Mind: An Essay on
Human Feeling

Charlottesville

November 1967

MIND: AN ESSAY ON HUMAN FEELING

VOLUME I

SUSANNE K. LANGER

THE JOHNS HOPKINS PRESS

BALTIMORE

*To them in whom I hope to live
even to the great World Peace—
my children and their children*

Acknowledgments

Volume I

MY FIRST and well-nigh inexpressible thanks go to the Trustees of the Edgar J. Kaufmann Charitable Trust for the support which made the very undertaking of this book possible. For a decade and more they have repeatedly renewed their generous grant, without demanding concrete evidence or so much as a formal account of my progress, although I had to spend nearly half of the time in preparation before the writing could begin. Such faith is rarely shown to a scholar, especially one whose work is not expected to save thousands of human beings or to kill millions. It is as much for their confidence in my aim and the tribute they have thereby paid to Philosophy as for their generosity that I wish I could adequately thank the members of the Board of Trustees.

Old friends, new friends and colleagues (in one case, a colleague's friend's colleague) have contributed to the making of this volume. Most of all I am indebted to Mr. Bruno R. Neumann, for several years my research assistant, without whose aid it would certainly not have gone to press when it did. More important than any practical help, however, was the intellectual spur of discussions with him, the give and take of ideas between a political economist and a much less socially or historically oriented thinker; so that his taking a government post in the far-off Virgin Islands has left me with a loss that cannot be made up.

Three persons, besides Mr. Neumann, have read the entire manuscript and given me the benefit of their reactions. Mr. Edgar J. Kaufmann, Jr., has read it chapter by chapter, sometimes even more than one version of a whole chapter, with the closest attention, as his constructive and often astute criticisms proved. Dr. Anna W. Perkins, my friend since our college days, has done likewise, and in many conversations raised challenging questions. Mr. Thomas Hughes Ingle not only brought the authoritative artistic views of a painter to the general support and frequent improvement of Part II, but read the other parts with equal interest and extraordinary critical competence. I thank all these readers heartily for the time and pains they have devoted to my book.

Dr. Karl Ernst Schaefer of the United States Research Laboratory at New London, Connecticut, has read several parts of the volume, especially the biological chapters. To him I owe the chart of diurnal rhythms, specially constructed in his laboratory for my purposes, as well as the material he personally provided for its interpretation. I thank him for this great kindness, and for all that I have learned from him. For, in entering upon fields far from my own—biology, physiology, genetics and evolution theory—I have sought a great deal of help from my friends, and for this I am grateful to him and to two of my colleagues, Professors Dorothy Richardson and Bernice Wheeler. They have generously given me valuable information that saved me many hours of searching for facts and current theories. Professor Wheeler has allowed me to audit two of her courses at Connecticut College, and both she and Professor Richardson have often alerted me to new material in the literature of their field.

My special thanks go to Dr. Frederick Snyder, of the Adult Psychiatry Branch of the National Institute of Mental Health at Bethesda, Maryland, for providing the chart of sleep rhythms, and to my colleague Professor Otello Desiderato, and his friend Dr. Julius Segal of the National Institute of Mental Health, for their kind offices in that connection. Several other illustrations, too, were especially made for me, wherefor I wish to thank their originators: Mr. Louis Darling, for his photograph of a house centipede, Miss Alice Dunbar, for her sketches of Greek medallions in the Archeological Museum at Athens, Mr. Henry F. Dunbar, for his photograph of beetle carvings in dead wood, and Professor Walter R. Henson, of the Yale School of Forestry, who went to considerable trouble to make some identification of the carver. I am also indebted to Mr. Philip A. Biscuti at Connecticut College for his help with the photographic work involved in several of the reproductions, to the Yale Medical Library staff for their constant courtesy and helpfulness, and to the staff of the Connecticut College Library, especially to Miss Helen K. Aitner, for locating and often borrowing research material.

Finally, I wish to express my warm gratitude to Mrs. James R. Dunbar, who assumed the arduous task of reading the proofs; and to the staff of The Johns Hopkins Press for their unfailing cooperation and affability in the course of launching this first volume on whatever is to be its career.

Contents

PART ONE:
PROBLEMS AND PRINCIPLES

Feeling characteristic of animals as against plants; feeling as anything felt; fallacy of assuming physical and nonphysical entities; "mind-brain problem" engendered by dual standard of reality; reality as datum and as matter; epiphenomenalism and vitalism scientifically useless; doctrine of "logical languages" fallacious; behaviorism an evasion of real problems; psychical phases of physiological events; organism a center of activities; impact and autogenic action; all mental phenomena modes of feeling; crises in natural history

Lack of development in "behavioral sciences"; false notions of scientific status; idol of technical language—jargon; idol of scientific method—prescriptive methodology; idol of controlled experiment—trivial research; idol of objectivity—rejection of direct evidence; idol of mathematical form—specious algorithms; self-criticism in current psychology; basic weaknesses philosophical

Prescientific but intimate knowledge the beginning of every science; no such knowledge of feeling among practical psychologists; obstacles to simple introspection; models from physical science misleading; need of images before models; guiding image and working concepts; artists as prescientific experts on forms of feeling; adequate image makes all auxiliary models safe

PART TWO:

THE IMPORT OF ART

[x]

List of Illustrations

List of Illustrations

Introduction

THIS essay has arisen out of a previous book, *Feeling and Form*, which in turn arose from a predecessor, *Philosophy in a New Key*. The "keynote" recognized and stressed in that first book is the concept, dominant in twentieth-century philosophy, of symbolism as the characteristically human element in cognition, and the great departure from animal mentality which symbolic expression and understanding have effected. The influence of particular symbolic formulations on the very data of knowledge led the most original thinkers of our era, in widely separate schools—Charles Peirce, Sigmund Freud and Ernst Cassirer, for instance—to contemplate the possible modes of symbolism itself. By the recognition of the new key in which those thinkers and several others pitched the compositions of their thought, I was led to the concept of art as the symbolic expression of an artist's knowledge of feeling (a very different thing from symptomatic expression of currently felt emotions), which is the pervasive theme of *Feeling and Form*. There I have treated the arts in some detail under the rubric of non-discursive and non-systematic symbolic expression, an epistemological concept proposed and developed by Cassirer in his *Philosophy of Symbolic Forms*.

But one fundamental question was not answered—indeed, scarcely raised—in *Feeling and Form*: Why must artistic form, to be expressive of feeling, always be so-called "living form"? That it must be was noted and discussed there in an early chapter; also that such "organic" or "living" form is a semblance, since works of art are not actually vital organisms. But why is that semblance necessary? Aristotle already declared that organic form was the most important feature of any composition, and that everything in art, however fantastic, must seem like life. He also observed that it need only seem, not actually be, life-like. But why?

To answer this question proved to be a large undertaking, because it not only involved the whole theory of artistic creation, symbolic illusion, abstraction and presentation already developed in *Feeling and Form*, but also required a new, fairly extensive study of actual living form as biologists find it, and of the actual phenomena of feeling, to which we have at present no scientific access. It was in the course of seeking an approach to this last-named field of inquiry that the wide and manifold implications of every step in the emerging answer came to light, such as quite new relations of feeling to thinking; the differences and connections between animal mentality and human mentality, which led into a theory of mind; the differences between animal "societies" and human Society, which broached vast social and moral problems; and, finally, the truly philosophical problems of knowledge and truth which any theory of mind entails. The essay, consequently, soon outgrew the apparently moderate scope of the question which engendered it, and fell into the six parts, more or less distinct though intimately related and consecutive, that now compose it: Problems and Principles, The Import of Art, *Natura Naturans*, The Great Shift, The Moral Structure, and On Knowledge and Truth.

Since the concept of mind developed in this book has arisen from the philosophy of art presented in *Feeling and Form*, I have to assume that my readers have some acquaintance with the ideas contained in that earlier work, for it is impossible to repeat the statement and demonstration of them here. Philosophical interpretations are systematic; they follow a logical sequence by developing the implications of an initial, generative idea. Relatively remote implications of such key concepts as that of presentational symbolism contained in *Philosophy in a New Key* or the concept of artistic illusion advanced in *Feeling and Form* cannot be understood without knowledge and acceptance of those previously established ideas, which are their premises. This makes it hard to read the later products of a long intellectual career, because they require some preparation on the reader's part; the late books no longer can stand alone, as the early ones did. But this difficulty is inevitable and is, after all, not peculiar to philosophy. One cannot understand Beethoven's late quartets or sonatas without having studied his early work, nor evaluate a new theory proposed by a scientist in a single presentation (Einstein's Special Theory of Relativity was stated, I believe, in four and one-half pages) without knowing something of his previous encounters with the puzzling facts which finally have brought him to a radically new position.

The central problem of the present essay is the nature and origin of the veritable gulf that divides human from animal mentality, in a perfectly continuous course of development of life on earth that has no breaks. For animals have mental functions, but only man has a mind, and a mental life. Some animals are intelligent, but only man can be intellectual. The thesis I am about to develop here is that his departure from the normal pattern of animal mentality is a vast and special evolution of feeling in the hominid stock. This deviation from the general balance of functions usually maintained in the complex advances of life is so rich and so intricately detailed that it affects every aspect of our existence, and adds up to the total qualitative difference which sets human nature apart from the rest of the animal kingdom as a mode of being that is typified by language, culture, morality, and consciousness of life and death.

The main task entailed by the undertaking of a new attack on the problem of mind in the context of natural history, without resort to metaphysical assumptions of non-zoological factors for the explanation of man's peculiar estate, is to keep the biological concept adequate to the greatness of the reality it is supposed to make comprehensible. Any natural mechanisms we credit with the functions of life and try to trace from their simplest manifestation in a culture of *Neurospora* to human brains conceiving poetry, must be great enough to account for the whole spectrum of vital phenomena, i.e., for our genius as well as for the mold on our bread. Theories that make poetry "merely" an animal reaction, favored by "natural selection" as a somewhat complex way of getting a living, really prove, above all else, that our basic philosophical concepts are inadequate to the problems of life and mind in nature. It is relatively easy to carry a biological principle, discovered in protozoan or even the lowest metazoan forms of life, through to higher and higher levels; but in the course of that advance the principle usually becomes ever less important until its manifestations, though still discoverable, are trivial. The simple taxes, such as phototaxis, which are major principles of behavioral control in lowly organisms, still have some effects in most mammalian species, but generally very little significance. What is interesting at the higher levels is to find the principles which eclipse the ones that are paramount in the little flagellate and still important in the earthworm. Such scientific advances are not quickly made, and our real understanding of life consequently cannot be reached by the hasty generalization of a few biological findings to constitute a doctrinal or methodological "ism." Yet I think the philosopher who gives up hope of constructing adequate interpretive concepts in the face of mere difficulty, slowness and occasional frustration is shirking the assignment automatically given to him by the advance of science.

The assumption of some non-physical ingredient in human life, or even in all life, has a different kind of inadequacy. It does not tend to trivialize the phenomena of mind, but by treating all vital phenomena (and finally, as a rule, non-vital ones as well) as so many appearances of one Essence, it never takes one beyond those appearances to any details of their relations to each other or to the Essence, which is covered by a single word—Mind, Soul, *Lebensgeist*—and is unanalyzable. Oddly, people who find this kind of view "deeper" than a scientific view have no means of delving below the surface manifestations. The exciting conception of spirit incarnated in matter imposes itself as readily on inanimate nature as on living organisms; "tongues in trees, books in the running brooks, sermons in stones" are all on a par. But to understand life means to discover the differences (which are sometimes not sharp, and may even be only statistically appreciable) between organic and inorganic nature, and neither reduce the former to the latter nor, contrariwise, treat the symbolically rich appearances of winds, waters, mountains and heavens as vital phenomena.

The things or events in nature to which vitality is thus erroneously ascribed are not instances of life, but symbolic images of it. Religious thought, whether savage or civilized, operates primarily with images, by the long-sanctioned "principle of analogy." That this approach to the problems of life and mind does not lead to any exact knowledge need hardly be argued today. Yet there is a value in images quite apart from religious or emotional purposes: they, and they only, originally made us aware of the wholeness and over-all form of entities, acts and facts in the world; and little though we may know it, only an image can hold us to a conception of a total phenomenon, against which we can measure the adequacy of the scientific terms wherewith we describe it. We are actually suffering today from the lack of suitable images of the phenomena that are currently receiving our most ardent scientific attention, the objects of biology and psychology. This lack is blocking the progress of scientifically oriented thought toward systematic insight into the nature of life and especially of mind: the lack of any image of the phenomenon under investigation, whereby to measure the adequacy of theories made on the basis of physical models. In borrowing models from physics, one is apt to borrow its image of reality as well; and that image derives from inorganic nature. It is becoming more and more obvious that it does not fit the forms of life very far above the level of their organic chemistry.

It was the discovery that works of art are images of the forms of feeling, and that their expressiveness can rise to the presentation of all aspects of mind and human personality, which led me to the present undertaking of constructing a biological theory of feeling that should logically lead to an adequate concept of mind, with all that the possession of mind implies. The fact that expressive form is always organic or "living" form made the biological foundation of feeling probable. In the artist's projection, feeling is a heightened form of life; so any work expressing felt tensions, rhythms and activities expresses their unfelt substructure of vital processes, which is the whole of life. If vitality and feeling are conceived in this way there is no sharp break, let alone metaphysical gap, between physical and mental realities, yet there are thresholds where mentality begins, and especially where human mentality transcends the animal level, and mind, *sensu stricto*, emerges.

An image is different from a model, and serves a different purpose. Briefly stated, an image shows how something appears; a model shows how something works. The art symbol, therefore, sets forth in symbolic projection how vital and emotional and intellectual tensions appear, i.e., how they feel. It is this image that gets lost in our psychological laboratories, where models from non-biological sciences and especially from intriguing machinery have taken the field, and permit us to analyze and understand many processes, yet lead us to lose sight of what phenomena we are trying to analyze and understand.

Here the image of feeling created by artists, in every kind of art—plastic, musical, poetic, balletic—serves to hold the reality itself for our labile and volatile memory, as a touchstone to test the scope of our intellectual constructions. And once a measure of adequacy is set up for theories, models of biological processes may be taken from anywhere—billiard balls, crystal formations, hydraulics, or small-current engineering. A philosophy of life guided by the vital image created by artists (all true artists, not only the great and celebrated ones) does not lead one to deprecate physical mechanisms, but to seek more and more of them as the subtlety of the phenomenon increases. The better one knows the forms of feeling the more there is to account for in the literal, sober terms of biological thinking, and the bolder such thinking tends to become.

It is not usual in any intellectual field to give long or strenuous thought to the elements that enter into pure semblances; and art—as I have argued at length in *Feeling and Form*—is the making of virtual forms, symbolic of the elusive forms of feeling. But to discover the phenomena revealed in music, painting or any other order of art, one has to know what problems the maker of the symbol encounters and how he meets them. Only then can one see new forms of vital experience emerge. The techniques of art are different from those of science, and studio thought is not that of the laboratory. This means, of course, that to make art illuminate a field of science one has to be intellectually at

[xix]

home in both realms. Art is just as comprehensible as science, but in its own terms; that is, one can always ask, and usually can determine, how the artistic semblance of life is made and in what it consists. This sort of analysis I undertook in *Feeling and Form,* and have subjected to some generalization in the present book, to make the significant traits of that semblance available as a measure of what we are seeking to describe in the systematic concepts and direct language of science. The great value of a permanent image is that one can resort to it to recover an elusive idea, and reorient one's intellectual progress, when enticing simplifications and reductions have turned it away from its long course into shorter alleys that do not really lead to the same goal. Under the aegis of a holistic symbol, the concept of life builds up even in entirely scientific terms very much like the vital image in art, with no break between somatic and mental events, no "addition" of feeling or consciousness to physical machinery, and especially, no difference of attitude, point of view, working notions, or "logical language" dividing physics and chemistry from biology, or physiology from psychology. No matter how far apart the beginnings of research in various fields may be, their later developments converge, and in advanced stages tend to dovetail, and close like the perfect sutures of our skull, which become well-nigh invisible in ripe old age.

The relationship between the making of the vital image and its use in discovering the significant problems of psychology, morality and related contexts is demonstrated by the fact that Part II of the present essay, dealing wholly with artistic issues, leads directly into Part III, which is biological in matter and method. Similarly, Part IV, concerned with the development of mentality and the "great shift" to human mental life, epitomized in the concept of "mind," rests squarely on the three foundational parts contained in the first volume; just as the social theory (Part V) is implicit in Part IV, to which the epistemological ideas of the final part also go back. The source of this continuity is philosophical. Knowledge becomes coherent only as more versatile and negotiable concepts replace the generalities with which all systematizing thought begins, such as matter and motion, or body and mind, then cognition, reason and emotion, good and evil, truth and falsity, or whatever basic concepts govern the first analyses that organize a universe of discourse. Philosophy has traditionally dealt in such general terms; and the reason for its proverbial uselessness as a guide to the sciences, once they are born from its mysterious womb, is that it has made general propositions not only its immediate aim, but also its sole material. They are, in fact, the true scientist's ultimate aim, too; but his inevitable preoccupation with research that can be carried out in *ad hoc*, provisional terms is usually so complete that he cannot survey other fields and give his best hours of thought to reinterpreting his statements in more widely variable but specifiable ways.

The philosopher's aim is generality, but his true method is not therefore to deal in generalities. There are two familiar ways of establishing general propositions. One is the traditional method of selecting a key idea—life, mechanism, polarity, mind or other such basic concept, which is not further to be analyzed—and reinterpreting all known facts in terms of this universally exemplified "principle." This practice leads to a uniform outlook, such as vitalism, physicalism, psychism or some other "ism"—the sort of world view usually called "a philosophy." But it leads to no revelation of relationships among detailed operations in nature, nor new directions in which to look for facts that suddenly become important. The other method is that of progressive generalization from concepts which prove fertile in a limited way, that is, concepts which tend to expand and gradually become applicable to more and more phenomena, sometimes on a different level of abstraction, but in just as much detail—or at least with as much promise of detail—as in its original context.

As a rule, generalizations of this sort, and the widening of scientific knowledge that goes with them, are made by scientists; yet they are philosophical in essence. They are usually recognized as products of "basic research," and all really basic thinking is philosophical. Philosophy is the construction of concepts to work with, not only in natural science, but in the appreciation and treatment of values, in self-knowledge, historical understanding, education of feeling, and recognition of the moral structure, the structure of society. The scientist's fundamental thought is necessarily limited to the shaping and sharpening of his own intellectual tools, as the educator's and the statesman's to that of theirs, respectively. Consequently, the greatest pioneers in any particular line of research, though not unduly specialized or narrow within their own field, stand helpless at the edges of a really very different one; they can only accept popularly current ideas of what is going on in that strange domain. So an admirable physiologist can do no more with "consciousness" than suppose that it is somehow "associated with" or even "attached to" many cerebral processes but cannot be identified in or among them; a highly competent mathematician may believe that the purpose of music is to stimulate emotion, and that poems are a poet's spontaneous sighs or exclamations. Some noted biologists find that they must concede the existence of something called "free will," though no one knows just what that is.

The serious philosophical need of our day is a conceptual structure that may be expanded simply by modification (not metaphorical extension) of definitions in literally meant scientific terms, to cover wider fields than physics and chemistry proper, so that the exploration of those problematical domains—biology, with its special areas of genetics, evolution theory, neurology, etc., psychology, already departmentalized into animal and human, normal and abnormal, educational, social, and so on, the complex disciplines, mainly economic, that deal with values, and whatever other fields claim to be future "ologies"—may proceed as so many developments of our most exact systematic knowledge. To construct such concepts is, I believe, the task of professional philosophers; it is too large to be done by other intellectual workers on a basis of incidental insights reflected on in leisure hours. It requires familiarity with philosophical ideas, both general and technical reading in many fields, and logical training to the point of a liberated logical imagination; competences which may be demanded of philosophers, but hardly of anyone else.

Most scientifically minded philosophers, e.g., logical positivists and some phenomenologists, will probably approve these methodological ruminations as a rejection of metaphysics, and many existentialists will condemn them on the same grounds. But I do not reject or deprecate metaphysics; only it seems to me to be the natural end, not the beginning, of philosophical work. A. N. Whitehead once defined metaphysical statements as "the most general statements we can make about reality." Such statements, to be valid, must be built up by processes of generalization of systematic knowledge, not made on a basis of preconceived generality; and being attained stepwise, they are not likely to be ultimate, but only to be our present furthest reaches of thought. Whether this essay attains to any such synoptic view—or even to glimpses of truly metaphysical value—will have to be judged at its conclusion.

One word more: I am making no attempt to prove the sole rightness of my approach to the central problem mooted in the following pages, the problem of conceiving mind as a natural phenomenon, a "natural wonder," and to us the greatest of all such wonders of nature. To convince the contrary-minded is a hopeless task and an arrogant undertaking; they have their reasons for thinking as they do, as I have mine for my way. The foundations of a theory cannot be factually proven right or wrong; they are the terms in which facts are expressed, essentially ways of saying things, that make for special ways of seeing things. The value of a philosophical outlook does not rest on its sole possibility, but on its serviceability, which can only prove itself in the long run, by its multifarious turns, its amenability to mathematical or logical development, its scope and its applicability to factual findings. So, all I am seeking to do is to explain why I hold the views presented in the following pages, and to contribute to the work of more or less likeminded thinkers, dealing with the paradoxes of experience that always harbor the seeds of new conceptions.

S. K. L.
Summer, 1966

Problems and Principles

1

Feeling

Gefühl ist der mütterliche Ursprung der übrigen
Erlebnisarten und ihrer aller ergiebigster Nährboden.
—Felix Krueger

WE HAVE human psychology and animal psychology, but no
plant psychology.[1] Why? Because we believe that plants have
no perceptions or intentions. Some plants exhibit "behavior," and have
been credited with "habits." If you stroke the midrib of the compound
leaf of a sensitive plant, the leaflets close. The sunflower turns with the
diurnal changes in the source of light. The lowest animals have not
much more complicated forms of behavior. The sea anemone traps and
digests the small creatures that the water brings to it; the pitcher plant
does the same thing and even more, for it presents a cup of liquid that
attracts insects, instead of letting the surrounding medium drift them
into its trap. Here, as everywhere in nature where the great, general
classes of living things diverge, the lines between them are not perfectly
clear. A sponge is an animal; the pitcher plant is a flowering plant, but
it comes nearer to "feeding itself" than the animal.[2]

Yet the fact is that we credit all animals, and only the animals, with
some degree of feeling. The sensitive plant may have its foreshadowings
of sensation, perhaps even as much as sessile animals, hydras and
sponges. But if it has sensibility, that is an anomaly among plants,

  [1] This general statement holds true despite the fact that some psychologists have
devoted an article or chapter to reactions of plants, e.g., Claude Bernard in *La
science expérimentale* (1878), which contains an essay entitled, "la sensibilité dans la
règne animale et dans la règne végétale," and in our own day two chapters in the
Comparative Psychology of C. J. Warden, T. N. Jenkind and L. H. Warner ([1935–
40], Vol. II: *Plants and Invertebrates*), which deal with plant reactions.
  [2] Some carnivorous plants go even further, not only attracting insects to their
"mouths" (or "stomachs"?), but springing an active trap, by sudden movements. See
Francis E. Lloyd, *The Carnivorous Plants* (1942), chap. xiv, "The Utricularia
Trap."

whereas among animals any merest trace of feeling is a beginning, something with a tendency to develop. The plant kingdom shows no trend toward the elaboration of nervous structures for either perception or control of the environment; what affects a plant must come in contact with it, and its only approach to food or water or light is by growth.

Feeling, in the broad sense of whatever is felt in any way, as sensory stimulus or inward tension, pain, emotion or intent, is the mark of mentality. In its most primitive forms it is the forerunner of the phenomena that constitute the subject matter of psychology. Organic activity is not "psychological" unless it terminates, however remotely or indirectly, in something felt. Physiology is different from psychology, not because it deals with different events—the overlapping of the two fields is patent—but because it is not oriented toward the aspects of sensibility, awareness, excitement, gratification or suffering which belong to those events.

The vexing question in the philosophy of the biological sciences is how something called "feelings" enters into the physical (essentially electrochemical) events that compose an animal organism. The presence of such intangible entities, produced by physical (especially nervous) activities, but themselves not physical, not occupying space, though (according to most theorists) they do have temporal character, is hard to negotiate in the systematic frame of anatomy, physiology or the more circumscribed, physiological study of nervous process, neurology. Feelings, considered as entities or items, are anomalies among the scientifically acceptable contents of the skin. Yet they are universally conceived in this paradoxical fashion, and either accepted or rejected as scientific data of just such anomalous character; whether they are to be admitted or not seems to depend on whether one cares more about empirical evidence or about theoretical conceivability.[3] Harlow and Stagner, for instance, elected to recognize them and made quite explicit

---

[3] An explicit statement and avowal of the latter attitude is made by J. S. Haldane on the first page of his excellent monograph, *Organism and Environment as Illustrated by the Physiology of Breathing* (1936), where he says: "Animal physiology deals with the activities observed in living animals, including men; but under certain limitations. It deals in the first place with all the activities which are unconscious, such as digestion, circulation of the blood, secretion, or the growth of tissues. It deals, also, with the unconscious element in conscious action. . . . Physiology deals, also, with the sensations, impulses, and instincts of all kinds which appear in consciousness; but does not deal with the meaning and conscious control which are attached to them. It does not deal with this meaning and conscious control for the very good reason that the facts relating to them cannot be combined with the other material of physiology into a homogeneous system of scientific thought."

[4]

that they conceived of them in the traditional way, presumably because there seemed to be no other way that fitted the experiential facts; so they concluded that "the existence of feelings as distinct and discrete psychophysical entities can hardly be questioned, since there is abundant evidence to show that they exist as independent conscious experiences and that they are attended by characteristic reaction patterns."[4]

Since that was written, not many psychologists have come out as frankly with their conception of "feelings" as "discrete psychophysical entities," yet there is every evidence that manifestations of feeling in the widest sense, including emotions, sensations and even thoughts, are taken as just such entities wherever their status is seriously debated. The same thing holds for the concept of "consciousness" and gives rise to the question of what is the function of that strange addition to the economy of the organism. F. C. Bartlett, for instance, in his study of recall of previous information,[5] writes: "An organism has somehow to acquire the capacity to turn round upon its own 'schemata' and to construct them afresh. This is a crucial step in organic development. It is where and why consciousness comes in; it is what gives consciousness its most prominent function. I wish I knew exactly how it was done." Whitehead, in *Process and Reality*,[6] speaks of "the initial operations of consciousness itself"; a difficult concept to reconcile with his doctrine of "vague feeling" pervading all operations of all entities whatever. Even William Russell Brain, in "The Cerebral Basis of Consciousness,"[7] after remarking at the outset that "consciousness" is not a "state in its own right, but a quality of states," which therefore are "conscious states" of cognition, emotion, attention, etc., says on the very next page that the perception of a prick is "pin-pointed as it were upon the body-image, and this process is something that happens to the raw material of the sensation of touch before it is *presented to consciousness*" (italics mine). In the course of the article "consciousness" figures as a receptor and even an agent.

The fact that we feel the effects of changes in the world about us, and apparently of changes in ourselves, too, and that all such changes are physically describable, but our feeling them is not, presents a genuine philosophical challenge. The oldest line of attack was to treat "the

[4] H. F. Harlow and R. Stagner, "Theory of Emotions" (1933), p. 184. The use of "feeling" here is in a narrow sense, viz., pleasure, displeasure, excitement and depression.
[5] *Remembering* (1932), p. 204.
[6] A. N. Whitehead, *Process and Reality. An Essay in Cosmology* (1929), p. 22.
[7] (1950), p. 466.

psychical" as a sort of shadow, produced by physical events, but running parallel to them without further systematic value, as an "epiphenomenon." Our apparent reactions to such psychical data as pain could be conceived as induced not by the pain itself, but by its physical counterparts. This theory really amounted to little more than saying that psychical entities lay outside the realm of physiological psychology, which was aspiring to the status of a science. The epiphenomenalists, who found it impossible to imagine a psychical entity pushing a physical one around, still did not question how physical events could have effects of an admittedly non-physical kind, that is, effects categorially incompatible with their causes—and, indeed, with any elements of the system that was supposed to engender them.

But the next generation did question this mystery, and one important and flourishing school, at least, capitalized on the answer, which was that the alleged causes and effects could not possibly belong to the same system, and consequently could not be what they were supposed to be with respect to each other. They must belong to different systems; these systems might bear some relation to each other. At this point, the new semantic orientation which philosophical thinking received from Cassirer, Russell, Whitehead and most forcibly from Ludwig Wittgenstein's *Tractatus Logico-Philosophicus*, published in 1922, presented a happy hypothesis: the two systems contain equivalent statements of the same natural facts, but in different "logical languages." The system couched in physical terms is greater than that which can be constructed in psychical terms, but the latter is equivalent to a subset of the former.

This notion of two scientific systems differently formulated, but referring to the same objects and facts, namely, mental phenomena, and equally capable of describing them, actually antedates the semantic studies which occupy much of the limelight today. Charles Peirce, who in the 1870's had initiated those studies in all quietness, had noted the possibility of very dissimilar-looking, yet logically equivalent, abstract structures; William James, without entering into the conceptual analysis itself (to which he was temperamentally averse), did seize upon the suggestion, and attempted the first application of that logical finding to the "mind and body problem" in his famous essay: "Does Consciousness Exist?"[8] The same objects, he held, namely, physical things, may function either as "things" in what he called "objective history," or as

[8] Published in *Essays in Radical Empiricism* (1912; 1st ed., 1904). Note that "physical things," though asserted to be only one way in which "one primal stuff" may appear, i.e., function in experience, are still taken to be the ultimate "real" objects.

"percepts" in "subjective" (personal) history; and concepts, similarly, may function as "abstract" entities in science or as "real" relations in perception and concrete imaginative thought. This begets the division of reality into thought and things; a similar division of mind into consciousness and its contents stems from the fact that "any single non-perceptual experience tends to get counted twice over, just as a perceptual experience does, figuring in one context as an object or field of objects, in another as a state of mind: and all this without the least internal self-diremption on its own part into consciousness and content. It is all consciousness in one taking; and, in the other, all content."[9]

There is no need of tracing the history of this idea from such early formulations to the latest ones, except to say that it is still with us, and has gained a special persuasiveness through the concomitant advance of studies in symbolism and its influence on conception, and through the peculiar popularity of "semantics." One of the clearest statements of its present methodological version occurs in a book by T. S. Szasz, where the two sets of elements conceptually established are designated (following Bertrand Russell) as "public data" and "private data," respectively, and the treatment of one and the same event in terms of one or the other is based on the requirements of different methods, each of which has its own "frame of reference," one "physical" and the other "psychological." "For example, information theory has elucidated certain aspects of the reception, transmission, and reproduction of 'information.' It has done so by adherence to a purely physical frame of reference. Psychoanalysis, on the other hand, addresses itself to problems of communication in terms of object relationship, transference, the communication of affects, and other psychological concepts. In either case, emphasis is placed on the precise nature of the scientific method (*e.g.*, physics or psychology) and not on the nature of the verbal symbol (*e.g.*, mind or brain)."[10]

But the hypothesis that the two incompatible vocabularies may be regarded as terms of two different "logical languages," equally applicable to the same empirical facts, is neither proved nor removed by shifting one's emphasis from "the nature of the verbal symbol" to the pragmatic reasons for preferring one language or the other. The claim still stands that the two sets of verbal symbols, one of which presents the facts of psychology in terms of physical process, the other in terms of non-physical or psychical "ultimate factors," are logically equivalent, although their respective elements and constructs cannot be exactly coordinated.

[9] *Ibid.*, p. 17.
[10] *Pain and Pleasure* (1957), p. 8.

[7]

This interpretation is based on the analogy of alternative systems of mathematics, equally descriptive of physical space, but starting with different "primitive concepts" and "primitive propositions," and therefore taking incommensurable forms. But the analogy is not really sound; for mathematical statements in the most diverse systematic terms are still all equally mathematical, and work with elements of the same metaphysical category, namely, mathematical elements; whereas nerves undergoing electrochemical changes, and contents of consciousness—emotions, sensations, ideas—as elements of different systems, are not of one metaphysical category. This is attested by the fact that they cannot be handled by comparable operations. Their incommensurability is not of the kind that can result from differences of logical formulation, but springs from two different concepts of reality: reality as "primary substance," or "matter," and reality as the "datum," or "immediate experience." Ever since Locke, in the epoch-making *Essay*, set up the latter without abandoning the former, we have had a double standard in psychology. The empiricist concept of mind is, in fact, a counterpart of the materialist concept of the world—a mosaic of ultimate "constituents," a pattern of "ideas," or psychical entities, corresponding to the atoms or other irreducible entities of classical physics;[11] and as psychology strove to become a natural science, and to connect mental phenomena with the physical phenomena of animate nature, the very fact that the concept of mind was modeled on that of the material world made for two orders of basic elements instead of one, and defied the attempt to relate them causally; for the elements of a model and the corresponding items in the object it represents cannot be causally connected. They are two possible "loci" for one formal pattern,[12] not one "locus" for two different-looking but logically equivalent patterns; the alleged semantic shift from one symbolism to another is really a metaphysical shift from one interpretation of the same abstract system to another. But two interpretations, or "loci," of one form cannot be causally linked in a larger pattern.

The impossibility of connecting the facts of one such conceptual order with those of another has been recognized and candidly accepted in a very interesting, albeit problematical, article, "On the Nature of Pain,"

[11] Bertrand Russell called his early theory of knowledge "logical atomism." See his "Philosophy of Logical Atomism" (1918), together with the somewhat earlier essay, "The Ultimate Constituents of Matter" (1915). Whether he still abides by it I cannot certainly say.

[12] This pattern is, of course, the simple geometric pattern of *res extensa*. As the pattern of present-day physics becomes extended and elaborated, it looks less and less applicable to ideas and their supposed manipulation by the mind. Locke's *Essay* appeared even before Newton's *Principia Mathematica*.

Feeling

by William Gooddy (1957). Dr. Gooddy assents to the prevalent idea that the incompatibility of physical and psychological statements lies in the difference of logical languages, but unlike most of his colleagues he draws the proper conclusion, that between events in the nervous system and the "data" of experience no causal connections can then be established. This is very disconcerting, since the thesis of his paper is that all painful sensations bespeak unusual processes in the nervous network that links the brain with the countless functional subordinate parts of the body, and after his main exposition, he observes (p. 123): "We are now . . . forced to ask the question so commonly avoided 'How does the detection of the abnormal pattern give rise to pain?' and the second question 'Where does the neurophysiological state break through to feeling and psychological states?'

"This insoluble problem is not a valid one. The confusion has arisen from the error of changing from one semantic channel to another. The two channels in question are, respectively, quantitative and qualitative. . . . No theory of semantics can provide a cross-channel from quantity to quality, as the scientist falsely suggests in his desire for a simple connection between mind and brain."

The first question is, indeed, a false lead (every question has to be a lead to its relevant answers); for there is no act of "detecting" the condition of the nerves, which then "gives rise" to pain. But the second one is not invalid. The one injudicious term in it is "neurophysiological state"; a state cannot "break through" to anything. But I think a neurophysiological process can be said to "break through to feeling," and how this can reasonably be said is the theme of my present chapter. That is a philosophical issue. At what point the psychical phenomenon, feeling, begins to occur is a question of fact, i.e., a psychological question, difficult but not invalid. On the whole, the relation of mental events—i.e., felt impingements and activities—to the rest of nature is the subject matter of psychology, which demands studies in neurology, genetics, and also careful introspection and clinical protocol analysis for its systematic understanding. If we are indeed speaking two "logical languages" and always tending to mix them, as recent immigrants in a new country mix its spoken language with their native tongue, we must settle on one or the other.

It soon becomes apparent, however, that it is not a choice of language that confronts us. If "psychophysical entities" occur in nature, the vocabulary of natural science must be able to designate them, and to accommodate itself to discourse about them; if this is not possible, it is not our language that is inadequate, but our basic assumptions. Some

[9]

eminent twentieth-century biologists and psychologists, convinced that the facts of animate nature are not describable in the present frame of scientific thought, take the stand of admitting a non-physical and therefore incalculable universe of discourse. To these thinkers the philosophical issue seems paramount, too great to be ignored even for the sake of systematic progress, which they deem to be in a false direction. Biochemistry, for them, runs too close to the inanimate realm to promise any understanding of life. They hold that to account for psychical phenomena one has to assume the existence of a metaphysically ultimate psychical reality, active but non-physical, which "uses" the physical apparatus—nerves, muscles and specialized organs—as a means to act on the natural world. This "use," however, is not a physical process; the agent has no spatial properties, and is sometimes said to have no temporal ones. It is essentially a soul, an Aristotelian entelechy, which gives rise, in developed organisms, to intellect, and may even become endangered by a preponderance of that derivative power.

This "vitalist" theory is linked first and foremost with the name of the biologist Hans Driesch, who stirred the scientific world by declaring that the autonomous and teleological behavior of organisms required the assumption of an "entelechy" in each one, which powered and directed it.[13] Twenty years after Driesch's *Philosophie des Organischen* presented his ideas in their final form, Ludwig Klages wrote *Der Geist als Widersacher der Seele* (1929), and carried the vitalist doctrine of entelechy (*Seele*) from biology into so-called "philosophical anthropology." His main theme was the excess of intellect (*Geist*) in human psychical existence and the danger it presents to the soul, the instinctual guiding power, on which it is grafted. His book struck a sympathetic chord in several minds, especially in Europe, where it has carried its share of grist to the mill of antirational *Existenzphilosophie*.[14] But its chief importance is not its antirationalism, which Bergson had already introduced as a philosophi-

[13] Driesch's first proposal of an immaterial factor in living systems goes back further than the present century; *Die Biologie als selbstständige Grundwissenschaft* appeared in 1893. But it was in the early nineteen hundreds that he published his most startling and influential books—*Die "Seele" als elementarer Naturfaktor*, 1903; *Der Vitalismus*, 1905; *Philosophie des Organischen*, 1909. Several works appeared after that, especially *The Crisis in Psychology* (1925), but Driesch shot his bolt in his early works, and did not follow it up with scientific discoveries or explanations to prove its value.

[14] See, for instance, Karl Jaspers, *Reason and Existenz* (1955); Ernst Mayer, *Dialektik des Nichtwissens* (1950); and William Barrett, *Irrational Man: A Study in Existentialist Philosophy* (1958).

cal ideal; its influence stems rather from the fact that here an eminent scholar declared the bankruptcy of "physicalism" and proposed the assumption of a purely mental agency as a principle of explanation in the science of mind, as Driesch had done in the science of life. Among the followers and corroborators of Klages are such psychologists as Palagyi, Rothschild, Buytendijk and, with some reservations, Weizsäcker—people whose opinions cannot be taken lightly.

The foremost difficulty with the postulation of an immaterial entity in an active organism, which is steering its activity, is to bring any evidence for it. Since neither the "entelechy" nor the manner of its operation can be observed, either directly or by means of indirect recordings, its influence on vital processes can only be postulated, and the postulate justified by demonstrating an intellectual need of it to supplement the limited explanatory powers of any and every causal system. Several of its protagonists, consequently, have tried to prove that mental phenomena, even on the lowest level of vital response to outward conditions, could not possibly be conceived as purely physical events arising from other such events. The most elaborate of these demonstrations have been presented by the American biologist R. S. Lillie, building on Driesch and Bergson, and by Melchior Palagyi, who stood close to Klages. Since both of these writers came to their philosophical problem by way of biological studies, they could not ignore their very considerable knowledge of facts and theories scientifically attained in those fields; their primary aim was not to interpret the world idealistically, but to show the existence of a "psychic factor," which is a mental activity—will, purpose, judgment, conscious or unconscious desire—and "not a physical event." Lillie advanced an ingenious argument making this factor look "spiritual or psychical," but unfortunately the reasoning involves some highly questionable substitutions.[15]

[15] The following passages from his *General Biology and Philosophy of Organism* (1945) will suffice, I think, to substantiate this criticism. There we read: "Experiment indicates that the constant physical factors which determine the limits of optical discrimination or visibility are not themselves infinitely subdivisible, since radiation is transmitted in units, or quanta, having discontinuous or atomic properties. The question if, supposing a quantum to be a spatial (or spatiotemporal) unit (and, as such, occupying volume), it is theoretically possible to regard it as subdivisible into smaller subquantal units . . . , again raises the question as to how far the formal geometrical conception of space can be regarded as an exact model of 'real' space, i.e., of spatial conditions as existent in an independent extended world, in the sense of Descartes" (p. 80). Finding that quanta defined *in a spatiotemporal system*, which he mentions parenthetically as though it were equivalent to the classical spatial system, cannot be spatially defined "in the sense of Descartes," he continues in Cartesian terms: "Indivisible can only mean nonspatial, nonextended. . . . But

Palagyi attempted a proof that an act of human consciousness must be a mental (*geistig*) factor that is neither physical nor psychical, since these categories both apply to spatiotemporal events, whereas an act of "the perceptual consciousness" takes place in a mathematical moment, i.e., in a zero span of time.[16] His demonstration of this odd proposition, involving a metaphysical premise which is established simply by fiat,[17] is too involved and unconvincing to be repeated here; for in a mathematical moment, a time of zero duration, nothing can "take place" at all. The argument suffers, furthermore, from the same fallacy as Lillie's, namely, that it borrows the right to speak of "atemporal" events from relativity physics, but claims such events to be "non-physical" by tacitly identifying physics with eighteenth-century object physics.[18]

---

nonspatial (nonextended) is equivalent to nonmaterial (since matter occupies space), and a nonmaterial principle of existence or activity which nevertheless is real can only be regarded as spiritual or psychical" (pp. 81–82). The confusion of two systems, the casual substitution of "principle of existence" for existent units and the unfounded assertion that non-material must mean "spiritual or psychical" cannot but strike any trained philosophical thinker as pathetic ineptitudes or else sophisms. Far better the frank defeatism of Driesch, who simply said we "must assume" a mysterious entelechy, because we cannot go further with physical science. If someone does go further, and far enough, the assumption can be dropped without ado.

[16] *Wahrnehmungslehre* (1925), p. 18: "Indem wir also feststellten, dass unser wahrnehmendes Bewusstsein keinen fliessenden, sondern einen intermittierenden Charakter habe, lag es schon in dem Sinn dieser Feststellung eingeschlossen, *dass ein Wahrnehmungsakt für sich betrachtet völlig dauerlos sein muss, d.h. nur im mathematischen Zeitpunkt stattfinden kann.*"

[17] *Ibid.*, p. 13: "Wirklich ist immer nur das, was im mathematischen Jetzt stattfindet. . . ."

[18] In a succinct little article of only eleven pages, "Atemporal Processes in Physics" (1948), Richard Schlegel presents the possibility of atemporal processes, their physical nature and the fact that *qua* atemporal they cannot act upon material processes. This would make the "act of consciousness" ineffectual if it were "atemporal," and deprive it of the virtue for the sake of which it is introduced by the vitalists—its supposed power to engender action.

"The movement of electromagnetic radiation in an otherwise force free region is an atemporal process. . . . That is, there is no time extension beyond one oscillation in the sub-universe or domain of the radiation being considered. This domain is isolated from the remainder of the universe. Whenever it [the radiation] interacts with the world outside of its domain it undergoes an energy change of some sort, and therefore its time extension increases, by virtue of the progressive change which has occurred in the system" (p. 26).

"A state which has neither progressive nor cyclic change of any kind, and which is completely isolated from the larger universe of change, is a state for which time extension is zero. This state could be considered a limiting case of the isolated, cyclic process, in which the extent of a cycle has gone to zero. Such a state, in which 'absolutely nothing ever happens,' would be an atemporal process. In the natural

Feeling

Buytendijk, writing out of a great fund of factual knowledge, and with full awareness of the rapid advance of neurology, histology, biochemistry and other relevant studies, is too cautious to base his doctrine of non-physical activity on the assertion that processes of one sort or another lie forever beyond the physicist's intellectual reach. He evades that dangerous ground by the forthright and radical assertion that mental acts like intending, willing, perceiving and feeling are not processes at all, but "functions" of a metaphysical subject which achieves their execution by means of "its" body. How "means" can be required where no process is to be propagated remains somewhat obscure. But here we have, at least, the identity of the mysterious "psychic factor": the metaphysical Subject.

The categories involved in this interpretation of vital phenomena are difficult to combine so as to let the act as an empirical event fit into them all. The metaphysical Subject has functions which are not processes, but the motions which it engenders in the body certainly are processes. The body and even its environment provide the opportunities for the subjective functioning. "Whenever a person experiences the floor as that upon which he is standing—and an animal, too, even a mutilated animal [preparation] is still capable of this—then the Subject disposes not only over its body, but over the floor as well, which is not given to it as a stimulus object [een prikkelgestalte], but as a condition and regulative basis for his standing as a form of behavior."[19] Reflexes, apparently, are natural processes, not acts of the Subject; for elsewhere Buytendijk says, "A jump can not be explained in terms of reflexes, it is not a process, but a genuine function...."[20] The precise relation of the Subject to "its" body presents some further anomalies which he frankly exhibits, for instance, where he says: "If someone touches my hand, he touches me, myself; if I reach for something, I move *my own* hand."

---

world such processes probably have no more than momentary existence, if even that." The only exception may be represented by the "reversible processes" of thermodynamics. But, Schlegel immediately observes, "the existence of reversible processes in nature is open to controversy; the only possibility of their existence is apparently in biological processes which are not at present understood" (p. 28). I do not know whether this statement is made in reference to the vitalists' hypothesis. Schlegel goes on (p. 29) to list the four propositions assumed by eighteenth-century physics and repudiated by its modern successor. They are all tacitly accepted in the arguments for "non-physical" acts.

[19] F. J. J. Buytendijk. *Algemene Theorie der Menslijke Houding en Beweging als Verbinding en Tegenstelling van de physiolische en de psychologische Beschouwing* (1948), p. 161.

[20] *Ibid.*, p. 92.

And again: "The infant kicking with his *own* legs is, at the same time, the one being excited. . . . 'Jede Erregung *ist* auch eine Regung,' says Klages, rightly. Autonomous movement is the reverse aspect of excitation and thus has the same significance, albeit the stimulus, as an incoming (virtual) movement has this significance for the individual, the movement as a centrifugal phenomenon for the outer world. . . ."[21] This and similar difficulties are not satisfactorily resolved by Buytendijk's vacillating argument, first from a basic "neutral" reality[22] (like the metaphysical "neutral stuff" conceived by James, Peirce and others of their generation) and then from a difference of metaphysical status, the Subject being transcendental, the body and all its psychological experiences empirical.[23]

The difficulties of a theory, however, do not necessarily condemn it; if its promise of conceptual power is great, if it gathers up piecemeal findings and tentative explanations, and makes many of them support each other systematically, its logical defects can probably be corrected gradually by more precise statement and more circumspect abstraction. But here is the most serious weakness of Buytendijk's hypothesis—a weakness he himself points out in the "vitalism" of Driesch,[24] yet does not himself escape: for his assumption of a non-physical factor in living creatures not only dictates no scientific procedures, just like Driesch's entelechy or Bergson's *élan vital*, but causes him to reject the most successful theoretical constructions based on empirical findings. For instance, he rejects Coghill's general principle of progressive functional maturation with bodily maturation, illustrated by the salamander's behavioral advance, Bethe's theory of muscular coordination in the act of grasping, Storch's analysis of tactual sensation which shows "directional sense" to be an isolable factor, Voigt's explanation of the sensory guidance of an act in progress, on no other ground than that they are "psycho-physical" and deal with biological mechanisms. Thus he says of

---

[21] *Ibid.*, p. 58.

[22] *Ibid.*, p. 38: "Onze beoordeling van de zin der bewegingen ligt dus in een psycho-physisch *neutrale* spheer, d.w.z. *in een gebied van ervaring, waarin de tegenstelling psychisch-physisch nog niet voltrokken, althans niet werkzaam is.*" Also, p. 130: "Het zich bewegen van het wiegekind is zeker een gedraging, maar de *differentiatie* in een der beide hoofdrichtingen is nog niet voltrokken."

[23] On p. 93 of *Algemene Theorie* he remarks "dat een physische verklaring der functionele zelfbeweging evenmin mogelijk is, *daar de levensfuncties een ander ruimte- en tijdsysteem bezitten dan de physische en psychische processen.*"

[24] *Ibid.*, p. 24: "Het vitalisme bleef altijd in het dualisme van stoffelijke processen en een besturende factor (entelechie) gevangen en was wetenschappelijk onvruchtbaar, daar het geen *nieuw* gebied van natuur-wetmatigheid ontsloot."

Coghill's results: "One could ascribe the primacy of the head-movements to a greater stimulability of the extended spinal cord, and this in turn to a more rapid histological differentiation; . . . but in this sort of explanation from structure one would be pursuing the mechanistic mode of thought. . . ."[25] The action of the metaphysical Subject, however, offers no alternative account of how vital activities are engendered, because this entity with its uncaused effects is essentially unamenable to conception, definition or description.[26] Such a doctrine, of course, removes the study of mind altogether from any scientist's docket.

No wonder, then, that psychologists who desire above all else to make their work scientific as quickly as possible will not accept an assumption leading to such results, and choose the other alternative, which is to deny that anything in nature is inexplicable in terms of physics. They are no more able than their "vitalist" colleagues to deal in such terms with the existence of the "psychophysical entities," i.e., feeling, intentions, acts of will or the "data of consciousness" given in sensation and memory; but they prefer to fall back on a methodological principle, which is to circumvent discussion of the embarrassing elements by eliminating words like "willing," "intending," "feeling"—that is, all words for introspectively known factors—from their vocabulary, proscribing reports of subjective states or events in their laboratories and ruling that all psychological facts are to be conceived as physical facts. That, too, is a radical remedy; but instead of extending assumptions to cover the subject matter, it limits the subject matter to what falls within the scope of safe assumptions.

The most ardent physicalists have gone so far as to treat all psychological data under the rubric of "transposition of matter in space."[27] But most of them have compromised on a distinction between organisms and inorganic entities, and on a special class of events involving organisms, known as the "behavior" of those organisms. All the "social sciences," from psychology to history, have even been dubbed "behavioral sciences," as though their only possible formulation depended on this somewhat vague concept.

[25] *Ibid.*, p. 405.

[26] *Ibid.*, p. 52: "Er bestaat geen begripsmatige formulering, noch een definitie of beschrijving van het eigen ik, nich van het subject, dat wij aan een ander mens toekennen . . . onze innerlijke ervaringen *verwijzen* slechts naar het eigen ik, naar de persoonlijkheid van andere mensen en naar het vitale centrum van het dier."

[27] See, for example, E. Augier, *La mémoire et la vie. Essai de défense du mécanisme psychologique* (1939), pp. 55 ff; A. J. Lotka, *Théorie analytique des associations biologiques* (1934), esp. pp. 29-31; and Otto Neurath, *Foundations of the Social Sciences* (1944), with instances scattered throughout.

Several different definitions of the term "behavior" have been given, yet the uses of the term in supposedly scientific contexts often overstep them all. J. Z. Young states that "the sum total of the action of the effectors constitutes the *behaviour* of the animal."[28] Nikko Tinbergen says, "By behaviour I mean the total of the movements made by the intact animal."[29] These are similar and usable definitions: behavior consists of the total movements, or sum total of actions, of creatures (though instead of "the intact animal" one might better say "the entire animal," since decerebrate, blinded, or otherwise prepared animals are certainly not intact, yet the experimenter's aim is to observe their "behavior"). But the word loses its usefulness when it is defined, as by Rosenblueth, Wiener and Bigelow, in such a way that "behavior" may be attributed to chairs, pebbles or even molecules as well as to organisms.[30] This is a hasty assimilation of "behavioral sciences" to physical sciences by applying the behavioral terminology to non-living entities—a purely verbal trick, easier but intellectually less genuine than the "physicalism" that seeks to describe vital phenomena as "transpositions of matter in space." The word "behavior" simply designates entirely different processes according to the entities to which it applies. Yet therewithal its employment sometimes goes even beyond the above definition, in contexts where one may question the existence of "entities" credited with behavior, as when Neurath assigns to the social sciences "research on the behavior of tribes, on the behavior of customs and languages . . . on the behavior of whole nations . . . and even on the behavior of markets and administrations," and speaks of "the behavior of artists, priests, statesmen" in exactly the same way as "the behavior of languages."[31] Are languages "entities" in the same sense in which priests, or even their hats and shoes, are entities?

The chief motive of this formulation is to establish psychology as a natural science distinct from physiology, biochemistry, or any other physical science, yet to give it the virtues of experimental research and so-called "objective" truth. But behavior has every appearance of being the macroscopic result of microscopic processes in the behaving organism; and no scientific research can restrict itself to any set of phenomena having causal connections with some other set, yet leave that

[28] "The Evolution of the Nervous System and of the Relationship of Organism and Environment" (1938), p. 185.

[29] *The Study of Instinct* (1951), p. 2.

[30] A. Rosenblueth, N. Wiener and J. Bigelow, "Behavior, Purpose and Teleology" (1943), quoted and criticized on this same score by C. J. Herrick in *The Evolution of Human Nature* (1956), p. 26n. The article, nevertheless, deserves attention.

[31] *Op. cit.*, p. 1.

other set alone. So the blessed word "behavior," while purporting to supply an "objective correlate" for the mental phenomena—feelings, thoughts, images, etc.—which are the original material of psychology, but stand condemned as "subjective," really creates methodological difficulties of its own. To these we shall return in the next chapter. At present, suffice it to say that "behaviorism" is essentially the last resort of psychologists, especially in America, who find themselves faced with the choice between "the physical" and "the psychical," and elect the former as their universe of discourse.

All these controversies spring from the desire to establish a clear and adequate concept of mind and of its relation to matter. The older ones had mainly an epistemological or cosmological aim, which goes back to Plato in our culture and considerably further in India; but the ones here considered all bear on the actual study of mind, and are the work of people conversant with the issues and difficulties of psychology. Physicalism and its opposite, the assumption of an immaterial "psychic factor" in living structures, are attempts to define the subject matter of psychology, in the belief that if we knew exactly what we are dealing with we could apply scientific methods to this material and thus find the basic laws which govern it, as physicists have done in their proper realm.

But it may be questioned whether this is really a profitable approach. The precise definition of the matter and scope of a science is more likely to become accessible in the course of intimate study, as more and more becomes known about it, than to be its first step. Physics did not begin with a clear concept of "matter"—that concept is still changing rapidly with the advance of knowledge—but with the working notions of space, time and mass, in terms of which the observed facts of the material world could be formulated. What we need for a science of mind is not so much a definitive concept of mind, as a conceptual frame in which to lodge our observations of mental phenomena. The field of inquiry should then define itself in various ways, by stages; but at every stage it should articulate with the rest of our scientific thinking, and especially with those fields which lie adjacent to it, biology on the one hand and ethnology on the other. It is our working vocabulary that has to be coherent and adequate. So far, it has never met one of these requirements without sacrificing the other.

This dilemma is clearly illustrated by the fact that people who handle psychological problems not under the conditions of the laboratory, where problems may be (and are) cut down to fit the capacity of the apparatus and the sanctioned procedures, but under the pressure of necessity—psychiatrists, neurologists, therapists of all schools, who deal with human behavior in crisis—can neither use the notions of non-material "psychical factors" among material ones, acts occupying a zero length of time, or "functions" which are not processes, nor can they abide by the prohibitions and dictates of behaviorism. They cannot forgo the use of introspective data, fantasies, dreams and alleged memories which cannot be empirically checked; above all, they have to deal with feeling, no matter how the admission of its existence confuses their philosophical notions. William Gooddy observed that, in studying voluntary control of muscles, "it will be with feeling that we shall be largely concerned, for sensory symptoms are so commonly a complaint of patients with dysfunction of voluntary movement. 'My hand feels funny when I try and move it.' Indeed, it was the observation . . . that the natural use of the words 'feel,' 'seems,' 'numb,' 'clumsy,' 'heavy,' 'helpless,' 'stiff,' was made to describe what turned out to be motor dysfunction, which suggested this present subject for elaboration."[32] Yet it was Gooddy who insisted elsewhere[33] on the incommensurability of physical and psychical phenomena despite their conjunction in scientific discourse.

It is sometimes said that therapists are essentially practicing physicians who have no need of philosophical coherence in their working theory. In a sense this is true; a good deal of clinical judgment and even research is possible in partial, not to say fantastic, theoretical frameworks. But such surface knowledge, which is developed in detail and under the stringent control of given conditions, with relentlessly changing probabilities of connections among them, leads into deeper problems and finally runs into the mysteries of unclarified basic conception. It is then that the greatest psychiatrists, neurologists, neurosurgeons or psychotherapists become aware of the shaky foundation which is carrying all their intellectual constructions. Being too busy and also technically unprepared for logical analysis and formulation, they usually accept, though perhaps reluctantly, one of the current answers to the "mind and body problem"; but they are certainly not blithely content with the solution. Wilder Penfield, for instance, has stated the old puzzle of the interaction between living matter and immaterial "mind-substance" in his famous address to the American Philosophical Society in 1954, saying: "It is obvious that nerve impulse is somehow converted into thought and that thought can be converted into nerve impulse. And yet this all throws no light on the nature of that strange conversion."[34] The fundamental

---

[32] "Sensation and Volition" (1949), pp. 316–17.
[33] "On the Nature of Pain" (1957) (see above, pp. 8–9).
[34] "Some Observations on the Functional Organization of the Human Brain" (1954), p. 297.

dualism of physical entities and mental entities, processes of thinking and elements of thought, is here accepted as simple empirical fact. An even stronger statement of this common-sense metaphysical assumption is made by J. C. Eccles in *The Neurophysiological Basis of Mind* (1953), where "the mind" is accepted as a genuine non-physiological entity, for he makes the forthright statement: "The special purpose of this inquiry is the nature of man, and it is the way in which the brain achieves liaison with the mind that is the essence of the problem . . ." (p. 260).

One thing every psychiatrist and clinical neurologist knows is that neither the principle of causal connection which we find in physical nature, nor the phenomena properly called "mental"—feeling, thought, sensation, dream, etc.—can be ruled out of his domain. As C. J. Herrick observed, "The two sets of data—the subjective and the objective—when examined separately seem to be disparate and incommensurable, yet actually we know that they are not because, if they were, a purposive action would be impossible."[35] That is putting it simply, even naïvely, but nonetheless sagely. Our so-called "objective" set of data—the elements of physics, relatively concrete or abstract, molar, molecular or atomic, according to the level of the physicist's problem—seems to fall naturally into a logical system of almost inexhaustible complexity without more than passing and minor loss of coherence. But "subjective" data do not. Whichever way you arrange such items, they do not compose a universe of discourse in which they can be operationally connected with each other, and still less a class of non-physical events to be intercalated among physical ones. So, presumably, our treatment of "subjective" data as elements categorially distinct from physical elements, but to be related to the latter, is untenable. Something is profoundly wrong with our conception of "the psychical"; some very fundamental notions need to be reformulated to give us more acceptable meanings, negotiable and illuminating, for the moot words "psychical," "conscious," "thought," "feeling" and others of the same shady character that fall under the present taboo in respectable laboratory parlance.

The basic misconception is, I think, the assumption of feelings (sensations, emotions, etc.) as items or entities of *any* kind, whether produced by physiological processes, or independent of them, non-physical "genuine functions" of a "life" or "soul" casually "making use of" bodily mechanisms, in the sense of Buytendijk. This is a genuine metaphysical fallacy; yet those theorists who have tried to treat it as semantic had an essentially right idea, for the conception of such psychical "factors,"

[35] *Op. cit.*, p. 22.

which is expressed in the question of how something called "feeling" can enter into physical processes, probably is essentially of linguistic origin. The fact that we call something by a name, such as "feeling," makes it seem like a kind of thing, an ingredient in nature or a product. But "feel" is a verb, and to say that what is felt is "a feeling" may be one of those deceptive common-sense suppositions inherent in the structure of language which semanticists are constantly bringing to our attention. "Feeling" is a verbal noun—a noun made out of a verb, that psychologically makes an entity out of a process. To feel is to do something, not to have something; but to "have" a feeling, a sensation, a fear or an idea, seems a perfectly equivalent way of conceiving the fact expressed by the verb. The supposed equivalence is given in the syntax that governs our intellectual processes. It is, perhaps, not as purely a product of language as current doctrines make it appear; there is a deeper reason, of

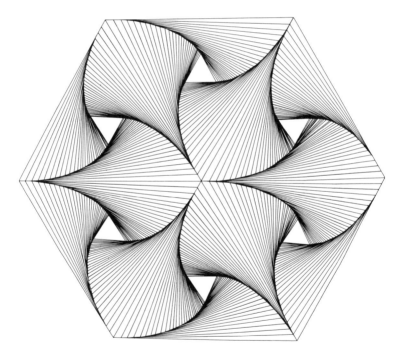

Figure 1–1. Mystery of the Vanishing Triangle

'in constructing the diagram, only triangles were drawn, yet the weird spade-like shape so dominates the result that the triangles pass unnoticed'

(Design by Rutherford Boyd, "Mathematical Ideas in Design," *Scripta Mathematica*, XIV [1948]. Reproduced by permission of Harriet R. Boyd.)

course, why language (despite considerable variations among different tongues in this respect) tends to hypostatize acts as entities. That reason should become apparent in a later chapter. Just now the effect, not the source, of the reifying tendency of our grammar presents the philosophical challenge. It is the concept of feeling—the modulus of psychological conception—that I propose to reconstruct.

In the first place, the phenomenon usually described as "a feeling" is really that an organism feels something, i.e., something is felt. What is felt is a process, perhaps a large complex of processes, within the organism. Some vital activities of great complexity and high intensity, usually (perhaps always) involving nervous tissue, are felt; being felt is a phase of the process itself. A phase is a mode of appearance, and not an added factor. Ordinarily we know things in different phases as "the same"—ice, water and steam, for instance—but sometimes a very distinctive phase seems like a product. When iron is heated to a critical degree it becomes red; yet its redness is not a new entity which must have gone somewhere else when it is no longer in the iron. It was a phase of the iron itself, at high temperature. Heat is not a thing, but an agitation, measurable in degrees, not amounts, and when the iron is no longer hot there will be comparable degrees of heat, or of some equivalent process or sum total of processes, outside the iron. But the redness simply disappears; it was a phase of the heated iron.

A striking demonstration of how constituents of one kind, brought together in a special combination, may seem to produce a new ingredient which is, however, a phase of their own occurrence, is given by Rutherford Boyd in the design shown on the page opposite (Fig. 1–1).

Unlike many other aspects of vital processes, which are propagated outward with the processes themselves beyond the organism as effects on its surroundings, the phase of being felt is strictly intraorganic, wherever any activities of life attain it. It is an appearance which organic functions have only for the organism in which they occur, if they have it at all.[36]

---

[36] This privacy of feeling was observed many decades ago by William James, who used "thinking" in the sense in which I use "feeling." (Remarking on the difficulty of finding a generic term for "mental states at large, irrespective of their kind," he said: "My own *partiality is for either* FEELING *or* THOUGHT," but he finally rejected "feeling" because it had kept particularly bad company in his day.) See *The Principles of Psychology*, Vol. I (1890), p. 186. In a later passage on "thought" in this broad sense, he wrote: "My thought belongs with my other thoughts, and your thought belongs with your other thoughts. Whether anywhere in the room there be a mere thought, which is nobody's thought, we have no means of ascertaining. . . . The only states of consciousness that we naturally deal with are found in personal consciousnesses, minds. . . .

Millions of processes—the whole dynamic rounds of metabolism, digestion, circulation and endocrine action—are normally not felt. One may say that some activities, especially nervous ones, above a certain (probably fluctuating) limen of intensity, enter into "psychical phase."[37] This is the phase of being felt. It may develop suddenly, with great distinctness of quality, location and value-character, for instance, in response to a painful stimulus; or similarly, only with less precise location in the organism, like a shock of terror; or a deeply engendered process may go gradually, perhaps barely, into a psychical phase of vague awareness— come and gone—a sense of weariness or a fleeting emotive moment. The normal substrate of "feeling-tone," from which the more acute tensions build up into specific experiences, is probably a dynamic pattern of nervous activities playing freely across the limen of sentience.

It is this transiency and general lability of the psychical phase that accounts for the importance of preconscious processes in the construction of such elaborate phenomena as ideas, intentions, images and fantasies, and makes it not only reasonable but obvious that they are rooted in the fabric of totally unfelt activities which Freud reified with the substantive term, "the Unconscious." There may be a describable system of functions that terminate in felt events, i.e., something that could fairly be called "the unconscious system"; but so far I do not think we have found more than a few lines of functional development, which may or may not

---

"Each of these minds keeps its own thoughts to itself. . . . No thought even comes into direct *sight* of a thought in another personal consciousness than its own. . . . The breaches between such thoughts are the most absolute breaches in nature" (p. 226).

James, in keeping with the psychological vocabulary of his age, reified thoughts, or he could hardly have set up as an undecidable question whether "a mere thought" were afloat in the room apart from any (equally substantive) "consciousnesses." But a much later psychologist, writing in modern terms and using "thoughts" in a somewhat stricter sense, remarks similarly on their privacy: "Because they are perceived only by the organ that produces them, they cannot even be detected in transit from point of origin to point of consumption" (Mortimer Ostow, "Entropy Changes in Mental Activity" [1949], p. 502). Felix Krueger, in "Das Wesen der Gefühle. Entwurf einer systematischen Theorie" (1928), p. 97, similarly noted the privacy of "feeling" in the strict sense.

[37] The use of "psychical" in this connection has been protested, especially by psychiatrists, who have given it a somewhat different meaning. "Sentient phase" would be a possible substitute, but meets with the objection that it seems to refer to sensation and to exclude other felt processes. In the end, one cannot avoid the use of all words that have acquired special meanings in professional contexts or even in currently popular discourse. The reader must be ready to accept words as defined, and accordingly put associations and even other definitions aside.

belong to a single system. In this respect the theoretical basis of classical psychoanalysis is overassumptive. But the inconceivability with which it has often been charged stems from a philosophical error that is remediable—the belief that desires, ideas or emotions cannot be psychologically engendered and psychologically modified if they are essentially physiological processes, so that physiological psychology and "dynamic" psychology are rival sciences. As soon as feeling is regarded as a phase of a physiological process instead of a product (perhaps a by-product) of it, a new entity metaphysically different from it, the paradox of the physical and psychical disappears; for the thesis I hope to substantiate here is that the entire psychological field—including human conception, responsible action, rationality, knowledge—is a vast and branching development of feeling. This does not mean that all reasoning is "really" rationalization, all judgment "really" emotional, all moral intentions specious, and so on. There is not some primitive form of feeling which is its "real" form, any more than a bird is "really" an egg or water is "really" a vapor. Emotion as we know it is not even a primitive form of feeling; it is not a rudimentary nervous process, such as fairly simple organisms might exhibit, in a psychical phase. Human emotion is phylogenetically a high development from simpler processes, and reason is another one; human mentality is an unsurveyably complex dynamism of their interactions with each other, and with several further specialized forms of cerebral activity, implicating the whole organic substructure. Our knowledge of neural functions is as yet very scrappy and tentative, but I think research has reached the point at which the understanding of these specializations becomes a scientific target rather than a piece of science fiction.

As there are many distinct nervous processes, some originating at the periphery of the central nervous system, others within it, especially in its core which is the brain, so there are many ways in which activities may be felt. The most important distinction within the realm of feeling is between what is *felt as impact* and what is *felt as autogenic action*.[38] The existence of these two fundamental modes of feeling rests on the nature of vitality itself. The pattern of stimulus and response, the guiding principle of most psychological techniques, especially in the laboratory of the animal psychologist, is a simplified schema derived from that natural division. In work with animals, the overt response has to take the place of events "felt as action," since we do not know what animals feel. We cannot know what impact they experience; there are some amusing cases on record where the behaviorist and the behaving

---

[38] The phrase "felt as action" will sometimes be used in opposition to "felt as impact," and is to be understood to mean "felt as autogenic action."

animal seem to have disagreed about what was the stimulus.[39] In human psychology the stimulus-response schema has only a limited value, because cerebral activity is so high that an isolated impact like a slight electric shock does not find the way clear to evoke a distinct correlative response. But human psychology can resort to protocol statements, and so approach its subject matter more directly.

What is felt is always action in an organism, but some of it is felt in a special way, as encounter; and, moreover, this mode is not entirely peculiar to events of peripheral origin, though it is their characteristic mode. There will be much more to say about this; for the present we may consider the two modes of feeling as given typically with the fundamental relation of living systems to the surrounding world. A closer look at that relation may serve to explain the difference between these two chief ways of feeling organic processes.

Many biologists and social philosophers have remarked the importance of the constant interchange between a living creature and its environment. There is a vast literature on adaptation and "fitness," some of it indispensable to philosophical insight as well as to biology proper, so I shall draw upon it constantly. Just now, however, we are concerned with the general and patent fact that an animal organism is always under the influence of the world around it, and in its turn always affects at least its immediate neighborhood, if only by using up oxygen, exhaling carbon dioxide, radiating warmth, even when it is not eating anything or excreting its waste. Plant organisms take nourishment all the time and transpire all the time unless they are deeply dormant.[40] A botanist has

[39] H. S. Liddell, in "Adaptation on the Threshold of Intelligence" (1949), tells of one of Pavlov's dogs that did not await the signal for food to which it was "conditioned," but salivated and licked its chops at the click of the metronome in the laboratory, and later—in the same session—at the sight of a bell clapper used in the experiment. He further cites similar cases of his own (p. 58).

Consider also the ambiguity of "stimuli" noted by N. L. Munn in "Pattern Brightness Discrimination in Raccoons" (1930): "Lashley had found that rats could discriminate complex visual patterns such as those used by Fields. . . . We obtained results similar to those obtained by him until we rigidly controlled the size-brightness factor, whereupon the rats could no longer discriminate" (p. 5). Probably every experimenter with animals has faced this difficulty. Wilson McTeer encountered it in applying a simple stimulus to human beings and measuring the response; he was measuring effects of an uninvited stimulus of which the experimenter was unaware. But this case will engage us later, in a more important connection.

[40] Some, in the form of seeds, seem able to suspend activity almost indefinitely in an undisturbed, unchanging environment; grains found in an Egyptian tomb are said to have been still in a state of suspended animation thousands of years after the burial. J. T. Bonner, apparently incredulous of such reports, nevertheless writes in *Size and Cycle* (1965), p. 66: "Although the extent of seed longevity has been

summed up the normal situation by saying that "the plant and its environment may be looked upon as two interacting systems, the plant system comprising a complex of organs, tissues, and cells, the environment system a complex of activating agencies."[41] C. J. Herrick, who is a neurologist, writes in a similar vein of animals that "the vital process is essentially a special type of mutual interaction between the bodily mechanism and the forces of the surrounding world, of the correspondence between the organism and the environment. . . ."[42] However different a plant and an animal—especially an animal with a nervous system—may be, they bear the same essential relation to their surroundings, a constant interchange of matter.

One trait of that interchange, however, is rarely if ever explicitly stated, perhaps because it is so familiar that it seems self-evident to every biologist: it is the asymmetrical character of the interaction. The exchange is not strictly speaking a relation between two systems, for the environment is not a system in the same sense as the organism is. It is not so much "a complex of activating agencies" as a conglomerate of them, some in short supply, some unlimited, but not systematically organized among themselves, nor meted out with any respect for the organism's uses. There may be barely enough food, and a thousand times more than enough air, water and light; and the surrounding space will usually take up excretions and exudations to any degree. The exchange of matter is, therefore, not really a mutual transaction, but one in which the inanimate world has the gross control, while the fine control rests with the organism. This asymmetry of the two factors in the many-sided, continuous pattern of exchanges makes the organism a center of activity.[43] The environment may also enfold other organisms, but to the organism in point they are parts, though perhaps rather special parts, of the surrounding world. In extreme cases the whole relevant environment

---

greatly exaggerated, there are known cases of seeds being successfully stored over three hundred years. This means that for this great period the embryo remained in what amounts to suspended animation."

[41] George E. Nichols, "The Terrestrial Environment in Its Relation to Plant Life" (1924), p. 37.

[42] *Introduction to Neurology* (1931), p. 18.

[43] Cf. K. F. Muenzinger, "The Need for a Frame of Reference in the Study of Behavior" (1954), 227–28: "Psychological events occur within the microcosm of a situation of which a single personality is the center . . . . behavior is always the behavior of one whole personality in its unique situation. . . ." For physics "there is a continuity of texture, so to speak, from one place of the universe to any other place. Behavior takes place in as many microcosms as there are personalities, animal or human" (I would say "individuals, animal or human").

may be another organism, e.g., for parasites living in the bodies of other creatures.[44] Even here, what makes the parasite an organism in an environment is that it has fine control of the exchange of matter, whereas the organism it has invaded has, with respect to it, only gross or indefinite control of the constant transaction. If the parasite's waste is non-poisonous to the host and its ingestion of matter is not excessive, the host may survive, i.e., remain an organism itself. The parasite within its own confines is part of its world, something to be dealt with. Such anomalies, which strain our categories and present "borderline" cases, are typical of the whole realm of life. All divisions we can make are somewhat fluid, and nearly all general rules have exceptions.

An organism is a continuous dynamism, a pattern of activity, basically electrochemical, but capable also of large, concerted forms of action with further principles of organization. Its round of functions is continuous, and every constituent action in it, every chemical transformation, osmotic exchange, contraction of muscle, discharge from glands, etc., is more or less finely adjusted to the immediate situation created by the environment. Many of these activities are internal to the organism, so the direct environment of the tissues engaged in them is composed of other parts and products of the same system. Environment is, in fact, a relative concept. The active structures in any metazoic organism are incredibly detailed, and its activities are usually composed of smaller and smaller complete cycles of action, in subordinate but functionally distinct structures. Each active unit, be it a single cell or an integral complex, has its direct environment as long as it exists, i.e., performs. All vital action, whether of the organism as a whole in its surroundings or of an organ internal to it, is interaction, transaction, in which the functioning unit has the fine control, and the medium in which it maintains itself has the gross control; that is, the latter determines what is given, the former what is taken.

The provision, usually not precisely measured for the needs of the agent, comes as impact, and has to be dealt with; as it makes action possible it also makes it compelling.[45] The organism, *in toto* and in every one of its parts, has to "keep going." Every act of a living unit transforms

---

[44] The young in the womb might also be cited, but as they are in process of individuation their status as organisms (though not as organic structures) is not complete.

[45] Hans Jonas remarked this fact in his article "Motility and Emotion. An Essay in Philosophical Biology" (1953), p. 119, where he concludes an exposition with the words: "Thus animal metabolism makes mediate action possible; but it also makes it necessary."

its situation and necessitates action under the impact of that new development as well as of any fortuitous changes coinciding with it.[46] This is what Whitehead called the "creative advance" of nature. It is certainly the pattern of life.

The structure of every entire organism reflects this basic functional law: whether it be a vertebrate covered with skin or a unicellular creature with a semi-permeable cell wall one or two molecules thick, its periphery is adapted to the exigencies of contact with the plenum of external events. The organism may or may not react with far-reaching effects upon its surroundings, but in any case it presents some mechanism which filters the impinging influences, letting only some substances or some propagated vibratory patterns (temperature, light rays, etc.) go through it for the rest of the body to meet. The periphery is specialized to handle emergencies. Its activities are normally of an improvisational character, adaptable, quickly responsive. If they rise to the level of being felt they are generally felt as impact; that is, without preliminaries, presenting themselves almost instantly, and with the semblance of coming inward from outside. This last aspect has generally not been recognized. Theories of perception which assume that feeling occurs only at the inner terminus of nervous activity, and in doing so does not express the transitional dynamic pattern, have therefore to meet the problem of how percepts are "projected" into an outer world instead of seeming to be inside us, like thoughts.

Sensations are a class of felt activities. The special organs of sense that develop as permanent structures in all but the lowest animal forms are peripheral areas highly elaborated to deal with impingements which the rest of the body covering filters out entirely, or admits in a general way that creates no elaborate pattern, and which often activate no organic processes which have any special psychical phase. Our sense organs are "processing" mechanisms, and also peripheral outposts protecting the active deeper parts which are less capable of improvising responses.[47]

---

[46] At the same time, its own state changes with every act. Compare I. D. London, "Some Consequences for History and Psychology of Langmuir's Concept of Convergence and Divergence of Phenomena" (1945), p. 184: "H. M. Johnson has . . . directed my attention to . . . the fact that living organisms constitute hysteretic systems, in which every response sets up mnemic traces in their structure, so that the next stimulus impinges on a system that may differ quite markedly from the original. Consequently, the act of producing a response may so change the system that the second stimulus, though externally equivalent to the first, produces a different response."

[47] Cf. J. Donald Harris, *Some Relations between Vision and Audition* (n.d., copyright 1950). Speaking of vision, Harris says: ". . . the sense organ acts as a

The high and varied elaboration of those receptor organs causes their activities to be felt not only as impact, but as qualitatively different kinds of impact, the incommensurable deliverances of sight, hearing, smell, taste and the various tactile senses, respectively. Sensations arise from within the body, too, but more typically the sources of sensations are peripheral; and they always carry, however vaguely, some indication of an impingement met and dealt with. Taken together they constitute a major department of feeling, sensibility.

The counterpart to this essentially centripetal action springs from the constant functioning of the central nervous system itself (or whatever precedes it in the lower forms of life). An act engendered predominantly from within, such as the "firing" of an assembly of neurons as a result of organic processes without specific external stimulation,[48] if it is felt, is differently felt from a response to environmental impact. Its psychic phase begins gradually, rising from a background of general body feeling and a texture of emotive tensions so closely woven that the separate strands of process in it are not distinct, but compose a "mental state," which includes the body feeling as a constant somatic factor, usually very near the threshold of the psychical phase, playing across it. Against this background, more specifically articulated acts are felt as envisagements, intentions, cogitations, insights, decisions, etc. They are felt to arise without the attack of a sensory impact, and to proceed from within us toward their termination in expression, which may be muscular motion, or the forming of an image apparent in the space between our eyes and their focus point, or an act of tonal imagination proceeding still more vaguely toward the outside world. Their neural paths are probably centrifugal from areas of most intense activity in the higher parts of

---

protecting layer, or buffer, between the stimulus and the nervous system.

"Something of the same sort of thing goes on in audition, in that the ear structures serve to protect the organism against the sometimes quite intense acoustic event . . ." (p. 11).

Heinrich Husmann, in "Der Aufbau der Gehörswahrnehmungen" (1953), p. 98, makes a similar observation, though he regards the transformation of the stimulus as a change to a more congenial form rather than a mere damping of the impact: "Das Ohr selbst ist ganz wie der Augapfel, insbesonders die Linse des Auges, ein den eigentlichen Empfangsorganen vorgeschalteter . . . Apparat. Er hat . . . die Aufgabe, die ankommenden Schwingungen so umzuformen, dass sie von den Empfangsorganen des Nervensystems möglichst einfach aufgenommen . . . werden können."

[48] The nature and extent of genuinely autogenic activity in higher organisms will be discussed at length in another connection, and some existing evidence of it will be adduced there.

the brain, stimulated from other cortical centers and from deeper centers, but not directly from the peripheral organs of perception. Since they are not acts of coping with constantly and ineluctably impinging forces, their structure is organic, developmental, rather than reactive and improvisational. Perhaps that is why they are felt more intimately as our own actions. Their rhythm is recognizable as vital rhythm. They are felt as autogenous action springing from the central parts of the organism, rooted in its emotivity; while perceptions are largely felt as impact, starting from the organism's peripheral organs as external forces play on its gamut of sensibility.

If one conceives the phenomenon of being felt as a phase of vital processes, in which the living tissue (probably the nerve or a neuronal assembly) feels its activity,[49] the problem broached but abandoned as hopeless by Dr. Penfield—namely, of how nerve impulse can be "converted" into thought and thought into nerve impulse, which he designates, rightly indeed, as a "strange conversion"—becomes a different sort of problem. The question is not one of how a physical process can be transformed into something non-physical in a physical system, but how the phase of being felt is attained, and how the process may pass into unfelt phases again, and furthermore how an organic process in "psychical phase" may induce others which are unfelt. Such problems, even if far from solved, are at least coherent with the rest of biological inquiry and logically capable of solution.[50]

The proposed new concept of feeling, furthermore, permits a new way of construing the greater concept of mind. Instead of accepting "mind" as a metaphysically ultimate reality, distinct from the physical reality which subsumes the brain, and asking how the two can "make liaison," one may hope to describe "mind" as a phenomenon in terms of the highest physiological processes,[51] especially those which have psychical phases. That is the purpose of this book. The construction of the

[49] Ostow (see above, p. 22n) also proposed that the activity is felt in the organ performing it, and a few years later, apparently independently, Bernhard Rensch expressed the same view in his *Psychische Komponenten der Sinnesorgane* (1952). But a primary center may not be sufficient. This issue will receive discussion later.

[50] It also makes the question which Dr. Gooddy (see above, p. 9) rejected as an invalid problem—"Where does the neurophysiological state break through to feeling and psychological states?"—a perfectly valid one; the "breakthrough" is not a transformation of a neural "state" into an ontologically different psychological "state," but the heightening of a neural process to the stage of being felt in the active organ itself.

[51] What is meant by a "high" physiological process will be discussed in a later chapter.

concept of "mind" is not simple, but it involves some major working notions and interesting sidelights: for instance, a cogent reason why only man can be properly said to have a mind, though all the higher animals (I do not know how far down the biological scale) have mental functions; what the radical difference is between animal and human societies; why the mind appears so forcibly as a separate entity, a "being" independent of the physical organism; and other incidental views and decisions emerging more or less unexpectedly from the main work, which is the formulation of basic concepts capable of scientific development, and adequate for a study of mental life in any direction and to any extent that we want to carry it. Instead of looking for a point of "liaison between the brain and the mind" we may look for a psychical limen in the rise and abatement of cerebral processes. But, in comparing the hypothesis of such a limen with Dr. Eccles' observation on the conditions for a mind-brain connection, I find the hypothesis so suitable to his problem that it might have been fashioned for it, when he says: "The term 'mind' will be restricted to 'conscious mind' in all its general operational field of perceiving, feeling, thinking, remembering and willing,"[52] and afterward: "Only when there is a high level of activity in the cortex . . . is liaison with mind possible. By activity is meant the active neuronal response, i.e., generation of impulses in the neurones of the cerebral cortex. It would appear that unconsciousness supervenes instantly the activity is lowered below a critical level. . . ."[53]

Such conditions of consciousness are certainly plausible enough if we view the mental phenomenon not as a product of neural impulses, but as an aspect of their occurrence. We might compare the phenomenon of being felt to a less problematical one in everybody's experience. When a tree leans over a quiet surface of water, its visible form is reflected by the surface under normal conditions of daylight. The tree's being reflected is an aspect of the whole natural situation; a complex situation, but of frequent occurrence. There is no ontologically real but non-physical "inverted tree" produced by the physical interaction of the upright tree and the water. With many objects and mirrors we can produce reflections at will, and place ourselves to see or not see them at will. We can turn the rear-view mirror in a car to abolish the reflection of lights following us, or shift our own position so as not to receive it. Also, in using the mirror for backing up, we learn that right and left in kinetic space are represented by their opposites in the peculiar visible phase of the mirror surface caused by the situation; but there are no "mirror things"

[52] J. C. Eccles, *The Neurophysiological Basis of Mind* (1953), p. 260.
[53] *Ibid.*, p. 265.

intercalated among the objects we negotiate as we back into a space. Yet the reflection is a genuine occurrence, a phase of the mirror, and, indeed, the only phase that makes the mirror important to us.

The changed concept of feeling, which calls almost at once for the distinction between a sense of autogenic action and a sense of impact, roughly establishing the realms of emotivity and sensibility, suggests a corresponding conceptual change with regard to the terms "subjective" and "objective." These terms are derived, of course, from the traditional dichotomy of subject and object, perceiver and perceived; the basic metaphysical dualism which Carl Stumpf declared to be simply given in the nature of things and ineluctable.[54] In the context of organic activities here assumed, that dichotomy is not simply given, and whether it can and should be theoretically constructed remains to be seen. Yet "subject" and "object" do mean something, whether their usual intension is acceptable or not; they are important words, especially in such derivative forms as "subjective" and "objective," "subjectified" and "objectified." Sometimes, in making a systematic framework, it is more profitable to define terms in a special form than to start with the grammatically normal; and it is the adjectival form that will be defined here. By "subjective" I mean whatever is felt as action, and by "objective" whatever is felt as impact. Cognate meanings will be logically derived as they become relevant.

The first consequence of these definitions is that one does not find a class of objective things, with which the scientist is concerned in his laboratory, and another class of subjective things which are scientifically embarrassing. Any felt process may be subjective at one time and objective at another, and contain shifting elements of both kinds all the time. "Subjective" and "objective" denote functional properties. Since organic functions have dynamic forms, which they build up and melt down again constantly, their identifiable properties are transient. The properties in question are two possible modes of feeling, i.e., of psychical phases of activity.

The real value of the concepts of subjectivity and objectivity here proposed can only become apparent in later parts of this essay, where they

---

[54] "Erscheinungen und psychische Funktionen" (1906), p. 14: "Statt Ausdehnung und Denken, sagen wir allgemeiner (aber den Intentionen Spinozas wie Descartes entsprechend) Erscheinung und psychische Funktion. In diesem Punkt ist in der Tat weder Spinoza noch einer der Späteren über den Dualismus von Descartes wirklich hinausgekommen. Das uns gegebene Tatsachenmaterial zeigt eben schon in der Wurzel ein Doppelantlitz, und was man auch weiter über Einheit der Substanz und der Realität, über Pansychismus, universalen Idealismus sagen mag: *diese* Zwiespältigkeit ist nicht wegzubringen."

serve to describe the structure of experiences so as to let us guess, at least, at some vital processes that may be involved in the occurrence of those experiences. The credentials of any definition are found in its implications and the constructions it permits. Such are, of course, the virtues of the definition of feeling, too, which cause me to adopt it: it can be intellectually exploited, it has more illuminating implications than any other within my ken. Not only does it make feeling conceivable as a natural phenomenon, and some valuable subordinate notions (like "subjective" and "objective") derivable, but it permits one to construe the more impressive forms of mentation—symbolic expression, imagination, propositional thought, religious conception, mathematical abstraction, moral insight—as functions of that most active and complex of all organs, the human brain, with intense and prolonged psychical phases.

Feeling stands, in fact, in the midst of that vast biological field which lies between the lowliest organic activities and the rise of mind. It is not an adjunct to natural events, but a turning point in them. There must have been several such turning points in the evolution of our world: the rise of life on earth, perhaps the beginning of irreversible speciation, the first true animal form, the first shadows of a "psychical phase" in some very active animal, and the first genuinely symbolic utterances, speech, which marked the advent of man. It is with the dawn of feeling that the domain of biology yields the less extensive, but still inestimably great domain of psychology.

That is why I make feeling the starting-point of a philosophy of mind. The study of feeling—its sources, its forms, its complexities—leads one down into biological structure and process until its estimation becomes (for the time) impossible, and upward to the purely human sphere known as "culture." It is still what we feel, and everything that can be felt, that is important. The same concept that raises problems of natural science takes one just as surely into humanistic ones; the differences between them are obvious, but not problematical. Society, law and institutions create their own issues to be appropriately resolved. An adequate concept of "psychical" should serve all psychological purposes.

2

Idols of the Laboratory

When half-gods go, The Gods arrive.
—R. W. Emerson

THE social sciences, originally projected by Auguste Comte in his sanguine vision of a world reformed and rationally guided by science, have finally come into recognized existence in the twentieth century. They have had a different history from the natural sciences which grew up chiefly in the seventeenth and eighteenth centuries under the name of "natural philosophy," and gradually took shape as formal and systematic pursuits. Astronomy, mechanics, optics, electronics, all merging into "physics," and the strange kettles of fish that became chemistry, had a free and unsupervised beginning; and the most productive thinkers of those maverick days ventured on some wild flights of fancy. Not so the founders of the "young sciences" today. They cannot indulge in fantastic hypotheses about the aims or the origins of society, the presence of sentience or intellect in anything but the investigator himself, the sources of fantasy, the measures of animal and human mentality. They began their work under the tutelage of physics, and—like young ones emulating their elders—they have striven first and hardest for the signs of sophistication; technical language, the laboratory atmosphere, apparatus, graphs, charts and statistical averages.

This ambition has had some unfortunate effects on a discipline for which the procedures of classical physics, for instance, the experimental techniques of Galileo, may not be suitable at all. It has centered attention on the ordering and collating of facts, and drawn it away from their own intriguing character as something distinct from the facts encountered by the physicist, and perhaps differently structured. The main concern of the early physicists was to understand puzzling events; each scientific venture grew from a problem, the solution of which threw unexpected light on other problematical phenomena. It was always in

[33]

such a light that the concepts of physical science were set up. But the chief preoccupation of the social scientists has been with the nature of their undertaking, its place in the edifice of human knowledge, and—by no means last, though seldom candidly admitted—their own status as scientists. For decades, therefore, the literature of those new disciplines, especially of psychology, has dealt in large measure with so-called "approaches," not to some baffling and challenging facts, but to all the facts at once, the science itself. Every theoretical thinker in the field set out to define and circumscribe this science and propose a strict proper method for its pursuit, until a rotating program committee seems to have had the lion's share in the whole venture.

At about mid-century, Egon Brunswik published a critical survey entitled *The Conceptual Framework of Psychology* (1952), reviewing the development of his subject from its inception in the time known as the Enlightenment, when the notion of mental phenomena as a department of "natural philosophy," i.e., as something to be empirically viewed and studied, had its beginning. The first formulations, of course, were inconsistent enough, mixing theological concepts with the Lockean concept of experience, and both with the traditional view of the physical world as a realm of passive "matter" moved by forces extraneous to it. Inevitably, in such an intellectual setting, the attempt to treat of the mind in "the new scientific manner" precipitated a battle of metaphysical tenets: materialism, idealism, mechanism and all intermediate hybrid doctrines. At long last, a fairly distinct pursuit called "psychology" took its place in the universities. This early history is well known, and Brunswik does not rehearse it, but passes quickly to the recent period, which Sigmund Koch has called "the Age of Theory"[1]—the twenties, thirties and forties of our century—roughly, the contemporary scene at the time of Brunswik's writing.

If, as Koch maintains, there has lately been some slackening of the pace in program-making, that is presumably because a saturation point has been reached. None of the "frameworks," so far as I know, has been expressly given up. Psychology as Brunswik surveys it still exhibits the same welter of "isms," with some two hundred years' growth behind it. Besides the classical "isms" already mentioned, and, of course, behaviorism, we find reductionism, emergentism, physicalism (indeed, thematic, emulative and methodological physicalism); introspectionism, and hedonistic introspectionism; elementarism, sensationism, nativism,

---

[1] In his "Epilogue to Study I," in Vol. III of *Psychology: A Study of a Science*, projected in seven volumes under his editorship, and so far represented by six published volumes (1959–63), a work to which frequent reference will be made.

intuitionism, intentionalism, situationism; behavioral environmentalism versus hereditarianism; micro-mediationism, and micro-mediational phys-iologism which leads to nomothetic encapsulated centralism. Despite his sound and often astute critique of various methodological failings which hamper the young science (see especially chap. iii, "Misconceptions of Exactitude in Psychology"), the major obstacles seem to be unremoved by his own circumvention of them, which is a program of "probabilistic functionalism."

In proportion to the effort spent on all these "approaches," the harvest of interesting facts and systematic ideas remains meager (which, of course, is not to say that there are none). By comparison with other biological sciences, both the method and the findings of laboratory psychology look extremely simple. Their essential simplicity is sometimes masked by a technical vocabulary, and even an algorithmic-looking form of statement; but when we come to the interpretation of the "variables," their values prove to be such elements as "somebody," "some fact," "some object," without any formal distinction between a person, a fact and an object, which would make it logically not possible to interchange them, as it would have to be to assure the formula any sense. We find, for instance, the definition: " 'Culture' $= \hat{y}$ (x learns y from z and $x \neq z$).' "[2] We need the gloss which tells us that y is a response and x and z are persons (here called "hominid individuals"), so that "x learns y from z" defines $\hat{y}$ as the class of responses learned by someone from someone else. Logically, of course, the relation "learns," which is unfor-malized, is all that tells us that if "x learns y from z," we cannot have "z learns y from x" with the same value of y, and, above all, that "y learns z from x" makes no sense. Symbols of the same category represent members of entirely different categories. There can, then, be no logical rule for manipulating them; the formula is purely specious. In another study, entitled "Verbal Behavior Analysis,"[3] formal behaviorist language is freely mixed with ordinary language to interpret it, so we find, for instance: "Many investigators have engaged in studies of verbal behavior using various stimuli to produce verbal activity, such as the clinical inter-view, the interpretation of proverbs, and Thematic Apperception Test cards." There is a table of elaborately classified stimulus words, wherein a general heading of "material objects" covers such subheadings as "Self," "Self-Other," "Others" and "Things"—all familiar enough,

[2] Mortimer Brown, "On a Definition of Culture" (1953), p. 215. The "defini-tion," with its interpretation, is quoted from O. K. Moore and D. J. Lewis, "Learning Theory and Culture," in *Psychological Review*, LIX (1952), p. 385.

[3] L. A. Gottschalk, G. C. Gleser and G. Hambridge (1957), p. 300.

informal ideas; but the responses of the experimental subjects are related to "internalized restrictions" on "desire to take in food," etc.[4]

To speak of "hominid individuals" instead of "persons" and of "verbal behavior" instead of "speech," of a clinical interview as a "stimulus to verbal behavior," and so on, is to translate ordinary thinking into a jargon for literary presentation. Jargon is language which is more technical than the ideas it serves to express. Genuine scientific language grows up with the increasing abstractness or extraordinary precision of concepts used in a special field of work, and is therefore always just adequate to express those concepts. It is not deliberately fixed (with the exception of Latin nomenclature in taxonomy), but may become completely technical if scientific thought moves very far away from ordinary thought. Jargon, on the other hand, is a special vocabulary for common-sense ideation. It is an Idol of the Laboratory, and its worship is inimical to genuine abstractive thinking. A sociologist or psychologist who will spend his time translating familiar facts into professionally approved language must surely have more academic conscience than curiosity about strange or obscure phenomena. We are often told that such exercises are necessary because the behavioral sciences are young, and must establish their formal rules, vocabularies and procedures. Not long ago chemistry was young, too; its modern history only began with Lavoisier, who died in the Reign of Terror. But did any chemist ever write an article to show how a recipe for fudge could be stated in proper chemical form, i.e., without any household words?

Another source of idolatry is the cultivation of a prescriptive methodology, which lays down in advance the general lines of procedure—and therewith the lines of thought—to be followed. According to its canons all laboratory procedures must be isolated, controllable, repeatable and above all "objective." The first three requirements only restrict experimentation to simple responses, more significant in animal psychology than in human contexts; but the fourth is a demon. The Idol of Objectivity requires its servitors to distort the data of human psychology into an animal image in order to handle them by the methods that fit speechless mentality. It requires the omission of all activities of central origin, which are felt as such, and are normally accessible to research in human psychology through the powerful instrument of language. The result is a laboratory exhibit of "behavior" that is much more artificial than any instrumentally deformed object, because its deformation is not calculated and discounted as the effect of an instrument. Here, indeed, is a

[4] *Ibid.*, p. 302.

[36]

critical spot where haste to become scientific destroys the most valuable material for investigation. It completely masks the radical phylogenetic change induced by the language function, which makes one animal species so different from others that most—if not all—of its actions are only partially commensurable with those of even the nearest related creatures, the higher primates. For the revelation of subjectively felt activities through speech is not a simple exhibit from which the observer infers the action of external stimuli on the observed organism; in a protocol statement, the dividing line between the observer and his object is displaced, and part of the observation is delegated to the experimental subject.[5] Language, in such a situation, does not belong entirely to the behavioral exhibit; the act of using it is a part of the psychological material, which may or may not be relevant to the intended observation, and so are the subjective phenomena to which it refers; but the semantic function of the words is part of the perceptual medium, the instrument of observation. The speaker, who is the experimental subject, automatically participates in the work; both he and the experimenter handle the semantic instrument. If his observation or power to report is inadequate the instrument is crude. Many vitally important data have to be treated as insecure because the experimental conditions did not permit a definitive view. This is as true of protocol statements as it is of one-way screens, tachistoscopes and galvanometers.[6]

The relativity of the subject-object division, which lets that division come now at the eye, now at the lens of the microscope, or—in psychological observation—now between the experimenter and the experimental subject, now between the latter's report and its matter of reference, is one of the serious instrumental problems for scientific psychology. But it is more than that; it is one of the most interesting phenomena characterizing human material itself. Its extraordinary interest and its exact

[5] The lability of that dividing line, and with it the difficulties of defining directness of observation, were first observed by Niels Bohr and were given serious philosophical consideration by Victor F. Lenzen in *Procedures of Empirical Science* (1938), pp. 28 ff.

[6] A good instance in point is the experimental material which R. G. Bickford *et al.* present as "Case I" in their neurological study, "Changes in Memory Function Produced by Electrical Stimulation of the Temporal Lobe in Man" (1958), p. 231, where they preface their discussion by saying: "In discussing the significance of our stimulation findings on this patient, we must emphasize that our observations were severely limited in scope owing to his intellectual deficiencies. He showed little interest in, or capacity for, subjective observation, with the result that we were unable to settle satisfactorily the nature of the 'hollering' that he heard during stimulation."

significance will become apparent later, in connection with the theory of symbolic activity in general. The fact that I wish to point out here is that to make a fetish of "objectivity" means to assume, in the first place, that some phenomena are intrinsically objective and others intrinsically subjective so that they can be accepted or rejected accordingly; it is one of the tacit assumptions which have frustrating metaphysical implications, and lead some great biologists and pathologists to accept strange philosophical doctrines as the only possible supports for those assumptions. In the second place it means that problems of the relationships between subjective and objective factors in mental activity are removed from the psychologist's proper sphere of investigation. These relationships, and the terms that develop in conjunction with them—symbols, concepts, fantasy, religion, speculation, selfhood and morality—really present the most exciting and important topics of the science of mind, the researches toward which all animal studies are oriented as indirect or auxiliary moves. To exclude such relationships for the sake of sure and safe laboratory methods is to stifle human psychology in embryo.

The emulation of physical science naturally leads to premature attempts at mathematical expression of known or presumed regularities in behavior. In the interest of the quickest possible mathematization the observed facts may be pared down to a few overt actions in situations which elicit the simplest and most invariable movements. An almost paradigmatic instance of this practice may be found in D. G. Ellson's article, "The Application of Operational Analysis to Human Motor Behavior,"[7] which he essentially reiterated and defended a decade later in Koch's collaborative survey.[8] In his first essay Ellson proposes to compute "the effects of stimulus variables (in this case, existing or conceivable design characteristics of machines) upon response characteristics (accuracy, speed, frequency, latency of movements, etc.)" of human operators. He adds an appendix presenting the basic equations required by the program, but it is written by another person, "who made this article possible by clarifying my ideas concerning the mathematical aspects of operational analysis. . . ."[9] On this slender stalk of mathematical knowledge is grafted the method of precalculating human motor behavior. "The method," Ellson claims, "is a direct parallel of the stimulus-response approach in psychology, which is concerned with the relationships between stimuli and responses rather than with intervening physio-

[7] (1949), p. 9.
[8] "Linear Frequency Theory as Behavior Theory" (Koch II, 1959).
[9] P. 9n.

logical mechanisms."[10] Input and output in a machine "correspond to the stimulus and response in a psychological organism . . . and they may be investigated without reference to intervening mechanisms."[11] But engineers do, in fact, understand the intervening machine, and their correlations of input and output would have no scientific value—though, perhaps, some limited practical use—if they did not. The "direct parallel" presupposes, furthermore, that input and output are known, and "that all individuals are linear systems"—that is, that each stimulus element will keep its identity through its transformation from input to output, so that the total response will be the arithmetical sum of those transformed stimuli. Unfortunately, however, this does not seem to be the case, and Professor Ellson is aware that it is not; but he is undaunted by adverse experimental findings. They may be wrong, or at least atypical; there may be circumstances in which the empirical facts would agree with the required assumptions, and then those circumstances could be set up as standard. By restricting the human "input" in a "man-machine system" to a few simple movements and manipulating other conditions one might make the input-output of such a system calculable.[12] The stakes he has put up warrant the long chance; for, as he puts it: "In requesting information concerning the frequency characteristics of human behavior, the systems engineer has in effect handed the psychologist a mathematical model and suggested coordinating definitions which are sufficient to convert the model into an S-R [stimulus-response] theory. . . . If counterparts of the relationships between input and output variables specified by this model can be observed consistently between stimulus and response of man under any single set of limiting conditions, then the model becomes an empirical law of human behavior. If relationships described by the model are found for many different situations, e.g., for many tracking responses, task inputs, and limiting conditions, and these relationships are adequately systematized, then the model becomes a theory of human behavior."[13]

There are many other scholars in the programmatic sciences, sociologists as well as psychologists, whose immediate objective is to "mathematicize" their findings as fast as they are found and even faster, evidently feeling that the dress of mathematics bestows scientific dignity no matter how or where it is worn. The most obvious and therefore most popular means to this end is the use of statistics, often based on as few

---

[10] *Ibid.*, p. 16.   [11] *Ibid.*, p. 10.   [12] *Op. cit.* (Koch), pp. 652–57.
[13] *Ibid.*, p. 648.

as a score of cases, if not fewer.[14] Another, more imaginative, practice is to apply the terminology of higher mathematics—relativity, quantum theory, topology, doctrine of aggregates—to vaguely conceived behavioral items, which seem thereby to be quantified and established. L. L. Thurstone, for instance, speaks of "vectors of the mind,"[15] which are such traits as perseverance, generosity, quick temper; but, of course, the vector symbols which represent these traits in his diagrams do not really express any measurable directions or quantities, because traits are not activities, which might be precisely symbolized by vectors and manipulated with respect to each other. In another project[16] we meet dimensions of an experimental continuum, a "limited number" of which are said to define "a semantic space within which the meaning of any concept can be specified." Three of these dimensions, independent of each other, are evaluation, potency and activity. The resulting geometry threatens to be difficult. In the same vein, the "indeterminacy principle" is interpreted by A. J. Lotka as "freedom of the will," in a book too competent and interesting to permit of such argument.[17]

The cult of borrowed mathematical terms is especially pernicious when it invades serious original thinking, where there are really fundamental psychological concepts in the making, which are obscured and turned from their own implicit development by the unessential though enticing suggestiveness of scientific words. In a sober but trenchant critical article,[18] I. D. London has shown even so influential and impor-

[14] See, for instance, S. Beneviste, H. B. Carlson, J. W. Cotton, and N. Glaser, "The Acute Confusional State in College Students: Statistical Analysis of Twenty Cases" (1959); or L. Petrinovich and R. Bolles, "Delayed Alternation: Evidence for Symbolic Processes in the Rat" (1957), presenting statistics on the performance of sixteen rats, four of which were eliminated because they did not perform at all.

[15] "The Vectors of the Mind" (1934).

[16] E. T. Prothro and J. D. Keehn, "Stereotypes and Semantic Space" (1957). The "dimensions" are adopted from C. E. Osgood and G. J. Suci, "Factor Analysis of Meaning" (1955).

[17] *Théorie analytique des associations biologiques* (1934). On p. 31, the author introduces "le 'principe d'incertitude,' d'après lequel la nature même des phénomènes nous rend impossible de jamais connaître l'état d'un système physique assez en détail pour pouvoir déduire de cette connaissance le cours des événements ultérieurs." On the basis of this conception, he further argues: "L'incertitude de ce genre se fait ressentir particulièrement dans les phénomènes à l'échelle moléculaire ou au-delà, et du fait que les actions dependent grandement des phénomènes à cette échelle, nous avons tiré la conclusion provisoire que le libre arbitre pourrait se ranger définitivement dans l'ordre de la nature à la faveur du 'principe d'incertitude.' "

[18] "Psychologists' Misuse of Auxiliary Concepts of Physics and Mathematics" (1944).

tant a venture as Kurt Lewin's "field theory" to be only verbally modeled
on the field theory of relativity physics, since Lewin's key concept, force,
has no analogue in real topology, and his psychological "field" conforms
to no known geometry. The result is that Lewin can use none of the
powerful principles of substitution that make topology reveal new facts
in physical science. "Topology concerns itself with the properties of sets
of points. . . . Sets may be added and multiplied in the Boolean sense. . . .
The definitions of these two operations in topology are practically the
same as those given by Lewin. However, although these operations have
all the formal properties of ordinary addition and multiplication—a fact
of great importance—Lewin nowhere makes any essential use of
them. . . ."[19] Also, "a set is made equivalent to a region and a point to
a part-region, that is, a subset, in spite of the fact that a point can be
only the ultimate element of a set."[20]

"The host of theorems that form the actual machinery of topology
should have been made to function and so to take over the work of
rigorous deduction. Lewin in reality does not utilize *one single theorem*
of topology."[21]

Here, I think, we have the central and fatal failing of all the projected
sciences of mind and conduct: the actual machinery that their sponsors
and pioneers have rented does not work when their "conceptualized
phenomena" are fed into it. It cannot process the interpretations that are
supposed to be legitimate proxies for its abstract elements. Lewin, for
instance, designates chewing and swallowing as "regions." But can such
terms be generalized so that it is entirely a matter of formal manipulation to prove that they are or are not identical, that they can or cannot
be simultaneous, or what is the product of either of them if combined
with sneezing?

In many cases, the mathematization of behavior is the work (or better,
the dream) of non-mathematicians to whom single expressions or
propositions encountered in an alien field look analogous to some conceptual constructs of their own. The analogy would soon break down if
they could carry their interpretation through all the algorithmic possibilities of a formal system. An outstanding exception to this general rule
is the work of one of the most ambitious formalizers, Nikolas Rashevsky,
who has systematically developed techniques for computations of human
behavior, from the realm of reactions of the organism as such in a ma-

[19] *Ibid.*, p. 283.   [20] *Ibid.*, p. 286.   [21] *Ibid.*, p. 287.

terial world[22] to interpersonal behavior,[23] and further to a theoretical treatment of the latter in terms of the former.[24] Such, at least, is the plan and purpose of his elaborate mathematical system. Its terms are quite genuinely abstract, and are to be applied to measurable units of living matter or of its motions, or to any identifiable items of experience that can be ranged in quantitative orders, so that relations among such items may be numerically expressed and precisely known. Rashevsky offers examples of applications selected from various fields of "behavioral science," ranging from hypothetical cerebral processes to "socio-economic dynamics" (in which the equations add up to a Marxian pattern of inevitable economic events). But the applications in every case require such simplifications of the facts that the approximation of the instance to the form is very tenuous. For example, in regard to "the mathematical biology of the nervous system," he admits that he has used "highly oversimplified assumptions," but adds: "The oversimplifications will hardly bother a reader who is familiar with the developments in mathematical physics. . . . Has, indeed, a greater over-simplification ever been made than to consider actual molecules, with all the complexity of their electronic and nuclear structures, as rigid spheres?"[25] Some readers may be greatly troubled; for the simplifications in Rashevsky's psychological chapters, e.g., "Satisfaction and Hedonistic Behavior," are not of the same sort as the representation of molecules by rigid spheres. If every organism were treated (let us say) simply as a point of origin of acts, in order to furnish a defining function for a class "acts of A," this operation would be comparable to ignoring the structure of molecules. But it is the pattern of acts that is simplified in ways not irrelevant. The trouble with Rashevsky's calculations for "hedonistic behavior" is that they fit a very doubtful psychology, and do not engender any developments, such as theorems that prove to be exemplified in formerly undiscovered facts, or resolve mysteries by showing up subtle fallacies.

So, here as elsewhere, the intellectual results are disappointing. Such a powerful machinery should make the behavioral sciences rocket to success. Instead of that, it has so far left psychology, sociology, anthropology, and ethical theories just where they were before. The reason for the failure of even this well-equipped expedition into the mathematical realm is that abstract concepts borrowed from physics, such as units of

[22] *Mathematical Biophysics: Physicomathematical Foundations of Biology* (1938).
[23] *Mathematical Theory of Human Relations: An Approach to a Mathematical Biology of Social Phenomena* (1947).
[24] *Mathematical Biology of Social Behavior* (1951).
[25] *Ibid.*, p. viii.

matter—even with the adjective "living" to qualify them—and their motions, do not lend themselves readily to the expression of psychologically important problems. The aim of the work is to make mathematics applicable to a given material. In genuinely scientific research the aim is to explain events (their prediction is primarily a check on the validity of the explanatory hypothesis), and mathematics is not employed unless it is needed; but it is quickly needed, because the terms in which natural processes are conceived were originally abstracted from the exemplifying concrete system by observation of those processes themselves, and may therefore be elaborated and manipulated to any extent without danger of absurdity or of only trivial meaning. They are not descriptive categories under which things and their accidents are ranged, but abstract concepts which must be manipulated in order to describe anything at all. Newton's famous apple would not have revealed any astronomical relations if he had only relegated its falling to a general category of "descending" (John Dewey would have had him say "descendings"), and treated its fall from a tree to earth as a special case of "descending," to be subsequently differentiated from that of a cat's jumping voluntarily off a wall, etc. The relation of two masses, i.e., of apple and earth, respectively, as masses, in terms of which Newton described the fall, was not a visible relationship at all; for neither the quantitative proportion nor the mutual attraction of the two objects was a sensible aspect of the event he witnessed. Bulk is visible, but mass is not; and even the earth's bulk is not phenomenally given to our perception and amenable to our eye-measure for comparison with the bulk of an apple. The only obvious way to express the concept of the gravitational force obtaining between two objects so diverse was in abstract terms, which applied to the moon as well as to the apple in relation to earth, and to the earth in relation to the sun, and served to explain, by a very easy calculation of masses and distances, why the apple did not gravitate toward the sun; and together with the first law of motion, already established, the relation of masses explained why the lesser heavenly bodies moved in orbits around the greater ones. Were it not for the concepts of mass and gravitation, which do not themselves denote any natural objects or events, but are purely conceptual terms, there would be no similarity at all between the fall of an apple to the ground and the motion of the moon round the earth—let alone the swing of the tides under the influence of moon and sun.

The terms of the behavioral sciences, on the other hand, are not abstractions, but concrete elements about which general statements are made. Stimuli and organisms are concrete things classified by their informally established properties; responses are changes which can be predicted of organisms, but not of (say) a Skinner box; just as "falling" can be predicted of apples but not of the ground, nor of the heavenly bodies held to their orbits by gravitational forces.

There is no justification for "postulate sets," "theorems," or other systematic forms unless one operates with abstract concepts in terms of which rats, food, boxes, lights, shocks and all other relevant factors may be described and distinguished, treated as members of classes established by systematically derived defining functions, and related one to another (or to others) purely as such. But for this sort of systematization the basic concepts are lacking. The "systems," consequently, are very simple codifications of informally conceived facts, some of which have measurable features: repetitious phases, changing tempo, concomitantly controllable factors. Such data should, of course, be measured and compared, varied and recorded, so that their regularities are set forth as simply and clearly as possible. But a few regularities observed in a special field of interest, as for instance the gradients of approach and avoidance in animal learning processes, do not provide the framework of a science in a general field that is probably more complex than any inorganic physical system. Even the more ambitious "factoring" of human behavior is as yet only a taxonomic approach to that immense subject, with somewhat sharpened and refined common-sense principles of classification, which may lead analysis further than ordinary observation. But all such treatments are far from any mathematically exact science.

The deceptiveness of simple achievements as models for highly complex ventures becomes apparent when one follows rather closely the processes of generalizing, extending, or otherwise transferring the "laws" found in the animal laboratory to human life. One of the most serious attempts to do this is Neal E. Miller's "Liberalization of Basic S-R [Stimulus-Response] Concepts: Extensions to Conflict Behavior, Motivation, and Social Learning"[26]—a brave but probably last attempt to carry the general propositions (i.e., propositions stating regularities or "laws" of nature) of a system which describes the behavior of rats in experimental alleys over to human beings faced with a choice of lovers. The rat-calculus is orderly and verifiable enough, within its narrow limits. After establishing a gradient of approach to food and a steeper gradient of avoidance of electric shock, Miller provides a subheading, "Tests of Simple Deductions," under which he proves the hypotheses (1) "that there is a gradient of stimulus generalization of the approach responses," that is, new stimuli which are very similar to the one to which the rat

[26] (Koch II, 1959).

was originally trained are more readily responded to than less similar ones; and (2) "that increases in drive raise the height of the entire gradient," i.e., hungrier rats pull harder toward both similar and dissimilar goals than less hungry ones. This is, roughly, the framework of the S-R system which is experimentally established. The next subheading reads "More Complex Deductions," and here the illustrative matter suddenly shifts to a purely imaginary girl and her sweetheart and the man she is likely to marry if her sweetheart dies. "Stimulus generalization" is traded in for "displacement," in the clinical, not the ethological, sense,[27] and the whole terminology is shifted from its sober literal meanings of movement in a measured space, a definite condition of hunger, cues such as right and left or light and dark, controllable hazards, and goals that conclude the experiment when they are reached, to the figurative senses in which one speaks of "approaching" health or "avoiding" an idea. As Miller says:

"The definition of nearness is extended to apply to any situation in which the subject can be said to be coming nearer to a goal in space, time, or some dimension of qualitative or culturally defined similarity of cues.

"The definition of avoidance is extended to apply to the responses producing inhibition and repression.

"The definition of approach is extended to the responses that are inhibited or repressed."

These are, actually, not modifications of definitions, but reinterpretations of "S" and "R." With such new meanings, however, the gradients of approach, avoidance and stimulus generalization no longer fit any observable facts, so the parameters on which the measurements are based have to be shifted too, and the author further remarks: "The analysis we have made of the reasons why the gradient of avoidance is steeper than that of approach suggests that it might be better to classify responses on the basis of the way the motivation involved is aroused."[28]

Here, of course, is the vital difference between the conditions prepared for the derivation of the "laws" which govern the miniature system, and the conditions the liberalized version is to meet: the motivations of human beings in moral and emotional conflicts springing from a more

[27] Were it applied, here, to the acts said to be substituted for frustrated instinctive acts, such as preening for avian sex activity, which Nikko Tinbergen calls "displacement activities" (*The Study of Instinct* [1951], pp. 114 ff), a true extension of the study of animal activity might be made; but the S-R system does not admit the ethologist's theories, which center on mechanisms and treat stimuli as "releasers" of highly patterned whole acts.

[28] *Ibid.*, pp. 226–27.

or less permanent, oppressive social situation are so different from the motivations of a rat which has not been fed for forty-eight hours that they do not offer any comparable patterns. The fact that an intervening variable, "motivation," may be postulated in both contexts is the only point of similarity. In the starved rat this intervening variable may reasonably be taken to be hunger, but in the human case the nature of the motivation arising from complexes of ideas, feelings of potency or impotency, desires and fears, all inscrutably molded by a rampant imagination, constitutes the key problem in the psychologist's research. So Dr. Miller really offers an exemplary piece of understatement when, reviewing his "extensions," he concludes: "It is obvious that the definitions stated above are a considerable distance from the empirical level. Therefore, it will frequently be more difficult to secure completely agreement on the application of these definitions to clinical phenomena than it has been to secure agreement on the application of the preceding ones to simple experimental situations."[29]

The reason for the lack of agreement on the applicability of the liberalized concepts is that their liberalization is not effected by generalization, but by metaphorical use. Generalization widens the denotation of a term without changing its connotation, as a trade name denoting a specific manufacturer's product sometimes becomes the common name for this sort of product after the registration lapses; but metaphorical use changes the connotation of terms in an unspecified though often extreme fashion, and consequently makes their denotation indefinite. The moot problem of what verbal element is paralleled by what empirical datum is, however, not the most serious result of the metaphorical practice; even more serious is the fact that the liberalized concepts cannot be related in the same ways as the literally designated ones nor, indeed, in any systematic ways. If, for instance, "nearness" may be either a spatial or a temporal relation of a subject to a goal, then velocity can no longer be defined in terms of distance and time, nor drive measured by speed of running. There is, furthermore, no reason why gradients for "approach" in one sense should hold for "approach" in quite another, nor why fear of social embarrassment should follow the same "curve" as fear of electric shock. Analogies are essential to thought, but they cannot be automatically used to pass from known to unknown domains of nature.

It is interesting to find that some excellent experimenting psychologists are aware of this and other fallacies in the many advocated "approaches" to their projected science, including the danger that too

[29] *Ibid.*, p. 227.

much programmatic steering of research may be detrimental to its actual progress. They seem, in fact, to be recognizing the Idols of the Laboratory one by one—Physicalism, Methodology, Jargon, Objectivity, Mathematization—and to be taking leave of them, sometimes with declarations that they never worshiped them anyway, sometimes with apparent regret and apology. The latter is perhaps the more common, for the realization that the *Novum Organum* as modernized in Chicago a few decades ago is a will-o'-the-wisp comes painfully. So it is apparently a hard conclusion for a social psychologist to say, "It is not possible, as a rule, to conduct investigation in social psychology without including a reference to the experiences of persons. The investigator must, for example, take into account what the person under investigation is saying; and such utterances have to be treated in terms of their meaning, not as auditory waves, or sounds, or 'verbal behavior.' One can hardly take a step in this region without involving the subject's ideas, feelings, and intentions. We do this when we observe people exchanging gifts, engaging in an economic transaction, being hurt by criticism, or taking part in a ritual. The sense of these actions would disappear the moment we subtracted from our description the presumed mental operations that they imply."[30] And in justification of this heterodoxy he further remarks, "Heider has pointed out that we observe persons to produce effects intentionally. . . . In the case of persons, a cause is not merely a preceding state of affairs; it is a state of affairs as known or understood by the actor; an effect is not merely a later state of affairs; persons make things happen or intend them. The movements of persons thus gain the status of actions."[31]

Anyone who has never been on the program committee of the behavioral sciences will probably find these propositions quite easy to accept; they may even sound ironic. Yet in their contemporary frame they present some sensitive elements. One is the reference to "mental operations"—a genuinely difficult concept which needs not only to be admitted but handled, i.e., defined and established; the other is the key concept of "actions," which demands the same sort of explication. There are some other really significant observations, of which the most important—coming, as it does, from a social psychologist—is the statement: "At this time, I know no way of describing the psychological and sociological happenings within a single conceptual formulation." Is the

---

[30] Solomon E. Asch, "A Perspective on Social Psychology" (Koch III, 1959), p. 374.
[31] *Ibid.*, p. 375.

long-standing debate as to whether sociology should be regarded as a "branch" of psychology, or *vice versa*, perchance a specious issue?[32]

In the same compendium from which the passages above are quoted, several other leaders in the behavioral sciences declare their defection from the academic tin gods. The most outspoken are E. C. Tolman and E. R. Guthrie. The former says in his very first paragraph:[33] "I think the days of such grandiose, all-covering systems in psychology as mine attempted to be are, at least for the present, pretty much passé." And further: "I myself become frightened and restricted when I begin to worry too much as to what particular logical and methodological canon I should or should not obey." It is in the same spirit that Guthrie remarks, "The scholastics systematized a world of unripe knowledge and may thereby have protected it from its enemies, but at the same time they denied it the chance for progress."[34]

One of the best essays is the editor's own "Epilogue to Study I," at the end of the third volume, wherein he surveys the general change of heart that marks the most recent history of psychology. His reflections on this new turn—which is, perhaps, not quite as strong yet as he deems it—are on the whole so fair and circumspect that one would like to quote them *in toto*. The next best thing, I suppose, is to recommend his fifty-odd pages to the reader's attention, and to select a few significant passages to convey their trend. After reviewing the doctrine of "intervening variables," i.e., such concepts as "drive," "means-end readiness," "perceptions" and other hypothetical conditions assumed to lie between the observable stimulus and the observable responses, Dr. Koch writes: "First introduced in the early thirties by Tolman as a modest device for illustrating how analogues to the subjectivists' 'mental processes' might be objectively defined, the concept of the intervening variable was soon after elaborated by Tolman and others into a paradigm purporting to exhibit the arrangement of variables which must obtain in any psychological theory seeking reasonable explanatory generality *and* economy." Among other aspects of this idea which made it appeal to the programmatic scientists was the aptness with which it lent itself to some harmless-looking metaphorical extensions. "The demand for explicit linkages with observables could be equated with *operational definition*. The statements interlinking the three classes of variables could, if one so desired, be asserted as *postulates*, thereby making place for the paraphernalia and

[32] Cf. E. C. Tolman, "Principles of Purposive Behavior" (Koch II, 1959), p. 96.
[33] *Ibid.*, p. 93.
[34] "Association by Contiguity" (Koch III, 1959), p. 193.

imagery of *hypothetico-deductive* method. The fervent drive towards the *quantification* of systematic relationships characteristic of the era could become the quest for quantitatively specified intervening variable functions. And so on."[35]

After some more detailed discussion, Koch declares with emphasis: *"The overall tendency of the study is to call the intervening variable paradigm and much of the associated doctrine sharply into question, and to do this in almost every sense in which questioning is possible.* Virtually every contributor has shown a disposition to qualify some aspect of the doctrine: in some instances only diffidence seems to prevent qualification in all aspects."[36]

Next to come under discussion are the tenets of behaviorism. It is, perhaps, not necessary to adduce these familiar articles of faith, except to point out that the doctrine in its narrow, "classical" form admitted no dependent or independent variables that were not observable by several persons at the same time and describable in physical terms, whereas neobehaviorism tends to admit as independent variables "certain intra-organismic 'states' of a sort wholly or partly unspecifiable in stimulus terms," provided that their indicators and functions be capable of such description.[37] What is important here is the editor's verdict on recent developments, namely: *"An outstanding trend of the study is the presence of a widely distributed and strong stress against behaviorist epistemology (both in the narrower and broader senses just distinguished).* It is evident from all sides. It is strikingly evident among behavior theorists who themselves have powerfully molded behaviorist tradition. . . . Many of those whose problems have *not* been set by the emphases of behaviorism seem more disposed than formerly to question the adequacy of its epistemology—even if they are not always ready to relinquish the 'objectivist clang' of the independent and dependent variable language which has seemed so necessary a condition of respectability in recent decades. . . ."[38]

The foible of using jargon is but casually mentioned, perhaps because only one of the authors—Tolman—has expressedly paid it his disrespects. For the most part this idol is still enthroned. Yet Dr. Koch has not quite missed it, but remarks, *en passant*, that "after reading Tolman's presentation it is more difficult than ever to avoid the impression that much of the power of his thinking derives precisely from his use of, or fidelity to, 'common sense' conceptual categories of a cognitive sort. One feels more strongly than ever that whatever the inadequacy of the assumptions

[35] *Op. cit.*, pp. 733–34.   [36] *Ibid.*, p. 735.   [37] *Ibid.*, p. 753.
[38] *Ibid.*, p. 755.

made in such a vocabulary, they will probably be in some sense less 'vicious' than assumptions which, because of a principled commitment to some simplistic vocabulary, are forced into abusing and distorting ontology."[39]

Yet there is one connection in which he does report some general suspicion, though not repudiation, of pretentious borrowed verbiage, namely, in dealing with the dream of mathematization, of which he says: "Perhaps the most passionate Age of Theory demand has been that for the mathematization of systematic relationships—preferably at levels of quantitative specificity at least comparable to classical physics." In that sanguine age, "it seemed to most a matter of course that the goal of science, and thus of psychological science, was over its *entire* range the statement of mathematical laws."[40]

Since I have already discussed the products of that movement I shall not retail his account, for our findings are the same. But it is to this source that he attributes the artificial language of the psychology laboratory; and surely he is not making an unfounded indictment when he says that "the strength of the Age of Theory autism for quantification led to some blurring as between aim and achievement. A kind of pseudo-mathematical jargon became common to all. This is indicated in the general fondness for the language of 'variables' and 'functions,' the incessant use of terms like 'parameter' and 'parametric' in purely metaphorical contexts, etc. Moreover, the intervening variable schema proved a ready milieu for facile talk in this idiom of wish."[41] Similarly, the procedures of mathematical logic offered an enticing model of systematization, and "the atmosphere became permeated with an 'imagery' of hypothetico-deduction—the use or presence of which often seemed interpreted as equivalent to hypothetico-deductive practice. A language of 'postulates,' 'derivations,' 'primitive terms,' 'defined terms' (and in more rarified cases, logical variables, constants, arguments, predicates, operators, functors and connectives) became the tongue of psychological commerce. . . . This language was, of course, one with the mathematical language of 'functions,' 'equations,' 'variable,' 'constants,' 'parameters,' etc. that we previously sampled."[42]

"*The trends of the present study are in distinct contrast to the state of affairs above described.*"[43] I must confess that I cannot see a change quite as distinctly as Koch does. But there are indeed some voices lifted against this most impressive, and therefore most seductive, phantas-

---

[39] *Ibid.*, p. 767.   [40] *Ibid.*, p. 769.   [41] *Ibid.*, p. 770.
[42] *Ibid.*, p. 777.   [43] *Ibid.*, p. 770.

magoria of science; the most forthright, as one might expect, the intrepid Professor Tolman's. Tolman's criticism goes deeper than an expression of no-confidence in mathematization; it charges that interest in this new game has given psychologists a preference for facts which can be easily systematized, and such facts are usually trivial. So, Tolman says—and Koch has quoted him—"to me, the journals seem to be full of over-sophisticated mathematical treatments of data which are in themselves of little intrinsic interest and of silly little findings which, by a high-powered statistics, can be proved to contradict the null hypothesis."[44]

Here we are faced with the real fiasco to which the programmatic sciences have come—that in worshiping their picture of Science they have rarely come to grips with anything important in their own domains. One notable exception is the quiet work done in the fields of sensory psychology, which has not escaped the thoughtful editor's notice. Another exception is the progressive research going on in psychoneurology, to be treated by Koch in the second study. But psychology proper, under the rubric of "the science of behavior," instead of slowly extending such work to its limits and letting it rest, for the present, at those points, has by-passed it and imitated more finished masterpieces instead of chipping away at its own. So it is not surprising, and indeed heartening, that Koch concludes, *"It can in summary be said that the results of Study I set up a vast attrition against virtually all elements of the Age of Theory code. . . .* There is a longing, bred on perception of the limits of recent history and nourished by boredom, for psychology to embrace— by whatever means may prove feasible—problems over which it is possible to feel intellectual passion."

All these insights and changed attitudes are entirely to the good, yet in themselves they are not enough to launch an intellectual enterprise. The basic need is for powerful and freely negotiable concepts in terms of which to handle the central subject matter, which is human mentality —properly, and not foolishly, called "mind." But such concepts are still missing, or at least unrecognized; and as long as they are missing there will always be some primitive, scientifically useless entity—soul, entelechy, metaphysical Subject or vital essence—ready to slide into the vacant place to work havoc with the incipient science. This ever-present danger creates a constant desire on the part of psychologists to fill that empty place somehow with borrowed concepts, or at worst to shut it off with a verbal screen such as the "physicalist" vocabulary of behaviorism. Meanwhile, however, our understanding of mental phenomena does not

---

[44] *Ibid.*, p. 771; quoted from Vol. II, p. 150.

progress except by inches. We have reached a point at which a sounder substructure is required, and the philosophical work of construing the facts in logically negotiable, intellectually fertile ways is imperative.

If a science is to come into existence at all, it will do so as more and more powerful concepts are introduced. Their formulation is often the work of empirical investigators, but it is philosophical, nonetheless, because it is concerned with meanings rather than facts, and the systematic construction of meanings is philosophy. Wherever a new way of thinking may originate, its effect is apt to be revolutionary because it transforms questions and criteria, and therewith the appearance and value of facts. Even basic facts in an older system of thought may seem unimportant in a new one. Ultimately—perhaps at very long last—they are rediscovered in terms of the new formulation. But then they may be so masked that only a very discerning eye will recognize them.

If a philosophical theory of mind can serve the mental and social sciences at all, its effects in those fields of research are likely to be radical. It is not a matter of giving new definitions to old terms, as one might substitute "verbal behavior" for "speech" or "muscular reactions" for "dance,"[45] and going on with former lines of thought and experimentation. It means going back to the beginnings of thought about mental phenomena and starting with different ideas, different expectations, without concern for experiments or statistics or formalized language; if the concept of mind is philosophically sound it should serve to define mental processes in ways that make one suspect connections and derivations, and should lead from odd facts to bold hypotheses and from hypotheses to verification by any possible means. The experiments should suggest themselves automatically, and techniques and language grow up apace. But this might take a generation, and meanwhile we would have no science of mind, only notions to work with.

The state of having turbulent notions about things that seem to belong together, although in some unknown way, is a prescientific state, a sort of intellectual gestation period. This state the "behavioral sciences" have sought to skip, hoping to learn its lessons by the way, from their elders. The result is that they have modeled themselves on physics, which is not a suitable model. Any science is likely to merge ultimately with physics, as chemistry has done, but only in a mature stage; its early phases have to be its own, and the earliest is that of philosophical imagination and adventure.[46]

[45] As, for instance, F. C. Bartlett does in *Remembering* (1932), p. 271.

[46] Cf. Koch's comments on the causes of this precocity of the "young" sciences, in his "Epilogue," pp. 783–84.

It is even conceivable that the study of mental and social phenomena will never be "natural science" in the familiar sense at all, but will always be more akin to history, which is a highly developed discipline, but not an abstractly codified one. There may be a slowly accruing core of scientific fact which is revelant to understanding mind, and which will ultimately anchor psychology quite firmly in biology without ever making its advanced problems laboratory affairs. Sociology might be destined to develop to a high technical degree, but more in the manner of jurisprudence than in that of chemistry or physics. Were that, perchance, the case, then the commitment to "scientific method" could be seriously inimical to any advance of knowledge in such important but essentially humanistic pursuits.

Whatever the future, let us not jeopardize it any further by denying to our researches the free play which belongs to brain children as well as to animal and human infants. The philosophical phase we have missed lies at the very inception of research; if we would build a sounder frame of psychological, ethical and social theory, it is to that incunabular stage that we must return.

3

Prescientific Knowledge

Nelle cose confuse l'ingenio si desta à nove inventioni.
—Leonardo da Vinci

TO START the study of mind with an inquiry into the nature of feeling is to start *in medias res*, not with the most elementary biological phenomena which may be relevant to that study, for they probably lie below the level of felt events, nor with anything properly called "mind," because only the highest development of mental activities really constitutes a genuine "mind." Feeling includes the sensibility of very low animals and the whole realm of human awareness and thought, the sense of absurdity, the sense of justice, the perception of meaning, as well as emotion and sensation. The advantage of starting from an intermediate point is that the scope of one's inquiry may expand in both directions. There is no need of assuming that all organic responses are felt; there is some evidence that even fairly high human cerebral functions, guiding behavior and influencing conscious mental activities, may take place without entering into a psychical phase themselves. A reasonable judgment about what animals feel or do not feel is not based only on analogy between their behavior and our own, but also on the phylogenetic relationship between their activities and those human ones which rise to feeling (this is a point the humanizers of "mechanical brains" and cybernetic models ignore). The substructure of psychology is part and parcel of its subject matter, though it may be pure chemistry. In the opposite direction lie complexities of perception and conception, emotional sensitivity and selectivity, logical and semantic intuition, abstraction, communication, and cognition—the phenomena we ultimately want to describe, understand, and relate systematically. Again, the scope of our study may expand as far as we see any avenues for it. All those developments from simpler forms of feeling which become so specialized that they are no longer called by that word compose the mentality of man, the mind, the material of psychology.

We all have direct knowledge of feeling—"knowledge by acquaintance," as Bertrand Russell has called it—but very little of what he termed "knowledge by description," or knowledge *about* feeling. Science is essentially "knowledge about." What is more, a science does not arise directly from pure experience: such experience is a backward check on ideas that have long become general, and in their generality fairly exact. That is to say, every science arises from prescientific knowledge about its subject matter, empirical and haphazard, but developed by practice to considerable detail and precision. A great deal of correct and applicable mechanics was known before any physics had been developed. Egyptians and Mayans and Aztecs moved enormous stones, but left no theoretical works on dynamics which would indicate that they knew—or even asked —how and why their methods worked just the way they did. Chemistry, too, has had its seedbed of practical knowledge. Historians of science usually point out that chemistry grew from alchemy and that some of our common pharmacological and chemical words, such as "spirit," "elixir," "affinity," still reflect the mystical mode of thought which governed its earlier phase. But alchemy was not prescientific knowledge, although a few chemical facts may have been discovered in the necromancer's kitchen. It was essentially theory, built on symbolic rather than physical properties of substances, and was regarded as "occult science"; so it was a pseudoscience rather than an accumulation of factual knowledge without any theoretical framework. Prescientific empirical knowledge of chemistry is older; it came from cooking, preserving, tanning, embalming, and medicine which all too often gave cause for embalming. The same is true of astronomy and its alleged forerunner, astrology, although more genuine data were discovered by the soothsaying sky-watchers than by the alchemists. Again, the theoretical aspect, which gave (and still gives) astrology its status as "occult science," contributed nothing at all to the natural science of celestial bodies; the use of instruments and charts was developed and practiced more expertly by navigators, who had no theory, but had to keep exact bearings, than by astrologers who attached good or bad characters to the planets and traced their passage through equally portentous houses. The interests in which divisions and connections are made express themselves in diagrams as in language.

Builders may know the basic facts of mechanics, cooks find out chemical properties, and sailors map the sky; but who has any such naïve yet expert knowledge of psychical phenomena? Who knows the essentials of feeling? Offhand, one might reply: those who have to deal with emotional responses and types of mind—advertisers, politicians, people who constantly deal with so-called "human nature." Such people, however, need not be concerned with the nature of feeling at all, but only with overt behavior, intelligent or emotional or dull, as the case may be. What is felt, what feeling is like, how human activity appears within the agent himself is of no consequence in their business, and is known to them only as it is to everyone else, that is to say, vaguely, except for the acute states which have names such as indignation, anxiety, relief, terror, pain, joy, etc. The real patterns of feeling—how a small fright, or "startle," terminates, how the tensions of boredom increase or give way to self-entertainment, how daydreaming weaves in and out of realistic thought, how the feeling of a place, a time of day, an ordinary situation is built up—these felt events, which compose the fabric of mental life, usually pass unobserved, unrecorded and therefore essentially unknown to the average person.

It may seem strange that the most immediate experiences in our lives should be the least recognized, but there is a reason for this apparent paradox, and the reason is precisely their immediacy. They pass unrecorded because they are known without any symbolic mediation, and therefore without conceptual form. We usually have no objectifying images of such experiences to recall and recognize, and we do not often try to convey them in more detail than would be likely to elicit sympathy from other people. For that general communication we have words: sad, happy, curious, nauseated, nervous, etc. But each of these words fits a large class of actual events, with practically no detail. There are some distinctions expressed by adjectives not quite synonymous, for instance "happy" may mean cheerful, merry, gay, jolly, which all express shadings of a happy general state; or glad, delighted, ecstatic, which refer to a specific response rather than to an emotional condition; or joyous, blithe, blissful, beatific, again denoting a state, but with an overtone of high importance, as in a religious context. Yet an adjective never presents the complexities of the intraorganic processes from which overt behavior arises, whereas feeling really begins far below the issue of such processes in visible or audible action. Words are intrinsically public, and refer to the public aspects of life; and to understand someone else's words we need not follow their implications very far into his private ways of feeling. To say "I'm sorry" may bespeak remorse, sympathy or frustration; we may understand it to mean "I'm sorry I said that," or "I'm sorry your tooth hurts," or "I'm sorry the rain spoils our picnic." In understanding the sense of the word, however, we usually do not imagine what the speaker feels like, but simply note his attitude, in ex-

[57]

pectation of his subsequent behavior, and take our attitude toward him accordingly. Our own is what we feel, and he does not.

For most practical purposes, the nature of feeling does not need to be known conceptually beyond the point to which language, voice, physiognomy and gesture will express it. But for the study of mind such conceptual knowledge is needed, because the dynamic forms of felt experiences are a major exhibit of the rhythms and integrations, and ultimately the sources, of mental activity. Feeling is the constant, systematic, but private display of what is going on in our own system, the index of much that goes on below the limen of sentience, and ultimately of the whole organic process, or life, that feeds and uses the sensory and cerebral system.

Unfortunately, the methods of introspection which were used by Wundt, Titchener, Külpe, Brentano, James and some other great nineteenth-century psychologists showed that direct observation without any conceptual frame is impossible; some schema is bound to impose itself on the findings in the very process of seeking them, and the data reflect the expectation of the observer to whom they are supposed to be purely "given." To turn "knowledge by acquaintance" into "knowledge by description" is not a simple procedure of reporting private experience, because the formal possibilities of language are not great enough to reflect the fluid structure of cerebral acts in psychical phase, by which the substructure below the threshold of sentience is suggested, guiding physical research on the highest of vital phenomena. The earliest introspectionists were, in fact, the most influenced by theory, because they worked uncritically with the epistemological model provided by British empiricism—the supposed mosaic of pure sense data linked by a sort of magnetic process, association, into compound entities that were stored away as memories, and could somehow be retrieved from storage for later use.[1] So Wundt and Titchener, looking into their own private experience (conceived as a non-physical entity, "consciousness," which received the data and automatically did the associating, storing and recalling), found nothing there but sense data tinged with pleasure or displeasure; while Brentano, operating with the religious concept of moral agency, saw the contents of his "consciousness" as acts. One of the most candid and direct observers was William James. His unsystematic temper, suspicious of logic and often too disdainful of theory, was at once an asset and an obstacle in his intellectual pioneering—an asset because it freed his thought from the distorting drag of preconceived categories,

[1] For a historical account of introspective method and "structural" psychology, see Edwin G. Boring, A *History of Experimental Psychology* (2d ed., 1950).

[58]

and an obstacle in that he never brought his insights into any single focus. The German psychologist Felix Krueger was a close second to James in his powers of psychological observation, and more than his equal in systematic thought. With such men a fruitful era of fact-finding might have begun. Most of the modern "classical" psychologists, however, were more anxious to simplify their findings so they could be systematized than they were to examine them patiently and empirically— with the results discussed in the previous chapters. Before we had any clear image of the phenomenon we call "mind" we committed ourselves to a model, the system of physical laws, whereby the material was automatically cut down to what the model could represent, and the very subject matter of psychology—the psychical phase of vital functions— was eliminated altogether, leaving only its overt record, behavior.

I said, "Before we had any clear image. . . ." An image is not a model. It is a rendering of the appearance of its object in one perspective out of many possible ones. It sets forth what the object looks or seems like, and according to its own style it emphasizes separations or continuities, contrasts or gradations, details, complexities or simple masses. A model, on the contrary, always illustrates a principle of construction or operation; it is a symbolic projection of its object which need not resemble it in appearance at all, but must permit one to match the factors of the model with respective factors of the object, according to some convention. The convention governs the selectiveness of the model; to all items in the selected class the model is equally true, to the limit of its accuracy, that is, to the limit of formal simplification imposed by the symbolic translation.

It is different with images. An image does not exemplify the same principles of construction as the object it symbolizes but abstracts its phenomenal character, its immediate effect on our sensibility or the way it presents itself as something of importance, magnitude, strength or fragility, permanence or transience, etc. It organizes and enhances the impression directly received. And as most of our awareness of the world is a continual play of impressions, our primitive intellectual equipment is largely a fund of images, not necessarily visual, but often gestic, kinesthetic, verbal or what I can only call "situational." The materials of imagination which I crudely designate here as "images" will be discussed at length in later parts of the book. Suffice it now to point out that we apprehend everything which comes to us as impact from the world by imposing some image on it that stresses its salient features and shapes it for recognition and memory.

Images, however, are abundant and often fragmentary, not single and coherent like a model. We use and discard them prodigally. Sometimes there is but one passing phase, one extraordinary aspect of a familiar thing or common-place event, suddenly revealed by an image of what we call "that sort of thing," but without the image we would never have conceived it as it now seems—or as henceforth it may seem. We constantly and spontaneously produce images of action, of objects, of pure motion or of moving things, and of audible events. Most people know the experience of having a tune "running in one's head." That is an auditory image.[2] Silent thinking, though we may believe it to be a succession of pictures, is really conducted largely in unspoken words. Some dream analyses uncover the most astounding verbal material even on that primitive level of ideation. No one knows how words which are certainly not articulated—sometimes, indeed, are illustrated, in a way that hides their effective presence from the thinker or dreamer himself— enter into cerebral processes; no one knows how far the various language functions permeate our mental life, but with every new research the soundings go deeper.

The high intellectual value of images, however, lies in the fact that they usually, and perhaps always, fit more than one actual experience. We not only produce them by every act of memory (and perhaps by other acts), but we impose them on new perceptions, constantly, without intent or effort, as the normal process of formulating our sensory impressions and apprehended facts. Consequently we tend to see the form of one thing in another, which is the most essential factor in making the maelstrom of events and things pressing upon our sense organs a single world (Fig. 3–1). In this way all the things which one image roughly fits are gathered together as instances of one conception. The image is not, I think, made of an accumulation of specific impressions, as many specific photographs, superimposed, constitute a composite photograph. The original image may have been derived in roundabout and irrecoverable ways. But it fits many impressions, even if somewhat imperfectly, nearly enough to permit their treatment as things of one kind; which is to say, it permits their interpretation in terms of the conception which the image expresses (many words, such as "conception" and "express," are loose and inexact here, but will receive greater precision in later discussion). I do not know what the cerebral function of fitting images to sensations is; but it is, or at least enters into, the act of interpretation, and is probably intimately related to the processes of concept formation, for which the scanning mechanisms of television and some devices on mathematical machines may furnish revealing models.

[2] Jean D'Udine calls it *"l'image phonique."* See *L'Art et le geste* (1910), p. 196.

 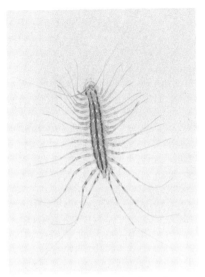

Figure 3–1. (*Left*) Carvings of Bark Beetle (probably *Scolytus*);
(*Right*) House Centipede, *Scutigera forceps*

we tend to see the form of one thing in another

(*Left*, made by a beetle of the family Scolytidae, species undertermined,
photo by Henry F. Dunbar; *Right*, photo by Louis Darling)

Where does one "kind" of thing, or class, end and another begin? How nearly must an image fit a new percept to gather it to its kind? The answer is, I think, as indefinite as most categorizing statements about vital phenomena: it must fit better than any other image (in the broad sense of envisagement, gestalt) available at the moment. We may hesitate to decide among several possible interpretations of an unusual phenomenon; or we may impose a form on the object (or act, or circumstance) at once, only to exchange it for another afterward.

As one gestalt may have many exemplifications, so one object may also exemplify various gestalten. Ordinarily, in the kind of thinking which civilized adults today call "common sense," every familiar physical object has a stable dominant gestalt according to which it is publicly classified, i.e., named; and how it is named largely determines the way we experience it even privately. One may fancy a witch riding on a broom, but one thinks of the object she rides on as a broom; it would be strange to think of sweeping the kitchen with the bristly tail of a witch's hobbyhorse. The standard designation expresses what is considered the object's "real" nature; if it is called anything else, which brings it under a different rubric, this is regarded as a metaphorical designation. The points of appearance, use, origin, etc., which make an object capable of being interpreted in an unusual way, that is, taken to exemplify some other than its accepted form, are its points of resemblance to things normally classified by virtue of those traits. We speak of the "leaves" of a book by metaphorical transfer from the leaves of a plant. The resemblances may, moreover, be purely relational, as in our speaking of the "flow" of time, or of the "leap" of thought from one topic to another. These elementary semantic functions have all been discovered and discussed by the philosophers of our century, so that there is a considerable literature dealing with literal language, figurative language, the reasons for circumlocution such as taboos on the literal designation, or the sheer limitations of language making roundabout expression necessary, and at the same time the growth of exact, economical, wholly conceptual language with the advance of scientific thought.

But in our semantic studies we generally assume that there is a clear distinction between the literal meaning of a word or a statement and its metaphorical extensions. This assumption, which is almost a premise of common sense to civilized human beings, may nonetheless be unsafe in reconstructing the part language has played in the articulation not only of ideas, but of perception, the making of the world out of the fragmentary findings which prompt our overt actions and covert felt responses. The primitive use of words may have been much less bound to specifiable objects than its present use, much richer in connotation and therefore more elastic in denotation, so that literal and metaphorical meanings were not distinguishable, but the same word simply meant a variety of things which could all symbolize each other. It has often been asserted that most if not all common words for mental attributes are "borrowed" from the physical realm: "bright," "dull," "clear," and "obscure" are examples. We speak of thoughts as "heavy" or "light," "shallow" or "deep," and of "bald" statement and "sharp" wits. Even terms for mental acts such as "insight," "foresight," "reflection" and "rumination" fall into the category of borrowings from the physical realm.

The linguistic fact on which this assertion is based is beyond dispute; most words for intellectual traits also signify physical qualities or acts. But the anthropological doctrine that primitive men knew the physical world before they were aware of mental activity, and borrowed words with physical meanings to refer metaphorically to mental states and functions, is open to doubt. On grounds of more general phylogenetic patterns it is at least as reasonable to suppose that the light of the sun and the light of reason or of joy were named by the same word because they were charged with the same feeling, and consequently taken as the same thing, and that their distinct characters only showed up in language in the course of its logical development. The physical meaning became the "literal" one because it was the most public and therefore socially the most negotiable one. Physical light is a permanent, ever-available symbol for everything that "light" may have meant in its earliest uses, which were probably quite spontaneous applications to a large vague class of things similarly felt. The sun can be pointed out when it is visible in the sky, and graphically represented when it is not; so it is naturally the image of all conceptions of light. Mithra, god of the Sun, is also the Truth, and Power, and Life. It is primarily as an image of these intangibles that the sun is adored, and they in turn are known only in its image, to which they seem to belong. Light, Truth, Power and Life are of one essence to the sun worshiper. He is not aware of light as a symbol and the several religious concepts as its meanings, but of one reality, divine Light, which the religious concepts characterize.

Knowledge begins, then, with the formulation of experience in many haphazard ways, by the imposition of available images on new experiences as fast as they arise; it is a process of imagining not fictitious things, but reality. Some French psychologists, notably Henri Wallon[3] and the late Philippe Fauré-Fremiet,[4] refer to this process as *"réalisation,"* the making of reality out of impressions which would otherwise pass without record. The role of imagination in direct perception, the essentially imaginative nature of our spontaneous, current interpretation of impinging "data," has been gaining recognition ever since Ribot,[5] just at the last century-mark, extolled scientific imagination as the highest form of the creative faculty; Jean Philippe,[6] at the same time, took a somewhat broader view, more directly suggestive of present developments. The depth and reach of the imaginative functions in the making of human mentality will be discussed in a later part of this essay, but their very early occurrence is important here. The imposition of imagery on all materials that present themselves for perception, whether peripheral or intraorganic, enters into the most naïve experience, and into the making of our "empirical" world. It is more primitive than the adoption of any "model." The use of a model belongs to a higher level of conception, the level of discursive thought and deliberate analogical reasoning. But the process of seeing things as exemplifications

[3] *De l'acte à la pensée* (1942).

[4] *La re-création du réel et l'équivoque* (1940).

[5] *Essai sur l'imagination créatrice* (1900).

[6] *L'image mentale (évolution et dissolution)* (1903).

of subjectively created images gives us the original, objective phenomena that theoretical reasoning seeks to understand in causal terms, often with the help of highly abstract working models.

To return, at long last, to the unanswered question: who has a naïve but intimate and expert knowledge of feeling? Who knows what feeling is like? Above all, probably, the people who make its image—artists, whose entire work is the making of forms which express the nature of feeling. Feeling is *like* the dynamic and rhythmic structures created by artists; artistic form is always the form of felt life, whether of impression, emotion, overt action, thought, dream or even obscure organic process rising to a high level and going into psychical phase, perhaps acutely, perhaps barely and vaguely. It is the way acts and impacts feel that makes them important in art; their material identity may be suggested in quite sketchy or distorted fashion, where it is wanted at all, for it serves artistic purposes only in so far as it helps the expressive function. This does not mean that it is always unimportant to the artist; representation may be the prime purpose of the thing he makes, which gives him the occasion to create a work of art. In the hands of a natural artist—professional or not, called "artist" or "artisan" or whatever else— almost anything may become a work of art: a bed, a doll, a scientific drawing, a photograph. Haydn wrote the bugle calls for the Austrian army, and they are music.

In the course of projecting the forms of feeling into visible, audible or poetic material, an artist cannot escape an exact and intimate knowledge of those passages of sentience which he succeeds in expressing. His range may be small, but if any of his work is good he has found some true expression of felt experience. He knows something of how feeling rises, develops, tangles or reverses or breaks or sinks, spent in overt action or buried in secrecy. He is not a psychologist, interested in human motivation and behavior; he simply creates an image of that phase of events which only the organism wherein they occur ever knows. This image, however, serves two purposes in human culture, one individual, one social: it articulates our own life of feeling so that we become conscious of its elements and its intricate and subtle fabric, and it reveals the fact that the basic forms of feeling are common to most people at least within a culture, and often far beyond it, since a great many works do seem expressive and important to almost everyone who judges them by artistic standards. Art is the surest affidavit that feeling, despite its absolute privacy, repeats itself in each individual life. It is not surprising that this is so, for the organic events which culminate in being felt are largely the same in all of us, at least in their biologically known aspects,

[64]

below the level of sentience. Yet in the highest development—the psychical phase of activities hard to observe even in ourselves, except through some focusing and arresting device—individual differences may be extreme. One cannot safely argue from uniformity at one level of vital processes to uniformity at another, that is, from physiological directly to psychological similarities among creatures of a species, especially the most highly individuated species, Man. That is why psychology is not a "branch" of physiology: there is no way in which physiology can put forth such a branch. But if we can find systematic access to the intricate and multifarious ways of feeling which build up the whole pattern of the mind in the course of human life, we may hope to trace them to their sources below the psychological level, and perhaps conjoin two sciences in a single system of facts, once we really have two sciences.

The direct perception of artistic import, however, is not systematic and cannot be manipulated according to any rule. It is intuitive, immediate, and its deliverances are ineffable. That is why no amount of artistic perceptiveness ever leads to scientific knowledge of the reality expressed, which is the life of feeling. What it gives us is always and only an image. But without this or some other image we cannot ask questions about the empirical data with which knowledge begins, because the image enters into the objectification of the data themselves. Unless they are objectively seen and intimately known we cannot formulate scientific questions and hypotheses about them. Here, I believe, lies the weakness of our present psychological researches: we do not really grasp the data we propose to deal with, because we are trying to transfer methods of observation simply and directly from physics to psychology, without taking account of the fact that the intraorganic character of the material presents a special difficulty and does not lend itself to those methods. But a special difficulty does not necessarily spell total frustration. It requires special treatment, some indirect approach to the facts.

Since art is an objectification of feeling, whereby men have perceived and recognized its rhythms and complexities and scope since time immemorial, it is not unreasonable to expect some formulation of psychological basic facts to come from art. But here we meet the second difficulty: what knowledge of feeling comes to us as the import of art comes through artistic perception, which is a form of intuition that does not lend itself to discursive analysis and synthesis as, for instance, mathematical intuition does. Artistic expression is essentially non-discursive; but to yield negotiable facts its homogeneously presented import has to be somehow analyzed and demonstrated. Experience has long convinced all scientifically inclined persons that no matter how exact and

[65]

sure their artistic insights may be there is no direct way of demonstrating them to other people, no basis for arguing their validity. So we require an indirect approach, some parallel phenomenon which can be manipulated.

The art symbol itself furnishes this parallel, for the symbol is a work, and its elements are analyzable. I say "elements" rather than "factors," because the components of the art symbol—of the virtual entity, to which import belongs—are created elements, whereas the factors are materials used, and are replaceable by others for most purposes. Elements, too, sometimes have alternatives; but their measure is expressive power, which pure materials possess only at second remove. Pigments, clay, things capable of producing sounds, the vocabularies of languages, the physique of dancers are materials of art; colors, volumes, accents, word uses, gestures are elements. Poetry can be made in any language and pictures painted with any pigment or pigments. It is easier to work with some materials than with others, but the scantiest ones have sometimes served for work of great scope and depth. Consider the meager materials that entered into Gregorian church music: no instruments of various tone colors, no modulation, no metric division, no polyphony, only men's and boys' voices in unison or alternation. Yet the music so created is complete and impressive.

It is with created elements, then, that we are dealing when we consider the components of the art symbol, which is always the entire work; and one way to understand what passages or aspects of feeling (always in the broad sense of anything that can be felt) are expressed in it is to analyze just what the artist has done and for what purpose. Here the basic principle that a good work of art is one hundred per cent expressive, so that nothing which does not serve to create or enhance the expressiveness of that virtual form can enter into it without cluttering and compromising it, proves to be a strong support; for it guarantees the close association of the import, which we know directly, with technical processes that do invite analytic study and objective demonstration of facts. There is still a wide range for interpretation of facts, but once the more detailed expressions of psychical activity are tracked down, there is less and less disagreement about their purposes. For instance, once you discover the means whereby a composer prepares an effect of unmistakable emotive value, you realize that such emotion does, in fact, arise from scarcely discernible beginnings, and that the preparation of the musical effect is a process of articulating the structure of the feeling expressed, so it does much more than convey some nameable mood or emotion; it may even be given different names, but best none at all, for no name such as "sorrow" or "joy" fits any actual feeling throughout its

[66]

course. Feeling is a dynamic pattern of tremendous complexity. Its whole relation to life, the fact that all sorts of processes may culminate in feeling with or without direct regard to each other, and that vital activity goes on at all levels continuously, make mental phenomena the most protean subject matter in the world. Our best identification of such phenomena is through images that hold and present them for our contemplation; and their images are works of art.

What makes a work important is not the category of its expressed feeling, which may be obvious or, on the contrary, impossible to name, but the articulation of the experiential form. In actual felt activity the form is elusive, for it collapses into a condensed and foreshortened memory almost as fast as the experience passes; to hold and contemplate it requires an image which can be held for contemplation. But there is no simple image of our inner dynamisms as there is of visually perceived forms and colors and of sound patterns. A symbol capable of articulating the forms of feeling is, therefore, necessarily presented in some sort of projection as an extraorganic structure that conveys the movement of emotive and perceptive processes. Such a projection is a work of art. It presents the semblance of feeling so directly to logical intuition that we seem to perceive feeling itself in the work; but of course the work does not contain feeling, any more than a proposition about the mortality of Socrates contains a philosopher. It only presents a form which is subtly but entirely congruent with forms of mentality and vital experience, which we recognize intuitively as something very much like feeling; and this abstract likeness to feeling teaches one, without effort or explicit awareness, what feeling is like.

"The art in painting," as Albert Barnes called it, or what some aestheticians of his generation called "significant form," is this image of feeling, the "life" of lines and colors themselves, which requires no pictorial record of people in emotional states—in fact, is rather endangered by such distracting elements of representation. It is the deepest principle the arts have in common; the forward movement in drama, not the particular "thing done"; the articulate transience of music that makes us call fragments of music "passages"; the self-defining masses of good architecture which possess life even if they happen to house nothing but lifeless things, or even death itself in a funerary shrine.

The art symbol is an image, not a model. An image abstracts the semblance of its object,[7] and makes one aware of what is there for direct

---

[7] The term "semblance" was introduced by C. G. Jung to denote what Friedrich von Schiller called "*Schein*"; a more detailed discussion of this concept may be found in *Feeling and Form* (hereafter cited as F.F.), chap. iii, *passim*.

perception. A model illustrates a principle of construction or function, quite apart from any semblance. Our first acquaintance with the material of any research has to be negotiated by images which organize and present the phenomena as such, for it is always phenomena that we ultimately wish to explain, and this requires detailed empirical knowledge. Such knowledge cannot be gathered without some systematic device whereby observations can be made, combined, recorded and judged, elements distinguished and imaginatively permuted, and, most important, data exhibited and shared, impressions corroborated. The symbol of feeling permits such manipulation, although feeling itself does not. If we want to study the phenomena of emotion, sensibility, ideation and especially the integral mental life in which they are all what Husserl's English-speaking disciples call "moments" (*Momente*), the most promising method is to study art, painstakingly and technically, always asking: "What for this doubled time? Why this elaboration? Why that emptiness, that color, that perspective?" If the answer does not suggest itself upon inspection, a simple device is to change or omit the element to see what is lost. Sometimes an alleged fault in a work proves to be a strong and essential element. The procedure of the so-called "New Criticism" in literature serves this purpose of understanding the making of the symbol, of following the author's undertaking through all his artistic problems and decisions as the surest guide to the import of the work, the idea expressed.

The image of feeling, however, does not provide us with any scientific concept such as psychological research requires. That is the function of a model. An image may be—and usually is—built up on entirely other principles than the phenomenal character of its object, and its own construction may be utterly different, while the created semblance confronts us like the phenomenon itself—sometimes in an extreme projection that we do not even note; for instance, the projection of a visible object as a thin gray contour line on a flat surface. A model, on the other hand, need not share any phenomenal traits with its object, but symbolizes its structure or function and is true to the original in every proportion and represented connection, to a stipulated limit of accuracy. A model is usually based on a single systematic abstraction which can ultimately be expressed in mathematical terms. Consequently it is a model, not an image, that one works with in science. At an advanced stage of scientific research resort to images revealing the original data becomes unnecessary; the macroscopic facts are sufficiently well known to be used as a check against the elaborate speculations to which the powerful abstract concepts lend themselves, and which are often illustrated by highly ingenious models. A play of imagery at that level only hampers invention, because it limits logical conception to the imaginable. But problems of this sort arise only with the full development of systematic knowledge in completely mathematicized fields.

The chief reason for our general hesitancy to undertake a serious study of psychological data themselves is that there seems to be no instrument to negotiate it; few people realize how excellent a presentation of such data is to be found in the arts. A few scholars have become aware of it—Koehler, Husserl, Fauré-Fremiet, Bertalanffy, Portmann and undoubtedly some others also.[8] But no one, as yet, has pursued the revelation to the point of finding problematical facts never presented before, or recognizing vital patterns in pure art which may be keys to essential relations in the life of feeling. There is a wealth of such material presented indirectly in the technical problems artists have to solve to attain and sustain expressiveness, which is the same as to create a work of art.

In solving such problems, an artist rarely thinks about feeling, at least in the way which we usually call "thinking about" something. He thinks about his work, about what is wrong, what seems right, where the "dead spot" really is, what can put "life" into it; he judges what is expressive and what is inexpressive, but very seldom formulates in verbal thought the idea to be expressed. He will know it when he succeeds in embodying it. Discourse tends to arrest it; if the embodiment is not perfect the work will be somehow out of kilter and make new technical demands. The essentially intuitive process of artistic construction is as intriguing a psychological problem as the process of artistic apprehension. Neither is understood as yet.

To learn the facts of feeling from art, the best general practice is to follow the evolution of a work from one technical problem to another. A student of art would presumably be interested above all in the solutions; a student of actual sentience, however, in the reasons for their necessity. Gradually, but not systematically, the image of living form builds up from the exigencies of its creation. The principles of such image-making are entirely different from those of biological growth and function; the image projects nothing but the empirical datum, but this it presents with a degree of precision and detail beyond anything that direct introspection is apt to reveal. In the arts, and especially in their technical triumphs, lies the store of prescientific knowledge from which a science of psychology may draw its first inspiration.

[8] Reference to the works of these authors will be made where they are discussed.

Part II

The Import of Art

4

The Projection of
Feeling in Art

THE theory of art from which this whole philosophy of mind arose is the theme of a previous book, *Feeling and Form*, wherein it has been developed chiefly with reference to the distinct basic abstractions and created dimensions giving rise to the respective great orders of art—plastic, musical, poetic, etc. The theory itself is systematically set forth there, and the chief implications of its tenets discussed in some detail, which is necessary if it is to be judged on its actual credentials, rather than accepted or rejected on a basis of a few catchwords whereby it can be relegated to an "ism," and thus quickly evaluated as current or obsolete. To reiterate that whole exposition here would involve us in chapters of discussion which would, after all, be largely repetitious; in presenting new studies built on former work one cannot possibly start from the beginning of the entire story every time. Some acquaintance on the reader's part with the foundations of the new ideas has to be presupposed. Instead of repeating myself, therefore, I shall make frequent reference to *Feeling and Form*, even as though it were a previous volume of the present book, because this practice seems the least onerous in a necessary choice of evils.

In some respects the treatment of art itself in this book transcends the compass of *Feeling and Form*. There my main concern was with the principles of creation and expression as they appear, often quite particularly, in each of the great orders of art, and with the equally particular problems and techniques which consequently arise in the different arts. The ultimate unity of art, which most philosophers seek by resolutely ignoring or denying any differences between music and painting, painting and literature, and so on, I expected to find precisely by tracing those differences as far as possible, to their vanishing point, where no more traits can be found that belong to some arts but not to all.[1] That there

---

[1] Cf. *F.F.*, p. 103.

are some fundamental principles by virtue of which we subsume so many distinct pursuits under the single word "art" is experientially hard to doubt, and in the last part of *Feeling and Form* those principles do come to light. The problems which engendered that book were general problems of art: what is created in works of art, and how, and to what purpose? The three major sections of the text[2] roughly correspond to those three issues; the complexity of the answer to the second one caused the second part to deal with the several arts seriatim, and to constitute the greatest portion of the work.

Now, in looking to the import of art for guidance in the study of mind, we shall find ourselves dealing almost entirely with problems of art as such, as a unitary phenomenon, and with the separate arts as branches of the same pith and fiber, merely reaching out toward different opportunities for the essential artistic functions. The questions dealt with in this book pertain equally to all the creative arts. What is a projection? How does an artist project an idea of feeling by means of his work? How does the idea become perceptible? And finally: what new empirical knowledge of the morphology of feeling can we derive from its image in works of art, and what light can this knowledge throw on the unfelt processes of life and the emergence of feeling, animal mentality, human experience and mind? These are far-reaching issues that invade many fields of special study. They bring strangely assorted facts together in new designs, and lend philosophical importance to a great deal of research already performed without any claim to such importance, in a modest spirit of clinical or experimental checking on factual detail. The amount of significant work that has been done becomes more impressive as the ramifications of the philosophical issue—the construction of a biological concept of mind adequate to the phenomenon itself—become explicit, and the consideration of them ineluctable.

To start, then, with the most elementary question: what does it mean to speak of "projecting" an idea? What is a projection?

Literally, of course, a projection is something that extends in one direction beyond the generally smooth contours of a mass. A bracket projects from a wall, or horns from an animal's head. In so doing, the projecting item stands out noticeably and presents itself most readily to perception. This is, I think, the circumstance that gives the concept of "projection" its metaphorical value. We speak of projecting a picture on a screen, which means to produce light, dark or variously colored areas

[2] Part I, The Art Symbol; Part II, The Making of the Symbol; Part III, The Power of the Symbol.

there, corresponding to the light, dark or respectively colored areas on a slide; the instrument effecting this projection is a projector. We project a plan of action, and call the projected undertaking a project. We map the global surface of the earth on a flat surface, in polar, equatorial or conical projection; psychologists have long called the process of referring our sensations to points of origin in the outer world projection of those sensations; and today they say that a paranoiac projects his own feelings into another person when he imputes them to that person instead of to himself. These many uses of "projection" seem to be widely disparate metaphorical applications of a word which literally means protruding from a mass or surface, and they do not even exhaust its figurative uses. Yet all those uses seem to go back to the enhanced perceptibility of a projecting object or feature; and I think it is safe to say that, in all metaphorical senses, a projection is a principle of presentation. The operation of such a principle may be intentional, as in a cartographer's work, or purely spontaneous, unintended, as in the outward reference of sights and sounds, or even unadmitted—perhaps vehemently denied, as in the case of paranoid projection of a patient's attitudes and motives. Graphic renderings, ordinary sense data, and the paranoiac's alleged observations all appear to be presented to us as impinging "data."

"Projection" is really a word-of-all-work; sometimes it is used to denote a principle, as I just used it above in saying that a projection is a principle of presentation. Sometimes it is applied to the act of making the presentation, i.e., setting up the symbol; the paranoiac's projection of his own feelings into someone else, who serves him as a symbol of himself, is held to be an act, and indeed a mistake on his part. And finally, perhaps most often, we call the symbol itself a projection of what it symbolizes. In this sense, art may be said to be a projection of the artist's idea into some perceptible form; the expressive object he creates is called a projection of the life, mood, emotion or whatever he makes it express. Ordinarily the sense in which the word is used is easily gathered from the context, so it would be pedantry to specify it by a modifier every time; but it should never be forgotten that "projection" is really an ambiguous term, which may have to be explicitly defined in analytic or speculative discourse where precision is all-important.

During the last forty years logicians and epistemologists have developed the concept of logical form, and discovered its relevance to the long-baffling problems of symbolic expression and communication. Evidently the process of projection rests on the recognition of one and the same logical form in different exemplifications, which are, therefore, different expressions of it.[3] Very frequently one of the examples is much easier to perceive and hold in view than the other; then that negotiable one is used as a symbol, to make the obscure or elusive one conceivable, especially if the latter is the more important to conceive. Hence our many devices for measuring insensible or only imprecisely sensible conditions in the world—temperatures, velocities, intensities, pressures, forces—by means of analogous visual phenomena; we project the rise and fall of temperature or of gas pressure in more easily observable terms by coupling its changes with visible changes in an instrument.

When we take, say, a thermometer reading, the stand of the mercury column is an indicative sign of the temperature actually prevailing. That physical covariance is what makes the thermometer a meter, an instrument, but even without its instrumental character it has an intellectual value which it retains though the bulb break and the tube be emptied; for besides being a device to determine actual temperatures, it is a symbol of the conceptual order in which all possible temperatures may be arranged so that any particular one may be unequivocally designated and any two may be distinguished and compared. All possible temperatures together form a one-dimensional continuum, which can be symbolically projected as a spatial one-dimensional continuum, i.e., a line; this visible line can be measured against a series of equal lengths marked off and numbered, a scale; so the continuum of possible temperatures appears in spatial projection as a series of degrees of heat. The glass tube in which the mercury rises and falls exemplifies the line, and is a visual representation of the conceivable range of temperatures (physically our thermometers are limited to part of the entire range, but theoretically, of course, they have no upper limit, and the lower is absolute zero); the mercury column in it is a device whereby the actual temperature at the moment may be found in terms of the symbolically established degrees.

This is one of the simpler examples of symbolic projection, in which the basis of isomorphy, i.e., sameness of logical form in the visible phenomenon and the more elusive one it represents, is obvious. It is adduced here as a stock example, to typify the conceptual process, which is an essentially symbolic process pervading all human mental activity—perception, appreciation, selfhood, emotion, as well as thought and dream. To understand its origin and offices is the chief purpose of human

---

[3] W. W. Skeat, in *An Etymological Dictionary of the English Language* (1910), points out that the word "like" had an original meaning of "form." See the entry, "*Like* (1) similar, resembling. (E) ME *lyk, lik*; . . . Dan. *lig*.; Sw. *lik*. . . . All signifying 'resembling in form,' and derived from the Teut. sb. **likom* a form, shape, appearing in AS *lic*, a form, body. . . ."

psychology, the study of mind. But the general semantic theory of projection, symbolic presentation and conceptual formulation has to be taken for granted here, where we are concerned with one of its special and very complicated applications—the analysis of artistic expression, or presentation of what artists call "the Idea" in their works.

The study of symbols is a philosophical interest of rather recent date. The eighteenth and nineteenth centuries were the era of "sense data" and their "associations," the search for ultimate units or "impressions" to make up the mosaic of knowledge. A few writers, especially in Germany, sensed the inadequacy of the prevalent empirical psychology to deal with art, imagination and many aspects of emotional life.[4] Even the highest intellectual functions seemed to them to entail something more than the recognized contents of the mind, and the simple activities of connecting, dividing and permuting those contents, that were deemed the only functions of the mental organ. But it was particularly the understanding of art that obviously involved some other factor than realistic cognition; and naturally enough, people who reflected long and seriously on the nature and import of art were the first to realize that here was a sort of significance that differed from the definable meaning of words, and a logic that was not like the logic of discourse, yet was a form of reasonableness. Here was something like a symbol but not a conventional symbol, and something that was more than a symbol—a form that contained its sense as a being contains its life. So F. T. Vischer wrote in 1863: "Form [in art] has intrinsic expression; everything is viewed as Form, the Form expresses its inward content, and so all Form is viewed as expression. The peculiarity of the phenomenon lies in a reversal: In ordinary reality a force, the Idea, comes first. It begets the Form. But in the sphere of the beautiful, perception goes backwards from the Form to the force, the Idea of which it [the Form] is the expression. Yet in the course of aesthetic contemplation the reversal is turned back again; feeling, and then judgment, too, have penetrated from the presenting form to its core, and from the core now retrace its configuration."[5]

The door to a wider conception of mind than the impressionable

---

[4] E.g., F. T. Vischer and his son (cited below); Eduard von Hartmann (*Philosophie des Unbewussten,* 1869); Konrad Fiedler (*Moderner Naturalismus und künstlerische Wahrheit,* 1881, and *Über den Ursprung der künstlerischen Thätigkeit,* 1887).

[5] *Kritische Gänge,* Vol. II, No. 5 (2d ed.; 1st ed., 1863), p. 86. Just a decade later his son, Robert Vischer, published an interesting little book, *Über das optische Formgefühl,* in which he refers to his father's concept of the aesthetic function as "*formsymbolik.*"

The passage quoted above will prove to be of special interest later, in connection with dialectic and functional shifts in vital processes.

tabula rasa of the empiricists, with its distinct faculties of logical manipulation, was opened by the Kantian analysis of experience into form and content and the emphasis on the imposition of form by the transcendental Subject. The effect of this doctrine on aesthetic thinking was not entirely direct.[6] In Kant's metaphysics mind and world, the knower and the known, are logically implied, transcendental ultimates. To speak of their manifestation, the phenomenal world, as resulting from an activity, an imposition of form by a Subject, *ex hypothesi* spaceless, timeless and causeless, on an Object of similarly negative character is to speak in a metaphor, and has to be taken in a metaphorical sense, since anything that can literally be called "activity" is a process, and processes—especially effective, transitive ones—belong strictly to the phenomenal realm. The transcendental Subject and Object are final terms of a purely conceptual analysis, terms which cannot be used as factors in any historical account or hypothesis. The fallacy of treating transcendental relations as generative causes begot the mythical—and non-Kantian—cosmologies of the post-Kantian Idealists. Most of the aestheticians who saw works of art as expressive forms setting forth "ideas" of some sort by direct sensuous presentation had experienced the liberating influence of Kant's thought on their powers of conception,[7] and though they generally swung under that influence to some romantic, not to say mystical, view of the human mind, this extravagant mood may have been precisely what they needed to venture on an entirely new line of thought—the candid study of symbolism, even in its furthest reaches where the symbol relations are manifold and complex as in art.

This study arose in several quarters of the intellectual world more or less independently during the nineteenth and twentieth centuries, and each of its starting points was the beginning of some major theoretical development. In the realm of mathematics, where it took its earliest rise, the concentration on semantic problems revealed the influence of particular symbolic formulations on the lucidity, combinatory power and even the degree of abstractness of propositions, and begot the modern science of symbolic logic. In another academic field, the study of lan-

---

[6] Kant's own aesthetic, presented in the *Kritik der Urteilskraft* in 1790, had an influence apart from that which his metaphysical principles exercised; some aestheticians who did not adopt his theory of art nevertheless built their own on their respective versions of his transcendentalism, as, for instance, Schopenhauer, Nietzsche and Hartmann.

[7] Not only in Germany, but in England, America and Italy philosophy of art received this vivifying influence; Coleridge, Emerson and later Croce were led by it to speculate on the conceptual and expressive nature of art. Cf. Helmut Hungerland, "The Concept of Expressiveness in Art History" (1944).

[78]

guage, it led from principles of verbal expression to those of literal conception—the basis of ordinary, traditional logic—and their fore-runners, the mythical and metaphorical forms of thought which seem to have preceded our sober literalmindedness. At the same time, outside the universities, in the domain of practical psychiatry, those primitive conceptual processes which express laws of imagination rather than verbalized thought have received concentrated attention, so another special meaning, and with it another evaluation of "symbol," has grown up. On the whole it is quite evident that concepts of symbolic presentation and its converse, the intuition of meaning, have effected something of a revolution in the basic assumptions of our psychology, epistemology, logic and various special fields of research; and the medley of half-baked, exciting ideas from all these different sources has given new life—howbeit in the way a deluge brings new life—to philosophical thought in our day.[8]

The disconcerting aspect of the sudden appearance of this rich material is that so much of it does not fit together. In logic every symbol stands proxy for one defined or initially accepted idea; the symbols may be assigned and reassigned simply by fiat—"let x equal this, y equal that" —and nothing but the definition of "this" and of "that" enters into the current meanings of the respective terms. Logical symbolism is the extreme of literal expression, and a logician's (or mathematician's) intention is to allow no leeway at all for alternative or additional meanings. In ordinary language, the precise connotation and denotation of a word usually represents only a core of its meaning in discourse. The core is of prime importance, being the conventional meaning that makes the phonetic pattern a word, wherefore we have dictionaries to preserve what is called the "strict" sense of each linguistic unit; but the dictionary, in many cases, lists also various other meanings which may be given to words and still fall within the vague confines of "proper" usage. Words are our most powerful symbols; their use is universal, constant, as natural to man as walking on only two limbs; nothing can hold him to an unchanging standard of verbal expression, as long as he finds that he can, in fact, use words in highly unusual ways and still be understood. So even the strict dictionary meanings change with time, and the senses admitted as "figurative" in one age are regarded as literal in another, and usages which no lexicographer would list as correct at all force themselves into

---

[8] This growing emphasis on symbolic functions is the "keynote" referred to in *Philosophy in a New Key*, where further discussion of its origins may be found, especially in the early chapters.

public acceptance by dint of popular tendencies to error that the psychology of language so far has not fathomed.

At the opposite extreme from mathematical expression stands the great phenomenon of artistic expression, the symbolization of vital and emotional experience for which verbal discourse is peculiarly unsuited. Epistemologically this sort of symbolic presentation has hardly been touched. The philosophers of science and mathematics have drawn a great black line between propositional language used to state facts as unequivocally and literally as possible, and all other kinds of expression and their various purposes, which are lumped together under one caption, "emotive."[9] This category embraces self-expression, the symbolization of wishes and fears in dream or fantasy, myth and other religious beliefs, and all artistic expression, which positivistic thinkers generally regard as an exhibition of the artist's own emotions, either in lyrical sighs and confessions, or by representation of things which evoke his emotive responses.

Like most sweeping and simple classifications, the treatment of all non-propositional symbols together as one kind with one function has thoroughly confused the study of human mentality as a whole, and especially of the crucial humanizing activity, symbolic projection. No crasser oversimplification could possibly be made than the assumption that symbolic processes are either concerned with receiving, handling and storing information, or with externalizing and working off emotions. The effect of symbolic expression is primarily the formulation of perceptual experience, and the constant reformulation of the conceptual frames which the cumulative symbolizing techniques—conscious or unconscious, but rarely altogether absent—establish, one upon another, one in another, one by negation of another.

The dependence of different modes of thought on respectively different symbolic forms has been treated at length, with admirable scholarship and awareness of its implications, by Cassirer in *The Philosophy of Symbolic Forms*. Since the publication of that pioneer work, a number

---

[9] The first proponent of this division, so far as I know, was Bertrand Russell, in one of his early essays, "Mysticism and Logic," originally in *The Hibbert Journal* for 1914, and many times reprinted, especially in a collection to which it gave the title (New York, 1918). Since then it has become an article of faith among semanticists in general and logical positivists in particular; see especially Rudolf Carnap, *The Logical Syntax of Language* (1937), p. 278: "The supposititious sentences of metaphysics, of the philosophy of values, of ethics . . . are pseudo-sentences; they have no logical [meaning "conceptual"] content, but are only expressions of feeling which in their turn stimulate feelings and volitional tendencies on the part of the hearer."

of books and articles have appeared which reiterate and apply its basic ideas,[10] but none, so far as I know, has gone on to develop the differences among the forms beyond those which Cassirer indicated. The whole progressive genesis of conceivability, the evolution of human thinking in all its complexity, lies in the divergences of those forms. The objection has sometimes been brought against the doctrine of distinct symbolic forms that it places all orders of thought—the most fanciful and the most scientific—on a par as far as truth is concerned. I do not think this criticism is valid, though the emphasis Cassirer gives to the intellectual value of mythical conception invites it, if we gauge the intellect by a direct relation to truth. But intellect (perhaps unlike animal intelligence) has many indirect relations, too, with that ideal attainment; there are many steps in the processes of concept formation before the measure of factual cognition becomes paramount, and the "truth value" of mythical formulations belongs to this preparatory stage.

Not so the truth value of art. Artistic conception, for all its similarities to mythical ideation and even dream, is not a transitional phase of mental evolution, but a final symbolic form making revelation of truths about actual life. Like discursive reason, it seems to have unlimited potentialities. The facts which it makes conceivable are precisely those which literal statement distorts. Having once symbolized and perceived them, we may talk about them; but only artistic perception can find them and judge them real in the first place.

What the many researches on symbolism published in the English-speaking world today all tend to skip is the nature of the symbolic projection itself. They are generally occupied with its effects; that is why they stand so ruggedly apart, with no clue to the reason why the most diverse sorts of elements may be called "symbols" (for instance, "symbols" in chemistry and in psychoanalysis). But the differences in the ultimate appearances of symbolic projections lie in their many ways and means; and the sphere where diverse means and very subtle ways of projecting ideas force themselves on one's attention is the sphere of art. The reason is not hard to find: for art has no ready-made symbols or rules of their combination, it is not a symbolism, but forever problematical, every work being a new and, normally, entire expressive form.

The demand for a rule of translation to make the symbolism of art an unambiguous means of communication has been made repeatedly, sometimes as a challenge expected to be frustrating, which is supposed to disprove the expressive character of art, and sometimes in the genuine hope

[10] These sources will be mentioned later as their contributions come under discussion. *Phil.N.K.* seems to have been the first one to appear in English.

of achieving an interpretive technique that would make art a true "language of feeling." This hope has been seriously entertained chiefly with regard to music, where the formal creation is less masked by representational functions than in plastic and poetic arts, and the appearance of pure emotional expression most evident. So common sense does not rebel when an aesthetician declares: "The problem confronting us in the psychology of music, as I see it, is to determine what musical organism corresponds to what living organism, and in what respects."[11]

The well-known work of Albert Schweitzer[12] on the emotional significance of Bach's music comes to mind at once as an acceptance and a step to solution of this problem. It is just such a hermeneutic as the demand quoted above seems to call for, to decide in what respects a "musical organism" might correspond to a living one—that is (I take it), what functions of living organisms are analogous to the dynamic forms presented in music. Schweitzer's systematic analysis shows plainly that particular musical forms can be paired with particular emotions, and that Bach made use of this principle in setting texts to music.[13] But it also reveals the limitation, and indeed the danger, of such semantic studies. Schweitzer's method is entirely empirical, using Bach's own treatment of emotionally cathected words as a sort of Rosetta stone to give him the meanings which he then transfers to wordless compositions, so his attribution of emotive contents to musical elements is restricted to

---

[11] H. Jancke, "Das Spezifisch-Musikalische und die Frage nach dem Sinngehalt der Musik" (1931), p. 109.

[12] The first edition of this work was his *J.-S. Bach, le musicien-poète* (1905). The latest to date is the English translation by Ernest Newman, with C. M. Widor's Preface, under the title *J. S. Bach* (1955).

[13] André Pirro, in *L'esthétique de Jean-Sebastien Bach* (1907), goes somewhat further in the same line of research, giving far more thought to dissonances, changes of key, etc., than does Schweitzer, who worked mainly with melodic figures; yet Pirro's text is essentially an exposition, almost word for word, of Schweitzer's problem and method, even to saying that once the "dictionary" of themes is derived from Bach's treatment of the texts, it can be applied to instrumental music, and that the vocal compositions would be interpretable in the absence of their texts once we had the key. Whole paragraphs are simply taken over. Therewithal, he mentions Schweitzer only three times, late in the book, to take issue with him on trivial points.

The same charge of unacknowledged borrowing may be brought against E. Sorantin, who, in his dissertation, *The Problem of Musical Expression* (1932), uses Schweitzer's technique of investigating melodic figures in song and opera with sad or happy words, and then interprets these same figures as they occur in instrumental music. Sorantin traces all the emotions he finds to joy and sorrow (on the basis of a pleasure-displeasure psychology of feeling) and their possible compounds.

More thorough research might uncover a much greater body of such modern hermeneutic inspired by Schweitzer.

affects which can be named or generally indicated by words. His herme-neutic really makes music a "language of feeling," with formally estab-lished unit phrases or motifs to "mean" joy, sorrow, anger, etc.; but this symbolism, being homologous with a small part of our verbal symbolism, namely, the words that serve in the "dictionary of themes," is subject to the same limitation as language in expressing concepts of feeling.[14] One might say, perhaps, that a musical figure, besides referring to a feeling as a word does, also serves as an auditory semblance of it, and in this way reveals what it is like. Anyone, however, whose musical intuition can perceive the emotive import directly does not need a key to tell him into what category to put it. He might even have good reason to put it into another, but the question is not likely to occur to him.

It appears to be a very common assumption that if music expresses emotions, feelings or states of mind, it must be possible to name each psychical phenomenon that is expressed, and to assign it to some circum-scribed part of the music. So, for instance, Albert Gehring, in a book[15] which contains as much sagacity and astute criticism as misconception and plain blundering, first builds up a hypothesis of music as an image of the mind, saying that if experience could be viewed "sweeping past at a quickened rate," simplified, in a vast perspective, "streams of ten-dency would come to view, surgings to and fro, obstructed efforts, pre-cipitate advances, victorious emergences, expectations, hesitations, and satisfactions," and asking rhetorically: "is not music admirably adapted to form a picture of all these things?"[16] But a little later he observes (truly enough) that one cannot prove that every step in the progress of a musical composition means a namable feeling, a memory or an idea, and (falsely enough) concludes: "Until this is done, we must deny that symbolization accounts for the essential charm of the art."[17] In proof of his finding he cites the Andante from Beethoven's piano sonata opus 14, no. 2, as "an exquisite piece of music, sparkling with beauties, in which almost every measure, like a separate gem, contains charms of its own," and asks the reader: "Where, however, is the expression of emotion? Does the emotion lie in the first measures? . . . does every part express a

---

[14] According to Pirro, some music teachers in the seventeenth century compiled handbooks, or textbooks, of "correspondences" between musical forms and certain words, and compositions were produced by simply following the words of a text according to the conventions of this fictitious "language," making musical sentences parallel to the verbal ones. See *op. cit.*, pp. 16–18.

[15] *The Basis of Musical Pleasure, together with a Consideration of the Opera Problem and the Expression of Emotions in Music* (1910).

[16] *Ibid.*, p. 75.

[17] *Ibid.*, p. 90.

different state of affection?"[18] and so forth. Finally he clinches his argument with the words: "Even the most extreme expressionists admit that the interpretation cannot be hunted down to the individual bars. Would not the conclusion seem to follow that the musical beauty, which adheres to the bars, does not depend on interpretation?"[19]

Almost every error one can make in criticizing the theory of expressive form is represented here; in the first place, the tacit assumption that what is expressed in art is always an emotion, when in fact it may be the mere feeling of vitality, energy or somnolence, or the sense of quietness, or of concentration, or any of the countless inward actions and conditions which are felt in the living fabric of mental life; in the second place, the constant confusion of feeling with the events or ideas that evoke it; and in the third place, the concept that if expressiveness is the essence of art, then beauty depends on interpretation. This last fallacy is the most serious. Artistic import requires no interpretation; it requires a full and clear perception of the presented form, and the form sometimes needs to be construed before one can appreciate it. To this end, interpretations of verbal material or representational compositions may be useful, even necessary. But the vital import of a work of art need not and cannot be derived by any exegesis. Such a process, indeed, destroys one's perception of import.

All the failings of musical hermeneutics (the only developed technique of interpretation we have, with the possible exception of Vitruvius' theory of the significance of architectural forms) rest on one cardinal error, the treatment of the art symbol as a symbolism. A work of art is a single symbol, not a system of significant elements which may be variously compounded. Its elements have no symbolic values in isolation. They take their expressive character from their functions in the perceptual whole.

Art has a logic of its own (and by "a logic" I mean a relational structure), which is very complex; it is largely by virtue of its complexity that

---

[18] *Ibid.*, p. 175.

[19] *Ibid.*, p. 190. In this same movement one finds such bars as:

This bar might occur in the most trivial music (e.g., "Three Blind Mice"). It is hard to see how beauty, any more than feeling, "adheres to the bars." This one happens to express a strong feeling of conclusion, coming to rest, or cadential feeling.

it can present us with images of our even more complex subjective activity. The only work that I know which centers on this topic, and to which I can subscribe almost without reservation, is that of Ivy G. Campbell-Fisher,[20] interrupted by her death before it was brought together in a book, but published as well as might be by her husband in three articles, two of which are immediately relevant here: "Aesthetics and the Logic of Sense,"[21] dealing primarily with the problem of logical form in art and its difference from discursive form, and "Intrinsic Expressiveness,"[22] by which she meant exactly what I mean by the symbolic projection of feeling, though she took exception to the term "projected emotion" because most people mean by this phrase the imputation of either the artist's or the beholder's own emotion to the object.[23] The prudent evasion of the supposedly dangerous concept of projection, like most evasions, caused considerable confusion and some inconsistency in her analysis, which she proposed to make in terms of "external form," i.e., the sensibly created object, and "internal form," or "the emotionalized thought"—professing thus to avoid the traditional but highly equivocal language of form-and-content. Without admitting the projective function of the "external form," however, she ran into great difficulties in determining what "combination" of the two forms, outer and inner, yields a work of art; "intrinsic expressiveness" is attributed to both kinds of form; and finally she said, "The link common to internal and external form is their emotional content."[24] So here we have the old form-and-content language after all.

Yet Mrs. Campbell-Fisher, despite these semantic weaknesses, was very sure of what constitutes artistic import and how the import of a work differs from meanings occurring in it, which she distinguished as "referential meanings." "What art invariably has," she said, "in addition to its organization of sense materials is its expression. Nonreferential as well as referential art has this. All sense materials and their organizations are intrinsically expressive. . . . One abstract painting has an intrinsic expressiveness different from another's, and one has seen dances having their expression without a shadow of referential, associational meaning."

[20] The chief exception is in regard to her theory of beauty, but that does not concern us here.

[21] Published in 1950. This article is the more solid and intellectually developed of the two, perhaps because her notes were more worked out and ready for presentation.

[22] Published in 1951. This article is immediately followed by the third, "Static and Dynamic Principles in Art," but the topic there discussed belongs to a later chapter in this book.

[23] "Intrinsic Expressiveness," p. 245.

[24] *Ibid.*, p. 268.

Her recognition of such essential artistic expressiveness (which I call the projection of artistic import) does not beguile her, as it does many other people, into regarding "referential meaning" as an encumbrance or impurity;[25] on the contrary, she took special note of the fact "that fine representational art has always been formal too. That Giotto's, Rembrandt's, El Greco's formal organizations are as great as those in any purely abstract canvas of Braque, Kandinsky, or Picasso. . . . The 'story telling' of the great artists never interfered with the formal use of their materials; indeed, these men performed the art miracle of fusing the two. . . ."

The "art miracle" is so normal to painting, sculpture, dance and all the poetic arts that "fusing the two" is probably not the right idea; there are not two purposes to be served, two interests to be adjusted to each other, because the artist has accepted a theme. Leonardo speaks of the pictured subject—whether history, portrait or allegorical figure—as "the artist's narrative," and among his scattered advices to painters in the *Treatise on Painting* there are directions for using the possibilities of the "narrative," just as for using the characteristics of pigments—the transparency of some and opacity of others, the purity of spectrum colors and the plastic function of mixed and especially blackish hues, the chiaroscuro effects of subtle transitions from one to the other and of bright details on a dark ground. He remarks on the opportunity afforded the painter by the transparent, bright skin color of young people and the weathered look of the old, but also on the support given to such contrasts by the contrasts in the motions depicted. "The variety of motions in man," he says, "are equal to the variety of accidents or thoughts affecting the mind, and each of these thoughts, or accidents, will operate more or less according to the temper and age of the subject; for the same cause will in the actions of youth, or of old age, produce very different effects."[26]

The secret of the "fusion" is the fact that the artist's eye sees in

---

[25] As Arnold Schoenberg, for instance, wrote in one of his essays (a very early one, to be sure) in *Style and Idea* (1950), p. 2, after having remarked on the pure immateriality of music: "So direct, unpolluted and pure a mode of expression is denied to poetry, an art still bound to subject-matter." In the same spirit, A. H. Barr, in his *Cubism and Abstract Art* (1936), p. 16, makes an excuse for the thematic materials in contemporary art, saying: "The cows and seascapes and dancers which lurk behind the earlier abstract compositions of Mondrian and Doesburg have no significance save as points of departure from the world of nature to the world of geometry."

[26] Leonardo da Vinci, *A Treatise on Painting*, trans. John Francis Rigaud (1877), p. 34.

nature, and even in human nature betraying itself in action, an inexhaustible wealth of tensions, rhythms, continuities and contrasts which can be rendered in line and color; and those are the "internal forms" which the "external forms"—paintings, musical or poetic compositions or any other works of art—express for us. The connection with the natural world is close, and easy to understand; for the essential function of art has the dual character of almost all life functions, which are usually dialectical. Art is the objectification of feeling; and in developing our intuition, teaching eye and ear to perceive expressive form, it makes form expressive for us wherever we confront it, in actuality as well as in art. Natural forms become articulate and seem like projections of the "inner forms" of feeling, as people influenced (whether consciously or not) by all the art that surrounds them develop something of the artist's vision. Art is the objectification of feeling, and the subjectification of nature.

The rationale of the ancient and almost ubiquitous practice of representation in painting and sculpture lies in this dialectical aspect of the arts. Our present cultivation of non-representational art, for all its importance, is episodic in history. It is a purification of vision that has become necessary, partly because people's conception of portrayal had degenerated to a point of complete misconception, and partly because, at just that same time, an entirely new demand on formulative, intuitive vision arose from the social stresses and disconcertments of a new age. A new mentality was crying for recognition and projection. The power of abstraction became a central need, and the elimination of referential meanings threw the whole task of organizing the work on the artist's imagination of sheer perceptual values. The great "abstract" painters and constructivist sculptors are clearing the way to a new vision; and when they have found and completely mastered the principle of its presentation, they presumably will turn to nature again for the same purpose as always before. And it will look different to their eyes.[27]

---

[27] There are some interesting essays, and more recently a book, *Peinture et société; naissance et destruction d'un espace plastique de la Renaissance au cubisme* (1952), by Pierre Francastel, dealing with the artistic organization of space and its influence on our perception of actuality and, consequently, by the dialectical exchange of imagination and perception on what appears "realistic" in art. See his "Espace génétique et espace plastique" (1948), and especially "Naissance d'un espace: mythes et géometrie au Quattrocento" (1951). The thesis of this interesting article is that the spatial conception of the Renaissance painters was not the discovery, made once and for all, of visual space "as it really is," but was a re-imagination of space necessitated by changes in practical and scientific conception, and was a gradual achievement: "Depuis quatre-vingts ans la mesure du monde a changé non dans

Two further facts which Mrs. Campbell-Fisher saw more precisely and therefore stated more convincingly than most other aestheticians who have seen them at all are (1) that the import of a work, if it is emotional (art can express other feeling than emotion), does not consist of ordinary emotion as we actually experience it, and (2) that many forms of feeling expressible in art have no names, but are nonetheless made perceptible and comprehensible through the "intrinsic expressiveness" of the work. Both statements occur in conjunction with each other, in a very fecund passage, as the following excerpts may show:

"It may well be that an artist never creates a work of art unless he is emotionally stirred; if so, it does not follow that this, his own, emotional excitement is what he portrays. . . . An artist may portray something quite independent of his own psychological processes. . . . He may go beyond the felt thing. . . . In what actual person have we ever experienced the inevitability of passionate pattern that Oedipus or Lear present to us? . . . If I could be as sad as certain passages in Mozart, my glory would be greater than it is. . . . The fact that I know as much as I do of the essence of pathos comes from meeting with such music. If those passages made me sad, which in fact they most often do not, that would be an extraneous and irrelevant detail. My grasp of the essence of sadness . . . comes not from moments in which I have been sad, but from moments when I have seen sadness before me released from entanglements with contingency. We have seen this in great beauty, in the works of our greatest artists.

". . . But what shall we call the emotions in *Medea* or a hundred other works of art? We . . . do not put or read into them some petty emotion we ourselves have experienced and can readily name. Rather we learn from them what an emotional reality of greatest stretch can be. . . . Or the calm, the spirit of eternal rest, that comes to birth in an Egyptian statue or a Mayan head—such are the emotions unnamed perhaps, but expressible and realizable in art. Those, and not personal emotional excitements, are what great artists give."[28]

---

l'abstraite mais pour chaque homme en particulier et dans chaque instant de son activité. . . . alors comme aujourd'hui, il s'agissait, pour une société en voie de transformation totale—et progressive—d'imaginer un espace à la mesure de ses actes et de ses rêves. L'espace plastique en soi est un mirage. Ce sont les hommes qui créent l'espace où ils se meuvent et où ils s'expriment. Les espaces naissent et meurent comme les sociétés; ils vivent, ils ont une histoire. . . . la société contemporaine est 'sortie' de l'espace créé par la Renaissance et à l'intérieur duquel les générations se sont, pendant cinq siècles, mues à leur aise. La mesure du monde a changé et, necessairement, sa representation plastique doit suivre" (pp. 3–4).

[28] "Aesthetics and the Logic of Sense," pp. 266–68.

Here we have a lucid statement of the difference between self-expression and artistic expression, with their respective complements, sympathy[29] and intuition. The intuition of artistic import is a high human function which so far both psychology and epistemology have completely by-passed. Yet its roots lie at the same depth as those of discursive reason, and are, indeed, largely the same. Probably the chief reason why it has never been recognized as a characteristic mental act is that the treatment of art as emotional self-expression and social communication is so much more familiar to common sense that it is axiomatically accepted, and once accepted, obscures the subtler phenomena of metaphorical presentation and insight.

But why should common sense find the concept of art as a direct expression of actual emotions easier to accept than the concept of the art symbol? It is because the latter idea requires a distinction which is not always easy to make and maintain. It takes some analytic effort to distinguish between an emotion directly felt and one that is contemplated and imaginatively grasped. In the case of outward events and objects, it is usually not hard to judge whether we are confronted with them here and now, or remember them as something in our own past experience, or have read or heard of them, or whether they are products of our own imagination. Especially the first case is normally easy to distinguish from any and all of the others, because perceptions come as impact, while all the others are subjective activities—recall, rumination or invention. In the case of emotions this ready means of distinction is not available. Emotions are felt as actions originating within us. An emotion, mood or disposition actually felt is as subjective as any thought about it; so the conception of feeling, and the contemplation of it as a part of the larger inward activity that characterizes human life, is not automatically distinguished from the actual occurrence called "having" that feeling. No wonder, then, that when an artist is said (by himself or someone else) to be engaged in expressing an emotion he is *ipso facto* supposed to have it and to be giving vent to it, as he might by shouting or gesticulating, and to be causing the same emotion in each person who sees or hears his self-expressive display. This assumption is made even more seductive by the fact that creative work always produces an actual excitement, which is colored by the feeling to be projected, and is sometimes more massive than the intended import. It is, I believe, this intel-

---

[29] Many people might be inclined to call the response to self-expressive acts "empathy" rather than "sympathy"; but empathy is a highly complex mental function which probably enters into both sympathy and intuition, and perhaps into all mental response. It will be a subject of serious discussion later.

lectual excitement, the feeling of heightened sensibility and mental capacity which goes with acts of insight and intuitive judgment, that the artist feels as he works, and afterwards evokes in those people who appreciate his creation. But—as Mrs. Campbell-Fisher remarked—this is not the import of art; what the created form expresses is the nature of feelings conceived, imaginatively realized, and rendered by a labor of formulation and abstractive vision. Their envisagement may be spontaneous and easy or very arduous and slow; the result need not show the way it was achieved.

There is, however, no basic vocabulary of lines and colors, or elementary tonal structures, or poetic phrases, with conventional emotive meanings, from which complex expressive forms, i.e., works of art, can be composed by rules of manipulation. The musical figures to which Schweitzer was able to assign a fixed association with certain emotionally cathected words may really be units with a natural feeling-value, and in Chinese poetry, where the material units seem to be a limited stock of available phrases rather than freely combinable words, it may be that those phrases have more or less fixed emotional significance; but neither in our music nor in Chinese poetry (as far as one can judge from translation) is there a rule for the composition of elementary feelings into great complexes. It is easy enough to produce standard cadences, manufacture hymn tunes according to familiar models and some experiential knowledge of standard alternative resolutions, etc.; but such products are at best mediocre, and their import too slight to show any articulation of feeling. The analysis of spirited, noble or moving work is always retrospective; and, furthermore, it is never definitive, nor exhaustive.

Now, that is certainly an odd enough logical characteristic to make any logician question the possibility of such a structure. I have pondered that question myself for many years, yet always with the conviction that the structure exists and has to be explained. The explanation of its peculiar resistance to systematic treatment lies in the nature of the symbolic projection effected in art.

The fact that different modes of thought derive from different principles of presentation, which is to say, that they operate with different projections of their subject matter, has long been recognized. At the beginning of our century, Jean Philippe published a very interesting book entitled *L'image mentale* (*évolution et dissolution*), which, for reasons whereon one can only speculate, has made too little impression on his colleagues and successors; one hardly ever finds a reference to it even in the literature of exactly its own subject, the material and processes of imagination. In American psychology this neglect is not surprising, because in his day our interest was already centered almost entirely on practical intelligence, survival values, social adaptation, etc., so the functions of recognition, generalization, memory and reasoning were usually treated as rather simple manipulations of "stored" and "reproduced" sensations. Philippe makes the character of images either in isolation, as sheer occurrences, or in their various operational contexts— that is, their functioning within all sorts of mental processes—his wide and general topic. Only William James would have been deeply interested in such problems. But why did *L'image mentale* receive so little attention in France? Perhaps the influence of Ribot, whose *Essai sur l'imagination créatrice* had but recently appeared and commanded great respect, made any serious study of imagery as such seem a purely academic pursuit. Ribot valued the scientific imagination above all other kinds and deprecated even art as a form of play (leaning on Groos,[30] who lists nine such forms); herein he followed the scientism which was the European equivalent of our pragmatism, and placed a distorting emphasis on imagination held in the frame of discursive thought. To this attitude of mind, Philippe's impartial, detailed structural analysis was uncongenial.

Ribot's essay itself, despite its bias, is full of important observations, as for instance of the emotional pressures under which the mental image arises as a spontaneous organic product,[31] and of the fragility and instability of such imagery;[32] of the part played by the recognition of analogy—i.e., the sameness of logical form in two diverse exemplifications—in the imaginative rendering and expression of ideas, which would

---

[30] In *Die Spiele der Thiere* (1896).

[31] T. Ribot, *op. cit.* (6th ed., 1921), p. 36 (all references are to this edition):

"Il y a des besoins, appétits, tendances, désirs communs à tous les hommes qui, chez un individu donné, peuvent aboutir à une création; mais il n'y a pas une manifestation psychique spéciale qui soit l'instinct créateur. . . . L'invention n'a pas une source, mais *des* sources." And further, p. 67:

"Toute émotion fixe doit se concréter en une idée ou image qui lui donne un corps, la systématise, sans laquelle elle reste à l'état diffuse; et tous les états affectifs peuvent revêtir cette forme permanente qui en fait un principe d'unité." This unity he describes below as "une synthèse subjective qui tend à devenir objective."

[32] *Ibid.*, p. 70: "La forme instable prend son point de départ direct et immédiat dans l'imagination reproductrice, sans création. Elle assemble un peu au hasard et coud des lambeaux de notre vie; elle n'aboutit qu'à des essais, de tentatives. . . . Citons par exemple les songes à demi cohérents qui semblent au dormeur contenir un germe de création. Par suite de l'extrême fragilité du principe synthétique, l'imagination créatrice ne parvient pas à accomplir son oeuvre et reste dans un stade intermédiaire entre l'association des idées simples et la création proprement dite."

remain amorphous without expression;[33] of the unconscious level at which that symbolic transformation takes place;[34] of the fragmentary, essentially imaginative mentality prevailing in infancy.[35] He cites Hippolyte Taine's dictum that myth is "not a disguise, but an expression,"[36] and Vignoli's characterization of it as "a psychophysical objectification of man in nature."[37] All these aspects should have intrigued him as much as the one which alone he valued—the function of imagination in scientific thought. Although he did not go as far as Lombroso, who denied artistic creation any importance whatever, Ribot considered it a luxury product born of the pure fun of creating, or else derived from outgrown mythology;[38] and he finally distinguishes three stages of mental development corresponding to three types of imagination: (1) initial ("*imagination ébauchée*"), which is transient, protean, irrational—dreams, fleeting images, at a somewhat higher level fancies, vague hopes; (2) fixed ("*imagination fixée*")—myth, speculative theories, art; and (3) objectified ("*imagination objectivée*"), or practical imagination, invention, truth. The second type he considers capricious, though not formless; the third has necessary form, verifiable by sense and reason.[39] That is the

---

[33] *Ibid.*, p. 22: "L'élément essentiel, fondamental, de l'imagination créatrice dans l'ordre intellectuel, c'est la faculté de *penser par analogie*, c'est-à-dire par ressemblance partielle et souvent accidentelle."

"Le travail créateur . . . est réductible à deux types ou procédés principaux qui sont la personification et la transformation ou métamorphose" (p. 23).

"La personification est le procédé primitif . . . ; il va de nous-mêmes aux autres choses. La transformation est la forme avancée qui va d'un objet à un autre objet" (p. 23).

"Entre l'imagination créatrice et la recherche rationelle, il y a une communauté de nature: toutes deux supposent la faculté de saisir les ressemblances" (p. 25).

"Le mécanisme psychologique du moment de la création est très simple. Il dépend d'un unique facteur . . . : la pensée par analogie" (p. 105).

[34] *Ibid.*, p. 48: "L'existence d'un travail inconscient est hors de doute. . . . *Théoriquement*, on peut dire que tout se passe dans l'inconscience comme dans la conscience, seulement sans message au soi, que dans la conscience claire le travail peut être suivi pas à pas, avec ses progrès et ses reculs; que dans l'inconscience il procède de même, mais à notre insu."

[35] See *ibid.*, p. 57.

[36] *Ibid.*, p. 105; from the *Nouveaux essais*.

[37] *Ibid.*, p. 101.

[38] *Ibid.*, p. 276; here he speaks of "le développement esthétique qui est une mythologie déchue, appauvrie"; and earlier (p. 114) he said, "Les dieux sont devenus des poupées dont l'homme se sait maître et qu'il traite à son gré."

[39] *Ibid.*, p. 150.

mature stage; and he remarked elsewhere that a man usually begins his creative activity with music and poetry and then rises to really important or "realistic" imagination.[40]

As for any precise observations on the processes and sources of imagination, they are scant in Ribot's work; the usefulness of the "inventions" was what mattered to him; so the problem he originally set himself, and defined as "a very clear and definite one: What are the forms of association that beget new combinations, and under what conditions do these take shape?"[41] is only superficially treated, because he does not delve into the nature of the things which are associated and combined, nor of the mental functions vaguely lumped under the term "association."

It remained for Jean Philippe, unhampered by any prejudice favoring one activity over others, to tackle the problems presented by imagination in all its forms, from fleeting images to useful inventions; and he found, at the outset, that in the "initial" type, *imagination ébauchée*, lay the essence of the whole self-expanding and self-elaborating process. This essence, moreover, he saw in the representational character of the mental image. Its involvements in recollection and invention are only uses to which it can be put, and these are variable, but representation is its primary function, which determines its structure.[42]

As he himself puts it: "Literally speaking, imagination is the activity which embodies in mentally perceptible forms the effects of our sensory impressions, present or past. Fashioned by this action, . . . the image is neither a memory nor an invention; it is a sheer representation, an *image* in the elementary and primitive sense of the word."[43] It is the modulus of imagination: "In the complexity of our mental organization it is a sort of living cell, which maintains its life through manifold and diverse transformations."[44]

Philippe's contribution to our understanding of the art symbol and its import—and, epistemologically, of the intuition which apprehends that import—is indirect, but great, perhaps basic. His work is a study of imagination, not of art. But it is a rare analysis of the forms through which images pass and the processes of their articulation, their ingestion of other products of mental activity—current percepts, older images, accents of feeling, verbalized or unconsciously held beliefs—and the cadence of their extinction. Despite the lack of recognition he encoun-

---

[40] *Ibid.*, p. 138. [41] *Ibid.*, p. 19. [42] *Op. cit.*, pp. 1–2.
[43] *Ibid.*, p. 3. [44] *Ibid.*, p. 4.

tered, the temper and direction of his thought were not entirely lost on the French psychologists who succeeded him. We find essentially the same approach to imagination taken by Henri Wallon and especially by Philippe Fauré-Fremiet, who saw its relevance to some problems of art. That relevance is actually greater and closer than he was prepared to claim.

The importance of Philippe's analysis of imagination for the study of art as a special symbolic projection lies in two fundamental tenets advanced and substantiated in *L'image mentale*: that the primary function of images is representation (which is to say that they are essentially and originally symbolic), and that their formation is a living process, and therefore as complex as all living processes are. The first of these two guiding propositions provides a key to the art projection: for the work of art is a constructed image, and should reflect the basic structure of the primitive and spontaneous image; if that structure is determined by the elementary function of representation, it must express first and always the natural laws of representation.[45]

In this connection we may draw on an observation made and used in the early part of the present century by Gustaf Britsch, but not published until after his death:[46] that the laws of representation are not the laws of optics which govern our physical vision.[47] According to Britsch, the essential task of art is *"a mental working up* ['geistige Verarbeitung'] *of the experiences of our visual sense.* This immediate—that is to say, non-conceptual—mental working up of visual experiences leads to intellectual connections within this field, to *judgments* and *cognitions* concerning experiences of vision."[48] The laws of representation which govern image-making are not laws of optics, but of visual interpretation, which begins

---

[45] A "natural law" is a logical form exemplified by such events as are said to be "governed" by the "law." Our conventional wording derives, of course, from the old conception of nature as an agent obedient to the dictates of a divine Will. The legislative connotation of "natural law" is unfortunate, but the term is too firmly established to be abolished now.

[46] Britsch himself presented his ideas chiefly in lectures. They were finally published in a systematic book, *Theorie der bildenden Kunst* (2d ed., here used, 1930), under his name, though the book was actually composed by a capable and faithful editor, Egon Kornmann. An excellent discussion, in English, of Britsch's work, with voluminous scholarly annotations indicating its relations to the theories of Fiedler, Wölfflin and others, may be found in Wayne V. Andersen's "A Neglected Theory of Art History" (1962).

[47] This proposition has been extensively used by Rudolf Arnheim in his *Art and Visual Perception: A Psychology of the Creative Eye* (1954).

[48] Britsch, *op. cit.*, p. 13 (from Kornmann's Introduction).

with the act of looking.[49] Therefore all aids to exact reproduction of what the eye actually takes in—from the surveying contraptions and "transparencies" of the Renaissance painters to the traditional device of measuring relative sizes of objects with a pencil held at arm's length—are irrelevant or inimical to plastic ideation and rendering.[50] Britsch was speaking only of images created in art, as Philippe spoke only of envisagements in imagination: but the two are so closely related that what Philippe says of the bodiless mental image rings true when it is applied to the plastic image, and corroborates what Britsch maintains, namely, that the work of art is not a "copy" of a physical object at all, but the plastic "realization" (*"Verwirklichung"*) of a mental image. Therefore the laws of imagination, which describe the forming and elaboration of imagery, are reflected in the laws of plastic expression whereby the art symbol takes its perceptible form.

Britsch is never done with condemning the evils of drawing so-called "foreshortened" figures by all sorts of devices which automatically produce a two-dimensional projection of a three-dimensional object.[51] He makes it abundantly clear that, and how, artistic representations are built up by means of visual conceptions, such as figure and ground, direction and change of direction, opposition, proportion; differentiation of color, and interaction of colors within a field of general coloration; volume, and visual changes of volume with torsion. This last-named principle creates great problems of representation, and is what has tempted artists—especially those with a flair for mathematics—again and again to seek some means of calculating the distortions of the traditional or "normal" forms of objects in all possible spatial relations. But all such devices seem to him a very contradiction of art. And he has good reason for his claim; for they lead to other than artistic achievements.

The fact is that the draftsman who is guided by a method such as Dürer advocated in his *Unterweisung der Messung* is not merely "cutting corners" in the process of making an image, but is making something else, namely, a model of the object, by a rule of symbolic translation, the projection of its three-dimensional geometric properties into a two-dimensional field. So Britsch is not exaggerating when he says that

[49] On the "intelligence of the sense-organs" there is a considerable and very interesting literature, physiological as well as psychological; but most of it will be of more importance in later contexts, and will be cited there.

[50] Britsch, *op. cit.*, p. 83.

[51] Britsch uses the word *Projektion* only in this sense of geometric projection; so his emphatic declaration (p. 105) that painting has nothing to do with projection refers to geometric methods (as Mrs. Campbell-Fisher's rejection of the term was also with respect to a special sense).

"foreshortening, in an artistic sense, i.e., rightly: variability of extension, has nothing, in principle, to do with the sort of foreshortening practiced by the 'man drawing a lute' and the 'man drawing a reclining woman.' All these geometric methods—whether they employ fixed viewpoints, grids, auxiliary geometric constructions (e.g., block figures), or merely the pencil for taking measure—stem from the difficulty of representing extensional variation, and replace this very advanced intellectual achievement, which requires a maximum of mental tension, by a geometric substitute that one can produce without any such tension."[52]

Why, then, did the great masters of the Renaissance who recommended such geometric methods not ruin their own art with them? Britsch says, because they were already too imbued with a truer artistic tradition.[53] At worst they played with their grids and transparencies. Their paintings are not constructed in any such projection.[54] Perhaps the very wrongness of the process kept it apart from their intuitive work, and let them indulge in their reproductive inventions as unharmed as Maillol or Brancusi would be by building a scale model of a ship.

The principal difference in the effects of calculated and imagined spatial forms, respectively, is that the former seem "dead," "empty," "unfelt."[55] Something about systematically executed models is so radically different from products of imagination that one is led to look for a basic difference in the whole symbolic projection. The art image has an irresistible appearance of livingness and feeling, though it may not represent anything living. A hat on a table, a pile of rocks, may have as much life as a dancing faun, and far more life than an "action photograph" of a dancer. It is this sort of vitality and feeling that constitutes

---

[52] Britsch, *op. cit.*, p. 82. The references in the text are to two illustrations in Dürer's *Unterweisung der Messung* (Nuremberg, 1525).

[53] *Ibid.*, p. 126: "Der historisch Denkende könnte dagegen einwenden, dass doch gerade unsere grössten Künstler, wie etwa Lionardo oder Dürer, mit diesen Theorien wie Perspektive, Proportion, Anatomie gearbeitet haben, . . . dass ihre Fruchtbarkeit also durch die Leistungen unserer besten Kunst erwiesen sind.

"Dem wäre zu entgegnen, dass sich bei genauem Zusehen die Auswirkung dieser Theorien immer als der Einheit der künstlerischen Leistung feindlich zeigen wird, so bei Dürer etwa in den konstruierten Köpfen, bei Lionardo durch das mangelnde Zusammengehen konstruieter perspektivischer Hintergründe und vorstellungsmässig gezeichneter Figuren. Man muss also sagen, dass diese Künstler nicht mit Hilfe, sondern *trotz* ihrer Theorien Einheitliches geschaffen haben, weil die Theorie ihrem Schaffen genügend weite Grenzen liess, um innerhalb dieser Grenzen—des Proportionsschemas z. B.—ihr Vorstellungsbild frei verwirklichen zu können, auch wenn sie selbst meinten, das Schema zu verwirklichen."

[54] Compare the statement of Pierre Francastel above, pp. 87–88.

[55] Britsch uses the terms "*tot*," "*leer*," "*unempfunden*."

the import of the work; it is conveyed entirely by artistic techniques, not by what is represented, for the represented objects could be recognized at once as "the same" if a photographic or made-to-measure record of them were put beside the drawing or painting in which they figure. Representation of objects is itself a device,[56] though such a natural one that plastic expression has grown up—as Britsch convincingly showed—on the principles of representation, which reflect the processes of visual imagination, the subjective envisagement of things and the increasing conceptual differentiation of their spatial properties. Those spatial properties, however, which are abstracted in the interest of artistic expressiveness are not essentially the geometric ones; they are noted and formulated chiefly with an eye to their dynamic, cohesive and other non-geometric aspects, because the space they are to organize is not actual space (the space of our actions, which is abstracted and refined by scientific thought), but virtual space. All presentation of the artist's idea—his conception of human feeling—is made through the expressiveness he gives to that virtual space (or time, or other substrate, according to the art), which he creates and fills with appearances.

Based on such abstractions and having such a purpose, the process of creating an art symbol is entirely different from that of making a model of an object. It is guided by imagination, and imagination is fed by perception; there lies the reason for all drawing from nature, as for all Aristotle's poetic "imitation." The process itself is a labor of sustained imagination, which exemplifies the laws of imagination, not of optics,

---

[56] This fact has often proved an Asses' Bridge of art theory. The interpretation of forms as objects gives the forms a meaning in the primary, discursive sense. Even if those objects have further symbolic values, those values are meanings; and both levels of meaning are elements in the construction of the art symbol, not separate parts of its import. The deeper level—the symbolic significance of objects such as Dürer's skulls and hourglasses, Picasso's Minotaur—is almost always confused with the artistic expression of feeling; even a critic who recognizes representation as a device in art may fall into this confusion. So, for instance, Ludwig Lewisohn, in *The Permanent Horizon* (1934), pp. 174: "All art . . . externalizes or expresses the processes of the human soul. Now the outer expression of the inner facts must be through forms, and these forms must borrow their substance from nature. . . . The artist uses as a symbol of his inner world details from nature which he first selects and next symbolically refashions so that they may express the passion or thought or idea that he desires to communicate." I find it hard to believe that Rembrandt's "Old Woman Cutting Her Nails" is a refashioned old woman, or even that the lights falling on her are details first selected and then refashioned, to express the artist's passions or thoughts. They are all created, not selected. Lewisohn seems, indeed, to know this, when he says on the very next page: "The heavy leaning on the fruits of exact observation as symbols is merely one technique, one method of projecting the artists' inner world, among others."

[97]

acoustics, phonetics or whatever scientific study pertains to the material employed. The material is, in fact, never entirely circumscribed, as it is in the sciences. Sound, for instance, is not an object of study in geology (though it might be used as an indicator), but it can be a true element in architecture; great bells in a tower bring the edifice to life, and their tolling belongs to the architect's conception of the tower. Or, to take an example from another art, one often finds in English folk songs a discrepancy between the speech cadence, which is given to the words by their sense, and the musical cadence, created by musical phrasing; this discrepancy is no ineptness of an untutored composer, but a strong device, making for an urgent forward movement somewhat like that of a syncopation.[57] But, although it is achieved by the use of word rhythms, it is not a poetic contribution to the song, for there is nothing comparable to it in poetry at all; it is a purely musical effect attained by the use of the rhetorical (not the phonetic, or "musical") structure of the text. Such foreign materials could never enter into acoustics, nor into music conceived as an acoustical pattern; vowels and consonants might, but not prosody.

The techniques of art are intricate, subtle and manifold beyond any prevision or accounting. And this brings us back to Jean Philippe's work, to his second thesis: that the forming of a mental image is an act, a vital process, as ramified and complex as all functions of living things are. The product of that act, i.e., the image itself, which presents with something

[57] A good example is offered by the familiar carol, "God Rest Ye Merry, Gentlemen." The pressure against the speech rhythm is established first of all by the unvaried march of even quarters that seem to stride right over the varying values of the words; and when that steady pattern breaks, the rhythmic tension, instead of being lost, is increased by the inserted little figure:

And its tid-ings of com-fort and joy (comfort and joy), and its tid-ings of com-fort and joy.

Another, even more striking example of the driving force of relentless even quarters is the time of "St. John's Steeple," which constrains the text like a straitjacket so that it really has to be sung on a single breath. There is no rest or break of any sort. And a further case of the disregarded verbal cadence occurs in "The Holly and the Ivy," where the last two phrases of the text are allowed no separation:

Oh the ris-ing of the sun, — and the run-ning of the deer, the sound of the

mer-ry or - gan, sweet sing-ing in the choir.

like the objectivity of a percept, still bears the stamp of the thing it really is—part of the cerebral process itself, a quintessence of the very act that produces it, with its deeper reaches into the rest of the life in which it occurs.[58] In the actual event this involvement with the whole vital substructure is simply given by the feeling of activities interplaying with the moments of envisagement. But works of art are not natural occurrences with the stamp of life upon them. They are constructed symbols, made in the mode of imagination, because imagination reflects the forms of feeling from which it springs, and the principles of representation by which human sensibility records itself. If a piece of art is to express the pulse of life that underlies and pervades every passage of feeling, some semblance of that vital pulse has to be created by artistic means. But of this, more afterward.

The great complexity of the imaginal processes which scientific psychologists are all too apt to miss (consider Ribot's statement, quoted above,[59] that the mechanism of creative imagination is "very simple" and depends on one mental factor!) is implicitly if not explicitly known to all intuitive artists, whose work is guided by an artistic idea, and not by a discursive one such as a program or prescriptive theory of art. So it is not surprising that Philippe's study of mental imagery received a further development at the hands of a psychologist who had spent part of his life as a musical performer. I do not know whether Fauré-Fremiet received a stimulus from his precursor, or found the same approach independently. It is, actually, not quite the same; for in his first book, *Pensée et re-création*, he starts from his artistic experience, assuming that it will reveal the ways, if not laws, by which images are formed, and traces their operation through more casual and fragmentary mental activities, the endless, varied improvisations of ordinary life in which imagination is more or less constantly engaged. This is, essentially, the path I am taking here, though with a further premise by which not only the processes but also the products of art become starting points for the understanding of

---

[58] So he says (*op. cit.*, p. 5) of the image, which he called the "living cell" of mental life: "Sous sa forme élémentaire, cette cellule psychique est en réalité aussi complexe que la cellule physiologique.

"Quoiqu'elle offre toujours les caractères atténués, d'une représentation objective, aucune image n'a été façonée de toutes pièces au moment où naquirent en nous les contours de son objet. Loin de là, c'est au contraire le résultat d'une double mise en oeuvre: d'un côté les éléments qui préexistaient en nous et pouvaient servir (ayant déjà servi) à nous présenter l'objet actuel; de l'autre, quelques éléments tout à fait neufs, produits . . . des sensations présentes, et qui donnent à la représentation son caractère de vie extérieure, et à la perception son objectivité."

[59] P. 92, n. 33.

mind, and, indeed, for fundamental inquiries into its entire evolution from primitive biological traits and tendencies. That premise (which was not a premise but a conclusion in *Feeling and Form*) is that works of art exhibit the morphology of feeling. I think Fauré-Fremiet would have agreed to it, though he did not really exploit its implications, which were blurred for him by the usual difficulty of distinguishing between the occurrence of feelings and the conception of them. Yet he was certainly aware that feeling in art exists only as what is expressed,[60] and that the matrix of a work is always an idea, a single idea whereof all apparently separate ideas in the work are further articulations (Britsch would say, further differentiations).[61]

What he means by the word "re-creation," which figures in the titles of both his books that are relevant here, is the symbolic projection of experience, whereby it is, as he says, "realized." All conscious experience is symbolically conceived experience; otherwise it passes "unrealized." But this symbolic conception can be practically instantaneous. In a work of art, the idea has to be embodied in a perceptible creation, worked out coherently as an organic form.

His later book is less concerned with art and more with life than *Pensée et re-création*, but it is in *La re-création du réel et l'équivoque* that he furnishes a key to many problems of the art symbol, namely, in the peculiarity of its logical projection, which makes it untranslatable and indivisible, but also makes it "organic," "living" and "emotive," though no emotion or even living creature capable of emotion be represented in it. The peculiarity he notes belongs to all imaginative construction: that its elements need not lie all in one and the same symbolic projection. Their respective transformations may be made on different

[60] *Pensée et re-création* (1934), p. 38: "L'expression est, proprement, la manifestation d'un sentiment, d'une pensée ou d'une émotion. Mais . . . quand je parle de l'expression, l'expression s'identifie avec le sentiment, la pensée ou l'émotion. L'une est inseparable de l'autre. . . . C'est pourquoi nous entendons par *expression* une disposition toujours plus ou moins créatrice de l'être, pensable instantanément, mais qui ne peut devenir efficiente et transmissible qu'en s'étalant dans la durée—ou dans l'étendue."

[61] *Ibid.*, pp. 59–60: "En pensant la pièce que je voudrais faire, je conçois *un accord harmonieux* entre les diverses scènes qui doivent se succéder. Telle situation me semble discordante et je la rejette—ou je la modèle pour la ramener à cette tonalité générale, à cette sorte de parenté sonore que l'on trouvera—si libre soit elle —dans une sonate, dans une symphonie, et qu'il en définit presque la première substance, comme la qualité de marbre ou de bronze d'une statue. Toute oeuvre humaine belle même moyenne, repose sur un projet constructif harmonieux—si audacieux soit-il—ou bien elle n'est rien qu'un contrefaçon."

principles.[62] In the ordinary, spontaneous envisagements involved in our consciousness of actual experience and of its precipitate in memory, different projections alternate as we "realize" fragmentary facts mixed with a more or less inchoate current of discursive thought, spoken or tacit. Images arise and fulfill their purpose, conjoining their functions in our thinking without any mutual adjustments of visual scale, clearness, degree of schematization or directness of reference.[63] We shift from one momentary vision to another without any sense of their incongruity because as a rule they are used only instrumentally, so as visual experiences *per se* they are simply not compared. Only their deliverances are compared.

It is different in art. Here the same multiplicity of logical projections prevails, but since a work of art is a single symbolic form presented to perception, it has to encompass all its elements without losing its unity of semblance and sense. That difficulty seems formidable enough to make one ask: why not limit the technique of art to one ruling principle of presentation? The answer is that the artist's idea cannot be

[62] *Ibid.*, pp. 15–16 (describing his own envisagement of a voyage): "Tantôt, en effet, je me contente de refaire un dessin mental du réseau . . . je recompose la carte Chaix et j'y représente mon train et moi-même comme ces petits drapeaux que l'on pique, en temps de guerre, pour symboliser une armée et les positions qu'elle a conquises; tantôt ce schéma linéaire s'élargit, s'étoffe. Il continue à me servir d'armature, . . . il m'offre des coordonnées, des points de repère, mais je sais bien qu'un pays n'est pas plat comme une carte. Pour peu que je le connaisse, je le recompose plastiquement, je modèle en moi une sorte de carte en relief qui n'est le decalque ou l'image d'aucune et que, déjà, je remplis de réalités objectives. Je retrace, non plus le cours linéaire du fleuve ou de la rivière, mais *l'expression* de la nature où l'eau apporte ses thèmes, son rythme, ses reflets, ses mirages, ses brouillards. C'est dans ce paysage que je me situe, paysage expressif, quasi vivant et, cependant, *toujours reductible au schéma de la carte Chaix.*" And later, p. 68: "En fait, nous pouvons toujours *glisser* d'un système réalisateur à un autre, même apparamment fort éloigné, et prendre point d'appui sur très peu de chose pour nous engager dans l'imaginaire."

[63] *Ibid.*, p. 132: "La maison lointaine n'a pas le même sens que *vue de près*, mais elle participe à la réalisation du paysage.

"Seulement, nous ne pouvons pas penser l'être lointain, la fourmis qui est maintenant devant ou dans cette maison, selon les proportions de la maison lointaine. Nous sommes obligés de le penser . . . dans ses proportions expressives, réalisatrices."

"Le sens commun ne s'étonne point et admet que nous *imaginons* la personne et ce qui l'entoure . . . en vertu de nos souvenirs. Sans doute, mais ces souvenirs . . . constituent des noyaux d'étendue que nous plaçons 'au dela de l'horizon' ou 'ailleurs,' nous ne savons pas bien où, quelque part dans l'espace. . . . La composition de ces noyaux d'étendue est autre chose que le découpage usuel des objets et des êtres opéré par l'intelligence. Celui-ci correspond à *un* système de morcelage du réel. Mais nous vivons couramment d'après plusieurs systèmes de morcelage . . ." (p. 133).

expressed by such simple means. Herein precisely lies the incapacity of verbal statement to form and convey conceptions of feeling. What can be stated has to be logically projectable in the discursive mode, divisible into conceptual elements which are capable of forming larger conceptual units by combinations somehow analogous to the concatenations of words in language. The principles of verbal concatenation are few enough to be formally known as rules of syntax and grammar; the combinable units are known, too; and though the meanings of those elementary symbols do have some range, only Humpty Dumpty can ignore their conventional sense entirely. Word meanings change, but only with time.[64]

The most salient characteristic of discourse is that its symbolization of concepts is held to one dominant projection, which enables users of words to "run through" elaborate combinations of them, building up meanings by accretion.[65] Other expressive devices may find their way into the pattern of discourse, but they are contingent to the basic pattern, and their sense is very aptly said to be "between the lines." The "lines" of discourse are propositional constructions; other accepted forms, which are not strictly propositional—interrogative, imperative, vocative—are auxiliary forms developed in use. The essence of language is statement.[66]

The single pre-eminent projection of language is both its power and its limitation. It makes language not only a means of symbolization, but a symbolism; in fact, it is the paradigm of symbolisms, so that every discursive form is usually, and rightly enough, called a language. The fundamental principle of such organization, which imposes itself on our tacit conceptual processes as well, is "logic" in its traditional sense, reflected in the various grammatical systems of particular languages.

The enormous power of language, whereby we are enabled to form abstract concepts, concatenate them in propositions, apply these to the world of perception and action, making it a world of "facts," and then manipulate its facts by the process of reasoning, springs from the singleness of the discursive projection. A symbolism to be currently manipu-

---

[64] The use of words in poetry is another matter. The elements of poetry are not words, but are created by words—albeit sometimes by just one word; but that is an unusual feat. Words, with their meanings literal or figurative, are materials of poetry; so are statements. Cf. F.F., chap. xiii, "Poesis," and Problems of Art, No. 10, "Poetic Creation."

[65] See above, Chapter 3, pp. 61–62.

[66] This subject belongs properly to Part IV below, but a brief treatment of it may be found in Philosophical Sketches, No. 2, "Speculations on the Origins of Speech and Its Communicative Function."

lated as a tool in practical life must have a single set of rules; and the units deployed by those rules, the words of the language, must be as unequivocal as possible. We all know that unequivocal verbal meanings are hard to secure and even harder to maintain; the necessity of scientific languages for very precise and abstract discourse testifies to that. But we do have ways of bringing words with unduly many meanings back to a "strict sense" by agreement, if only in a given context.

The same characteristics, however—its atomistic structure and singleness of projection—set limits to the expressive power of language. It is clumsy and all but useless for rendering the forms of awareness that are not essentially recognition of facts, though facts may have some connection with them. They are perceptions of our own sensitive reactions to things inside and outside of ourselves, and of the fabric of tensions which constitutes the so-called "inner life" of a conscious being. The constellations of such events are largely non-linear, for where sequences occur they normally occur simultaneously with others, and every tension between two poles affects (evokes, modifies, cancels or precludes *ab initio*) many concomitant ones having other poles. The tensions of living constitute an organic pattern, and those which rise to a psychical phase—that is to say, felt tensions—can be coherently apprehended only in so far as their whole non-psychical organic background is implied by their appearance. That is why every work of art has to seem "organic" and "living" to be expressive of feeling. Its elements, like the dynamic elements in nature, have no existence apart from the situations in which they arise; but where they exist they tend to figure in many relationships at once.

This multiplicity of functions is reflected in any symbolic form that can express the morphology of feeling—a fact which makes discursive expression a poor candidate for that office. But another product of human mentation, the spontaneous image, has exactly this character, which psychologists designate as a tendency to be "overdetermined" in significance. It is this overdeterminateness that Fauré-Fremiet meant by *"l'équivoque,"* and found inherent in all *"réalisation,"* or formulation of subjective experience.[67]

The only adequate symbolic projection of our insights into feeling

[67] *La re-création du réel et l'équivoque* (1940), p. 2: "Disons tout de suite que l'analyse de la pensée du réel nous convainc non seulement de sa complexité, mais de son incurable, de son irréductible équivoque. Nous sommes tentés de croire que l'homme est affligé d'une sorte de strabisme mental, qu'il voit à la fois au moins deux univers, qui sont le même, et dont les images différentes, dans l'état actuel de sa pensée, ne peuvent jamais coincider exactement."

(including the feeling of rational thought, which the discursive record of rational order has to omit) is artistic expression; and the material for this is furnished by the natural resources of imagination. But artistic expression has an enormous range; for what we usually call a mental image—a visual fantasy—is only one kind of figment produced by our brain, prodigally, in sleep as well as waking, apparently without effort. Forms of sound and of bodily movement and even envisagements of purposeful action are similarly engendered, and have the same sort of plasticity and the same tendency to take on symbolic functions as visual forms. With these diverse materials, variously endowed people create works of art in the several great orders—music, drama, painting, dance, and so on. The potentialities of the imaginative mode seem to be endless.

As literal language owes its great intellectualizing power and its usefulness for communication to the relative simplicity of its logical structure, which also sets its limits of expressiveness, so the non-discursive structure of artistic presentation prevents art from ever being a symbolism which can be manipulated by general rules to make significant compositions, but at the same time is the secret of its great potentiality. Its elements are all created appearances which reflect the patterns of our organic and emotional tensions so ostensively that people are not even aware of speaking figuratively when they speak of "space tensions," "harmonic tensions," the "tension" between two dance partners, or even between two unrelated events in a poetic work, for instance, the entrance of the drunken porter in *Macbeth* and the scene which precedes it. A sculptor, Bruno Adriani, adduces "tension" between values as an architectural device, without, apparently, the slightest consciousness of its virtuality.[68] The illusion of tensions is the stuff of art.

The creative processes which build up an artist's image of subjective acts are numberless, because they can work in combination by many projective techniques at once, mingling several principles of presentation. A work of art is like a metaphor, to be understood without translation or comparison of ideas; it exhibits its form, and the import is immediately perceived in it. So far, I have always called its characteristic symbolic mode simply "non-discursive"; but there are other non-discursive symbols, such as maps and plans, which have not the organic structure or the implicit significance of art. One might well call a work of art a metaphorical symbol.

[68] *Problems of the Sculptor* (1943), p. 44: "The sculptor . . . establishes a balance between movement and static values. The tensions between these opposite tendencies—the hidden dynamics instead of lively exterior motion—is one of the fundamental conditions of monumental plastic style."

It is, however, not a simple metaphor, but a highly elaborated one, even though it may seem the simplest of compositions; for its elements are articulated to various degrees, their sense is one thing in one of their internal relations and something else in another.[69]

Every symbolic projection is a transformation. There are projections which are merely transpositions, such as the projection of a picture from a slide to a screen, but they are not symbolic. The function of a symbol is not only to convey a form, but in the first place to abstract it; and this requires transformation, because it is the sameness of logical structure in experientially different loci that makes it apparent. In the simplest use of symbols, such as ○, ⊙, and ◎ on a map for cities of various classes, we say that cities "appear as" circles with distinguishing modifications. Such marks are variable and usually explained by a key. They continue to look like mere signs. It is somewhat different with signs we manipulate all the time, and especially those which stand for immaterial entities, as numerals do. In most people's thinking, as well as in our written records, numbers appear as numerals. Yet we may translate our numerals into those of a different system, for instance, Arabic into Roman, "7" into "VII." Then the number seven—a purely mathematical concept—appears as "VII," at least after a little habituation. A still more radical symbolic transformation, psychologically speaking, takes place when we watch changes of a perceptually inaccessible sort by means of a device, e.g., the passage of time by the advance of the hands in a clock, or the fluctuation of prices on the stock market by the progressive action of the ticker tape. Ordinarily, in such contexts, we disregard the symbol and consider ourselves in direct touch with the reality it conveys; yet we know, upon very little reflection, that a symbol intervenes, and that its physical character is merely so irrelevant to its meaning that we ignore it. Its elementary meanings are assigned by convention and could be changed by agreement.

The art symbol, however, does not rest on convention. There are conventions in art, and they do change, but they govern only the ways of creating the symbol, and not its semantic function. When they change, it is not by any agreement but by individual departure from them. The

[69] Compare G. Anschütz, "Zur Frage der musikalischen Symbolik" (1928), p. 116, speaking of the expression of moods and emotion by musical articulation: "Die gleichen oder ähnlichen musikalischen Formen erhalten auf diese Weise bald den einen, bald den anderen Symbolwert. Es wäre verkehrt, hierin eine Beschränkung musikalischen Ausdrucks zu erblicken. Gibt es doch innerhalb unseres gesamten Bewusstseins und Geisteslebens nirgends ein Element, das nicht durch Verknüpfung mit einem anderen zum Träger einer neuartigen Bedeutung werden könnte."

import of art inheres in the symbol, which has no dispensable or change-able physical character at all, because it is an image, not an index; its substance is virtual, and the reality it conveys has been transformed by a purely natural process into the only perceptual form it can take. Feeling is projected in art as quality.

What I call "quality" here is, of course, not what the British empiricists meant by that word when they distinguished primary and secondary qualities, and spoke of simple qualities and their compounds—mixed colors, for instance, or mingled odors. There is a kind of quality that different colors, or even a tonal form and a visual one, may have in common; even events may have the same quality, say of mystery, of portentousness, of breeziness; and a word like "breeziness" bespeaks the qualitative similarity of some moods and some weathers. Homer refers to "the wine-dark sea," although Greek wine is red, and the Mediterranean is as blue as any other sea water. But the translucent blue in the curve of a wave and the glowing red in a cup of wine have a common quality.

It is quality, above all, that pervades a work of art, and is the resultant of all its virtual tensions and resolutions, its motion or stillness, its format, its palette, or in music, its pace, and every other created element. This quality is the projected feeling; artists refer to it as the "feeling" of the work as often as they call it "quality." The image of feeling is inseparable from its import; therefore, in contemplating how the image is constructed, we should gain at least a first insight into the life of feeling it projects.

5

The Artist's Idea

*Bei jedem Kunstwerk, gross oder klein, bis ins
kleinste, kommt alles auf die Konzeption an.*
—Goethe

FEELING is projected in art as quality"; where has that been heard
before? The proposition sounds familiar, and with its familiarity
there comes to most people a conviction that they are through with it.
Presently the old phrase echoes in memory: "Beauty is pleasure objectified." Beauty was the quality, and pleasure the feeling thus projected.
Santayana, of course.

The cardinal ideas with which I am working can almost all be found
in the literature, but usually not in each other's company. So, in a sense,
the thesis that artistic quality is an objectification of feeling was set up
by Santayana even before the turn of the last century;[1] only it is the
sense that makes all the difference. Both "objectification" and "feeling"
have such radically different meanings in his theory and mine that there
is little relationship, even historically, between our respective assertions,
although the same words occur. The process of objectification he called—
as I would call it—"projection"; but by "projection" he did not mean
symbolic projection, he meant the supposed process of "projecting" sensations actually occurring in our own bodies onto an outside world as
qualities of "matter." This error to which mankind was presumed to be
irresistibly prone, which explained our naïve belief in "secondary qualities" and justified Hume's skepticism, is extended in Santayana's aesthetic theory to another level of psychical response than the level of
sensation, namely, that of feeling. Just as colors and sounds are objectified sensations, beauty is the objectification of our pleasure in having
those sensations, and is wrongly imputed, like them, to an outside

[1] In *The Sense of Beauty* (1896).

[107]

object.[2] This doctrine is "classical" enough to need no exposition; the points at issue here are that for Santayana (1) the feeling objectified in an aesthetic object is a special and somewhat superficial one, the pleasure of agreeable sensation;[3] (2) the "projection" whereby the pleasure is objectified is exactly the same process of erroneous imputation which is supposed to operate in producing the illusion of secondary qualities;[4] and (3) the projected feeling is one which the beholder of the beautiful object is undergoing as he sees or envisages the object. On all three counts the similarity in the wording of Santayana's theory and the one expounded in *Feeling and Form* is deceptive.

Santayana's psychology of art (it is really not a philosophy, since he had rather little interest in the essential nature of art, but only in the aesthetics of our perception of art among other pleasing sights or sounds), going back some three generations now, sounds peculiarly "dated" even for its age. It makes one suddenly aware that a revolution has occurred in philosophical thinking since his day, which has deeply and rather specially affected our understanding of art—a revolution which was creeping up on us even then: the shift from a concept of human mentality built up in terms of sensation and association to a concept built (a little shakily still, but broader based) on principles of symbol and meaning, expression and interpretation, perception of form and import.

This new orientation changes the meanings of some terms which Santayana himself introduced into aesthetics. Old words with new meanings are treacherous, chiefly because of the academic habit of classifying theories by schools or "isms" based on words, not on the various senses in which words are used, which senses are then erroneously assumed to be the same. Yet finding a word which stands while its meaning gradually changes has an essential function in the development of philosophical thought. The right word for a difficult concept is often recognized

---

[2] The first part of *The Sense of Beauty*, called "The Nature of Beauty," ends with the words: "Thus beauty is constituted by the objectification of pleasure. It is pleasure objectified" (p. 32).

[3] That Santayana never found anything deeper than this in the plastic arts is attested by his statement in a much later and probably definitive essay, his reply to critics entitled "Apologia pro Mente Sua," in *The Philosophy of George Santayana* (2d ed., 1951), where he wrote: "In my own case, aestheticism blew over with the first mists of adolescence; . . . nor has my love for the beautiful ever found its chief sustenance in the arts. If art *transports*, if it liberates the mind and heart, I prize it; but nature and reflection do so more often and with greater authority" (p. 501).

[4] *The Sense of Beauty*, p. 49: "It [i.e., our pleasure] becomes, like them, a quality of the object. . . ."

as such before the concept is defined; the definition grows up under the constant suggestion of the word, by virtue of its etymology and all its linguistic relations, its traditional ambiguities, its poetic and even slangy uncommon uses. The thinker who leans on a promising word hopes to find his apt and enlightening concept as a possible meaning for that word; where logical invention flounders the term with its suggestiveness holds him to his line of thought. A strange condition indeed—where one could say without irony:

> *Denn eben wo Begriffe fehlen,*
> *Da stellt ein Wort zur rechten Zeit sich ein!*

The word "projection" has had such a history of intellectual service. I remarked in the previous chapter on the many senses in which it has been used. Santayana's usage was, of course, that of empiricist sensory psychology. By the time we meet the term again in a context of aesthetic theory, in Bernard Bosanquet's *Three Lectures on Aesthetic* (1915),[5] it has undergone a change, not so much in explicit definition as in implicit meaning, due to the direction in which the theory headed. Bosanquet, coming from a different metaphysical school, did not have the notion of projected sense impressions to distract him from the concept toward which he was working, as Croce and Collingwood[6] also were: the concept of symbolic projection.

Yet that concept was slow to take shape. Aesthetics, as its unhappy name bespeaks, started as an analysis of sensibility, though it was aimed from the beginning at making some systematic description of the pleasures of the higher senses, vision and hearing. These pleasures were soon found to be easily destroyed or at least eclipsed by most people's proclivity for using their eyes and ears only to identify things and receive information about them, to orient themselves and steer their own courses of action. To enjoy the deliverances of sense as such appeared to be, at least in our practical-minded society, an unusual indulgence; and so the aesthetic attitude became the first requirement for the experience of aesthetic pleasure, and sometimes seemed even to be the active source of it.

In Bosanquet's three published lectures, the beauty of a work of art is still the projection of an actual feeling, and that feeling is the beholder's

---

[5] In his *History of Aesthetic*, which appeared first in 1892, it occurs only incidentally, without theoretical significance.

[6] Croce and Collingwood, especially the latter, have been discussed at some length in *F.F.*, chap. xx, "Expressiveness." That entire chapter is relevant here; I can but refer the reader to it for a more systematic treatment of the subject as a whole and especially for their contributions to it.

pleasure,[7] but at least the projection is not the "error" of taking something subjective, which occurs in his own mind, for a property of the object. The projection is intentional and its purpose is to make the mood or emotion perceptible like the object which "embodies" it. Since the projected "pleasant feeling" is the percipient's, Bosanquet, like Santayana, starts with the "aesthetic attitude"; but it transpires almost immediately that his real problem is not what is felt at sight of the work, but what is there to be seen. With almost every paragraph the schism between the "embodied" feeling and the beholder's reaction widens, and the "embodiment" rather than the actual feeling moves into the focus of interest:

"The aesthetic attitude is that in which we have a feeling which is so embodied in an object that it will stand still to be looked at, and in principle to be looked at by everybody."[8]

"Feeling becomes 'organized,' 'plastic,' or 'incarnate.' This character of aesthetic feeling is all-important. For feeling which has found its incarnation or taken plastic shape cannot remain the passing reaction of a single 'body-and-mind.' "[9]

"The aesthetic attitude is that of the feeling embodied in 'form.' "[10]

To call "feeling embodied in 'form' " an attitude is certainly straining the traditional terminology. And, furthermore, as soon as Bosanquet's attention veers from the famous attitude to center on the embodiment of the feeling, it appears that this feeling need not be pleasurable at all. His first example is, in fact, a sorrow or pain which the artist is presumed to have undergone, and which he projects in visible form by creating an object to "embody" it. Now we are well on the way to the art symbol, the work of art as an expressive form. He speaks of "transforming" the feeling, "submitting it to the form of an object" (i.e., giving it objective form), "developing" it by manipulating the form; and in the end, of the feeling "as finally expressed," which is not simply the artist's hard experience (no sensible person would strive to perpetuate that), but "a new creation."[11]

---

[7] At the beginning of his first lecture, after a short preamble, he says: "The simplest aesthetic experience is, to begin with, a pleasant feeling, or a feeling of something pleasant—when we attend to it, it begins to be the latter" (*Three Lectures on Aesthetic* [1915], pp. 3–4).

[8] *Ibid.*, p. 6.   [9] *Ibid.*, p. 7.   [10] *Ibid.*, p. 13.

[11] *Ibid.*, pp. 7–8: "In common life your sorrow is a more or less dull pain, and its object—what it is about—remains a thought associated with it. There is too apt to be no gain, no advance, no new depth of experience promoted by the connection. But if you have the power to draw out or give imaginative shape to the object and material of your sorrowful experience, then it *must* undergo a transformation. *The*

The Artist's Idea

The reason why the feeling can be developed according to the "laws of an object" is that the aesthetic object is composed entirely of appearances, abstracted from physical nature and manipulated to serve the artist's purpose. It is, as Bosanquet says, "a semblance."[12] And in manipulating this semblance the artist may even construct forms with an emotive import which does not stem from any moment of ordinary emotion at all. For, in the end, "You have not the feeling and its embodiment. The embodiment, as you feel it, is the aesthetic feeling."[13]

Now, "the embodiment as you feel it" is certainly a difficult notion, unless you let "feel" stand for "perceive" or "apprehend" or some such cognitive term. Then the aesthetic feeling becomes nothing more nor less than artistic perception, or intuition of artistic import, and the "embodiment" of feeling clearly means expression, or symbolic projection of a purely conceived passage of feeling, developed, like all conceptions, by developing the symbolic form. Bosanquet never committed himself, in the *Three Lectures*, to the admission of the word "symbol," and perhaps rightly so, since the meaning of that word in his day was still closely limited to the designation of objects such as religious or heraldic symbols.[14] Even today the "art symbol" is commonly misinter-

---

feeling is submitted to the laws of an object. . . . The values of which the feeling is capable have now been drawn out and revealed as by cutting and setting a gem. When I say 'of which the feeling is capable,' I only record the fact that the feeling has been thus developed . . . and the feeling as finally expressed is a new creation, not the simple pain . . . which was felt at first."

[12] The concept of semblance is developed in F.F., where the term was borrowed from C. G. Jung (see chap. iv, "Semblance," especially p. 48 f). It could have been borrowed nearer home, from Bosanquet, who presented the idea precisely in the words: "Anything in real existence which we do not perceive or imagine can be of no help to us in realizing our feeling. . . . Nothing can help us but what is there for us to look at, . . . which can only be the immediate appearance or the semblance. This is the fundamental doctrine of the aesthetic semblance" (*Three Lectures on Aesthetic*, pp 9–10). And later: "A cloud, *e.g.*, we know to be a mass of cold wet vapour; but taken as we see it with the sun on it, it has quite different possibilities of revealing aesthetic form, because its wonderful structure is all variously lit up, and so lit up *it* is the object or semblance which matters" (*ibid.*, pp. 17–18). The concept of the developed semblance is essentially the same as Adolf Hildebrand's "articulation of space" by the "architectonic method" discussed in F.F., pp. 72–75.

[13] *Three Lectures on Aesthetic*, pp. 8–9.

[14] Yet it had been boldly used even earlier by J. Sittard, in a serious and significant review of Conrad Lange's *Das Wesen der Kunst* which appeared in *Die Musik* (1903), under the title: "Die Musik im Lichte der Illusionsaesthetik." Here Sittard introduces the concept of "*Scheingefühle*," feelings apparent instead of actual, and of "expression" as presentation rather than catharsis, and finally says: "Die Formen der Musik sind . . . kein wirklicher Gefühlsausdruck, sondern symbolische Darstellung eines solchen" (p. 248).

[111]

preted as symbolism employed in art—iconography, or the dream material of the Surrealists, or the personal metaphors of the French Symbolists. Yet he wrote one passage which serves to move the notion of artistic expression of emotion squarely into the newer frame in which, for instance, Edward Bullough or T. S. Eliot would place it:

"I have avoided, indeed, throughout this lecture, the word which I myself believe to be the keyword to a sound aesthetic, because it is not altogether a safe word to employ until we have made ourselves perfectly certain of the true relation between feeling and its embodiment. But to say that the aesthetic attitude is an attitude of expression, contains I believe if rightly understood the whole truth of the matter. Only, if we are going to use this language, we must cut off one element of the commonplace meaning of expression. We must not suppose that we first have a disembodied feeling, and then set out to find an embodiment adequate to it. In a word, imaginative expression creates the feeling in creating its embodiment, and the feeling so created not merely cannot be otherwise expressed, but cannot otherwise exist, than in and through the embodiment which imagination has found for it."[15]

"An attitude of expression," which does not mean a physical attitude seen to be expressive, is a peculiar turn of phrase; "a process of expression" seems to be what it means. The old vocabulary of "the aesthetic attitude" died hard. Yet the radical change from "a preoccupation with a pleasant feeling embodied in an object"[16] to a perception of artistic import was really very nearly complete at this point. "Embodiment" obviously meant to Bosanquet the symbolic expression of feeling—and he makes quite plain that this feeling was not first "had," but was essentially conceived through the symbol and created with it.

Gradually the concept of art as a formed image revealing the life of feeling has been worked out by many thinkers, especially a fair number of articulate and keen-minded artists. Most of those theorizing artists began with the assumption that their work was a direct outward expression or "objectification" of emotions they were actually undergoing as they composed it, and ended with the realization that they were expressing their knowledge of subjectivity rather than their momentary moods or passions; that this knowledge of "felt life" stems from a person's own experience, but always from the whole of it, so that even if raw emotional upheaval may have furnished a maelstrom of new feeling and exerted a strong pressure to formulate and transmute it, yet that feeling is not all that feeds the work; an artist's long-developed background of

[15] *Three Lectures on Aesthetic*, pp. 33–34.
[16] *Ibid*. This is his definition of the aesthetic attitude given—tentatively—on p. 10.

emotional conception, his individual sensibility, his own ways of articulating and projecting assimilate the feeling to his entire art before it
finds expression.[17] What he holds in mind as he works on a new expressive form is not an upheaval, but an idea.

In the poetics of one of our foremost artist-critics, T. S. Eliot (whose
theoretical work, I now think, was not duly appreciated in *Feeling and
Form*), there is no more talk about contagious self-expression giving
pleasure to those who are infected with the poet's emotion. Eliot wrote,
quite early in his life: "The more perfect the artist, the more completely
separate in him will be the man who suffers and the mind which creates;
the more perfectly will the mind digest and transmute the passions
which are its material."[18] Neither did he, even in his youth, ever regard
the music of words as a purely sensuous beauty to please the listening
ear. Sound and structure, imagery and statement all go into the making
of a vehicle for the poet's idea, the developed feeling he means to
present. With Eliot, the conception of art as the impersonal expression
of feeling (he always spoke of it as "emotion," but I think he meant,
as I do, whatever can be felt) was not only entertained, but avowed,
discussed and singlemindedly adopted.

At once its wide implicative power becomes apparent, and offers solutions of some of the "chestnut problems" of older philosophies of art:
how a work can express more than its maker knew, why it may be read
differently, even in logically incompatible ways, yet correctly by different

[17] See, for instance, Arnold Schoenberg's essays collected in *Style and Idea* (1950);
in an early essay (1912) he declared: "Indeed, a work of art can produce no greater
effect than when it transmits the emotions which raged in the creator to the listener,
in such a way that they also rage and storm in him" (p. 8). But in 1941 he wrote:
"Alas, it is one thing to envision in a creative instant of inspiration and it is another
thing to materialize one's vision by painstakingly connected details until they fuse
into a kind of organism." Such work is not done in a moment of raging emotions.

Similarly, the German dramatist Friedrich Hebbel, at the age of twenty-two, wrote
in his diary: "Ist dein Gedicht dir etwas anderes, als was anderen ihr Ach und ihr
Oh ist, so ist es nichts. Wenn dich ein menschlicher Zustand erfasst hat und dir
keine Ruhe lässt und du ihn aussprechen, d.h. auflösen musst, wenn er dich nicht
erdrücken soll, dann hast du Beruf, ein Gedicht zu schreiben, sonst nicht" (see
Bernhard Münz, *Hebbel als Denker* [1913], p. 132). At forty-five, however, criticizing his own early drama *Genoveva*, he wrote: "Der Hauptfehler war, dass ich zu
früh an diese Riesenaufgabe kam. Sie verlangte die grösste Reife des Geistes und ich
hatte noch zu viel mit dem lieben Herzen zu tun. Denn warum leugnen, was schon
mancher Kritiker herausgefühlt hat: ich selbst steckte in einer gar heissen Situation,
als Golo entstand" (*ibid.*, p. 407; from a letter to Franz Dingelstedt). Instances
could be multiplied, but these are typical.

[18] "Tradition and the Individual Talent," in *The Sacred Wood: Essays on Criticism and Poetry* (1920), p. 48.

people without therefore being only a mirror of their own subjective states, how it can be "original" although materials, techniques and any number of stylistic conventions which have gone into it are familiar and the artist was not afraid to follow a tradition or repeat a theme; such problems come crowding as soon as their solutions are in the offing.

Eliot has put into words some of those emergent explanations; for instance, that the art symbol (i.e., the work) may present more than the artist was aware of, because in the process of manipulating its elements all sorts of possibilities of form appear that he recognizes as organically motivated, inherent in the style of the piece or dictated by what he has already done, as its natural continuation; the artist is the first person to see a new quality arise, which he develops as best he can, capturing a new feeling which he could not have conceived before. Just as we may, in discourse, state a fact of which we are aware and then find that we have stated, by implication, further facts of which we were not aware until we analyzed our assertion, so an artist may find that he has articulated ideas he had not conceived before his work presented them to him. He may even continue to construe the expression he has created, and see new significance, just as any other beholder of the work may. And because of the "over-determined" character of art, he may see more, or less, or simply a different import from what someone else just as truly sees.[19]

---

[19] Compare Eliot's observation in "The Music of Poetry" (*On Poets and Poetry* [1957]), p. 23: ". . . the meaning of a poem may be something larger than its author's conscious purpose, and something remote from its origins. . . . A poem may appear to mean very different things to different readers, and all of these meanings may be different from what the author thought he meant. . . . The reader's interpretation may differ from the author's and be equally valid—it may even be better. There may be much more in a poem than the author was aware of. The different interpretations may all be partial formulations of one thing; the ambiguities may be due to the fact that the poem means more, not less, than ordinary speech can communicate."

A very similar observation was made by Hermann Nohl in his book, *Stil und Weltanschauung* (1920), carrying the condition over from literature to painting: "Wie aber jedes Buch eine stärkere und rationalere Durchbildung zeigt als das freie Gedankenleben des Autors jemals besessen hat, so stellt dafür das Bild auch die Steigerung der Gesichtsvorstellungen des Malers dar und ist darum grösser als sein Schöpfer" (p. 22).

Oscar Wilde, too, remarked on this objective character of the work, saying: "The beauty of the visible arts is, as the beauty of music, *impressive* primarily, and it may be marred, and often indeed is so, by any excess of intellectual intention on the part of the artist. For when the work is finished it has, as it were, an independent life of its own, and may deliver a message far other than that which was put into its lips to say" (quoted by E. F. Jacques in "The Composer's Intention" [1891–92], p. 43).

A work of art—in Eliot's discussions always a poetic work—is thus treated as an objective entity, which anyone, including its own maker, can contemplate, appreciate, probe and construe, receiving from it whatever intuitions of subjective reality it may yield to him. Whether the import he finds in it is what the artist intended his poem to convey is not the definitive measure of its artistic success. That means, of course, that the essential function of art is not communication; for in communication the first requisite is always that the meaning which the recipient finds in the mediating symbol is the meaning its user intended. Naturally the artist intended to convey an idea, and if his work is good, his idea is expressed in it; but since the perception of import is intuitive, and there is no such machinery as language commands—paraphrastic restatement, logical demonstration and what not—to induce an unready intuition, some other import, incidentally presented, may eclipse the intended one for a differently predisposed mind. And the incidentally articulated reality may even be the greater and more vital one. It may stem from the limbo of the artist's unavowed knowledge, and may some day confront him in its objective presentation, too, when he revisits his own work. Such discoveries cause a moment of self-estrangement: "Did I compose that?—I wonder whether I could do that now." Most artists have experienced such impacts.

The prime function of art is to make the felt tensions of life, from the diffused somatic tonus of vital sense to the highest intensities of mental and emotional experience, "stand still to be looked at," as Bosanquet said, "and, in principle, to be looked at by everybody." That brings us face to face with the problem of objectification.

This issue, too, Eliot has attacked, and his attack on it has become famous. "The only way of expressing emotion in the form of art," he said in one of his most frequently quoted passages, "is by finding an 'objective correlative'; in other words, a set of objects, a situation, a chain of events which shall be the formula of that *particular* emotion; such that when the external facts, which must terminate in sensory experience, are given, the emotion is immediately evoked."[20] Yet I think there is an error in this prescription—an error which rests on a confusion, subtle (or its author could not have fallen into it), yet crucial: the mistaking of a principle of poetic construction for a principle of art.[21] The making of the art symbol, which is the work itself, is very complex, and in it other types of projection are involved, which have to be dis-

---

[20] "Hamlet and His Problem," in *Selected Essays, 1917–1932* (1932), pp. 124–25.
[21] This confusion to which theorizing artists are ever liable is treated more fully in *Problems of Art*, Lecture 8, "Principles of Art and Creative Devices."

tinguished from the expression of the import of the whole. One of these other types is the "objective correlative." It is, as he said, a formula which the poet finds for the rendering of a particular emotion; but it is for an emotion presented in the work, not by the work. The feeling of the work itself appears as a perceptible quality, and there is no formula for the creation of that effect, which is not a correlative of feeling, but its transformed appearance. Where and how that transformation is made remains the problem in hand.

But first, a word about Eliot's finding, to explain why I do not regard it as a principle of art as such: the "objective correlative" is a general device for depicting emotion occurring within the virtual world of the story told or dramatized. It is a principle of representation, creating the emotions of the fictive persons by putting the concrete image of their motivating objective conditions before us; but the emotion is "evoked" in those persons, not in us. It was, indeed, in his criticism of Shakespeare's practice in *Hamlet* that Eliot's recognition of that basic dramatic technique was formulated, and all the examples he consequently adduces illustrate this same technical purpose; as soon, however, as it is extended to the situation as it looks to a *persona dramatis*, and then even further to the author's situation in relation to his material, both the emotion and the correlative shift their meanings, so that the concept of objectification becomes blurred. Consider, first, the purely structural use of the principle set forth in the instances on which the theory seems to have been initially based, given in this passage:

"If you examine any of Shakespeare's more successful tragedies, you will find this exact equivalence; you will find that the state of mind of Lady Macbeth walking in her sleep has been communicated to you by a skillful accumulation of imagined sensory impressions; the words of Macbeth on hearing of his wife's death strike us as if, given the sequence of events, these words were automatically released by the last event in the series. The artistic 'inevitability' lies in this complete adequacy of the external to the emotion; and this is precisely what is deficient in *Hamlet*. Hamlet (the man) is dominated by an emotion which is inexpressible, because it is in *excess* of the facts as they appear."[22]

"Hamlet (the man)"; so far, our tacit supplementations to the somewhat ambiguous terms "inexpressible" and "as they appear" may keep the sense quite clear: Hamlet's emotions are inexpressible to the audience, being in excess of the "facts" as they appear to the audience. To motivate Hamlet's actions we must impute to him emotions which are not created for us with the actions. The formula, in this case, should

[22] *Op. cit.*, p. 125.

serve to make visible the feelings of a person in the play, and such feelings are as virtual as the situations and overt acts of that person himself. The "objective correlative" here is a principle of dramatic presentation.

With the next step, however, the dramatic principle is changed to a psychological one, and applied, within the play, to Hamlet's inability to understand his own feeling because "he cannot objectify it, and it therefore remains to poison life and obstruct action."[23] This is certainly a different issue. The protagonist's inability to understand himself must rest on the dramatic condition that the facts as they appear to him do not let him explain his emotions to himself. This condition might, however, be quite clear to the audience, as it is in *King Lear*, where the king's actions have created a situation which motivates the defection and hypocrisy of his daughters for us, but not for him, because he does not see his own part in it.[24] If Hamlet (the man) fails to understand his emotions for want of an objectifying symbol, this virtual situation may be analogous to our actual want of a symbol in *Hamlet* (the play) to convey them to us, but the two failures occur on entirely different semantic levels, and cannot involve the same function of an "objective correlative."

Finally, at another remove, the principle is applied to the play itself, and the critic tells us that "the supposed identity of Hamlet with his author is genuine to this point, that Hamlet's bafflement at the absence of objective equivalent to his feelings is a prolongation of the bafflement of his creator in the face of his artistic problem."[25] This last issue, the artistic problem, is really an entirely different matter; to treat the feeling conceived, the import of the play itself, as something in need of an "objective correlative" to represent it seems to me to bespeak a misconception of the artist's task. The expressive form, the work, is a symbol created, not found and used. One finds motifs and devices, and represents characters and their emotions, but the poetic import of the composition is not represented—it is presented, as the perceptible quality of the created image.

So we are back to the problem that I broached above and left suspended: how does the artist transform his idea of feeling—which must be vague, elusive, amorphous before it appears in any projection—into the objective datum, the perceptible quality of a poem, a musical piece,

[23] *Ibid.*, below.
[24] The first critic, as far as I know, who remarked that Lear himself was responsible for his tragic fate was Friedrich Hebbel, in a diary note of 1843. See Münz, *op. cit.*, p. 245.
[25] *Op. cit.*, p. 125.

a painting or whatever else he gives us? How does he envisage that quality before he or anyone else has seen it?

The answer is, I think, that he has seen it; it is this apparition that he tries to re-create. His idea is initiated by experiences, or perhaps even one isolated experience, of actuality colored by his own way of feeling (rather than by some emotion of the moment), and the image he creates is of the way things appear to his imagination under the influence of his highly developed emotional life. There is a beautiful statement of this condition and its origination made by J. Middleton Murry in *The Problem of Style*: "The literary artist begins his career with a more than ordinary sensitiveness. Objects and episodes in life, whether the life of every day or of the mind, produce upon him a deeper and more precise impression than they do upon the ordinary man. As these impressions accumulate . . . they to some extent obliterate and to a greater extent reinforce each other. . . . Out of the multitude of his vivid perceptions, with their emotional accompaniments, emerges a sense of the quality of life as a whole. It is this sense of, and emphasis upon, a dominant quality pervading the human universe which gives to the work of the great master of literature that unique universality which Matthew Arnold attempted to isolate in his famous criterion of the highest kind of poetry—'criticism of life.'. . . We have, however, to remember . . . that a great writer does not 'criticize' life, for criticism is a predominantly intellectual activity. . . .

"The great writer does not really come to conclusions about life; he discerns a quality in it. His emotions, reinforcing one another, gradually form in him a habit of emotion. . . . This emotional bias or predisposition is what I have ventured to call the writer's 'mode of experience.'. . . The greater the writer, the more continuous does that apprehensive condition of the soul become."[26]

What any true artist—painter or poet, it does not matter—tries to "re-create" is not a yellow chair, a hay wain or a morally perplexed prince, as "a symbol of his emotion," but that quality which he has once known, the emotional "value" that events, situations, sounds or sights in their passing have had for him. He need not represent those same items of his experience, though psychologically it is a natural thing to do if they were outstanding forms; the rhythm they let him see and feel may be projected in other sensible forms, perhaps even more purely. When he finds a theme that excites him it is because he thinks that in

---

[26] *The Problem of Style* (1960), pp. 23–25. The many elisions are only in part for abbreviation; I have also skipped, deliberately, his intercalated comments on lyric expression, which seem to me less justified, and are actually modified later in his book, but would only confuse the issue on which I quote his words here.

his rendering of it he can endow it with some such quality, which is really a way of feeling. That process is the abstractive process of art. It is true that, as one aesthetician after another has declared and stressed, an artist at his work does not think abstractly; but he thinks abstractively. He does not use the ready-made general concepts of ordinary discourse to reason verbally about his work—which is what they mean by "thinking abstractly"—to any great extent; but he is entirely absorbed in making a new abstraction, without the aid of that spontaneous, progressive generalization which is the normal machinery of intellectual abstract conception.[27]

Yet a person engaged in painting or composing, building up a novel or a scenario, a dance, or a monument—whatever may be his art—is not consciously making an abstraction, either. He is working toward the quality he wants this particular piece of his to have. If he is young, he is apt to have seen it in someone else's art rather than in other domains of life; that is, to have seen it created, where the more obvious techniques encourage him to make a start, with some confidence that his own individual vision will develop before he is through, and give his work a new value.[28] But, young or old, great talent or small, to make any work of art the maker must have an idea to realize.

The ways of getting and holding an artistic idea are extremely various and unpredictable. Some artists begin with emotion, pure and simple; some with a concrete symbol or even a sober fact, and a phrase or poetically perfect line to say about it. A line which rings true is usually the embryo of a poem. But how that line arose in the poet's mind cannot be generally described. Some musical composers start with a dramatic feeling, an adumbration of the scope and movement of a new work, and not much more; even with a visual image.[29] Brahms reported that

[27] A more detailed discussion of abstractive methods may be found in *Problems of Art*, the appendix entitled "Abstraction in Science and Abstraction in Art," and in *Philosophical Sketches*, No. 3, "On a New Definition of Symbol." See also Chapter 6 below.

[28] One of the frustrating circumstances in our public reception of art is the professional reviewers' habit of always determining, first of all, where a new artist got his initial insight: "in the style of ———," "influenced by ———," "———esque composition." That is usually easy to see, and fills the required line of print without demanding any attention to potentialities and goals.

[29] C. E. Montague, in *A Writer's Notes on His Trade* (1930), p. 97, says that Ibsen's basic ideas "began coldly and scrappily," and that only after he had fully plotted a piece "the divine accident happened. The magical interaction of technical effort and imaginative insight commenced." Tolstoi started with a massive idea and eliminated details to carve out its form. "To the one brooding mind the essential part of its task is an effort of amplification. To the other, all is denudation. And what comes of the one may be just as good as what comes of the other" (p. 105).

to him the whole "idea" of a piece was embodied in the theme, so that when this was found he could suspend his thinking, write letters or take a walk, without fear of losing his creative start.[30] Others find that the unfinished piece is vulnerable, the idea in suspension, or slow to take form, almost from beginning to end of their work on it. An interruption can destroy it, as the entrance of "a person from Porlock" ended the writing of *Kubla Khan*. Apparently the idea of the poem, conceived in an opium trance, had not registered itself in Coleridge's memory by any generative image apart from the words he was impulsively, quasi-automatically putting down. He never was able to recall or reconstruct it. Lucky the artist for whom the flash of apprehension has left its imprint on a readily recoverable image, or can be conjured up again by some personal trick like Proust's balancing on cobblestones, or any other psychological aid, no matter how absurd.[31] All that the stimulus has to do is to bring back vividly a feeling the artist has once known that imposed itself on things about him like a special light, and gave them the quality he wants to impart to the image he is making.

Yet this is only one type of creative experience. Many an artist begins his work simply with excitement about a new plot, even the suggestiveness of a new material, a seen form, an accidental effect that starts his imagination of new forms with expressive possibilities, and a tacit assumption that he can gradually work out a special feeling which may be mysteriously familiar, like an experience of *déjà vu*, or may be as new to him as to a stranger. His guide is not memory at all, but a conviction that the quality he cannot clearly envisage yet will emerge, and that he will know its pure expression when he sees it.[32] There is enough of it in the first formal conception to beget a sense of congruence or incon-

---

[30] Max Graf, in *Die innere Werkstatt des Musikers* (1910), p. 75, quotes a statement by Brahms: "Wenn ich die erste Phrase eines Liedes gefunden habe, kann ich mein Buch zuklappen, spazieren gehen und die Arbeit vielleicht nach Monaten erst wieder aufnehmen. Aber es ist nichts verloren. Wenn ich das Thema später wieder aufnehme hat es schon Gestalt angenommen und ich fange an, daran zu arbeiten."

[31] In Wilhelm Kienzl's "Die Werkstätte des Komponisten" (1886), there is an inventory of strange stimuli to musical invention, from noise and uproar to Wagner's necessary touch of velvet, and from weird lights to the smell of stew.

[32] The "click" of recognition has often been reported or remarked, e.g., by Roger Sessions in his essay, "The Composer and His Message," in *The Intent of the Artist* (1941), pp. 126–27, and by Robert Edmond Jones in *The Dramatic Imagination* (1941), pp. 75–76. Both authors have been quoted on that subject in *F.F.*, so I shall not repeat their words here. Leo Spitzer, in *Linguistics and Literary History* (1948), p. 26, speaks of the "click" of insight into artistic forms; the artist's own recognition of fittingness is probably such an insight when the form confronts him.

gruence in most of his further moves (sometimes it takes some trial and error). There is some analogy to this in our discursive, speculative thought, in that a thinker often sees a promise of new ideas, a potential use, in a general proposition he has not analyzed and deductively exploited or even safely established yet.

The achievement of artistic quality is the first, last and only aim of the artist's work. That achievement is expression of "the Idea"—an idea of human emotion and sentience, of which the work is a projection. There is no compromising with standards of beauty such as regularity, balance, decorative color, etc.—many eminent aestheticians to the contrary notwithstanding. If regularity does not serve to create the emotive quality, it is tiresome. If balance does not spring from the rhythm of feeling, such balance looks frozen. Color does not seem decorative if it does not enhance the appearance of livingness; and if it does it is expressive.

Here, I regret to say, I cannot share the view of Sir Herbert Read, with whom I do agree on some fundamental views he has expressed in the past. The opposition he professes to find between "vitality" and "beauty" seems to me to bespeak an unduly limited sense of the latter term, making it equivalent to what Bosanquet called "easy beauty," as against the "difficult beauty" of less canonical art. To argue that paleolithic art is not beautiful because the painters did not consciously aim at making it so (which we cannot know), and did not compose their representations into abstractly conceived designs,[33] equates "beauty" with regularity and symmetry of form.[34] In these confines, indeed, no

[33] *Icon and Idea*, the Charles Eliot Norton Lectures of 1953, subtitled "The Function of Art in the Development of Human Consciousness" (1955), p. 29: "Beauty . . . is not a necessary postulate of the prehistoric period. If we read beauty into . . . the Altamira bison—it may be that we are bringing to its appreciation eyes long accustomed to the stylized conventions of historical art." And further: ". . . before a sense of beauty could arise, there had to be a conscious *abstraction* from the process of nature. There is no evidence that such a conscious act of abstraction was made by paleolithic man."

[34] Cf. F. Mirabent, "Émotion et Compréhension dans l'expérience du beau," *Revue d'esthétique* (1948), 254–55: "Le beau est esthétique; mais tout ce qui est esthétique n'est pas beau. . . . L'art même, qui est presque exclusivement défini comme l'aptitude à créer beauté—définition, à notre avis, abusive et triviale—, nous donne souvent l'exemple d'artistes qui recherchent autre chose que le beau, et c'est avec raison que Joseph Warren Beach dans son livre récent A *romantic view of Poetry* signale avec perspicacité que le caractère et l'énergie sont les mobiles de Rembrandt et de Vélazquez, comme la nature humaine est le mobile de Shakespeare, tandis que pour Manet, Degas, Renoir, et surtout Cézanne, ce sont les structures. Dans tous ces buts à atteindre, ce n'est pas la même chose qu'ils soient beaux et qu'ils soient esthétiques."

This limiting usage of "beauty" and "beautiful" is old, but it has apparently

great range of feeling can come to expression. Paleolithic art, according to Sir Herbert, is naturalistic to such a degree that he believes it to have been traced from an eidetic image of the animal projected by vivid visual memory on the dark cave wall.[35] Yet "the effect of the drawing, if not the intention, is to enhance the vital potency of the animal."[36] In spite of this expression of vitality, however, he claims that these pictures have no style, because the postures of the animals do not depart from nature, and "stylization is always such a departure from natural attitudes."[37] If, therefore, they have any "abstract values," those values are accidental. "We have to wait for another cycle of human development . . . to see the emergence, in all its independence, of geometrical intuition, and with it, of that ideal category of representation which we call *beauty*. What the prehistoric period establishes, in all *its* independence, is an instinctive mode of representation[38] possessing vitality." So he comes to the general conclusion that " 'aesthetic' covers two very different psychological processes, . . . the one tending to an emphasis on *vitality*, the other discovering the still centre, the balance and harmony of *beauty*.

---

always troubled people whose sense of beauty responded to other than the standard "beautiful" forms. Goethe, for instance, wrote in his essay "Von deutscher Baukunst": "Die Kunst ist lange bildend, ehe sie schön ist, und doch so wahre grosse Kunst, ja wahrer und grösser als die schöne selbst" (*Sämtliche Werke*, Cotta ed. [1866–68], Vol. XXXIII, p. 10).

[35] *Op. cit.*, p. 23. If it were possible to imitate an actual eidetic memory, the result should look like an animal, as the *trompe l'oeil* figures do that continue mounted museum groups into a painted background, and not like pictures in unbroken profile with triangular hoofs pointing downward like the feet of medieval saints.

[36] *Ibid.*, p. 24. The cardinal point here is the difference between "livingness," which is an artistic achievement, and lifelikeness or direct mimicry. Oddly enough, the style and beauty of such primitive art is recognized in Sir Herbert's earlier book, *The Meaning of Art* (1936), p. 70, where he wrote: "The representations . . . show no attempt at perspective: the purpose is rather to represent the most expressive aspect of each element in an object—the side view of the foot, for example, being combined with the front view of the eyes. In other respects, too, the art of these primitive peoples is not naturalistic. There is a definite abandonment of detail in favour of what we may call symbolism." That does not sound like copying an eidetic memory image.

[37] *Ibid.*, p. 29.

[38] There is an astonishing piece of evidence in favor of the claim that the mode of representation is primitively intuitive (not instinctive) in an article by Anne Schubert, "Drawings of Orotchen Children and Young People" (1930). The children of this far northern Russian tribe had never seen a drawing, but once shown how a pencil will mark, they spontaneously drew what interested them—animals, especially reindeer and elk (*without* ritual motivation) strikingly like paleolithic pictures in their outlines and attitudes.

"We may imagine an ideal work of art in which vitality and beauty are both fully present and perfectly balanced; perhaps such a moment occurred when the Altamira bison was painted [Fig. 5–1]. . . . But it was always a precarious balance, difficult to achieve, impossible to sustain. In general, the artist has had to choose between the path of vitality and the path of beauty."[39]

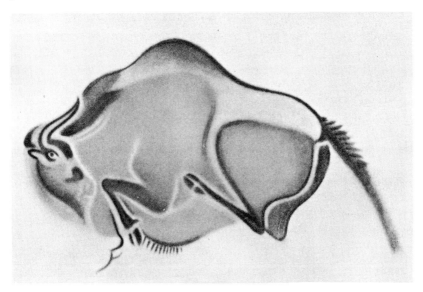

Figure 5–1. Recumbent Bison, Altamira Cave, Santander, Spain

'perhaps such a moment occurred when the Altamira bison was painted'

(Eckart von Sydow, *Die Kunst der Naturvölker und der Vorzeit*
[Berlin: Propyläen Kunstgeschichte, 1923].)

To select, as the paragon of artistic achievement, the Altamira bison, which he previously characterized as devoid of style and seeming beautiful only because we read beauty into a purely accidental composition, certainly calls into question the premises on which his criticism of "vitalistic" art rests—that beauty is a category, a special art concept, which has to be discovered and abstractly conceived before it can be used in art (his second chapter in *Icon and Idea* is entitled "The Discovery of Beauty"); and that it is indeed "used," not inevitably created wherever an artist is successful. The dichotomy between beautiful art

[39] *Icon and Idea*, pp. 32–33.

and expressive art is a firmly established principle of modern criticism; Nietzsche's contrast of Apollonian and Dionysian, Curt Sachs's "ethos" versus "pathos" art,[40] Read's conflict between beauty and vitality, all denote the same fundamental duality,[41] so one can hardly deny that such a duality exists.

Yet a genuine phenomenon may be misinterpreted, and its structure misconceived accordingly. The two tendencies in art, formalization and expression, are generally taken to serve two different aims, the one to please the "higher senses," the other to relieve emotional pressures; and the ideal artistic attainment, consequently, is supposed to be a perfect compromise between these two aims, doing the highest possible justice to each, as Sir Herbert proposed in the last passage quoted above.

I think it far more likely that every artist has only one artistic aim, whatever non-artistic interests he may also find opportunity to satisfy by his work. The sole artistic intent is to present his idea of some mode of feeling in the nameless but sensible quality which shall pervade his nascent creation. There may be many reasons why he wants to realize and present this particular import; it may preoccupy him, torture him, so he has to objectify and face it; it may merely intrigue him, because he has but partly known it, and thinks his own work would complete the revelation; or—very often, perhaps normally in working on commission or by command—the material and the stipulations themselves suggest a form congenial to the artist's "habit of emotion," which is the tenor of his artistic imagination. Such reasons have little to do with the seriousness of his intent once the work is undertaken, the motif accepted, the general feeling adumbrated in the first structural conception.

The developed rhythms of life appear in our traditional artistic rhythms, visual, musical or poetic, which are intrinsically "ornamental" —that is, sensuously organizing any material on which they are imposed, assuming its spontaneous transformation into a virtual field in which forms arise for pure perception. The first effect of formal design is to animate a surface; if the surface is curved, design lends its curves continuity by expressing their directions as flowing lines: if it is angled, borders or symmetrical forms emphasize the angles as conjunctions of areas, like meetings of segments in complex organisms, which individuate the parts and at the same time establish their mutual limitations. Even the visible expansion of an unbroken plane is peculiarly enhanced and

[40] See *The Commonwealth of Art: Style in the Fine Arts, Music and the Dance* (1946), *passim*.

[41] See also W. Worringer, *Abstraktion und Einfühlung* (1908), published in English as *Abstraction and Empathy* (1953); and Hermann Nohl, "Die mehrseitige Funktion der Kunst" (1924). The literature contains many other sources.

brought to account by evenly deployed forms, which may be the simplest geometric figures or very intricate, just so long as they occur in rhythmic repetition. The second great function of design, which may be the more important, at least if we measure its importance by its influence on the further potentialities of art, is the establishment of symmetry, or correlation of counterparts, which creates the axis as a structural element. At once we have right and left of a median; the projection of body feeling into symmetrical forms meeting along a straight line is such an elementary act of visual intuition that the forms are actually seen to "balance" the moment that line is upright. Our sense of space is gravitational as much as visual; the similar and opposed figures seem to exert the same pull on the axis so that the apparent tension is resolved only when the axis is erect in our field of vision. Balance establishes the vertical, and implies the ground line, the horizontal. Here is the simplest principle of monumental as well as pictorial organization; but what gives a symmetrical deployment of figures this formalizing power is that it expresses our deepest vital feeling, equilibrium, the resting tonus of the whole organism.[42] Its import is simple, but immediately received; the safest device to achieve living form is symmetrical composition. Sir Herbert aptly designates such balance as "the still center"; but it is the still center of life, not of geometric figure, which *qua* geometric is no more still at the center than anywhere else. To oppose the basic equilibrium of vital form to "vitality" itself, and to regard the latter as an interest on the part of the artist different from that of achieving "that ideal category of representation which we call *beauty*"—so that he is forever compromising his ideal, trying to eat his cake and yet keep it—makes the appreciation of art a constant juggling of two standards, and critical judgment an equally constant weighing of excuses.

An interesting analogue to this bifurcation of the artist's intent—perhaps a consequence of it—appears in Sir Herbert's theory of poetry as he has stated it in *The True Voice of Feeling*.[43] Here he argues that drama and "poetry"—by which he means lyricism—are antithetical literary principles, happily combined in what he calls "poetic drama," but really, intrinsically, opposed to each other. Drama is all action, whereas "poetry is essentially a quality, an abstract quality evoked by concrete images, and Shakespeare is an imperfect dramatist precisely because the poetry

[42] Cf. D'Arcy Thompson, *On Growth and Form* (1942), Vol. I, p. 67: "Gravitation influences both our bodies and our minds. We owe to it our sense of the vertical, our knowledge of up and down; our conception of the horizontal plane on which we stand, and our discovery of two axes therein, related to the vertical as to one another; it was gravity which taught us to think in three-dimensional space."
[43] The first half of the book contains a series of lectures delivered in connection with a seminar held at Princeton in 1951, which I was privileged to attend.

keeps breaking in, suspending the action for its own moment of existence."[44] He quotes Aristotle's words, that drama is "a mode of action,
not a quality"; but very brief quotations are dangerous, as in this case,
for instance, where Aristotle has just made it quite plain that "quality"
refers to the moral quality of depicted characters, not to the essence of
the composition. "Quality" in the latter sense Aristotle certainly does
regard as the prime achievement in drama as anywhere else in art; this is
evident from a statement toward the end of the *Poetics*:

"Tragedy may produce its effect even without movement or action
in just the same way as epic poetry; for from the mere reading of a play
its quality may be seen."[45]

Drama is a form of poetry, and action is the organizing principle of
the dramatic projection.[46] But the artist's intent is still the achievement
of the quality which characterizes the play—its "livingness," its tensity
and pace, depth, romance or realism, irony, its sweep and grandeur, or
perhaps its lightness and lilt, a miniature stature, whim, humor and
complication of theatrical intrigue, its lizard-like movement from one
absurd moment to another. It is this kind of "quality" that may almost
as properly be called the "feeling" of the work.[47] To treat it as a special
product peculiar to lyric poetry seems to me an unhappy narrowing of a
key concept of art to a particular mode and style. Even more so is the

[44] *Ibid.*, p. 149.   [45] 1462ª.

[46] In *The True Voice of Feeling*, p. 147, Sir Herbert quotes from an article of
mine that "*I believe the drama is not literature at all.* It has a different origin, and
uses words as utterances." He adduces this passage as a footnote supporting his own
statement: "Mr. Eliot may be right: there may be an incompatibility between poetry
and the modern theatre with its prosaic audience." But I certainly never said drama
is not *poetry*, nor meant by "drama" the prosaic stuff of some present fashion. My
contention is that not all poetry is "literature"—i.e., essentially verbal composition. It
is whatever makes the poetic "primary illusion." I hope the fuller explanation of my
statement, given in *Feeling and Form* (which was not published when his lectures
were held), has cleared up that point.

[47] In *The True Voice of Feeling*, p. 110, is a quotation from T. E. Hulme's *Notes*,
in which Hulme speaks of "the creation of new images hitherto not *felt* by the poet"
(italics mine).

Katherine Dreier, in the *Three Lectures on Art* which she edited (1949) and to
which she contributed her own piece, " 'Intrinsic Significance' in Modern Art,"
discusses the fluidity achieved by El Greco through his use of light, and remarks:
"You can *feel the same qualities* in many of Kandinsky's paintings . . ." (p. 22)
(italics mine).

The tendency to identify feeling with its sensuous analogue is not a modern
phenomenon. In Hüller's "Abhandlung von der Nachahmung der Natur in der
Musik" (1754), for instance, one finds the remark: "Ein Ton also, von dem Gefühl
des Herzens erzeugt, ist das Gefühl selbst. Es wird so gleich dafür erkannt, und
gelanget unmittelbar und ohne Umschweif zu dem Herzen."

claim that "organic form" is "the specific form of modern poetry, in so
far as this poetry is specifically modern."[48] But any disputation on that
point must needs become so far-reaching that it had better be shelved
at present, in favor of the apodictic statement of my dissenting view that
every artist is interested in achieving what Fauré-Fremiet called the
"quality of expression"[49]—whatever may be the immediate source of his
interest; to make the image holy, or potent, or illustrative, high-lighting a
story, or simply to produce a decorative design, perhaps without any
representational motif. This interest in sensuous or poetic quality,
whether avowed or unconscious, is what makes a person an artist, and
his product a work of art. As such it may be good, or poor, or even bad,
but it is art; whereas anything, made for any purpose, with perfect indif-
ference to the "quality of expression" (though the maker may aim to
please by falling in with fashion), is not art at all.

It is this quality that constitutes beauty in art. Every kind of art is
beautiful, as all life is beautiful, and for much the same reason: that it
embodies sentience, from the most elementary sense of vitality, individual being and continuity, to the full expansion of human perception,
human love and hate, triumph and misery, enlightenment, wisdom. And
it seems that all sorts of feeling can appear in art, which is to say, can
appear as peculiar qualities; so G. S. Dickinson speaks of "a purposeful
quality in the aesthetic pace" of music;[50] Rudolf Arnheim, commenting
on the early childish practice of picturing all objects as roughly circular
forms, remarks that the child's circle does not render shape at all, but
"the more general quality of 'thingness.' "[51] L. H. Sullivan—the great
architect who delved with true philosophical courage into the principles
of expressiveness in the most practical, material-bound, economically and
socially controlled of all arts—declared that the expressiveness of architectural form lies in a sameness of quality, which he called "an organic
quality," pervading every aspect and part of a fine building.[52] This
quality is primarily an expression of vitality: "a building should live with

---

[48] *The True Voice of Feeling*, p. 9.

[49] *Pensée et re-création* (1934), p. 37.

[50] "Aesthetic Pace in Music" (1957), p. 317.

[51] *Art and Visual Perception: A Psychology of the Creative Eye* (1954), p. 140.

[52] *Kindergarten Chats* (1947), p. 47: "If the work is to be organic the function
of the part must have the same *quality* as the function of the whole; and the parts,
of themselves and by themselves, must have the quality of the mass. . . .

"I can go on and consider my detail as of itself a mass, if I will, and proceed
with the regular and systematic subdivision of function with form, as before, and I
will always have a similarity, an organic quality . . . descending from the mass down
to the minutest subdivision of detail. . . . The subdivisions and details will descend
from the mass like children, grandchildren and greatgrandchildren, and yet they will
be, all, of the same family."

intense, if quiescent, life, because it is sprung from the life of its architect." But that aspect of art which Sir Herbert would trace to a distinct,
even antagonistic principle—ornament, which certainly aims at "beauty"
in the narrowest sense—Sullivan treats as an intensification of the life
and emotion expressed by the whole building:

"A building which is truly a work of art . . . is in its nature, essence
and physical being an emotional expression. This being so, ... it must
have, almost literally, a life. It follows from this living principle that an
ornamented structure should be characterized by this quality, namely,
that the same emotional impulse shall flow throughout harmoniously
into its various forms of expression—of which, while the mass-composition
is more profound, the decorative ornamentation is the more intense. Yet
both must spring from the same source of feeling."[53]

If one conceives art to be a projection of feeling by means of a transformation of subjective, intraorganic realities into objective, though virtual, forms of directly perceptible quality, the problem of artistic import
loses some of its mysterious (not to say mystical) aspect. At least it
breaks up into more manageable, distinct issues, which are of general
epistemological interest and even motivate some psychological queries.
Of the latter sort is the whole complex problem of immediate sensory
experience, on which some work has long been in progress: how, and to
what extent, do we detect qualities not correlated with some known
function of a specific sense organ? What do we really receive, at various
stages of life and under its various conditions, in the mode of direct
sensory impact?[54] But the other sort of inquiry it engenders, the epistemological, centers on the meaning of "intuition"; and as this is of prime
importance to the whole theory of "presentational" or "metaphorical"
symbolism, it has to be at least briefly introduced here, although it is
slated for more thorough treatment in connection with a later issue,
the emergence of "mind" from animal mentality.

Intuition is the basic intellectual function. The word has been popularly used to denote some alleged possession of information without any
demonstrable source—foreknowledge of rationally unpredictable events,
factual knowledge without any access to the facts, etc.; it is also used as a
synonym for instinct, which it certainly is not. I am using it here in the
strict sense which it was given by Locke in his *Essay*. In that sense,
"intuition" is direct logical or semantic perception; the perception of (1)
relations, (2) forms, (3) instances, or exemplifications of forms, and

[53] *Ibid.*, p. 188.
[54] The chief work on these topics emanates from the laboratory of Heinz Werner
at Clark University, and will be discussed at length hereafter.

(4) meaning.[55] Locke, who had reason to be wary of the word because the arch-opponents of his empiricism, the churchmen, rested their moral and theological beliefs on "intuition," often used the term "natural light" in its stead. It is not until the end of his fourth book that he uses the moot term quite freely, confident by that time that its sense in his vocabulary had been clearly enough established.[56] It still seems to have several senses, but I think they are all specifications of one general sense, referring to the same sort of mental act, sometimes called "insight," although that term may have a somewhat wider range. Animal psychologists have reported evidences of what seems to be quite properly called "insight," certainly in primates, and perhaps in other animals; but whether this really bespeaks a rudimentary intuition, or is still within the range of less questionably animalian functions which are convergent in effect with our intuitive procedures, is a difficult problem, not to be hastily settled.

The development of intuition, if not its very beginning, appears in the evolutionary theater as a hominid specialty; wherever else its roots may reach and show some budding life, it is in the human stock that it has had a steady growth, and made a radical difference in the entire design of that species, shifting its operational basis from directly stimulated instinctive action to more or less planned activity. By virtue of a symbolic envisagement of the world, human action has its symbolic rendering or conceptual form, too, which fits into the envisaged world (usually narrowed down to a situation) as a dynamic element, a potential change anticipated in imagination, so its performance or non-performance becomes a true option. This is, of course, the most obvious and momentous effect of the ability to use symbols, i.e., of logical and semantic intuition.

[55] Locke added the "intuition of ourselves," which is related to exemplification, but is probably not a direct intuition. He thought, like Descartes, that awareness of one's own perceiving was given with every perception, of one's thinking with every thought, etc., which is not borne out by psychological researches. See *An Essay on Human Understanding* (1690), Bk. IV, chap. ix, sec. 3.

[56] The first two chapters of Bk. IV are largely devoted to intuitive knowledge as the ultimate ground of all real knowledge. In chap. ii, sec. 1, there is an explicit statement of what Locke means by "intuition": ". . . sometimes the mind perceives the agreement or disagreement of two ideas *immediately by themselves*, without the intervention of any other: and this I think we may call *intuitive knowledge*. For in this the mind is at no pains of proving or examining, but perceives the truth as the eye doth light, only by being directed towards it. . . . *It is on this intuition that depends all the certainty and evidence of all our knowledge.* . . . Certainty depends so wholly on this intuition, that, in the next degree of knowledge which I call demonstrative, this intuition is necessary in all the connexions of the intermediate ideas, without which we cannot attain knowledge and certainty."

But it is not the only one—not even the only one of first magnitude. Perhaps the deepest change which the dawn of "natural light" has made in man is the vast expansion of his emotional capacity, under the more or less constant stimulation provided by the play of significant imagery and spontaneous ideation that have become his nature. Sheer conceptions evoke emotions, emotions focus and intensify attention, attention eventuates in symbolic expression that formulates more conceptions and sustains or reshapes emotion;[57] so the conceptual frame in which we feel our own activity and the impingements of outward events grows larger as long as the emotive and intellectual processes keep pace with each other in a dialectical advance, rhythmically self-sustaining like all major organic functions.

Intuition lies at the base of all specifically human mental functions, and even deeply modifies those which we still share with other creatures, so that most of our truly instinctive acts appear vestigial beside the complexity and adaptedness of theirs. The ordinary intuitive acts, such as the recognition of similar formal structures in sensuously dissimilar things, the reception of one as symbolic of the other, the spontaneous production of new perceptible entities—from private images to publicly objectified things or events—to present all sorts of abstracted forms *in concreto*, are as natural to us as the motions of our limbs, uncoordinated at first, but patterned and integrated in the normal course of maturation, and later developed selectively to various degrees by employment. We need not undertake such acts deliberately: symbolic projection and interpretation are spontaneous responses which we may become aware of and cultivate, or always perform at the unremarked level of what Coleridge called "primary imagination."

This brings us back to paleolithic art, and the opinion of several important critics that it is not beautiful because its makers had no notion of beauty, that it is not expressive because they were not seeking expression, and even that it is not art because they had no artistic intention. The appreciation of expressive form seems to be primitive and immediate in man, and long before things natural and supernatural, real and imaginary, physical and ideal could have been sorted out for him as so many categories of experience, they appear to have served as opportunities for creating symbols of his intuitive realizations. It is a prevalent fashion today to say that ancient art was produced without interest in any but utilitarian values. That opinion rests on a modern distinction

---

[57] In subsequent chapters I hope to show the intimate relatedness of thought, emotion and symbolic expression as more or less differentiated mental acts making up the partial individuation of "mind" within the human organism.

between efficacy and aesthetic appeal which does not occur to the unsophisticated mind. It is reflected also in the frequent misconception of magic as a prosaic activity performed just as one would perform a necessary chore. Magic is always mysterious power, and is viewed with awe as a special gift granted more freely to some persons than to others. Its employment is "practical" in the same sense as prayer; it is often built into purposive action; but its essential function is the ritualistic expression, or "realization," of the act of mind as a power among the symbolically conceived powers of nature. It is always mystical, fervent and, in its more advanced stages, religious; and although its exercise may be coupled with efficacious methods, their success becomes a demonstration of its power—not simply a result of the rite, but its consummation.

The same consideration applies to the modern anthropological tenet that primeval human societies must have been entirely practical, as animals appear to be, and had no artistic interests. The most categorical statement of this assumption that I know occurs in Arnold Hauser's *The Social History of Art*, where, apropos of prehistoric cave painting and sculpture, he writes: "We know that it was the art of primitive hunters . . . who believed in no gods, in no world and life beyond death. In this age of purely practical life everything obviously still turned around the bare earning of a livelihood and there is nothing to justify us in assuming that art served any other purpose than a means to the procuring of food. All the indications point rather to the fact that it was the instrument of a magical technique and as such had a thoroughly pragmatic function aimed entirely at direct economic objectives. This magic apparently had nothing in common with what we understand by religion. . . . It was a technique without mystery, a matter-of-fact procedure, the objective application of methods which had as little to do with mysticism and esoterism as when we set mousetraps, manure the ground or take a drug."[58]

I do not know on what authority Professor Hauser tells us of the caveman's sober pragmatic attitude and lack of belief in spirits, gods and mysterious forces, but he seems to have inside information, for there is no tone of conjecture in his statements, and he does not hesitate to say: "Any other explanation of Paleolithic art, as for example, decorative or expressive form, is untenable."[59] As it happens, the magic intent of the ancient animal paintings seems highly probable, but this does not rule out the artistic motive, nor even make it a secondary, extraneous, superadded interest. The projection of human feeling into visible forms may

[58] (1951), Vol. I, pp. 25–26.   [59] *Ibid.*, p. 27.

be a completely unconscious process, not only at the beginning but even throughout the work, yet gather momentum steadily with the growth of the form and guide the consciously intended act of pictorial representation. It is not as mammoth or reindeer that the deep inward rhythms of life are objectified, but as the flowing unity of lines and the distribution of accents, the tensions inherent in the strokes and virtual volumes, not in the represented animal's anatomy (Fig. 5–2). And I believe this kind of expression is elementary, prepared in the early phases of human vision.[60] Ontogenetically it has been observed by psychologists who call it the "physiognomic" stage of perception, in which over-all qualities of

Figure 5–2. Reindeer, Font-de-Gaume Cave, Dordogne, France

'the artist who painted the Reindeer . . . has given expression to his innate feeling for the beauty of line itself'

(Johannes Maringer and Hans Bandi, *Art in the Ice Age* [New York: Frederick A. Praeger, Inc., 1953].)

[60] This opinion is corroborated by Helen Gardner's judgment in *Art through the Ages* (1926), p. 7: "A constant characteristic that we have noticed in this art of the Reindeer Age is the keen observation of animal life and the ability to represent it in its most significant aspect with the greatest economy of line. But the artist who painted the Reindeer [at Fonte-de-Gaume] . . . showed furthermore that he was

fearfulness, friendliness, serenity, etc., seem to characterize objects more naturally than their physical constitution.[61] Whether this ontogenetic phase of our mental development has a phylogenetic counterpart is, of course, problematical, but there is some support for the hypothesis that human mentality as a whole passed through a period of much more active physiognomic seeing than it exhibits in most of the world today. Expressive form, or beauty of design, need not have been a deliberate concern added to representational intention, to vindicate the theory that representation was the only conscious intention and aimed at magical power. If those early painters saw essentially in the physiognomic way, they would inevitably paint physiognomic images to render what things looked like, and would be filled with awe at the emotional expression. They could not paint inexpressive forms first and then decide to make them artistic.

That, however, is just what some writers on prehistoric art maintain that its creators did. G. Baldwin Brown, for instance, has even set up a law-like formula: "*Every artistic activity and every form of artistic production is preceded by similar activities and similar forms of production that are not artistic.*"[62] If one measures artistic quality by the expressive-

---

sensitive to the beauty to be found in the animal world. In the instantaneous pose that he has caught as the reindeer is about to bend his head, he has seen the beauty of the curve of the back; and in the wide sweeping line with which he has painted the antlers he has given expression to his innate feeling for the beauty of line itself."

[61] Cf. *Phil. N.K.*, p. 123.

[62] *The Art of the Cave Dweller* (1928), p. 11. He introduces his formula with the words: "It is true, of course, that artistic feeling must be latent in man from the first, but it does not declare itself, and still less does it assert itself spontaneously as a yearning as if it needed expression. . . . It does not create art, but gives an artistic form to activities and products *that had a previous existence quite independent of art*" (italics mine). Yet on the next page Brown grants that "it often happens that in the act of representation an artistic element comes in so soon to modify and improve what is at first mere delineation, that on a superficial view the artistic element may be held to be present from the very first." I do not know what deeper view of Magdalenian art could reveal its origin in inexpressive forms which nonetheless delineated the animals, and how, without such an exhibit, one can assert that such pictures without artistic value had had an actual existence. The earliest scratched drawings (e.g., at Les Combarelles) show the swing and surety of lines and the spatial proportions that bespeak a developed artistic sense. Perhaps Professor Brown means that the original motive was not artistic; but that motive alone seems to have led to no previous product.

That practical activities *may* engender ugly things and only later be seen to offer opportunities for artistic expression is proven, however, by the history of our own industrial architecture, as well as of our factory products. But such a phenomenon is more likely to occur in a sophisticated, eclectically confused society, where intuition is corrupted, than in a primeval state of culture.

ness of lines and spaces, accents and rhythms, Professor Brown's formula sounds highly peculiar, for the most striking characteristic of the earliest representations we know is their amazing expressiveness, the vitality of their lines and volumes. But that is, of course, not his measure. According to his canons, pictorial art exists only where several figures are "composed," i.e., related to each other, or one figure is consciously related to a bounded surrounding space. The most beautiful figure alone on an indefinite wall space, or painted without reference to other figures on the same wall, is not art. "The presentation of a single object," he declares, "whether it be an 'intuition' or mental image in the artist's brain, or a copy of something in nature, just by itself as a single thing out of all relations to anything else, is not in strictness a work of art. . . . It is obvious that a transcript of an object of natural history may be a miracle of imitative skill, whereas a study of some object of natural history, with a background and accessories, may be artistically quite crude and faulty. None the less according to the strict canons of aesthetic philosophy the latter would rank as a work of art, the former, for all its intrinsic merits only as a work of art in the making. The object with a background and accessories is in relation with these."[63]

How a genuine work of art can be *artistically* inferior to something that is not really art at all presents a problem. One is tempted to question the "strict canons of aesthetic philosophy" on which this peculiar evaluation is based. He does not tell us what authority has set up these canons, but it is true that they are confidently invoked by many critics, or simply taken for granted in passing judgment on the works of all ages and peoples. Only one writer, as far as I know, has had the freedom of thought to ask whether spatial balance is really the sole possible aim of artistic composition, and whether such balance is itself an intrinsic value in art or an instrumental one—a major expressive device, but a device nevertheless, which could be dispensed with and yet let art be art. Max Raphael, in his *Prehistoric Cave Paintings* (1945), proposed that artists might wish to create something other than balance and symmetry, something that belongs most naturally to the essence of cave art, which Sir Herbert has aptly designated as "vitality," namely, the life and power within the herd as a whole (Fig. 5–3). So, speaking of the ceiling at Altamira, Raphael observes that there is a unified plan realized by precisely those features which go against our canons of composition:

"The dissolution and complication of the dimensions and directions, the repeated use of parallels and steps, the function of the bison leader

63 *Ibid.*, p. 23.

Figure 5–3. Ceiling of the Great Hall, Altamira Cave, Santander, Spain
'the internal movements of the herd: its crowding together and dispersion'
(Tracing by Breuil, in Paolo Graziosi, *Paleolithic Art* [Florence: G. C. Sansoni, 1956].)

as the focal point, the fact that the distance between the two hostile leading animals is exactly equal to the size of the hind, the corresponding positions of the wild boars, the accumulation of 'anthropoids' and signs at definite places—devices that can all be explained as representing the internal movements of the herd: its crowding together and dispersion, which had fundamental importance for the paleolithic people's physical and spiritual life . . . pure horizontals were avoided, asymmetry and successive storeys were preferred, and an axis to balance the left and right was not introduced. Perhaps it is the absence of such an axis that leads many to believe that the picture has no composition; for under the pressure of both the Renaissance and ancient art we see in every composition an attempt to "set right" in accordance with the rules of balance around an axis, a world that is 'out of joint' " (p. 45).

If the movement of animals past or against each other, and especially the appearance and disappearance of individuals in a herd, impressed the paleolithic painters, the superimposition of images, which has often been taken to prove that they were drawn without reference to each other,[64]

[64] See, for instance, Gyorgy Kepes, *Language of Vision* (1944), p. 69: "Primitive man had a limited understanding of space. For him, each experience was confined to

Figure 5–4. Marc Chagall, "I and the Village"

*no one, I think, would hold that Chagall . . . had used the
same canvas twice . . . or that he had been interested in
making separate imitation cows or little model cottages*

(Collection of The Museum of Modern Art, New York,
Mrs. Simon Guggenheim Fund.)

appears as an obvious artistic device, that might well be used by a modern artist—indeed, the more modern, the more likely; and no one, I think, would hold that Chagall (who superimposed forms for yet another purpose) had used the same canvas twice without painting out the older picture before putting on a new one, or that he had been interested only in making separate imitation cows or little model cottages (Fig. 5–4).

Nothing could show more clearly the power of a symbolic projection than the fact that hard-headed civilized people can see a pictorial or sculptural representation of an object as an artificial replica of the object, "a miracle of imitative skill." Yet it is by the doctrine of depiction as imitation that Professor Brown upholds his formula despite the fact that he rejects the theory of magical intent in prehistoric art, claiming that "the primary impulse is the impulse to imitate, and imitation is not in itself an artistic act."[65] Marcellin Boule declared that primitive art "is only a special manifestation of the general spirit of imitation, already highly developed in the apes."[66] Hauser maintains that to primitive men such imitation was an actual duplication, and that they never discovered the difference between their painted creatures and a real quarry. He even goes so far as to impute this sort of delusion—which would really indicate a dangerous degree of mental incapacity—to the civilized and highly advanced artists of the Far East.[67]

---

its own space-time life, without reference to past or future or wider spatial relationships. . . . His visual representation was limited, for the most part, to single spatial units. . . . He was little concerned with backgrounds, weight, top or bottom. Each element lived its own life in complete spatial independence. Represented figures were only transitory on the picture surface. . . . This unframed character of the picture surface . . . may be the reason that prehistoric artists frequently overlapped with new paintings the work of their predecessors. Children's drawing reveals a similar attitude." (The illusion that the forms are "only transitory on the picture surface" bears out Mr. Raphael's idea of the artistic intention, and speaks for the painter's achievement of it.)

[65] *Op. cit.*, p. 12.

[66] *Fossil Men* (1923), p. 257n, quoted by Brown, *op. cit.*, p. 182. The reference to the apes' imitative tendencies serves better to stress the distinction between imitation and projective rendering than to identify them; for the apes imitate only our gross movements, but never our works; and those apes that have lately set the world agog with their painting never make "imitation things"—that is, they paint no images. Cf. Desmond Morris, *The Biology of Art* (1962).

[67] *Op. cit.*, pp. 26–27: "When the Paleolithic artist painted an animal on the rock, he produced a real animal. For him the world of fiction and pictures, the sphere of art and mere imitation, was not yet a special province of its own, different and separate from empirical reality. . . . There is evidence of a similar approach when the Chinese or Japanese artist paints a branch or a flower and the picture is not

Most aestheticians regard imitation as a vice in art; others see it as the artist's essential task, the measure of his virtue.[68] But the sort of "imitation" they wrangle about is, in fact, very rarely even attempted. Respighi, in his "Pines of Rome," inserted a nightingale's song, actually imitated by a gramophone record. The effect (to my ears) is tricky rather than musical, not because it is mechanical—how it is produced does not matter—but it breaks the musical illusion with the delusion of a natural sound.

True imitation in art is an internal affair. It is one of the principles of "organic" construction, and one of the most directly comparable to an actual organic principle—the self-imitation of the organism in its basic activities. In animal life, self-imitation is usually the line of least resistance, confirmed as such by countless analytic studies of learning and habit formation. In growth—animal or vegetable—it is reflected in the general isomorphy of parts articulated within an individual, "where not only the leaves repeat each other, but the leaves repeat the flowers, and the very stems and branches are like un-unfolded leaves"[69] (Fig. 5-5).

intended to be a summary and idealization, a reduction or correction of life like the works of Western art, but simply one branch or blossom more on the tree of reality."

Again, I do not know where the author obtained these strange facts. I have talked with a good many painters at the Society for Aesthetics in Tokyo, and in studio conversation at Kyoto, but none ever showed signs of confusion as between actual flowers and pictured ones, nor offered me painted plums to eat, even if they were painted on a plate.

A still more astounding mental weakness is imputed to primitive man by J. C. Biederman in his *Art as the Evolution of Visual Knowledge* (1948), p. 40: "Abstracting from admittedly scanty data on visual evolution . . . it seems reasonable to speculate that the earliest human specimens sometime during the initial transitions from the animal to the human stage, had not yet learned to *project* upon the outside world, where the visual phenomena originated, the image reports which occurred in their eyes. That is, to the earliest humans the images may have appeared to be within the eyes themselves, where, incidentally, they do exist. . . . [So] it is not innate or natural that we are able to *project* the visual image, or at least make it appear to be projected. This has to be learned and acquired." This should certainly put the half-human creature at a great disadvantage, since he would tend to poke out his own eyes instead of casting his spear at animals in the outside world.

[68] See, for the purest statement, Biederman, *op. cit.*, chap. vii, *passim*; toward the end of the chapter, he remarks (p. 134): "The Pompeian painters had not yet achieved the ability of the sculptors contemporaneous with them to record details and individual characteristics, such as warts, wrinkles and facial expressions."

[69] Basil de Selincourt, "Music and Duration," *Music and Letters*, I (1920), 288. The passage (after a brief deletion) continues: "To the pattern of the flower there corresponds a further pattern developed in the placing and grouping of the flowers along the branches, and the branches themselves divide and stand out in balanced proportions, under the controlling vital impulse. . . . Musical expression follows the same law."

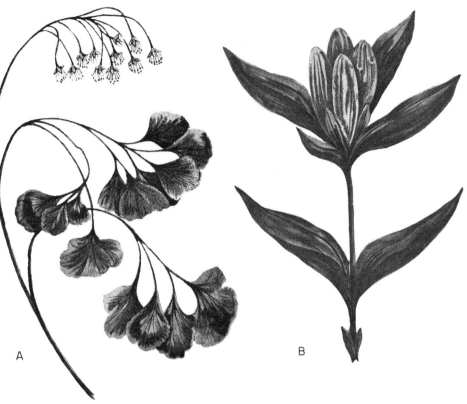

Figure 5–5. A, Early Meadow-Rue; B, Closed Gentian
*'not only the leaves repeat each other, but the leaves repeat the flowers,
and the very stems and branches are like un-unfolded leaves'*

(A, *Thalictrum divicum*; B, *Gentiana andrewsii*. From *Wild Flowers
of America* [New York: G. H. Buck & Co., 1894].)

In most, if not all art, such internal imitation is a fundamental principle,
underlying repetition, variation and other major devices for creating the
semblance of "living form," which is the subject of another chapter. But
imitation of anything outside the work is rare; the nearest to it is the
imitation of observed gestures in acting and pantomimic dance, but even
there the appearance of direct duplication is apt to be deceptive. Savages
performing a dog-dance may have studied their dogs very closely, but
men cannot really behave like dogs, and all their motions are suggestive
of canine actions rather than simply imitative. Dramatic gesture is a
rendering of behavior, as depiction is a rendering of objects; it abstracts

the salient forms of actual life to form an image that concentrates and presents the artist's idea.

Expression of conceived feeling seems to me the real criterion of art; and if such expression is intuitive, then artistic activity may occur without any verbalized intention, without preconceived standards and under any circumstances which induce people to produce objects, rites or utterances of distinctive form, no matter for what purpose. By all other measures which have been set up—emotional self-expression, recording of personal attitudes (which Sir Herbert calls "sincerity"[70]), a conscious purpose to create beauty,[71] to intensify sensory experience[72] or to realize standards of perfection which nature only approximates (i.e., to satisfy the canons of composition, such as unity in diversity, for lack of which both Professor Brown and Sir Herbert deny to the Pyramids the status of architectural works), symmetry, the relation of separate forms to each other, etc.—some things which competent people generally accept as examples of art are ruled out of the sacred precinct; and my own response to the juried exhibit which remains usually is: "This, too, is art."

Yet the measure I would apply—the measure of created form with vital import—sets its limits, too. One has already been remarked; where there is no evidence of any intuitive use of the material for the embodiment of feeling, we are not dealing with art, good or bad. There also are products which present artistic problems, but rarely emerge, at their completion, as total works of art: dress designs, too stringently governed by "the whirligig of taste"[73] to realize their maker's true ideas (though traditional costumes, scarves, kimonos often do so, and then are genuine art works), or discursive writings, which are full of literary challenges, but produce, in the end, no literary image of mental experience, no illusory process of growing thought, but only the literal report or discursive argument itself, to be analyzed, judged and, if it is philosophical, eagerly refuted. History, essays, even scientific expositions can be true literature, but that achievement, which consists in a complete use of the

[70] In *The True Voice of Feeling.*

[71] A measure still popular, as may be seen from some of the quotations in the text, and especially from Mirabent's remark given above on p. 121n.

[72] This view, that the artists sees and makes us see "what things really look like," is expressed in Roger Fry's *Vision and Design* (1925), and by John Dewey in *Art as Experience* (1934), where he wrote: " 'Art' does not create the forms; it is their selection and organization to enhance, prolong and purify the perceptual experience" (p. 391).

[73] The phrase is the title of a book by E. E. Kellett (his Hogarth Lectures on Literature [1929]).

undistorted actual discourse as a motif for artistic composition, is so rare that its instances are always masterpieces.

Perhaps the most controversial limitation, however, which the concept of "art" as a symbolic projection of feeling puts upon the extension (denotation) of the term is that it excludes so-called "animal art." The symbolic transformation and objective presentation of ideas is characteristic of human thought, being an intellectual process that involves the spontaneous, abstractive perception of forms and formal identities, and correspondingly of exemplification and significance; all of which are acts of intuition probably peculiar to man. Intuition is the basis of intellect. Animals often show intelligence, but not intellect. It is the intuitively grounded, symbolic activity—spontaneous and prodigal in man, and very rare if not quite absent in animals—that makes the difference between intelligent response to signals from the actual environment, and the intellectual functions that characterize a mind.

The treatment of art as self-expression would put it, in principle, not beyond the horizon of animal mentality. Self-expression can become very elaborate. People who credit animals with true artistic accomplishments generally start from the assumption that art is a venting of actual emotion stimulated by the presence of other individuals who watch or listen, and is intended to produce an aesthetic effect and win their admiration, or to arouse the same emotion in them that inspired the artist. Few protagonists of animal art will go as far as Étienne Souriau, who claims that cavorting reindeer are carrying out a public rite,[74] and that birds certainly hold ceremonial dances;[75] and, furthermore, that dancing animals always feel themselves engaged in a magnificent, noble and eminently respectable activity;[76] and that howler monkeys and gibbons vocalize with mystical fervor.[77] But a great many people, faced with some

[74] "L'art chez les animaux" (1948). After quoting a description by "un litterateur, Curwood," of the gambols of one animal that ended in a concerted pursuit by the whole herd, he concludes: "ce qui est remarquable, c'est que cet exercice est organisé socialement, et prend la forme d'une sorte de rite collectif et spectaculaire" (p. 224).

[75] *Ibid.*, p. 226: "Les oiseaux nous donnent quelques exemples incontestables de danses cérémoniales. Les choréographies en salutations et félicitations protocolaires des pingouins, des manchots, de tous les oiseaux sociaux de la banquise, sont bien connues. . . ."

[76] *Ibid.*, p. 228: "Volontiers je dirai que l'animal qui danse croit toujours plus ou moins faire quelque chose de magnifique et magnifiant, et véritablement de transcendant; ou tout au moins de noble, de conforme à quelque rite, nécessaire ou traditionnel, essentiellement respectable; quand il ne s'agit pas d'une improvisation admirable."

[77] *Ibid.*, p. 234: "Un mot pourtant sur les 'chants' des singes; . . . résultat esthétique médiocre ou déplorable (selon des oreilles humaines); mais caractère

of the more extraordinary courtship displays of animals, such as the bower-bird's preparation of his domain, or with the mimicry of heard sounds by some birds, especially in captivity, which really pose baffling psychological problems, are not content to leave the problems open for a long and complicated course of reasoning. Such reasoning would have to include weighing the implications of explanatory theories as well as the evidence supporting them, for an easy and obvious explanation of biological phenomena may cause havoc in the rest of our systematic thinking about life and mind.

Amateur naturalists are not usually troubled by such further demands, nor aware that the nature of art is a philosophical issue; to credit birds with musical interests and beavers with conceptions of architectural form seems good common sense to the simple nature lover. But scientists and especially philosophers do generally hold some theory of the nature of art, and also of the human mind and its relation to animal mentality. They have to square their aesthetics with their psychology and both with their biological concepts, evolutionary, Aristotelian, idealistic or whatever their persuasion. If animals are to be admitted as artists, then either they must be thought to share, to some degree, our mental powers of symbolic expression, as M. Souriau believes they do, or art must be conceived as an activity not requiring symbolic expression—as emotional demonstration, or as pure pleasurable sensory stimulation.[78]

Both hypotheses seem to me untenable; the former because if animals were capable of ritual acts, practiced ceremonials or had standards of respectability, they would have the rudiments of culture, so their groups would be clans, not flocks or herds. The latter assumption, which would make animal artistry quite plausible, is inadequate to the human phenomenon. In animal life, the art-like activities are spectacular culminations of behavior which up to the point of issuing in these characteristic patterns does not seem to contain any factors other than instinctual re-

---

cérémonial frappant; allant jusqu'à un point où il n'est peut-être pas interdit de parler d'une sorte de ferveur mystique." Such anthropomorphisms are particularly astounding after three pages of avowals that only strictly scientific findings are to be accepted. Their acceptance here is based, without any intended irony, on behaviorism: "J'étudierai seulement ici les activités propres, les comportements pratiques de l'animal; . . . (je demande simplement à dire: voilà des rennes qui dansent, voilà des singes qui chantent, exactement comme je dirais: . . . voici des enfants qui dansent, voici des passants qui chantent; simple caractérisation des comportements)" (ibid., p. 220). The psychological problem is safely packed away in the verb, which names the behavior, and grants the interpretation sub rosa.

[78] The latter interpretation is implicit, if not quite explicit, in Charles Hartshorne's article, "The Monotony Threshold of Singing Birds" (1956).

sponses to the actual situation—seasonal, parental or simply the quotidian round of self-support. The songs, dances, theatricals and building projects of animals are impulsive though stereotyped. Some individuals perform better than others, and there is some variability within the species-specific patterns, but there is no historic advance, no new art, stylistic development or criticism; above all, there is no visible relationship between the alleged arts and other activities that might be of a cultural nature—at least below a simian level, and a fairly high one at that.[79] Yet the most elaborate "works of art" are produced by lower animals than apes.

In human life, art may arise from almost any activity, and once it does so, it is launched on a long road of exploration, invention, freedom to the limits of extravagance, interference to the point of frustration, finally discipline, controlling constant change and growth. Among animals there are no schools of art, no primitive, classical and decadent periods. Human primitive art is intimately allied with religion, magic and all public activities, so it ranges from the carved doorway of the Men's House to the songs and drums of war. It is the spearhead of culture; before men have clothing, before they can count, their dances may be grotesquely splendid, their bodies adorned with lines that enhance their motions and postures. In this they suggest the nuptial displays of bowerbirds and the communal excitements of caribou. But in man, the first formalized gesture, the first stone shaped to fit the hand and seeming magic by virtue of its form, initiates a new era of mental life. That is because the gesture not only may be a movement of complicated and perfect form, but—however simple or imperfect—is made in order to realize the form. The shaped stone, treated as primitive implements usually are—with concentration on appearance—is not only a tool, it also presents itself at once, through tactual satisfaction and by its visible shape, as an intaglio image of the grasp; this semblance is a translation of touch into sight, bringing the feeling of the human hand to visual expression. Here is "expression" in the conceptual sense, a projection of the feeling, whereby it is conceived, not actually felt, through the appearance of the stone. It was thus gradually conceived as the maker of the tool worked with hand and eye. It is darkly but more steadily seen, by him and other people, now that it is embodied in an object, "objectified."

Animal art and human art are convergent phenomena, widely diverse

---

[79] The "paintings" of apes will be discussed later; also my own observation of a chimpanzee whose "dancing" could hardly be called "prancing," it showed such perfect form.

activities with overt results that look alike, and are therefore popularly equated.[80] But animal "art," even when it is elaborate, is self-expressive, and in so far as it is addressed to a spectator or auditor it is as a threat or a sexual lure, and is abandoned when the partner yields to it. Human art, whether serving for self-expression, self-stimulation (like war dances), literal communication (bugle calls, pictorial records, some African drumming), magic practices or worship, is art in so far as its producer uses the opportunity offered by his practical activity to project an idea of subjective realities—in simple, everyday things, such as the shapes of utensils and geometric decorative designs, the sense of rhythm, the dialectic of need and gratification, vitality, growth; on a nobler scale, in conjunction with religion and other high cultural interests which are naturally charged with emotions, the feeling patterns of the acts, objects and beliefs belonging to such contexts. The visible and audible forms, usually repeated to the point of intimate familiarity, become significant for an imaginative mind; the fact that all our perception has a "physiognomic" element in it, which may be heightened so that it dominates vision or lowered until it becomes negligible, makes it possible for the artist to see actuality as an expressive image, endowed with a "quality of feeling," which serves to hold his idea and let it haunt him before he has formulated any projection of it. This, I presume, is why representation is so common a practice that one cannot regard it as artistically unimportant, and even as an irrelevant, historical accident, as some aestheticians do

[80] The fallacy of equating convergent functions or forms has been noted repeatedly and criticized by scholars in many fields. In a context close to our present one the philosophical concept of "mind," Harriet Babcock has written: "All the results reported [in psychology] have been obtained by recording responses. It does not follow however that because mental activity is incomplete without response of some kind, the science of psychology is of necessity concerned only with responses. . . . The fact is that responses which are outwardly alike may result from entirely different kinds of mental activity" (*Time and the Mind* [1941], p. 232). In the field of biology, W. K. Gregory supplies many examples of convergent phenomena, but observes: "When comparisons are made between convergent resemblances of widely unrelated animals, . . . such convergent resemblances are but limited and imperfect" (*Evolution Emerging: A Survey of Changing Patterns from Primeval Life to Man* [1951], Vol. I, p. 60). And even in physics, what seems to be a common, single phenomenon calling for a single explanation may prove to have many possible sources. Werner Heisenberg, in his *Philosophic Problems of Nuclear Science* (1952), points out how the phenomenon of color may arise from the atomic structure of chemicals, from refraction of light or from retinal or cerebral processes, and says in conclusion: "We see here a general characteristic of nature. Processes appearing to our senses to be closely related often lose this relation when their causes are investigated" (p. 64).

who have found that it is not indispensable.[81] The represented thing—be
it a prehistoric game animal, or the dancing Siva, the perfect cone of
Fujiyama, the Mother of God—is the natural "motif," the source of the
artist's excitement which literally "motivates" his work; it was the first
organizing form for his idea, that gave him his start. In sculpture and pictorial art the boldest innovations and advances, which required the most
intensive conception, have, as a matter of historical fact, almost always
been achieved in works with an arresting and organizing subject matter.[82]

The most amazing animal "artists," notably the great apes that have
been led to imitate our technique of painting on paper and the bowerbirds that decorate their constructed dance stage by painting twigs with
berry juice, never try to produce an image. Neither do they interpret
images. They are the perfect Naïve Realists among philosophers. The
little rhesus monkeys reared in solitude treat the "mother-surrogate" as
a reality, or else ignore it;[83] a good many animals—apes, cats, decoyed

[81] F. B. Blanchard, for example, in *Retreat from Likeness in the Theory of Painting* (1945), says of any painter who achieves truly plastic values: "He looked at
whatever he painted as though it were still-life, caring not at all whether a roundish
object before him might be a head or a pumpkin" (p. 154).

Artists themselves, however, have frequently expressed a contrary opinion. Jackson
Pollock, for instance, commenting on the labels which had been attached to his
way of painting, said: "I don't care for 'abstract expressionism,' and it's certainly not
'nonobjective,' and not 'nonrepresentational' either. I'm very representational some
of the time, and a little all the time. But when you're painting out of your unconscious, figures are bound to emerge" (quoted by Selden Rodman in his *Conversations
with Artists* [1961], p. 82). Patrick Heron, in *The Changing Forms of Art* (1955),
expressed closely similar opinions: "Painting of whatever school is not understood
except through an extension of one's awareness of the visual . . . in discussion of the
iconography, symbolism or 'meaning' of a modern picture we have only . . . a discussion of that sort of meaning which . . . is not unique to painting.

"This literary content in painting is not to be ignored, however. It plays its part,
which is, pictorially, a subterranean one." And further: ". . . what at first appears
non-figurative (i.e., devoid of reference to objects external to the work itself) does in
the course of time begin to *take on* a figurative function. It is . . . as though all
forms become, willy-nilly, invested by the spectator with the property of symbols,
or signs, or images which overlap with those of reality" (pp. ix–xi).

[82] Compare A. Clutton-Brock, "The Function of Emotion in Painting" (1907–8),
p. 24: "In literature the statement of facts is only a means to an end. In painting
the imitation of reality is only a means to the same end. . . . No facts of reality are
essential or unessential in themselves for the painter any more than for the poet."
Each artist's handiwork, he holds, is likely to express a *habitual mood* even where
such a personal basic feeling is not really congruous with the subject. ". . . But in
the greatest works . . . the subject is congruous with an habitual mood of the
painter, provokes and intensifies that mood, and provides it with a perfectly suitable
machinery for expression."

[83] See H. F. Harlow, "Love in Infant Monkeys" (1959).

ducks—may be fooled by illusory objects, but once disillusioned, they show no further interest in them. The interpretation of a visual form as an image, just like the interpretation of a phonetic form as a word, is strictly human.

To us, every distinct form that is recognized as unsubstantial seems imaginal. Even a non-representational picture is an image—not of physical objects, but of those inward tensions that compose our life of feeling; almost as quickly as we construe a pure appearance confronting us, we interpret it, which is to say, we have an immediate intuition of meaning, and "realize" that intuition—to use Fauré-Fremiet's term—by giving the form a locus in our universe.

That is an entirely different function from self-display, release of actual emotion or the aesthetic pleasure birds may be taking in their own calls or apes in their deployments of color. It is an intellectual function which involves no reasoning. That sounds paradoxical, if not absurd; yet there is no contradiction in it. "Reasoning" is the process of building up insight into relations which are too complex to be grasped by direct inspection of the highly elaborate exhibit or statement, in which many terms are implicitly connected with others in several directions at once, and perhaps are even determined in their nature by such relations—that is, are functions of other terms. Reasoning is the use of logic to make implicitly given conditions explicit; and logic is a fundamentally simple yet powerful machinery for getting from one intuition to another, systematically, successively, without losing any member of the series.[84] The sort and degree of complex relational pattern which can be understood directly, without discursive analysis and technical process of reasoning, varies widely from one individual to another. There are, for instance, persons to whom numerical relations are immediately apparent which other people have to deduce by a long train of concatenated intuitive judgments. Both procedures are equally intellectual.

The perception of the formal aspects of concrete realities makes logical projection, or rendering of such realities in symbolic terms, possible and their recognition in those terms intuitive. Intuitive processes are not always immediate, and they may be selectively evoked, blocked or modified. The understanding of language is of this sort. It dawns at some time within the first two or three years of one's life under conditions of hearing speech, although the meaning of every phrase or word has to be conventionally determined. We have no intuition of what this or that word means; our intuition is, rather, of the fact that the articulated utterances

---

[84] This agrees exactly with Locke's concept of "demonstrative knowledge." See above, p. 129, n. 56.

of man are not chirps and cries, but speech—that is, we have an intuition of significance as such. Where this semantic perception is missing, there may be high animal mentality, but human mentality—mind—cannot reach a normal development.

Neither can it where feeling is a formless welter. If emotional reaction is sporadic and indistinct, though perhaps violent when it occurs, sensibility is likely to be crude, and its interpretive element limited to the practically essential minimum.

A highly developed mind grows up on the fine articulation of generally strong and ready feeling, both subjective—that is, autogenous—and objective, aroused by peripheral impacts. Where the process of interpretation is constantly elicited both by sensory impacts and by an active central production of images, inward verbalizing and variable emotional tone, that process becomes sure and flexible, and tends to enter into detail; for a mind so endowed, the semblances of most situations are complex, but complexity can rise to a high degree without becoming confused. And, furthermore, situations are not necessarily of the moment or of the outside world; symbolic activity begets its own data for constant interpretation and reinterpretation, and its characteristic feelings, especially of strain and expectation, vagueness and clearness, ease and frustration, and the very interesting "sense of rightness" that closes a finished thought process, as it guarantees any distinct intuition.

What makes this "sense of rightness" and the correlative "sense of wrongness" interesting for theory of mind is that these feelings are really the ultimate criteria whereby we judge the validity of logical relations. Once we see that a given proposition, A, implies another proposition, B, it is impossible to deny the validity of "$A \supset B$"; the sense of its rightness is absolute. Only a change of interpretation affecting the meaning of A or B could remove our immediate sense of conviction. Logical conviction is such a pin-pointed feeling that it has, in itself, none of the widespread and involved character of emotion; it seems the very opposite of emotion, although all sorts of highly cathected ideas may gather around it, and make it a tiny firm center in a maelstrom of fantasies. This intensive and exclusive focus on a distinct, discursively rendered concept, such as a proposition, is a structural characteristic of the feeling known as "logical conviction"; it makes that feeling easy to isolate from the matrix of sense and emotion in which most of our mental acts are deeply embedded. And, furthermore, it leads to the peculiar social circumstance that it is relatively easy to confront different individuals with the same challenge to feeling, unimpaired by the usual modifications due to personal context. This makes for a unanimity in logical convictions that has few if any parallels in the realm of human feeling, and gives to logical perception an air of "objectivity," i.e., of coming as impact upon us, not because we receive it with our peripheral sense organs, but because it is the same for all normally constituted people, to an even greater extent than any sense datum.

As a result of this formal peculiarity, logical conviction seems so different from other autogenous felt actions that it and they have traditionally been regarded as deriving from radically different respective sources, and meeting as antagonistic powers in the human mind, which is their battleground throughout life. Reason and feeling, logic and emotion, intellect and passion or however we name the incompatible pair, are most commonly treated as two opposed principles of motivation which can be reconciled only by striking a balance between them, sacrificing much passion, sympathy and gratification of desire to reason, and a little rational judgment to the admitted "natural affections." The alternative to this classical view is the prevalent modern one of rationality as a superficial process of "rationalization," not really determining action, which can be traced entirely to passional motives, but hiding those motives under a veneer of alleged "higher" aims and pseudo-logical plans.

The weakness of both doctrines is, I think, the same: they both treat reason as something intrinsically different from other mental functions, something not subject to spontaneous excitation, like fantasy or vivid memory, which are apt to be activated by emotion, but rather a constant, standard competence available for use by the organism, like an instrument. The concept of the brain (simplified by being stripped of all its interpretive and emotive functions) as a computer reinforces that view.

The principles of logic are exhibited both by the "mechanical brains" of systems engineering and by human thought. George Boole, who formulated the so-called "algebra of logic," called his famous treatise *The Laws of Thought*. But there is much more to rational thinking than the highly general form which may be projected in written symbols or in the functional design of a machine. Thinking employs almost every intuitive process, semantic and formal (logical), and passes from insight to insight not only by the recognized processes, but as often as not by short cuts and personal, incommunicable means. The measure of its validity is the possibility of arriving at the same results by the orthodox methods of demonstrating formal connections. But a measure of validity is not a ground of validity. Logic is one thing, and thinking is another; thought may be logical, but logic itself is not a way of thinking—logic is an abstract conceptual form, exemplified less perfectly in our cerebral acts than in the working of computers which can outdo the best brains a thousandfold in speed, with unshakable accuracy.[85]

There are many indications that rational thinking is a highly specialized phase of that constant symbolization and symbol concatenation which seems to be a spontaneous activity of the human brain, brought into close contact with the outward senses, and under their influence, by virtue of a cardinal intuitive function, the recognition of instances, or specific contents for conceived forms. The truly intellectual phase of symbol using, or "thinking" in a strict sense, is a late development concomitant with a fully articulate use of language. It was probably preceded in all societies by a period of riotous imagination, stemming from a chronic overload of emotional responsiveness.

These are all hypothetical propositions to be argued in a later part of the present essay (Part IV, "The Great Shift," on human mentality).[86] The reason for adducing them, tentatively, at this point is to suggest that the wide discrepancy between reason and feeling may be unreal; it is not improbable that intellect is a high form of feeling—a specialized, intensive feeling about intuitions. There are corroborations of this idea in the psychological literature, which frequently notes, at least, the similarity of form between intellectual processes and emotive ones.[87] But my reason for entertaining the hypothesis of the derivation of all forms

[85] The classical instance of the psychologistic misconception of logic is offered by John Dewey in his *Reconstruction in Philosophy* (1919), where he said that "logic is both a science and an art; a science so far as it gives an organized and tested descriptive account of the way in which thought actually goes on; an art, so far as on the basis of this description it projects methods by which future thinking shall take advantage of the operations that lead to success and avoid those which result in failure" (p. 135).

Jean Piaget, in *The Psychology of Intelligence* (1950; original French ed., 1947), tries to keep logic and psychology apart, yet chides logicians for treating "operations" as something else than a mental action. See chap. ii, " 'Thought Psychology' and the Psychological Nature of Logical Operations," especially p. 20, where the equivocation of terms and contexts is complete.

[86] A merely suggestive statement of the general hypothesis here referred to may be found in *Philosophical Sketches*, No. 1, "The Process of Feeling."

[87] Among psychologists, Wolfgang Koehler, Felix Krueger and Paul Schilder have made such observations, but these had better be reserved until we come to the study of the phenomena which serve as their context. Artists and critics, however, have discovered those resemblances in their own way. So. D. Dorchester, Jr., in "The Nature of Poetic Expression" (1893), wrote: "In the exercise of Imagination the mind experiences its tensest action, and feeling imparts the tension." And he quoted George Eliot's words from *Middlemarch*: " 'To be a poet is to have a soul . . . in which knowledge passes instantaneously into feeling and feeling flashes back as a new organ of knowledge' " (p. 89).

of human experience—self-awareness, *Weltanschauung*, mental suffering and joy, social consciousness or what you would name—from primeval feeling is the image of feeling created by art throughout its long, ramified history; that image seems to be capable of encompassing the whole mind of man, including its highest rational activities. It presents the world in the light of a heightened perception, and knowledge of the world as intellectual experience. Rationality, in this projection, is not epitomized in the discursive form that serves our thinking, but is a vision of that thinking itself, of a vital movement outstripping the sure, deep rhythms of physical life, so its tensions against those rhythms are felt as keen and precarious moments within the very limits of supportable strain; the sense of rationality appears as brilliance, perfection of form, a semblance of the tersest economy (which may be achieved with or without an actual restriction of means), or of great daring in the certainty of equally great competence (where the daring may be purely virtual, the competence that of a machine; there are beautiful designs in our contemporary art which depend for their feeling quality on their milled precision of form, and yet—without deluding anyone—create an impression of consummate skill).[88]

[88] The artistic projection of intellectual feeling often yields a great satisfaction of sense, and even appears as "easy" beauty; it is a different thing from so-called "intellectual art," a product of discursive thinking about principles of construction, relation to material, communication and especially the "message" of the work in the making. "Intellectual art" does not express rationality, but illustrates it. Cf., for instance, F. Heinrich, "Die Tonkunst in ihrem Verhältnis zum Ausdruck und zum Symbol" (1925–26), p. 79n: "Ich bezeichne als Geistiges in der Musik nichts anderes als das Beziehen gehörter oder vorgestellter Töne aufeinander, so wie wir das in sprachlicher Form auftretende Geistige, das Denken, ein Beziehen von Vorstellungen aufeinander nennen." Such a high intellectuality of procedure, which need not be reflected at all in the created quality, marks the work of Arnold Schoenberg; see *Style and Idea*, chap. v, "Composition with Twelve Tones" (pp. 102–43).

Cf. also the critical principle advanced by M. de Zayas and P. B. Haviland, *A Study of the Modern Evolution of Plastic Expression* (1913), p. 11: "The artist of today does not appeal . . . to our emotions but to our intellect." A couple of pages later the authors admit sadly, ". . . we do not think that art has yet reached a stage where it can be considered as a pure scientific expression of man through the elements offered by science and art."

A similar statement occurs, more surprisingly, in *Reason in Art*, the fourth volume of Santayana's *The Life of Reason* (1905), p. 111: "If a poet could clarify the myths he begins with, so as to reach ultimate scientific notions of nature and life, he would still be dealing with vivid feeling and its imaginative expression." The poet-philosopher goes on to speak of "that half-mythical world through which poets, for want of a rational education, have hitherto wandered," and concludes: "A rational poet's vision would have the same moral functions which myth was asked to fulfil, and fulfilled so treacherously; it would employ the same ideal faculties which myth expressed in a confused and hasty fashion."

Between all different levels and modalities of feeling, however, there are no breaks, as there have always seemed to be between the traditionally assumed realms of sense, emotion and intellect. There are graduated changes, and sudden shifts of rhythm; proliferations of detail, and sweeping simplifications; but the import of art is one vast phenomenon of "felt life," stretching from the elementary tonus of vital existence to the furthest reaches of mind. All psychical phases of human nature may furnish the "ideas" of art.

The expression of such ideas, however, reveals the nature of what is expressed, in a direction that is not open to actual experience: the unfelt activity underlying every event that enters the state of feeling. The work of transforming and projecting our concepts of psychical data forces their characteristic form on our notice, because a symbolic projection made for immediate intuitive apprehension has to be highly articulated; and, as a matter of long experience, we know that every form which seems to be charged with feeling also appears "organic," and makes the impression of "livingness," though it may not even remotely suggest any sort of living creature. This semblance of organism is implicit in the artist's "Idea." He probably thinks and does nothing explicit to attain it, because he thinks beyond it; but as soon as it is lacking and the work is "dead," it is emotionally inexpressive too.

It is this circumstance which really led me to think of feeling as a phase of vital process itself under special conditions, instead of as a new substantive element produced by such a process; as the incandescence of a heated wire, under better-known circumstances, is a condition of the wire itself and not an added entity. This led in turn to more and more intimate studies of the artistic projection, the ways of abstracting and organizing, individuating and deepening or etherealizing the virtual form. Those ways are legion, and constitute the artist's technique. The aims they serve are equally rich and diverse, and are, of course, the more significant aspect of his work, but they are not directly accessible. The readiest way to find them is to give attention to the means of their realization, i.e., to study the devices by which the expressive image is constructed and developed, and continually ponder what is the purpose of any problematical move. And here the power of the symbol really comes to light; for the great examples of any art exhibit—not in their physical structure, but in their virtuality, their perceptible form—an image of life that suggests some new basic concepts for biology and psychology. The aspects of vital functions that appear when one views them from their highest points, where felt impulses and conscious activities largely prevail, emphasize general characteristics of animate nature which are usually disregarded at lower levels of life, but are the very traits that make its generic continuity apparent. The necessity of "living form" for any rendering of psychical events rests simply on the fact that such events are the very concentration of life, acts in which the deeper rhythms of the organism, mainly unfelt, are implicated so that the dynamic structure of the individual is reflected in the forms of feeling as it is in the form of every voluntary movement of the body. But in the development of cerebral activity to the human level, some characteristics inherent in all such activity become highly specialized and finally transformed so they have no close analogues in the mentalities of other creatures: imagination, intuition and the whole gamut of new powers these engender, primarily of course speech and reasoning. These characteristics become paramount in forming the emotional patterns of man and even his perceptions, which are shot through and through with conceptual elements, so human experience is a dialectic of symbolic objectification and interpretive subjectification. But that is an anticipation at this point.

What is in order here, however, is the fact that an image of mind is that of a living process, and therefore entails the projection of "living form" in a symbolic transformation. The basic transformation in art is from felt activity to perceptible quality; so it is a "quality of life" that is meant by "livingness" in art. This vital appearance is exhibited, of course, by living forms in nature, which—as that great morphologist D'Arcy Thompson[89] has observed—are almost all records of growth, i.e., of biological activity; yet the most convincing images of such forms have often resulted in art where no natural model furnished the motif, and the shape seems to have sprung directly from the symbolic intuition of the artist. The principles of life are reflected in the principles of art, but the principles of creation in art are not those of generation and development in nature; the "quality of life" in a work of art is a virtual quality which may be achieved in innumerable ways. Yet it is in noting the differences between biological exemplifications of living form and the ways of creating its semblance in art that one finds the abstractions of art which emphasize the obscure, problematical aspects of life that are destined to develop into or to underlie higher activities, felt as emotion or sensation or the spontaneous ideation that is the intellectual matrix of human nature, the mind.

[89] On Growth and Form (1942), Vol. I, p. 57: "I have called this book a study of Growth and Form, because in the most familiar illustrations of organic form . . . these two factors are inseparably associated, and because we are here justified in thinking of form as the direct resultant and consequence of growth . . . whose varying rate in one direction or another has produced . . . the final configuration of the whole material structure."

6

A Chapter on Abstraction

THE problems of abstraction in art have never been philosophically surveyed and analyzed. They arise in practice, and people who
meet them there solve them practically, piecemeal, often without even
putting them consciously into any category. They occur as questions of
what to do next, how to handle a disturbing or commonplace passage,
how to concentrate or unify an impression. Those artists who have reflected more generally on abstraction either decry it, or praise it as the
aim and acme of their art. In the first case, of course, they mean the
kind of abstraction made in literal discourse, in the second some sort of
non-discursive appeal to intuition. But what sorts of the latter there may
be is not the creative artist's concern. He will find the ones he needs. It
is the philosopher's business to recognize their variety and reflect on their
functions, their relations to each other and their implications for the
concept of mind.

Scientific concepts are abstracted from concretely described facts by
a sequence of widening generalizations; progressive generalization systematically pursued can yield all the powerful and rarified abstractions
of physics, mathematics and logic. The process of establishing them may,
therefore, be designated as "generalizing abstraction."[1]

---

[1] After choosing this term purely on its merits, I find it, to my gratification, already
in use both in English and in German writings. It seems to have been introduced by
Sara C. Fisher in 1916, when she published a long and detailed monograph, "The
Process of Generalizing Abstraction and Its Product, the General Concept." Herein
she noted the difference between generalization and the abstraction of formal concepts by means of generalization, and also the tendency to recognize no other kind
of abstraction and consequently to assert of abstraction as such what is true only of
this one mode (see p. 2n). The product should, of course, have been called "the
abstract concept" in her title.

Another, much more recent adoption of the term occurs in Rudolf Schottländer's
"Recht und Unrecht der Abstraktion" (1953). Schottländer distinguishes *"generalisierende"* from *"isolierende abstraktion"* (herein following Wundt), and claims
that only the former really yields concepts. See p. 221: "Einen abstrakten Begriff
erzeugt nur die generalisierende Abstraktion." Thereupon he identifies "abstract"

The sort of abstraction, however, which artists mean when they use the word approvingly is of a different sort, and its procedures have never yet received any systematic study. Pointing out that they are not based on generalization and are not carried on by discursive thought tells us only what they are not, but provides no notion of what they are. Artistic abstraction is, in fact, of many kinds; some of these are peculiar to art, or at least unimportant in other contexts, and some are common to many mental activities and occur even in the ordinary use of language for social communication. Semantic intuition plays such a great role in human life that it is not surprising to find it elicited by many means, and as abstraction is involved in all symbolic functions, it also might be expected to occur in various ways and have several different forms.

The several different forms involved in the arts, however, are so different that one cannot arrange them in any order with respect to each other. They seem to have no such respect. The recognition of each one opens a new beginning in the analysis of whatever work of art one happens to find it in, and when one's analysis in terms of the given kind of abstraction has gone as far as it can go, it has not yielded the secret of how the artist's "Idea" is brought to expression. Some other abstractive principle seems to be at work, something stemming from an entirely different source. There are, in fact, at least four or five independent sources of abstractive techniques, and the interplay of logical projections which they engender creates the semblance of irrationality and indefinability which is the delight of artists and the despair of aestheticians.[2] It also begets the irrationality of many utterances made with warmth and conviction by artists expressing their aesthetic views; "abstraction" is alternately praised and decried, according to what the word means to the speaker at the moment; and with so many meanings to choose from and only the vaguest idea that the word has more than one, the moments when new meanings are carried into the fray succeed each other with bewildering speed. Some writers oppose "abstract" not only to "concrete," but also to "living," "expressive," "intuitive," or "real"—that is, to anything of which they approve. Others, however, or even the same writers at other times, name it as the pure essence of art. So we find:

---

with "general"—according to "the theory that dispenses with abstraction" this should be legitimate—and consequently makes the opposite of "abstraction," "specification," instead of its proper opposite, "exemplification." Perhaps this difference in their respective converse functions shows the non-identity of "abstract" and "general" in epistemology even if not in logic proper.

[2] Cf. the discussion of Philippe Fauré-Fremiet's work in Chapter 4 above, especially pp. 100–1.

"An abstraction is an individual with a life of its own,"[3] or as the title of an essay: "An Abstraction Is a Reality."[4] Roy Harris has said of his own art, "Music can only exist in the time stream of its own continuity. It can not be concrete. It is abstract."[5] In all these contexts, abstractness certainly figures as something laudable, not removed from life, reality, expression and intuition. One author even remarks: "Music being the most abstract and instinctive of the arts, musicians are least given to theorizing. . . ."[6] But the most disconcerting changes on the theme of abstraction are certainly rung in Erich Benninghoven's little article on Hindemith, where we find on one page: "Music, too, is no abstraction, but is somehow connected with physical reality," and on the next: "On this ultimate, purely abstract formulation the work of Hindemith is built up."[7]

The impatient or even angry tone in which most artists speak of abstraction when they mean the product of generalization springs from a perfectly sound conviction that the kind of thinking to which generalizing abstraction belongs is not only foreign to art, but inimical as well. The limitations inherent in verbal conception and discursive forms of thought are the very *raison d'être* of artistic expression; to surpass those limitations requires the abandonment of the activity which entails them, and which tends to interfere with the more precarious process of implementing formal intuitions of another kind than those usually called "logical": the process of perceiving and rendering the forms of feeling which are not amenable to generalizing abstraction. Because our verbal forms of thought are supported by conventions, they are incomparably easier to hold and to organize than the crowding, chaotic materials which sensuous or poetic imagination provides without any accompanying directions for use. Discursive thinking, once started, runs on in its own loosely syllogistic pattern from one proposition to another, actually or only potentially worded, but with prepared forms of conception always at hand. Where it seizes on any material—sensations, memories, fantasies, reflections—it puts its seal of fixity, categorical divisions, oppositions, exclusions, on every emerging idea, and automatically makes entities out of any elements that will take the stamp of denotative words. By virtue of its habitual exercise, it has an easy victory over any other

---

[3] F. B. Blanchard, *Retreat from Likeness in the Theory of Painting* (1945), p. 152.

[4] Grace Clements (1944).

[5] "The Basis of Artistic Creation in Music" (1942), p. 24.

[6] Michael McMullin, "The Symbolic Analysis of Music" (1947), p. 26.

[7] "Der Geist im Werke Hindemiths" (1929), pp. 719, 720.

process of conception and expression that competes with it; and similarly its mode of abstraction overrides the subtler abstractive techniques of art.

So far, the sort of abstraction that underlies artistic expression has not been given a name to distinguish it from generalizing abstraction. Some terms have been proposed, as for instance "isolating abstraction" by Wundt,[8] and "abstractive apperception" by Kuno Mittenzwei.[9] But "isolating" is too specialized, though quite correct for one important method, namely, suppressing or cancelling all obscuring factors so the intended form comes to light; and for this procedure I shall use it. Taken in a broader sense of revealing a form or trait, it applies to all abstraction whatever and therefore does not serve to make a distinction. Mittenzwei's term names the psychological process of apprehending forms, but again, this is the same process that the discursive method of generalization implements. In art this process has to be differently induced; but it is the reason for the difference of approach, rather than the type of approach itself, that must furnish the defining function of a whole class of abstractions not attainable by way of generalization. Such are all the abstractions which can be made only by presentational symbols; and perhaps the term "presentational abstraction" will serve most readily as the counterpart of "generalizing abstraction" to mark the main distinction here in question.

Presentational abstraction is harder to achieve and a great deal harder to analyze than the generalizing form familiar to scientists and recognized by epistemologists. It has no technical formula which carries the entire pattern from one level of abstractness to another, as progressive generalization of propositions does when it is exercised simultaneously on all the terms or all the constituent relations of a given order in a system. It has, in fact, no series of successive levels of abstractness to be reached by all elements in the complex of a symbolic projection at the same time. For purposes of logical analysis, art is unsystematic. It involves a constant play of formulative, abstractive and projective acts based on a disconcerting variety of principles. Presentational abstraction, consequently, has many subspecies, often related to particular creative devices, which differ from one art to another even if they show some general analogies. Those several subordinate kinds of abstraction allow many different logical projections to mingle in the making of the one

[8] Cited by Sara C. Fisher, *op. cit.*, p. 4.
[9] "Über abstrahierende Apperzeption" (1907). Also adduced by Fisher, *ibid.*, pp. 5 ff.

complex symbol, the created image that presents the artist's idea, the work of art, which consequently is not analyzable in any single set of terms. To comprehend its nature one has to take its author's mingled procedures one by one and consider what each of them contributes to the creation of the single expressive form; this sort of analysis is laborious, but it reveals aspects of feeling projected in the work—sometimes even in a second-rate work—that are not discoverable by any more systematic means. If then we look at art as a whole, through the ages, and at all its kinds in all cultures, it seems to be capable of expressing the entire range and complexity of human experience as I have just pictured it in the foregoing chapter; and it is not hard to see why that projection should demand such a multiplicity of means, and why presentational abstraction should have so many forms.

The artist's most elementary problem is the symbolic transformation of subjectively known realities into objective semblances that are immediately recognized as their expression in sensory appearances. Such a basic transformation is made at the outset by the establishment of the primary illusion, which creates the main substance (in the sense of *substantia*, not of matter) of every piece; it makes the most direct sort of presentational abstraction. The further development of the vital image, however, to the degree where its internal rhythmic relations appear more than just organic, more like the free play of thought, its immediate qualities like the warmth of emotion, its newness like an advancing awareness of its own,[10] requires indirect and subtle orders of abstraction: isolating, metaphorical, secondary, transcending and perhaps others for which one could invent suggestive names. Let us begin with the most recognized and essential.

When competent artists or critics speak of the ultimate values they find in finished works, they speak of quality, feeling, expressiveness, significance and (if they are old-fashioned, or if they dare) of beauty, apparently meaning essentially the same thing by all these words. But when they talk about works in progress, or about completed ones analytically, they talk in different terms; then they are likely to speak of tensions and resolutions, and all their language shifts to dynamic metaphors: forces in balance or imbalance, thrusts and counterthrusts, attraction and repulsion, checks and oppositions. Different arts favor different meta-

---

[10] Such effects are undoubtedly what Le Corbusier meant by "spiritualizing" when he wrote: "Architecture . . . brings into play the highest faculties by its very abstraction. Architectural abstraction has this about it which is magnificently peculiar to itself, that while it is rooted in hard fact, it spiritualizes it" (*Toward a New Architecture* [n.d.], p. 47).

phors, but tension and resolution are the basic conceptions in all of them.[11]

There is, moreover, one other peculiarity of diction that goes with the use of such kinetic terms: artists of all sorts—painters, poets, dancers, musicians—speak of "producing," "creating" or "establishing" tensions, and of "resolving" them, but usually of "achieving" or "getting" the ultimate qualities. "Tension" is a studio concept. "Quality" is a jury concept—the permanent jury being the public, present and future, for whom the work exists. "Aesthetic quality," as it is often called, is the beholder's touchstone, the measure of the artist's success. It is, of course, the latter's final aim (he is the first and most concerned beholder); but his immediate problem, as long as the work is in progress, is always to create and manipulate elements that build up and develop the expressive form.

The most fundamental elements seem to be tensions; and upon closer inspection, tensions show some peculiarly interesting traits. By their very occurrence they immediately engender a structure. They act on each other in a great variety of ways—they can be handled so as to intersect without losing their identity, or contrariwise, so that they fuse and compose entirely new elements. They can be intensified or muted, resolved either by being spent or by being counterbalanced, modified by a touch, and all the while they make for structure. This appears to be true in all the great orders of art; in every one of them, a general range of tensions is set up by the first element—line, gesture or tone—which the artist establishes. In performed works this immediate effect of a single, first-presented element is often apparent to the audience as well as to the

---

[11] To corroborate this statement one needs only to skim the literature of the arts and quote from the most serious and reputable sources. Heinrich Schenker, for instance, in *Neue musikalische Theorien und Phantasien*, Vol. III: *Der freie Satz* (1935), says: "Am Ende wird sich als gemeinsames oberstes Merkmal sämtlicher Künste feststellen lassen: das Gesetz von der inneren Spannung und der entsprechenden Erfüllung nach aussen, das aber in verschiedenem Stoff sich verschieden auswirkt" (p. 9).

I. K. von Hoeslin, in a valuable article, "Die Melodie als gestaltender Ausdruck seelischen Lebens" (1920), speaking of poetic images (*"Bilder"*), says: "Diese Bilder wirken dann . . . derart, dass zwischen ihnen musikalische Intervalle entstehen, die als melodische Intervalle gefühlt werden und als solche nichts anderes als Spannungen sind—aber keine . . . motiverzeugte Spannungen, sondern Spannungen reiner, formaler Art" (p. 253).

In a very similar vein, Ernst Toch has written, in *The Shaping Forces in Music* (1948), p. 156: "It is the right distribution of light and shade, or of tension and relaxation, that is formative in every art, in music as well as in painting, sculpture, architecture, poetry."

author. A perfect example is the definition of form achieved in the opening measure of Schubert's Impromptu, opus 90, with its unison dominant chord and dynamic contrast:

The rise of a curtain in the theater, even on an empty stage, as for instance in Philip Barry's *Tomorrow and Tomorrow,* is another. In Martha Graham's *Primitive Mysteries* the curtain is lifted on black darkness; the whole tensive frame of the piece is given in that moment.

In the plastic arts the creation of the decisive tensions unfolds in such an obvious way only for the artist, but to him it is none the less clear. Hans Hofmann has described the commencement of composition in a way that leaves no doubt about the making of the first visual abstractions, "space tensions."

"Take a piece of paper," said Hofmann, "and make a line on it. . . .

"The fact that you placed one line somewhere on the paper created a very definite relationship between this line and the edges of your paper. . . . By adding another line you not only have a certain tension between the two lines, but also a tension between the unity of these two lines and the outline of your paper. . . .

"Within its confines is the complete creative message. . . . The outline becomes an essential part of your composition. . . . The more the work progresses, the more it becomes defined and qualified. It increasingly limits itself. Expansion, paradoxically, becomes contraction.

"Expansion and contraction in a simultaneous existence is a characteristic of space. Your paper has actually been transformed into space.[12] A sensation of movement and countermovement is simultaneously created through the position of these two lines in their relation to the outline of the paper. Movement and countermovement result in tension. Tensions are the expression of forces. Forces are the expression of actions. . . . Your empty paper has been transformed by the simplest graphic means to a universe in action."[13]

Lines are not the only source of space tensions. Such tensions arise

---

[12] Both the "characteristic" and the "transformation" indicate that Hofmann was speaking of virtual space, which is created by the first line or color spot on the paper, by virtue of the tension that ensues between that mark and the contour of the sheet.

[13] *Search for the Real* (1948), pp. 47–48.

from any operation on the blank ground, for instance, the introduction of a spot, or spots, of color. "Color is a plastic means of creating intervals. Intervals are color harmonics produced by special relationships, or tensions. We differentiate now between formal tensions and color tensions. . . . Both, however, . . . aim toward the realization of the same image."[14]

So, of course, does sculpture, and Hofmann goes on to a discussion of volume, as spatial form created by the virtual movement of its related planes. It is interesting that what I call "virtual motion" the artist designates simply as "motion"; there is no other kind in his world. Only as an afterthought, he remarks: "Movement in sculpture does not actually exist in point of time, but the experience of movement is sensed in the limitation of the medium."[15] Here the system of tensions produces a new metaphorical abstraction, the experience of movement without anything that actually changes place. And more than that: it creates a virtual interior of the statue, which cannot be thought of, or felt to be, either the lump of material that actually constitutes it, or the bones and viscera of living creatures. Yet the planes are not like picture planes, but have "depth," mutual relations through the center inside. The center is a matrix of tensions and movements that seem to express themselves toward the surface;[16] actually, of course, the surface is all that has shape, and the "kernel" is purely implicit in the outward form.

The impression that space contracts or expands under the tensions of form is made with equal if not even greater force in architecture. Open, outdoor space, without limiting contours of hills or shore lines, is many times larger than the hugest edifice, yet the sense of vastness is more likely to beset one upon entering a building; and there it is clearly an effect of pure forms. Everyone who has stood in St. Peter's in Rome knows that its actual great size is somehow reduced, that the space contracts, simply by reason of its perfectly resolved proportions.[17] The

---

[14] *Ibid.*, p. 51.   [15] *Ibid.*, p. 57.

[16] Again, Hofmann speaks with the immediate sense of the artist, saying: "The basic forms either expand or contract depending on the various depth relations. In this way surface tensions are created—they express the inner life of an existing or created form" (*ibid.*, p. 56).

[17] Compare the observation of J. S. Pierce, in his "Visual and Auditory Space in Baroque Rome" (1959), p. 66: "There is no expansion in a centralized Renaissance church because there is a balance of forces. . . . But the full Baroque interior expands unequivocally. The effect is extraordinarily sudden and direct when the entrance is placed on the short axis of an oval as it is at S. Andrea al Quirinale or the piazza before St. Peter's where upon entering space seems to explode to either side."

Taj Mahal makes a division between earth and sky that seems to lift the building off its ramparts and hold it in the air. However such effects are achieved, the first move makes the abstraction of formal relationships, given to the eye as visible forces and setting up the limits and scope of an "organism" that is the framework of projected feeling. To speak once more with Hofmann, "The planes must be related to create the life-giving tension in an object and the force-impelled spatial tension outside and surrounding the object. The surface tension breathes the inner life of a form, the spatial tension is the life of the plastic unit."[18]

The existence of a tension pattern in music requires no demonstration to anyone who can hear sound as music at all, and to the few who cannot it is undemonstrable. In western music it is most obvious in the resolution or withholding of resolution of dissonant chords, but in purely melodic successions the movement toward their natural conclusions is no less resolvent. What is actually done with tensions in the making of a symbolic form is most readily analyzed in musical composition, where their presence is so obvious that not only artists in other fields but even the "gestalt" pyschologists resort to the vocabulary of music all the time to describe virtual or actual dynamic patterns.[19]

An original and contributive discussion of the structural use of the tensions set up by musical means, and incidentally of their setting-up as well, may be found in a little article by George Dickinson, "Aesthetic Pace in Music" (1957).[20] As one might expect, so basic a concept, when it is really explored, is found to contain more negotiable ideas, broader or

[18] *Op. cit.*, p. 57.

[19] In speaking of colors, Hofmann constantly makes use of musical terms; in one passage (p. 73) he speaks not only of color intervals but also of "symphonic painting." Compare his use of "interval" for color differences with I. K. von Hoeslin's employment of the same metaphor for relations between poetic images, which the latter author even maintains are *felt as melodic intervals* (see above, p. 158n). The words "harmony" and "harmonious" have been traditionally applied to color relations, as by Leonardo, and long before him by Chinese masters.

Among psychologists, Wolfgang Koehler and Kurt Koffka come immediately to mind, as the writers who really introduced the practice. See Koehler's *Gestalt Psychology* (1929), especially pp. 248–49. For a more detailed discussion see *Phil. N.K.*, pp. 226 ff.

[20] The article has some theoretical failings—chiefly, the many references to "stimulations" and their physiological effects, which seem to rest on questionable hearsay, and the comments on literary styles and their materials, which are old clichés —a prevalent weakness in artists making comparisons between the art they know and others which they do not. But these shortcomings make the contribution in the short paper no less important.

narrower, as the case may be, that soon require expression in special terms. So in Mr. Dickinson's treatment "tension" is made to yield the concept of "tensity," which applies to pervasive states, varying as a whole, as well as to distinct elements, i.e., particular tensions. The key words in his title, "aesthetic pace," denote a phenomenon which emerges from the interaction of two fundamental elements in music: temporal pace, which is simply the metronomic tempo of a passage, and the musical pattern, expressed in the score. From these two sources, the musical movement arises; and it is this movement that possesses its own "aesthetic pace." For it is by the interaction of the actual rate of succession of temporally spaced accents and the nature of these accents—their preparation, their melodic, harmonic or dynamic origin—that the flow of changing tensities is made.

Tensities, the author asserts, form a hierarchy, ranging "from a maximum *de*tensity to a maximum *in*tensity, with innumerable gradations between the extremes" (p. 312). Here the word "hierarchy" for what he intends to describe might be questioned. Certainly all tensities might logically be ranged in a single graduated series; but what we really find as the tacitly understood abstract framework is not so much a hierarchy as a plenum of potential tensities, with several gradients developing at the same time in different directions. In that frame the music not only progresses at its individual aesthetic pace, but expands and attenuates, always in keeping with that autonomous movement, so that any number of growing and of declining tensities may be simultaneously present. This creates the appearance of multidimensional motion, which is more than locomotion: cycle, breathing, life.

The chief interest of Mr. Dickinson's concept of "tensities" lies in what he is able to construe in terms of it: the image of feeling that is an articulation of forms in virtual time. The "hierarchy of tensities" in which the particular pattern arises is what I have called "the musical matrix";[21] and his "pattern" has the characteristics which Aristotle already termed "organic." These characteristics he traces to the interaction of created tensities.

"From the departure point of the initially established tensity," he says, "a pattern of structural tensities evolves within the potentialities of whatever overall design is contemplated. . . . The inner signs of continuity [of structure] are a closeness, cohesiveness, and persistence, in the progression of the structural logic itself. This, in manifold forms, is essentially development. Development may assert its power both locally and

---

[21] See *F.F.*, chaps. vii ("The Image of Time") and viii ("The Musical Matrix").

extensively, and has a particular influence on changing tensity through its capacity to progress with different degrees of insistency or rates of accomplishment. . . .

"In the flux of tensities, there resides—as both a cause and a consequence of their manipulation—a purposeful quality in the aesthetic pace, collateral to that of progression, which gives the progressing motion a sense of direction. The objective is an implied point or points of reference. . . . This phenomenon may be termed orientation: modality, tonality, and atonality, are merely individual manifestations of the wider concept of orientation. . . .

"There are . . . as many patterns of aesthetic pace as there are musical works. Each work is individualized in aesthetic pace and consequently possesses its own unique expression."[22]

If now we turn to the works of other musicologists, to books in which the analysis of the constructive process is much more detailed and frequently illustrated by specific passages, we find striking vindications of the doctrine which is but briefly sketched in the article from which the exerpts above have been gathered. So, for instance, Ernst Toch defines form in music as "the balance between tension and relaxation," and goes on to say: "The more equilibrated this balance is, the better will be the form of a musical piece. Thus it becomes clear that form will always be a matter of feeling, not to be pinned down like the signature of a key or the *established forms*. . . . when we speak of balance we by no means mean a half and half distribution of the contrasting elements. On the contrary, . . . the proportion, if expressed in measuring terms, will always favor the tension segment as against the relaxation segment."[23] More than twenty years earlier, W. Harburger, in a very interesting book on the expressive powers of various structural principles[24] (e.g., homophonic as against polyphonic principles), also observed that most of the different creative processes are just the many ways of building up and elaborating and resolving tensions.

"Pure form," he wrote there, "does not exist as material, but as polarity, as lines of force, as tensions, perhaps as longing. . . .

"The whole begets polarities within itself. . . . These polarities beget tensions. (Or is tension, perhaps, the original fact, the deepest source, which produces the polarities in the first place, giving birth to them out of itself?) . . ."[25] There follows a disquisition on tensions in living mat-

---

[22] *Ibid.*, pp. 316–18.
[23] *Op. cit.*, p. 157.
[24] *Form und Ausdrucksmittel in der Musik* (1926).
[25] *Ibid.*, p. 75.

ter, as the source of cell division, etc., whereupon he continues: "Music, likewise, is primarily tension. Not only in the most general, psychical [*seelischen*] sense. (For this asserts nothing particular; all life is tension, and so is all psychical experience.) But in a more special, musical sense. It engenders . . . that fundamental principle of polarity which operates in harmony as tonic and dominant. Just as in melody (in homophonic writing) it functions by way of keynote and leading-tone. Or in the domain of rhythm, as rise and fall of the rhythmic curve."[26]

From these and a great number of similar statements by artists and theorists in all the arts, it is fairly patent that the establishment and organization of tensions is the basic technique in projecting the image of feeling, the artist's idea, in any medium. They are the essential structural elements whereby the "primary illusion" of the incipient work is established, its scope and potentialities given and its development begun.

The pattern of tensions inherent in a work of art reflects feeling predominantly as subjective, originating within us, like the felt activity of muscles and the stirring of emotions. To regard the projection of this pattern as the whole being of the work leads to a "subjectivist" theory of art, which steers close to the concepts of direct emotional expression all too commonly applied to music and lyric poetry. The isomorphy of actual organic tensions and virtual, perceptually created tensions is so close that if the creation of the latter constituted the whole art process, our reception of art might really be simply empathetic or even sympathetic. But a true work of art—certainly any great work—is often above sympathy, and the role of empathy in our understanding of it is trivial. Art is an image of human experience, which means an objective presentation.

The need of its objectification has traditionally been met by a different principle of abstraction, a principle naturally inherent in perception itself, which organizes the impinging sensations spontaneously into large units: the tendency to closure of form, to simplification, known as the gestalt principle.[27]

This tendency is native to the perceptive apparatus of many of the higher mammals. The eye is particularly selective in its reception, favoring those photic factors that the rest of the visual apparatus (including the entire optic tract, the thalamic centers and primary cortical radiation) can compose into distinct retainable images. The abstraction of

[26] *Ibid.*, pp. 76–77.
[27] The literature of this subject, psychological and philosophical, is too extensive to be even suggested here.

form here achieved is probably not made by comparison of several examples, as the classical British empiricists assumed, nor by repeated impressions reinforcing the "engram," as a more modern psychology proposes, but is derived from some single instance under proper conditions of imaginative readiness; whereupon the visual form, once abstracted, is imposed on other actualities, that is, used interpretively wherever it will serve, and as long as it will serve.[28] Gradually, under the influence of other interpretive possibilities, it may be merged and modified, or suddenly discarded, succeeded by a more convincing or more promising gestalt.

This principle of automatically abstractive seeing and hearing deeply affects the potentialities of art; for it provides another and quite different means of constructing forms, whereby tensions, always created in the process, are subordinated to the unity of a substantive element. Instead of starting with the expression of linear forces which make points of arrest by their intersection, or with points that beget lines by their motions and volumes by expansion, the first productive envisagement may be of pre-eminent bounded shapes, carved out of the total virtual space of the work.[29] This is the basic conception of Hildebrand, discussed at some length in *Feeling and Form*,[30] and of Britsch.[31] It was also the conscious approach of Leonardo, who rated contour drawing as

[28] In this belief I find myself supported by Rudolf Arnheim, who observed, in an article entitled "Perceptual Abstraction and Art" (1947), that in dim light or brief tachistoscopic presentations, "patterns result which represent the structure of the model in a simplified way, by means of regular, often symmetrical forms, which quite frequently were not contained in the model [the reference here is to experimental sources]. It is as though the margin of freedom yielded by the lessened stimulus control enhanced a tendency of the receiving sense organ spontaneously to produce simple, regular form. . . . Perhaps perception consists in the application to the stimulus material, of 'perceptual categories' . . . which are evoked by the structure of the given configuration" (p. 69).

"Our assertion is that the individual stimulus configuration enters the perceptual process only in that it evokes a specific pattern of general sensory categories which stand for the stimulus . . ." (p. 70).

For my part, I would not assert that abstraction is never made by comparison nor by reinforced impressions, but only that such methods imply special conditions. Generalizing abstraction may involve either or both of those assumed processes. Such complex functions as abstractive intuition usually have a variety of sources and mechanisms with convergent effects.

[29] What is said here in terms of space holds for all other virtual dimensions.

[30] See chap. v, "Virtual Space."

[31] Cf. above, Chapter 4, pp. 94–96.

the first requisite in the making of a picture, more important even than shading and lighting, though these techniques of *relievo* were of particular interest to him.[32] Titian, a master of color, nevertheless is reported to have said: "It is not bright colors but good drawing that makes figures beautiful."[33] The outline created the volume and gave it its form, and therewith the semblance of an object, surrounded by free space of potential movement.

Herein lies the chief and immediate virtue of representation in art. Not duplication of things which are already in existence, but the gathering and projection of their forms, for their expressive and compositional values, is the artist's intent. His imagination draws on nature; in looking and noting he learns the potential growth and expressiveness of forms, and the continuous order of their distortions without loss of their basic identity. In this way differences in directions seem to have come from motions, and differences in volumes still imply an "absolute size." That implication was used in Renaissance painting chiefly to create the optical third dimension; but in other traditions it has been employed to center interest on one figure, which is superimposed by its mere relative size on a rhythmic mass of lesser figures, as the large, clear forms of the Christ on his worshipers, so that the small images, which have a perfectly distinct pictorial organization, assume something like the function of a textured ground (see Fig. 6–1); or it serves to intensify an attitude by repeating it without "realistic" meaning, like some figures in savage and folk art, decorated with other images, and even with little repetitions of themselves (see Fig. 6–2); or to telescope two rhythms into one, sometimes letting the dominant one barely connote the other (see Fig. 6–3).

Representation, far from being a non-artistic competing interest, is an orienting, unifying, motivating force wherever it occurs at all in the early stages of an art; it is the normal means of "isolating abstraction," or abstraction by emphasis. It provides terms in which a visual structure

[32] See *Das Buch von der Malerei*, trans. H. Ludwig (1882) (Italian and German given in parallel), Part II, pp. 133–34: "Due sono le parti principali, nelle quali si divide la pittura, cioè lineamenti, che circondano le figure de corpi finti, li quali lineamenti si demanda disegno; la seconda è detta ombra.

"Ma questo disegno è di tanta eccellentia, che non solo ricercha l'opere di natura, ma infinite più che quelle, che fa natura. . . . a tutte l'arti manuali, anchora che fissino infinite, insegno il loro perfetto fine.

"Sia con somma deligentia considerato; termini di qualonque corpo et il modo de lo serpeggiare; le quale serpeggiature sia giudicato, selle sue uolte participano di curuita circolare, o'di concauita angulare."

[33] Quoted in Robert Goldwater and Marco Treves, *Artists on Art* (1945), p. 77.

Figure 6–1. Tympanum, Central Portal of Façade, Cathedral of Autun

one figure, which is superimposed by its mere relative size
on a rhythmic mass of lesser figures

(Denis Grivot and George Zarnecki, *Gislebertus: Sculptor of Autun* [New York and Paris: Orion Press and Trianon Press, 1961].)

Figure 6–2. Ancient Fish of Hammered Gold

figures . . . decorated with other images, and even with little repetitions
of themselves

(Collection of the Staatliche Museen, West Berlin.)

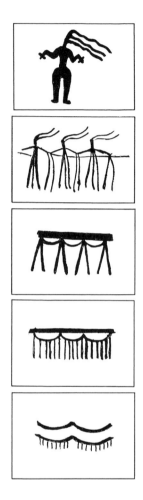

Figure 6–3. Garland of Dancing Women
to telescope two rhythms into one
(After André Parrot, *Archéologie Mésopotamienne* [Paris: Albin Michel, 1946–52].)

may be seen at once as a whole, and its parts as articulations of the whole. This character usually imparts itself to the entire work, so it comes with a single impact. The semblance of objects serves to objectify the total expressive form.

In European art the imitation of nature became an obsession, beginning with the Renaissance and culminating in the present popular standard of so-called "photographic truth to nature." But the great masters, even while they wrote about "imitation of nature," always knew that the abstractive power of representation lay not so much in giving virtual forms the semblance of being objects, as in seeing and using their resemblance to objects. So Leonardo, in a famous passage of his *Libro di Pittura*, advises painters to gaze at the texture and cracks of old walls, "where you can see all sorts of battles and swift actions of strange figures, facial expressions, and dress, and numberless things, which you can then render in complete and good form; which appear in walls and other such mixtures, just as with the sound of bells, you can hear in their tolling any name or word you fancy."[34]

The abstraction of gestalt from an actually given object by seeing it as an image of some entirely different thing—a plant, a roof, a boat, a human or animal figure—is a very ancient source of representational art. In a cavern at Commarque, in France, there is a paleolithic sculpture of a horse's head which was obviously suggested by the shape of the protrusion which served as the block.[35] The same sort of visual reinterpretation may be seen in the animal forms the Huichol Indians of Mexico produce, by a few slight touches, from uprooted bamboos, with their sharply recurved main root and lesser, opposite roots broken off short; Carl Lumholtz, in his exhaustive study of the symbolic objects fashioned by that tribe, has pictured the convincing deer heads on the batons of their nature goddess, "Grandmother Growth," which nevertheless are still perfectly recognizable roots (according to his account, the batons are her deer with plumes, also serpents, also her arrows).[36] The same obvious transformation was made thousands of years ago by

[34] *Op. cit.*, p. 66.

[35] S. Giedion, *The Eternal Present: The Beginnings of Art* (1962), p. 390.

[36] "Symbolism of the Huichol Indians" (1907), p. 50: "The roots of the bamboo sticks, having frequently three prongs, assume, with very slight exercise of the imagination, the shape of some animal with snout and ears or horns (or plumes), the cane accordingly forming the body. The prongs have natural transverse markings suggestive of snake-scales, and excrescences which might be taken for eyes or teeth; but the suggestive appearance of the stick is, besides, often improved on by cutting, painting, and adorning."

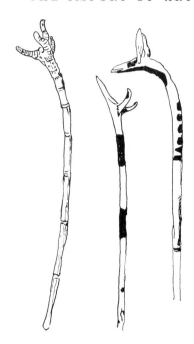

Figure 6–4. Batons of Grandmother Growth

*the animal forms the Huichol Indians of
Mexico produce, by a few slight touches,
from uprooted bamboos*

(After Carl Lumholtz, "Symbolism of the Huichol
Indians," *Memoirs of the American Museum
of Natural History*, III [1907], 91.)

the civilized craftsmen of Egypt, to produce similar animal-headed ceremonial staves and scepters for their divine personages[37] (cf. Figs. 6–4
and 6–5).

In the realm of plastic art, quite apart from symbolic intent, the
intuitive seeing of one thing in another is an invaluable means of abstracting not only shapes, but nameless characteristics. The conception
of one thing consciously or even unconsciously held in mind serves as a
scaffolding for the envisagement of the other, so the main lines of
representation of that other borrow their motivation from both; the
resulting gestalt "is and is not" its avowed object. But instead of giving

[37] William C. Hayes, *The Scepter of Egypt; a Background for the Study of the
Egyptian Antiquities in the Metropolitan Museum of Art* (1953), Vol. I, p. 285.

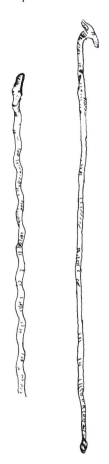

Figure 6–5. Egyptian Scepters

the same obvious transformation was made
thousands of years ago by the civilized
craftsmen of Egypt

(After William C. Hayes, "Ceremonial Staves and
Scepters, 54½–61¾ in.," *The Scepter of Egypt*
[New York: Harper and Brothers, 1953]. Used
by permission of Harper & Row, Publishers.)

it a profusion of meanings, as the religious symbol-user tends to do, the
artist sees the gestalt emerge as something in its own right; and if he
imposes another interpretation on it, he does so to see it undergo some
further transformation, until it yields elements of pure design. Such

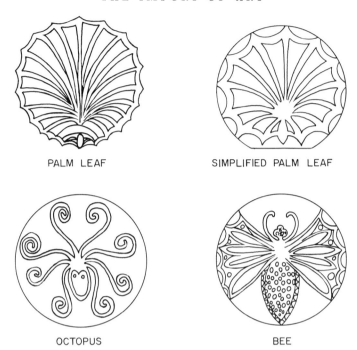

PALM LEAF   SIMPLIFIED PALM LEAF

OCTOPUS   BEE

Figure 6–6. Greek Medallions of Hammered Gold

(After sketches made by A. W. Dunbar in the Archeological Museum in Athens. Reproduced by permission of the *Journal of Aesthetics and Art Criticism*.)

elements, then, can be developed through a wide range of motifs, or used without any representational intent (see Fig. 6–6).[38]

In Europe, the ascendency of Greek art and letters gave Greek aesthetics, too, an unchallenged sway, so that its key concept of *mimesis*, translated literally as "imitation," became generally accepted as the obvious aim not only of plastic art but of literature and even music. Representation of objects and actions was measured by its compliance with the laws of optical projection instead of by its service to more fundamental artistic ends. In other parts of the world, however, where the arts also have firm traditions, the use of natural forms especially in painting is differently conceived. Its purpose is understood to be the

[38] Compare Henri Focillon, *The Life of Forms in Art* (1948; original French edition, 1934), p. 4: "By copying the coils of snakes, sympathetic magic invented the interlace. The medical origin of this sign cannot be doubted. . . . But the sign itself becomes form, and in the world of forms, it gives rise to a whole series of shapes that subsequently bear no relation whatsoever to their origin."

abstraction of elements of design which may be found in the most diverse contexts in nature, by representing objects in terms of lines, large and small dots variously produced, etc., derived from the envisagement of other objects. In the Japanese canon, for instance, there are eight methods of painting rocks, each based on the use of a highly adaptable unit of design, gathered from some phenomenon where it is clear and striking, such as axe strokes on a tree, the shape of alum crystals, hemp leaves, the wrinkles in a cow's neck. Henry P. Bowie, in his book *On the Laws of Japanese Painting* (1951), after listing these standard devices, said of them: "These eight laws are not only available guides to desired effects. . . . They are symbols or substitutes for the truth felt. Nothing is more interesting than such art resources whereby the sentiment of a landscape is reproduced by thus suggesting or symbolizing many of its essential features."[39]

Such conventions, based on the principle of seeing one thing figured in another, are often really amazing in their organizing power. So he adduces, for instance, "the fish-scale pattern used in painting the clustered needles of the pine tree or the bending branches of the willow; the stork's leg for pine tree branches; the gourd for the head and elongated jaws of the dragon. . . ." And the borrowing goes on from image to image, and comes back full circle to "the dragon's scales for the pine tree bark" (Fig. 6–7). From there it passes from the realm of objects altogether, to the recognition of pure line motifs, as he says further: "In addition to these, certain of the Chinese written characters are invoked for reproducing winding streams, groupings of rocks, meadow, swamp, and other grasses and the like.

"Of course the exact shape of the various Chinese characters here referred to must not be actually painted into the composition but merely the sentiment of their various forms recalled."[40]

The principle of gestalt or articulation of forms has intimate relations with the principle of dynamic structure or tensive design in all the arts. Either may predominate in evidence, but the life of every design springs from some interaction of these two creative processes. They are not opposed to each other as "motion and rest," for tensions arise from the very existence of closed forms,[41] from within them and from their outward relations, and rest or resolution may result from balance or con-

[39] P. 55. The author of that excellent little book had a thorough Japanese academic training in art.
[40] *Ibid.*, p. 58. There follows a discussion of the alleged derivation of Japanese painting from Chinese calligraphy, which Bowie calls "this rather wide-spread error."
[41] Cf. the last paragraph quoted from Hofmann, above, p. 158.

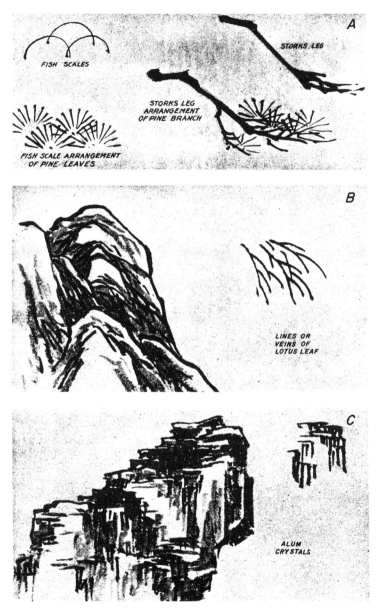

Figure 6–7. A, Motifs for Painting Pine Trees; B and C, Motifs
for Painting Rocks

*the abstraction of elements of design which may be found
in the most diverse contexts in nature*

(Henry P. Bowie, *On the Laws of Japanese Painting*
[New York: Dover Publications, Inc., 1951].)

vergence of tensions. They are aspects, abstracted from the actual sense of life in different and incommensurable ways. Intellectually we can conceive them only by turns, though perhaps very quick turns; but in the visual arts we see them, in the poetic art we understand them, in music we hear them, simultaneously. That is a fundamental fact of artistic structure, and one of its differences from discursive form. Within a work of art this sets up a level of deeper tensions than those which we perceive as such: a permanent tonicity, which pervades the work and is the most elementary source of its apparent life, or "livingness."

Gestalten, whether they are created by deployment of tensities or by an initial conception of figure, are multivalent elements, serving many purposes at once. Even the one interest we are pursuing in the present chapter, namely, abstraction, they serve in more than one way; for besides their "isolating" abstractive function, and besides their dialectical interplay with dynamic elements in a work, they have a character which has been called the "physiognomic" aspect of presented forms, and which is one of the most interesting puzzles in the psychology of perception today. It seems to be a primitive sort of "intrinsic expressiveness." Volkelt described this phenomenon as follows: "The feeling-content emanates from the perceptual content. By virtue of the existence of the perceptual content, the feeling-content exists, too."[42] What he specially remarks here is that some percepts convey ideas of internal feeling, without being cathected by association with any emotive experiences. Why they should do so, and how their so-called "feeling-content" is received, has been a bone of contention among students of human mentality, especially social philosophers, since the beginning of our century. Like all psychological problems of feeling, this issue has been clouded by two philosophical obstructions—the simple, classical concepts of sensory reception and its intracerebral effects (data being associated, infected with emotion in the process, stored away and occasionally retrieved as memories), and the influence of ethical concerns, which stresses the social uses of feeling-perception before the process itself is understood, and in this way slants its investigation and interpretation toward premature, special problems and axiologically rather than scientifically suggestive wordings. The resulting distortion of the basic psychological issues is evident in the writings of Klages, Allport and Scheler—to mention only the most important—where the question of "physiognomic perception" becomes entirely one of interpersonal understanding.

---

[42] Johannes I. Volkelt, *Das aesthetische Bewusstsein; Prinzipienfragen der Aesthetik* (1920), p. 93; quoted by Klaus Holzkamp, "Ausdrucksverstehen als Phänomen, Funktion und Leistung" (1956), p. 299.

Klaus Holzkamp summarized the arguments centering about direct feeling-perception in his valuable article,[43] wherein he pointed out the difference between the recognition of physiognomic self-expression in other persons and the perception of "intrinsic expressiveness" in sensory presentations as such: that in the former, feeling is always understood as actually felt by someone else, but in the latter it is experienced as "impersonal," not actually besetting anyone, not even oneself.[44] How this is possible remains unsolved. The well known Lippsian theory of empathy (*Einfühlung*) probably contains a good deal of truth; but it really does not go beyond the recognition and naming of a phenomenon. The somewhat older hypothesis,[45] that the perception of emotional expression leads the observer to an unconscious, merely incipient imitation of the fleeting act, and that the resulting faint tensions involve an equally faint feeling by which he understands what is passing in the other person, may also prove applicable, especially in some intriguing contexts in which we shall meet it later. The most conservative doctrine, i.e., closest to the orthodox empiricist standpoint, is that of experientially learned symptom interpretation without any element of empathy, sympathy or emotional contagion. It is also the weakest; its chief recommendation is that it does not strain any accepted modes of thought.[46] All these theories treat of the interpersonal appreciation of self-expression; as Holzkamp puts it, the "functional key-problem" for all of them is: "How can one's own experiences appear to belong to somebody else?" This question, as it stands, is unanswerable; none of the theories devised to meet it directly has any real constructive power. None of them, therefore, has gained general acceptance, or found its way into the psychological laboratory.

It is different, however, with the question of "physiognomic perception" of visible or audible gestalten as such: shapes, patterned sounds (bird calls, mechanical noises, brooks, also tunes and verses). Feeling perceived in this fashion is not usually imputed to the thing that conveys it. It does not seem like the self-expression of another being, except in the case of bird songs or animal features, and in some other cases, per-

---

[43] *Op. cit.*   [44] *Ibid.*, pp. 300–1.

[45] Holzkamp (*ibid.*, p. 305) traces it to William Benjamin Carpenter's *Principles of Mental Physiology* (1874).

[46] Holzkamp, whom I have followed here in surveying the major theories of feeling-perception, attributes the last one to Gruhle. It is also in keeping with the even more sweeping piecemeal-learning theory of D. O. Hebb, whose brilliant neurological hypothesis, set forth in *The Organization of Behavior: A Neuropsychological Theory* (1949), is followed by an unimpressive psychology.

haps, to children; and where such imputation is made, it is not essential. The veery's note is one of the most moving sounds in nature; yet it loses none of its emotional character when we realize that the bird probably is putting no other feeling into his call than the cheery robin or the scolding jay.

Several psychologists—Klaus Conrad, O. Kroh, Friedrich Sander and most notably Heinz Werner—have reflected and experimented on the conditions and causes of the non-social phenomenon of "physiognomic" perception, which often precedes or even replaces perception of physically describable sensory forms. So Werner reports that according to Sander's findings ("Über Gestaltqualitäten," 1927), "perception is global first, in contradistinction to a later stage at which the parts become increasingly more articulated and integrated with respect to the whole. Furthermore, much of the initial perceptual quality is dynamic, 'physiognomic'; feeling and perceiving are little differentiated, imagining and perceiving not clearly separated."[47] Conrad, checking Sander's work in his own researches, found that in peripheral vision, too dim light or too brief tachistoscopic exposure, a figure of bright lines on dark ground "loses its structure, . . . but gains a sort of physiognomy (Werner). Certain physiognomical qualities dominate the structural qualities."[48]

Under normal circumstances of adult life, the passage from an initial impression of "intrinsic expressiveness" to perception of "primary and secondary qualities" has become automatic and practically instantaneous in most people. But in childhood that process is slow, and in some perceptual experiences may not reach completion at all.[49] According to Oswald Kroh, the spontaneous interpretation of objects as expressive forms belongs to an early level of experience, the time of learning to distinguish and organize the data of the outer world, when autogenic activities still mingle freely with peripherally engendered ones, so that mental functions are not yet felt sharply as subjective or objective. Furthermore, they are transformed into a presentational datum which "mirrors" their dynamism and appears as its expression.

Here, certainly, the concept of symbolic transformation is employed, and all but named. The "transformation" operating spontaneously and

---

[47] "Microgenesis and Aphasia" (1956), p. 347.
[48] "New Problems of Aphasia" (1954), p. 495.
[49] See Holzkamp, op. cit., p. 301: "Nach Kroh gehört das physiognomische Wahrnehmen zu einer 'Frühschicht' . . . , in der das Zuständliche unserer Erlebnisweise sich in eine anschaubare Gegenständlichkeit wandelt, die den Zustand noch einmal ausdruckshaft spiegelt. . . . Das sinnliche Bild und der Sinn sind noch nicht von einander getrennt." Holzkamp's reference to Kroh is to an unpublished paper of 1942.

involuntarily at a mental level of sheer perception is precisely the projection of feeling—vital, sensory and emotive—as the most obvious quality of a perceived gestalt. To take up this sort of emotive import is a natural propensity of percepts in childhood experience. It tends, also, to persist in some people's mature mentality; and there it becomes the source of artistic vision, the quality to be abstracted by the creation of forms so articulated as to emphasize their import and suppress any practical appeal they would normally make. This is the just ground for the frequent assertion that an artist must see and feel as a child; and it is, I think, the only ground for that widely misused statement. For he must not "feel as a child," and project childish feeling, but only translate feeling into perceivable quality, by intense concentration on the potentialities of forms to symbolize it even for people who no longer see actual things "physiognomically."

Klaus Conrad has made an interesting though incidental observation on the process of artistic projection, prefixed by an account of various phases which occur in the genesis of a closed and objectified form in natural perception, as experiments with gradual illumination, gradually increased tachistoscopic exposures, etc., have shown. A figure thus slowly apprehended seems at first to move and vary; "it shows a certain flickering and wavering life, which is mostly surrounded by a more compact contour. Thus, a compact contour with a flickering core is set up. This tendency to restlessness we call fluctuation. In tachistoscopy . . . on being exposed for the second time, the same figure is seen as totally different from the impression received during the first exposure." This figure, furthermore, "appears unfinished and incompletely evolved. . . . In the subject, such an experience gives rise to a feeling of tension, and the tendency to give the figure finality. Then, again, one cannot grasp such a figure at will . . . but has to accept it as it inflicts itself upon one. One's mode of experience is changed from the critical to the receptive. I call this, the loss of the degrees of freedom. Only when the conditions of excitation have reached such a degree that the figure appears final, structured and finished, contrasting clearly and distinctly with its background and ceasing to vary in a fluctuating manner, does this tension slacken. The figure detaches itself, so to speak, as an object from the subject, it is released from a much stronger subject relationship, faces us coolly and remotely, and only then are we in a position to perceive the figure freely and at will."

A little further on, but still in connection with the above passage, he comes to speak of the artistic process. "One is bound to assume," he concludes, "that a creative artist . . . conceives at first a kind of bud of

the gestalt. The artistic achievement is thus not fully given and structured in the beginning, but seems to be at first a process pregnant with
possibilities, without structure but with a strong physiognomy, fluctuating and without definite shape, not clearly detached. The subject [i.e.,
the artist] is then charged with an impulse to elaborate the process, but
without complete freedom, as [he would be] in confronting the finished
creation. Only when the work stands fully finished in front of its creator
all this has changed, it is fully structured, clear and remote, is experienced with finality and can be perceived at will, with all degrees of freedom. The . . . *vorgestalt* has evolved into the *endgestalt*."[50]

The essence of the artist's task, however, is not only to create a form
"fully structured, clear and remote," but in so doing to hold in that final
objectified *endgestalt* all the phases of the evolving vision: the core
of "flickering and wavering life" within the compact contour, the
tension, the ambiguity, the sense of potentiality and non-finality, above
all the "strong physiognomy" and the cathexis, which appears in an
actually emerging gestalt as subjective involvement, incomplete detach-
ment. The artist's "realized" form has to retain all these experiential
aspects which an ordinary perceptual datum gives up as it reaches its
full objective status; because the ordinary percept becomes a thing for
the percipient, but the artist's creation becomes a symbol.

Since art is a symbolic expression, every aspect of life which it can
render has to be transformed in terms of its complex abstract presentation; any mode of abstraction that the human brain has evolved, therefore, may be drawn into the processes of our self-comprehension. Even
the discursive mode is not necessarily excluded, though its misuse is
such a constant danger that its happy employment as an artistic device
bespeaks a very sure expressive aim. The sort of conception that guides
the primitive art impulse is versatile and unfettered, and finds symbolic
possibilities in practically all aspects of actual experience. Just as the
simplest given element sets up tensions in its surroundings, so everything
that enters into a work has some physiognomy or at least the seed of
physiognomic value; not only the gestalt that emerges in acts of perception, but simpler elements, recognizable colors, sounds, tangible
surfaces, heat, warmth, coolness, iciness, light and darkness. There is a
reflection of inward feeling in the most typically outward, objective
data of sensation; their subjectification is practically started with their
very impingement on the specialized organ that receives them. Their
character is never as fixed and simple as the distillations our conventional store of qualifying adjectives has made from them.

[50] *Op. cit.*, pp. 495–96.

In previous chapters there was already some brief mention of the way physical attributes or events are metaphorically employed to describe mental and moral ones and of colors which can stand in for each other.[51] Obviously there are aspects of subjective feeling which are similar enough to some sensory qualities to seem inherent in presentations of those qualities, and, indeed, to inhere in more than one sort of sense datum, wherefore some sensations have equivalent values, and can stand proxy for one another. This fact has been recognized by many aestheticians, and broadly generalized to prove the basic unity of all the different arts. The proof, however, requires a previous assumption: namely that the art symbol is constructed essentially by selection and organization of such genuinely "aesthetic" elements, so that the pattern of feeling it is said to convey (more often, to "stimulate") is an integration of the natural emotive values of colors and color combinations, jagged or smooth lines, or high and low tones and the timbres of instruments, or poetic rhythms and rhymes, consonants and vowel sounds—in short, the exciting and soothing powers of those materials, peculiar to the art in question, which David Prall designated as the constituents of its proper "aesthetic surface."[52]

Besides having stimulative or inhibitive effects on our feelings, sense data within any particular realm tend also to modify each other, so each may gain or lose by any addition or shift of elements. Some colors advance and others recede, some intensify each other by contrast, some clash, some cancel each other.[53] Musical tones in low registers are intrinsically voluminous, those of high pitch seem thin and more penetrating; and the mutual influences among tones are, if possible, even greater than those which operate among colors. So artistic composition, conceived as the combination of sensory elements, appears to spring from the nature of the chosen materials just as the feeling content does, and the finished work is frequently regarded as an elaborated stimulus pattern evoking a corresponding emotional experience.

If this sort of composition is possible in any sensory mode, and the various modes have equivalent elements, then it should be possible to translate works of painting into music, music into dance (which Dalcroze attempted), dance into poetry, etc. The first step, however, must be to determine the equivalents which elements in one domain of sense have in some other domain. That is a psychological problem,

---

[51] See Chapter 3, pp. 62–63; Chapter 4, p. 106.

[52] For a discussion of this concept see F.F., pp. 54–59.

[53] Such relations among colors, and their uses in painting, are discussed by Leonardo in the *Treatise on Painting*, esp. Bk. II (most recent English translation, 1956).

and was properly undertaken in psychology laboratories. The investigation of sensory parallels has been carried on for some decades now, and certain similarities have become stock examples; the likeness of bright red and the tone of a trumpet, for instance, supposed to have been remarked in the eighteenth century by a boy cured of congenital cataracts, has become a byword in the literature (almost as banal as the philosopher's Ivory Tower).[54] For my own part, I have never been able to corroborate any such direct resemblance. Coming suddenly out of dark woods upon a cardinal flower in sunlight, one is certainly struck and arrested by the flaming red, yet it does not hit like a blast from a trumpet. There are uses of color, however (and not only bright red), which make "flaring" and "blaring" seem very closely related qualities. We do not know what it was about the boy's impression of redness that reminded him of a trumpet tone; it may have been richness quite as easily as startling impact.

The results of controlled experiment have not provided any usable parallels for a more scientific construction of "color organs" or "tone poems," although there are many records of so-called "synesthetic" responses.[55] Some significant relationships certainly exist between data of different sensory orders, yet I doubt the evidence of almost all the synesthesia experiments; the wording of the protocols usually shows all too clearly that a large dosage of suggestion was administered with the instructions to the experimental subject, or with the questions. When more than half of a group of subjects find various musical combinations "like peaches melting in the mouth,"[56] or when people have to be

[54] Charles S. Peirce even declared that "everybody who has acquired the degree of susceptibility which is requisite in the more delicate branches of reasoning . . . will recognize at once so decided a likeness between a luminous and extremely chromatic scarlet . . . [and the blare of a trumpet] that I would almost hazard a guess that the form of the chemical oscillations set up by this color in the observer will be found to resemble that of the acoustical waves of the trumpet's blare" ([1931], Vol. I, p. 155, Phenomenology, sec. 7, "The Similarity of Feelings of Different Sensory Modes").

[55] A. Argelander, in her monograph, *Das Farbenhören und der synaesthetische Faktor der Wahrnehmung* (1927), presents a bibliography of 466 titles to that date.

[56] See E. M. Edmonds and M. E. Smith, "The Phenomenological Description of Musical Intervals" (1923). The observers were given instructions to make a phenomenological report on perceptions of simultaneously sounded tones, and all sooner or later described their sensations in terms of touch and taste. The experimenters remark that in the course of the tests, "In general there was an increase in the number of analogies and in particular of analogies of the taste-touch order" (p. 289). Above, they reported: "All the Os had difficulty at first either in understanding the instructions or in assuming the phenomenological attitude. Eventually, however, after repeated trials, four of the five succeeded . . . in finding terms that characterized every one of the eight intervals . . ." (pp. 287–88).

deliberately trained to register synesthetic responses,[57] the value of their reports seems to me highly dubitable.

Aestheticians, however, are generally content to accept the notion of parallels not only between colors and sounds, sounds and gestures, gestures and words, but even between techniques belonging to different arts, without further investigation.[58] So, for instance, I. A. Richards simply posits the analogy of color relations and tonal (harmonic) relations,[59] and says categorically: "Representation in painting corresponds to thought in poetry."[60] De Witt Parker, likewise, did not hesitate to compare the elements of different arts, taking the equivalence of their respective functions for granted; and those elements are certainly conceived in the most obvious fashion when he makes his observation: "A picture and a poem are . . . essentially the same, so much so that it is strictly accurate to call a poem a picture in words, or a picture a poem in color and line. And it is this way of conceiving a picture that I wish to adopt at the outset."[61]

"Just as any poem, when you abstract from its meaning, may be

[57] Karl Zietz, in an article: "Gegenseitige Beeinflussung von Farb- und Tonerlebnissen" (1931), which contains many important ideas and some findings, also invited doubt of the synesthetic ones by saying: " 'Schwerfällige' Vpn. [Versuchspersonen] müssen erst eingeübt werden, ehe es zu positiven Ergebnissen kommt. Bei ihnen finden sich daher die erfolgreichen Versuche erst in der Mitte oder am Ende der Versuchsarbeit . . ." (p. 314). Since the practice consisted of constant requests to "identify" with the tones and colors and relate them to each other, the experiments may have been better tests of suggestibility than of sensuous connections.

[58] So, for example, W. Schmeer, in a little book entitled: *Farbenmusik und die Übernahme der Gesetze der Musik in alle Farbenkünste; mit Schlüssel und Farbakkordgreifer* (1925). And Jean D'Udine, whose *L'art et le geste* (1910) contains many careful distinctions, yet says with assurance: "Les lignes jouent, dans les arts plastiques, exactement le même rôle que le rythme dans l'art des sons" (p. 102); and further: "Un ornement qui tourne, une colonnade qui s'incurve, un motif qui se répète, une spire qui augmente ou diminue de rayon comportent exactement les mêmes qualités et les mêmes propriétés expressives qu'un motif musical rythmique, qu'un dessin 'ostinato' d'accompagnement ou que l'altération d'un mouvement sonore par un 'rallentendo' ou un 'accelerando' " (p. 106).

[59] *Principles of Literary Criticism* (2d ed., 1926), p. 156. Yet he declared just as pontifically only twenty pages earlier: "There are no gloomy and no gay vowels or syllables, and the army of critics who have attempted to analyze the effects of passages into vowel and consonantal collocations have, in fact, been merely amusing themselves."

[60] *Ibid.*, p. 160. The whole of chap. viii, "On Looking at a Picture," rests on the acceptance of this and other "obvious" parallels; the one quoted above proves very quickly to be deceptive when the *creative use* of representation in painting and the *created image* of thought in poetry are compared.

[61] *The Analysis of Art* (1926), pp. 65–66.

regarded as a pattern of words, so every painting, when you abstract from representation, may be considered as a pattern of color and lines."[62]

These judgments are not the result of an analysis of art such as the title of his book promises, but are made—as he says—"at the outset." Being based on the most superficial observation of the materials with which the various arts usually deal, and which consequently anyone can check with his own experience to the very slight degree required for illustration, they appeal readily to common sense.[63] But the real test of an isomorphy extending over the entire structures of two compared works is, of course, the equivalence of their respective artistic achievements; and here the parallel always breaks down. The visual renderings of musical compositions by the "color organ" are not works of art, and the greatest pieces of music "translated" into pictorial terms yield images that are not remotely their artistic equivalents.[64] An extension of this observation to the domains of poetic and graphic arts is justified by the exhibits in a rare little book by Marjory B. Pratt, which contains renderings of some of Shakespeare's sonnets in terms of black rectangles of various lengths and positions on a white ground. The figurations are all coherent and fairly interesting designs, and in several cases it is not hard to recognize which sonnet is reflected in a particular pattern; but the poetic value of the originals is not transferred (Fig. 6–8).

---

[62] *Ibid.*, p. 73.

[63] Artists, especially, have generally favored this doctrine, because they are not called upon to demonstrate it, and it falls in well with their natural tendency to see all other works of art in terms of their own. Rudolf von Laban, for example, calls music "sonorous gesture," and calligraphy "the dance of the hand" (*Die Welt des Tänzers; fünf Gedankenreigen* [1920], p. 15). Robert Schumann wrote in his *Dicht- und Denkbüchlein*, "Die Aesthetik der einen Kunst ist die der anderen; nur das Material ist verschieden" ([1914], Vol. I, p. 26). And a much more recent artist-aesthetician, Aleksandr Sakharov, in his *Reflexions sur la danse et la musique* (1943), declares: "Il existe une musique des mouvements, tout aussi bien qu'une musique des couleurs, une musique des parfums, des lignes, des formes.

"La peinture, la sculpture, la danse, ne sont-ce point la musique des couleurs, des formes, des mouvements? Je n'ai jamais su distinguer un art des autres que par le moyen d'expression dont il se sert" (pp. 15–16).

[64] Cf. Max Eisler, "Das Musikalische in der bildenden Kunst" (1926), p. 321, discussing the color patterns offered as "translations" of music in Rainer's *Musikalische Graphik*: "Ein selbstständiger, unmittelbar einleuchtender und ergreifender Kunstwert kann diesen Graphiken kaum zugesprochen werden . . . weil auch das Reifste von seiner Beziehung zur Musik lebt und so, soll es vom Betrachter verstanden werden, die musikalische Begleitung fordert. Man müsste es im Konzertraum . . . während das Orchester spielt, in leuchtender oder dämmernder Transparenz an die Wand bringen, um die Gleichzeitigkeit des Farbenhörens zu geniessen." Walt Disney did as Eisler advises, but *Fantasia* is still no work of plastic art.

Not mine own fears, nor the prophetic soul

Of the wide world dreaming on things to come,

Can yet the lease of my true love control,

Supposed as forfeit to a confined doom.

The mortal moon hath her eclipse endured,

And the sad augurs mock their own presage;

Incertainties now crown themselves assured,

And peace proclaims olives of endless age.

Now with the drops of this most balmy time

My love looks fresh, and Death to me subscribes,

Since, spite of him, I'll live in this poor rhyme,

While he insults o'er dull and speechless tribes:

    And thou in this shalt find thy monument,

    When tyrants' crests and tombs of brass are spent.

Figure 6–8. Syllabic Design of Shakespearean Sonnet

(Marjory B. Pratt, *Formal Designs from Ten Shakespeare Sonnets* [privately printed, 1940].)

A Chapter on Abstraction

There are, in the main, two methods of "translating" works of art into another medium: one is to ply the materials of the second art while enjoying the piece to be translated—painting without preconceived plan while listening to music, dancing under the same conditions, or improvising music while watching a dance, etc. The "translations" made by this method of personal exposure are usually amorphous products, because the structural virtues of the piece used as a stimulus object are not preserved in the emotional response that records itself directly in the expressive reaction. What the reaction expresses is not the nature of the presented piece but of the mood it induces, which at that point is unformulated, so there is no structure to reflect—let alone rendering that of the original work. The other method is the one adopted by Dr. Pratt in regard to poetry and by several people similarly in regard to music: namely, to find elements in the given work which may be represented by comparable elements of another sensory order. That means, of course, that a parallelism of elements must either be found to exist, or must be conventionally established. Dr. Pratt's configurations are conventional; they serve to abstract one structural factor—the syllabic pattern;[65] the product is schematic, showing the verbal skeleton, and makes no pretense of being anything but a projection of the word pattern of the sonnets. The development of the poem—the "curve," as Hulme called it[66]—is not rendered. But most of the translators are more ambitious; their hope is to reproduce the work itself in another medium. This hope has never been realized, even in serious efforts to take many structural elements into consideration, as for instance Kandinsky's proposals for a graphic translation of musical works in terms of dots to express rhythmic groupings, lines for melodic direction, their varying thicknesses for differences of timbre, their sharpness or "brilliance" for degrees of loudness.[67]

Kandinsky's principles of projection illustrate one of the major difficulties of such undertakings: the interpreter conceives works technically in only one of the two arts he compares. In this case, he sees plastic

[65] A letter from Dr. Pratt provided the following explication of her procedure: "I simply decided quite arbitrarily that each vowel sound (not letter, of course, but the phoneme) should be represented by a simple shape—triangles turned various ways, etc. Then I made the designs ten across for the syllables and fourteen down, the dimensions of the sonnet form. Just for fun I let the syllables run into each other, the solids in the case of the designs on the inside of the book and the outlines in the case of the one on the cover."

[66] "A series of images is the curve of a poem" (T. E. Hulme, quoted by Herbert Read in his Princeton lectures [1953]).

[67] *Point and Line to Plane* (1947; original German edition, 1926), p. 43.

works with a professional eye, but what he hears in music is what an amateur hears: rhythmic figuration, timbre, general levels of pitch and dynamics. Even the first of these items—the only musically essential one —he pictures only as a rhythmic motif, with no notice of its musical use. His rendering of the opening figures of Beethoven's Fifth Symphony (Fig. 6–9) conveys nothing of the impact produced by the minor third in the repetition of the figure, nor the suspense in the three eighth notes due to the fact that the metric accent (itself an up-beat) is on the second one, i.e., that the group is not a triplet. Very tiny differences in the sizes of the dots could render that accent graphically, but would certainly create no such effect for the eye as a tiny difference of accent does for the ear.

This last-mentioned discrepancy is the really fatal weakness in the alleged parallelism of the arts; there are no parallel elements that correspond in any regular fashion. There are, in fact, no elements that keep their character within the art to which they belong, or even in a piece where they recur under different circumstances. The structure of a work of art is nothing as simple as an arrangement of given elements by half a dozen, or even a dozen, combinatory operations. The techniques of abstraction and projection are largely derived from the opportunities offered by the material, often on the spur of the moment, in a situation that may never be repeated. Those are the highlights of expressive power. But even the most familiar ways and means require other principles of projection than combination of sense data.

Does this mean, then, that there are no relations between different arts except their basic sameness of purpose, to present the morphology of feeling? No; I think there are other relations, and even such as rest ultimately on sensuous equivalences, but not of whole categories of sensations to other whole categories, permitting of systematic substitutions. Such correlations as that of musical timbre with thickness of lines, and loudness with "brilliance," are largely if not wholly made by fiat. One could certainly as well pair thickness of line with loudness of tone, and let "brilliance" correspond to pitch; then the graphic work made on one principle of translation would represent, on the other, quite a different musical composition.

Despite all these difficulties, however, it is hard to shake off the conviction that "sensuous metaphor" does play some important role in art. The fact that we misconceive its functions and falsely generalize its rather special and striking occurrences does not militate against its reality; only we have to determine what different sensations really do have an emotive character in common, and how artists actually exploit

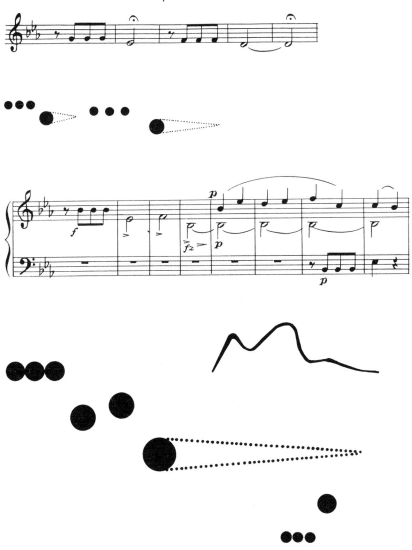

Figure 6–9. Wassily Kandinsky, Opening Measures of
Beethoven's Fifth Symphony

*very tiny differences in the sizes of the dots . . . would certainly create no
such effect for the eye as a tiny difference of accent does for the ear*

(*Top,* the first measures translated into points; *Bottom,* Theme 2 translated into points
and line. From Wassily Kandinsky, *Point and Line to Plane* [New York: Museum of
Non-Objective Painting, 1947]. Reproduced by permission of the
Solomon R. Guggenheim Museum.)

[187]

this fact. So it occurred to me, many years ago, that such parallels might be revealed by a study of the words which artists in different fields borrow from each other, when, for instance, musicians speak of the "color" of the woodwinds, and painters "harmonize" the "tones" of either monochrome or polychrome canvases. We have all heard of the "orchestration" in (say) Braque's or Picasso's collages; Le Corbusier says, "The cathedral is not a plastic work; it is a drama; . . ."[68] he calls the arrangement of masses a "rhythm,"[69] and architecture itself "a matter of 'harmonies.' "[70] Most of us understand his language. We are used to these raids on neighboring vocabularies, from all our reading of musical program notes, art magazines and even exhibition catalogues. Assuming, then, that the practice was an intuitive carrying over of words from their literal meanings to more general ones, i.e., a metaphorical extension of their sense, bespeaking an implicit analogy, it seemed reasonable to seek the logic and the limits of sensuous metaphor in such spontaneous studio language, and to expect a wide survey of it to establish the parallelism of all the arts on a factual basis.

This meant, first of all, to trace the habit through many writings, new and old. The literature of music is especially rich in examples; for, as one writer on music remarked, the vocabulary of music criticism is "all borrowed from other arts."[71] In C. P. E. Bach's famous essay on piano playing[72] the metaphors certainly flow freely; visual, tactual, kinetic, geometric, mental and moral characters are attributed to the effects of musical performance. Tones are colorful (*"bunt"*), lustrous (*"glänzend"*), keen (*"scharf"*) or peevish (*"verdriesslich"*); an excessively loud trill is impudent (*"frech"*); proper rendering should be "now caressing, now scintillating";[73] the pianoforte, however, induces a habit of "playing in a single color" (*"in einer Farbe zu spielen"*). We meet with "walking" and "leaping" musical ideas.[74] Sometimes the intersensory transfers reach a danger point, as in the warning that a passage will

---

[68] *Op. cit.*, p. 30.

[69] *Ibid.*, p. 49: "Arrangement is an appreciable rhythm which reacts on every human being in the same way."

[70] *Ibid.*, p. 19.

[71] J. Estève, "La pensée musicale" (1926), p. 238. His examples, however, are "warm" and "cold"; from what *art* could they be said to be taken?

[72] *Versuch über die wahre Art das Klavier zu spielen* (1753). References are to a reprint of the second (1759) edition.

[73] *Ibid.*, p. 27: "Man bringe bald eine schmeichelnde, bald eine schimmernde Manier an."

[74] *Ibid.*, p. 13. Bach points out "die Verschiedenheit der Gedanken, vermöge welcher sie . . . bald gehend, bald springend sind."

"sound unpalatable" ("*unschmackhaft klingen*") unless special precautions be taken.

The borrowing is not limited, however, to musical criticism and theorizing. Ruskin, for instance, was obviously using the current studio language when he said, "Every hue throughout your work is altered by every touch that you add in other places; so that what was warm a minute ago, becomes cold when you have put a hotter color in another place, and what was in harmony when you left it, becomes discordant as you set other colors beside it."[75]

In very similar fashion, Joseph Conrad, in his prefaces[76]—classics of literary comment—not only uses sensuous metaphors as freely, but chooses them more definitely from other arts, especially music and painting. So he says that *Nostromo* "will always remain my largest canvas"; and of all his stories, "Experience in them is but the canvas of the attempted picture." In the preface to *Almayer's Folly*, he wrote: "The picture of life, there [in the South Seas] as here, is drawn with the same elaboration of detail, coloured with the same tints. Only in the cruel serenity of the sky, under the merciless brilliance of the sun, the dazzled eye misses the delicate detail, sees only the strong outlines, while the colours, in the steady light, seem crude and without shadow. Nevertheless it is the same picture" (p. 37). Here he built up a deliberate figure of speech, but often he mingled his borrowed terms so they cannot be taken as allegorical, but only as direct ways of saying what is meant; for instance: "*Youth* is . . . a record of experience; but that experience, in its facts, in its inwardness and outward colouring, begins and ends in myself. *Heart of Darkness* is experience, too; but . . . there it was no longer a matter of sincere colouring. It was like another art altogether. That sombre theme had to be given a sinister resonance, a tonality of its own, a continued vibration that, I hoped, would hang in the air and dwell on the ear after the last note had been struck" (p. 73).

In view of the large literature that furnishes our stock of studio metaphors, it seemed a hopeful undertaking to trace their ancestries to the earliest writings and try to discover the most frequent intersensory transfers. But the quest soon came to an end; for the practice of borrowing or extending technical vocabularies goes back very little beyond the sixteenth century, and instead of being a primitive mode of expression, increases with the development of critical writings. It really gained momentum only in the eighteenth century, and is rampant in our

[75] *The Elements of Drawing in Three Letters to Beginners* (1889), p. 138.
[76] Collected by Edward Garnett in *Conrad's Prefaces to His Works* (1937), from which all the quotations here are made; page references are to this book.

descriptive and "appreciative" columns and captions today. In Leonardo's *Treatise on Painting*, for instance, there are hardly any metaphorical expressions. If one compares his text with the first English translation, made in the eighteenth century,[77] a little attention reveals the effects of the prevailing fashion even on such strictly limited aesthetic discourse. Leonardo refers to red as "fiery" (*"foccosa"*), but never as "warm" (*"calda"*); where his translator may speak of "deep" shadows, he says "dark" (*"l'ombre oscure"*). And where the English says that "a shower in falling, darkens the air, and gives it a precarious tincture," he said "a livid tincture" (*"livida tintura"*).

The development of music as one of the "fine arts," independent of words, ritual or dance, falls largely into recent centuries, so that its vocabulary grew up in the heyday of metaphorical usage. Perhaps that is the reason for its great wealth of borrowed terms. In the earliest treatises it is not apparent. Thomas Morley, in *A Plain and Easy Introduction to Practical Music* (1597), does not demand the achievement of high lights or shades, contours or colors. The use of loan words from any and all other arts was evidently, from its beginning, not a matter of spontaneous generalization, but of fashion; and as the fashion was born of the doctrine that all the arts are of one blood, differing only in material, but all doing the same things in directly analogous ways with their respective means, one cannot base a theory concerning the interrelations of the arts on the evidence of studio metaphor.

But the study of words, which the unsuccessful venture entailed, opened a different approach to the problem of sensuous metaphor and its significance for artistic expression. That approach starts from the etymological dictionary, and instead of proving the kinship of some or all the arts, it leads to a further understanding of artistic abstraction, as it occurs within a particular art and a particular work. The etymological phenomenon in question is the change of sense which words undergo in the course of linguistic history. The fact that they do change in meaning has, of course, been noted long ago, and some types of change recorded, such as the narrowing of originally wide meanings to special

[77] Reference here is to the first English translation, which was made by an anonymous translator, published in 1721 and reprinted in 1796, from a poorer copy of the original than the Codex Vaticanus Urbinas version discovered later; but as Rigaud, in 1802, used the same copy as the eighteenth-century translator, and his translation of the passages in question fits the Vatican version, there was presumably no discrepancy between the Italian texts. The more complete (Vatican) text is given with the German translation of 1882 (*Das Buch von der Malerei*), and in facsimile as Vol. II of A. P. Macmahon's translation published in 1956; I have not seen the other.

ones, either by some paramount application, or by moral association; also the opposite process, the extension of word meanings, terms being carried over from their original uses to variously related ones. But the interesting exhibit is the shift of meanings in words denoting sensory qualities, which seems to be based on a principle of truly spontaneous sensuous metaphor, and finally to take a turn which reveals a mode of abstraction developed only in works of art.

Words for colors are, at least in the Indo-European languages, among the most ambiguous and unstable. So, for instance, our word "blue" (German *blau*), is derived from *blavus*, the Middle Latin form of *flavus*, which means "yellow."[78] According to Walde,[79] the Russian word *Krasá*, "beauty," "ornament," in Old Slavic had the adjectival meanings "beautiful," "pleasant," "dressed in white," and is cognate with Russian *krásný*, "beautiful," "red." He lists the derivatives of the Indo-Germanic root *mel* as Greek *mélās*, "black," Latin *mulleus*, "reddish" or "purple," Breton *melen*, "yellowish," and Lithuanian *melsvas*, "bluish." Also, *madhro* or *modhro*, "blue," as cognate with Icelandic *mad̄ra*, "redness," and Old High German *matara*, "red dye"; and the various forms of *ĝhel-* (also *ghel-*), *ĝhelē-*: *ghle-*, *ĝhlō*, *ghlə*, as "shining," "scintillating," "yellow," "green," "gray," or "blue."

The basis of these apparently erratic shifts is suggested, in the last series, by the non-chromatic senses, "shining," "scintillating." Among other words derived from *mel* is the Middle High German *mal*, a spot, and Greek μωλωψ, a bruise or "black-and-blue" spot. The lexicographer remarks that color names from the root *mel* usually denote dark, impure, dirty colors, and that the "root metaphor" seems to be "discolored." The shift from Latin "yellow" to the "blue" of the northern tongues apparently followed some such route. Murray[80] connects the Middle Latin *blavus* with Old Spanish *blavo*, which means "yellowish gray"; the application to any other tint of gray is not hard to understand. A similar ambiguity exists even today in the use of the word "purple"; in English we have settled fairly well on the meaning, "a mixture of red and blue," but the German cognate word *purpur* is translated in the current edition of Heath's dictionary[81] as "purple, crimson, deep red," and *purpurrot* by the same words with the addition of "scarlet." "The purple" is used

[78] This derivation is attested by all authorities—Skeat, Walde, Pokorny, Murray, Paul and others.
[79] Alois Walde, *Vergleichendes Wörterbuch der indo-germanischen Sprachen* (1930).
[80] J. A. H. Murray, *A New English Dictionary* (1888).
[81] *Heath's New German and English Dictionary* (1939).

for a royal robe, and whatever the prince wears, that is purple. If he is a cardinal, "the purple" is scarlet.[82] The real sense of the word is "the color of majesty," the effect of grandeur. We speak figuratively of "purple passages" where literary style obviously aims at grandeur. Richness, dignity, power, also the vital warmth of blood, form the connotation of the word.

The same principle that governs the shifting of color words from one specific hue to another, namely, that all the hues it may denote seem to convey the same feeling,[83] also guides further migrations of terms to sensations of entirely different orders. The German word *hell*, "bright," "light," "illumined," is related to *hallen*, "to resound." It was used only with reference to sounds until about the time of Luther, who still usually modifies it as *hell licht* (*"am hellen lichten Tage"*).[84] Today, the expressions *heller Ton*, *helle Stimme* ("bright tone," "bright voice") are regarded as metaphorical. We would be inclined to speak, in English, of a "clear" tone or voice. "Clear" is derived from Latin *clarus* (German *klar* and French *clair*), which originally meant "shining," not "transparent" or "unsullied," as it later came to mean.[85] The intellectual sense of "comprehensible" is a still further extension, motivated by the modern

[82] *Cassell's New French Dictionary* (1934) translates *pourpre* first in the sense of "the purple," more particularly "the Cardinalate," and then gives as a special sense, "purple (colour)."

[83] A further, striking corroboration of this view is provided by Jacques Gernet in a paper read at a symposium on problems of color, and entitled "L'expression de la couleur en chinois" (1957), wherein he said: "Il y a un mot chinois, de caractère *ts'ing*, qu'on traduit d'ordinaire tantôt par bleu, tantôt par vert. Mais le caractère indéterminé de ce terme n'est qu'apparent, car il semble ne s'appliquer qu'à certains types de vert et de bleu. En effet, il existe, parallèlement à ce mot, des termes pour vert (*lu*) et pour bleu (*lan*) qui correspondent à des nuances de vert et de bleu qui ne pourraient en aucun cas être qualifiées de *ts'ing*. . . . Le terme s'applique spécialement au vert des arbres et au bleu du ciel *à une certaine époque de l'année*. Mais l'extension de ce terme est plus large, puisqu'il peut désigner des gris-bleus ou des bleu-noirs: le pelage d'un animal, les plumes d'un oiseau noir peuvent être qualifiés de *ts'ing*. Dans ce cas, ce n'est ni à un bleu ni à un vert caractérisés qu'on a affaire, mais à d'autres couleurs également particulières.

En fin de compte, il semble que ce terme, qui s'applique à des couleurs différentes mais en même temps définies, note l'impression ou l'ensemble d'émotions qu'elles provoquent" (p. 295).

[84] See Hermann Paul, *Deutsches Wörterbuch* (1935), s.v. "hell."

[85] Murray, *op. cit.*; also Hermann Paul, *op. cit.* In Italian, *chiaro* still retains the original meaning of *clarus*, even while acquiring all possible other senses. *Enenkel's English-Italian Dictionary* (1908) translates *chiaro* as "clear, bright; evident; illustrious, honest; true, loyal; limpid, distinct." As a noun, "brightness, clearness; lustre. *Chiaro di luna*, moonshine."

meaning. As *hell* has become a literal term for "light" or "bright," so has "clear" for a loud and pure sound.

According to Walde,[86] the original meaning of *krā-s*—from which he derives the words for "beautiful, white-clad," and "beautiful, red," already discussed—was "hot." To an ear attuned to American slang, the transfer presents no mystery.

The most recent edition of Webster's dictionary[87] defines "blare" as "(1) a loud, trumpet-like sound. (2) brilliance or glare, as of color." Murray lists the obsolete word, "brill: to make a sharp vibratory sound, as an insect by the rapid vibration of its wings," and as a variant of the same word [he marks it "(2)"], "brille: to shine."[88] Paul points out that *zwitschern*, to twitter, was formerly written *zwitzern*, "which also had the meaning 'to sparkle,'" and that *flink*, now meaning "quick," "nimble," is a low-German word that once meant "shining."[89] (Heath[90] still gives "bright" as its primary meaning, but unless he intends "bright" to mean "clever," usage is against him.) Skeat, after "*Tingle*, to thrill, feel a thrilling sensation," remarks: "*Tingle* is merely a weakened form of tinkle. . . . To make one's ears *tinkle* or *tingle* is to make them seem to ring; hence, to *tingle*, to vibrate, to feel a sense of vibration as when a bell is rung."[91]

These are only a few examples of the unconscious use of sensuous metaphor by which language expands as expansion is needed. Apparently the words which lend themselves most readily to metaphorical uses are those which denote light, heat, movement, or faintness, dullness, also pain and threat (which we shall come to presently). All these words either have direct application to mental states and acts, or have close cognates that are obvious extensions to psychical phenomena. It appears that light, smoothness and especially movement are the natural symbols of life, freedom and joy, as darkness and immobility, roughness and hardness are the symbols of death and frustration.[92] Those perceptual

---

[86] *Op. cit.*

[87] *Webster's New World Dictionary of the American Language* (1960).

[88] *Op. cit.*

[89] *Op. cit.*: "*Zwitschern* aus älterem *zwitzern*. Dieses erscheint auch in der Bedeutung 'funkeln.'"

"*Flink*, aus dem Nd [Niederdeutschen] in die Schriftsprache gekommen, bedeutet ursprünglich 'glänzend.'. . ."

[90] *Op. cit.*

[91] *A Concise Etymological Dictionary of the English Language* (1924).

[92] H. Paul, for instance, writes: "*froh* = *frô*, in der jetzigen Bedeutung nur deutsch, as., mnl. [*mittelniederländisch*] (das entsprechende altnordische *frar* bedeutet 'hurtig')." P. Persson, in his *Beiträge zur Indogermanischen Wortforschung*

impressions which are intuitively received as expressive lend their names quite spontaneously to conceptions of feelings, and at the same time exchange them among themselves.

It seems, then, that qualitative words have a wide range of meaning which freely cuts across the boundaries of various domains of sense. But this is not strictly true; for the metaphorical use of a qualitative word never has the specific character of its literal use. Ernst Kurth, in his *Musikpsychologie* (1931), observed this distinction, especially with reference to the very important and widely discussed phenomenon of "musical space." Commenting on that phenomenon he says: "It is not visible, not palpable, and really hardly imaginable; for . . . in transition to clearer conception the phenomenon is destroyed . . ." (p. 119). After considerable discussion of spatial, material and dynamic impressions, he remarks (p. 136) that musical space is really dynamic space (*"energetischer Raum"*), born of the kinetic forces of the psyche. Dynamic space is the primary illusion of sculpture. But in sculpture it has a permanent and complete development. The interesting aspect of Kurth's remark is that all space in art is plastic space, all event is poetic event, all time is musical time, etc. Primary or secondary, all dimensions in art are virtual.

Kurth goes on to the phenomenon of "color hearing," of which he says: "One of its main characteristics is the difficulty encountered in trying to render expressions borrowed from the visual realm in purely 'tonal' terms; the vocabulary of our language fails us here, and even if one resorts to descriptions of the impressions, a veritable flight of thought ensues, to all sorts of auxiliary concepts from other realms, as we can see . . . in such words as 'brightness,' 'clouding,' 'clearness,' 'glare,' 'high color.' Indeed, the expression 'toning down' in music seems like a re-borrowed loan word from painting! All this indicates how deep in the unconscious the obscure connection with alien sensory realms takes place. And for this reason it always remains imprecise and ambiguous, i.e., its indistinctness, like that of [musical] space, always remains something implicit in the *essence* of the element, which one must not coarsen to the point of real perceptibility" (p. 242).

"It is in keeping with the vague reverberant character of such concomitant effects that they do not connote specific colors; such ideas as green, red, yellow, or mixed colors would over-illuminate the conception

---

(1912), defines the Middle Dutch *blide* as "strahlend, klar, fröhlich, glückselig, erfreut," and renders Anglo-Saxon *blide* by the English words "pleasant, gentle, kind, friendly, clement, mild, sweet"; adding, "Wie mir scheint, liegt es auf der Hand, das von 'glänzend, klar, heiter' als Grundbegriff auszugehen ist."

and destroy it. Nothing like that is involved in the phenomenon, but rather just color *per se*" (p. 243).

That final observation reveals what the "concomitant effects" really are: symbolic elements that intensify the feeling of the "spatial," "colored" or "dramatic" passage. They are products of the composition, not given with its material, but created by the development of the material. That is why laboratory experiments, which require definite stimuli and responses, have never yielded any insights into artistic experience; they can test only correlations of simple isolated impressions. Genuine synesthesia is a highly individual matter, probably established in childhood. Few people who have specific color associations with vowel sounds, for instance (and I am one of them), would agree with Rimbaud's pairing of hues and vowels.[93] But sensuous metaphor is public enough to be a creative force in language, and secondary illusions can be deliberate achievements, because any specific sensation which may be connoted in one's experience of, say, a "musical" effect in poetry, not made by intonation, or a "colorful" symphonic passage, is unessential. Whatever the context evokes will probably clinch the emotive abstraction.

The conceptual processes which the metaphorical extensions of words reveal are manifold. The discovery that basic forms of subjective feeling relate many objectively disparate sensations is not the only formal discovery to which philology leads us. There is another abstractive principle that comes to light if the study of sensuous metaphor is pushed just a little further, and focused on one peculiar phenomenon, which appears superficially like an extreme case of such extension, but actually rests on a different logical intuition and consequently involves a different abstraction.

Many words or "roots" denoting qualities give rise not only to all sorts of derivative terms involving those qualities, but by further departure produce cognates wherein the original connotation is more and more attenuated, ending up in its exact opposite. So we find in Skeat the derivation of "gloom," which means darkness, from "glow"; and that of "swoon," to faint, from *swogan*, "to sound, resound, make a noise."[94]

[93] Arthur Rimbaud, *Le Sonnet des Voyelles* (1871):
    A, *noir; E, blanc; I, rouge; U, vert; O, bleu: voyelles,*
    Je dirai quelque jour vos naissances latentes.

[94] The passage from "resounding" to "fainting" goes through two entries in Skeat's *Etymological Dictionary*; the meaning of Anglo-Saxon *swogan* is given under "sough, a sighing sound, as of wind in the trees," which he connects also with "swogh, as in Morte Arthure, . . . where it has the sense of 'swaying motion'; formed as a sb from the A.S. verb *swogan*, to sound, resound, make a noise." Under another entry, "swoon," he says: "The descent of *swoon* from A.S. *swogan* is certain."

A similar about-face may be seen in the history of the word "wan," which he defines as "colourless, languid, pale," and derives from the Anglo-Saxon *wann* or *wonn*, meaning "dark, black." In this connection he says: "It occurs as an epithet of a raven, and also of night; so that the sense of the word appears to have suffered a remarkable change; the sense, however, was probably 'dead' or 'colourless,' which is applicable to black and white alike."[95] A similar derivation of "black" and words designating "white" from the same root appears in Webster, if we compare two entries: "*blank* [ME.; OF. *blanc* . . . OHG. *blanch*, white, gleaming . . . IE. base *bhleg-*, to shine, gleam; see *Black, Blink* . . .]" and "*black* [ME. *blak, blakke*; AS. *blaec*; akin to OHG. *blah*; IE. base *bhleg-*, shine, gleam. . .]."[96]

Latin *altus* means both "high" and "deep." Walde gives the basic meanings of *skar* as "light" and "shadow," and those of the root *kel-*, as: (1) "freeze, cold" and (2) "warm." Webster lists the meanings of "lurid" as (1) "deathly pale," (2) "glowing through a haze . . ." and (3) "vivid in a harsh or shocking way. . . ." From deathly to vivid!

There are many homonyms in our language with diametrically opposite meanings. They are accepted as different words, and their etymologies run back through different lines, but those lines ultimately lead to a common source. So, for instance,

Cleave:[97] (1) to split apart; (2) to cling to.

Farrow:[98] (1) to bear young (of pigs); (2) to be without young (of cows).

Souse:[99] (1) the act of rising in flight; said of a hunted bird.

(2) the act of swooping down on a hunted bird; said of a hawk, falcon, etc.

(3) in falconry, to swoop down.

The guiding principle of such changes is an aspect of conceptual thinking which no conventional symbolism can express: the fact that every primitive concept arises and exists in an area of relevance, ranging from its own logical domain to its converse domain, and including all conceptions lying between these extremes.[100] The roots of language usually convey ideas of felt experience, i.e., either of action or of impact,

[95] Skeat rejects the connection with the Anglo-Saxon *wan*, "wanting," from which Wedgwood derives it. Skeat is generally more reliable than Wedgwood.

[96] "*Bhleg-*" is adduced as "hypothetical," like most Indo-European "roots."

[97] F. G. Fowler and H. W. Fowler, *American Oxford Dictionary* (1935).

[98] Webster, *op. cit.*   [99] *Ibid.*

[100] Sigmund Freud noted the prevalence of opposite meanings in words, and wrote a little article, "Über den Gegensinn der Urworte," in which he unfortunately treats

and feeling is good or bad, pleasant or unpleasant, with a continuum between them, which, taken from either end, somewhere (not necessarily halfway) breaks over into its opposite. Sensuously, the change of cathexis usually corresponds to changes of intensity, from too faint to satisfy, to too intense to bear. In every sensory experience there is the threat of evanescence and the threat of intolerability, and the precarious balance between them is implicit in every moment of perfection. A sensible quality, therefore, gives into the artist's hand the whole range of feeling it can express, even the existence of that range itself.

The following chapter will deal with the constructive practices based on these numerous, more or less incongruous, yet often intersecting principles of abstraction in the making of virtual living form. The astonishing complexity which may be found in some apparently simple compositions—small lyric poems, line drawings of one figure or little piano pieces every amateur dares to play[101]—results from the countless elements a good artist can create, interweave and fuse, with so many dimensions of symbolic projection at his disposal. In the course of considering the most significant aspects of the art symbol, by virtue of which it presents us with an image of life and mind, we may encounter even further modes of abstraction, which cannot be discussed apart from their uses; in the study of expressive form, abstraction and creation are not always separable. This chapter has only done some of the spadework toward a survey of the prescientific knowledge of feeling, recorded by those people whose intuitive vision has put them in possession of it.

Egyptian as an archaic language and commits other unscholarly blunders, but proposes the gamut of meanings that may arise from different cathexes attaching to the same datum.

[101] A good example may be seen in Reti's analysis of Schumann's *Kinderszenen*, in *The Thematic Process in Music* (1951), chap. ii.

7

On Living Form
in Art and Nature

AT LAST we can really come to grips with the problem which engendered the whole study of mind contained in the subsequent parts of this book: why must every form symbolic of feeling appear to be "living"? What is "living form" in art? A figure of speech, certainly; a figure of thought, a powerful metaphor. What justifies the employment of that metaphor, and its development to such lengths as the whole literature of art exhibits?

The answer—for it is really one answer to a many-faceted question—has, I think, already been given: if feeling is a culmination of vital process, any articulated image of it must have the semblance of that vital process rising from deep, general organic activities to intense and concerted acts, such as we perceive directly in their psychical phases as impacts or felt actions. Every artistic form reflects the dynamism that is constantly building up the life of feeling. It is this same dynamism that records itself in organic forms; growth is its most characteristic process, and is the source of almost all familiar living shape.[1] Hence the kinship between organic and artistic forms, though the latter need not be modeled on any natural object at all. If a work of art is a projection of feeling, that kinship with organic nature will emerge, no matter through how many transformations, logically and inevitably. What is more, many aspects of life that never rise to feeling may appear in the art symbol; and they appear there as though they could be felt. Growth is one of them. The essentially somatic formative influences sometimes termed "organic memories" are another;[2] so are the activities that constitute

---

[1] Cf. the corroborative statement by D'Arcy Thompson quoted in Chapter 5 (p. 152n) above, to which one should here add, however, his further words: "But it is by no means true that form and growth are in this direct and simple fashion correlative or complementary in the case of minute portions of living matter."

[2] Eugenio Rignano called them "biological memories." See his book, *Biological Memory* (1926; original Italian edition, 1925).

our common biological heritage, known as "instincts." The image of feeling may encompass humanly unrealized potentialities, which in nature are realized by altogether different kinds of life. We shall meet instances of such extensions of biological conception as we follow feeling to its roots.

So far, our discussion of art has been concerned with its symbolic character, the logical peculiarities of its projection, the objectification of feeling as quality, the complexities of artistic abstraction; now, in the present chapter, we come to the point and purpose of introducing the art symbol into a philosophy of mind, i.e., to the image of mind which it projects. In the artistic projection, human mentality (which is "mind" in a strict sense) appears as a highly organized, intricate fabric of mental acts emanating more or less constantly from the deeper activities, themselves normally unfelt, that constitute the life of an individual. There is no simple dyadic relation which one could call "the body-mind relation"; a "psychophysical" organism is one in which some acts have psychical phases. All acts, including purely cerebral ones, are elements of a life. They arise in it and take shape and have some sort of termination, either in processes propagated outward beyond the organism, or else by spending themselves internally, reaching their own point of rest, or being taken up into other acts as constituents. This means that acts have characteristic dynamic forms; and one of the first revelations gained from works of art is that "living form" in art has those same characteristics.

In a work of art, however modest, the peculiar character of life is always reflected in the fact that it has no parts which keep their qualitative identity in isolation. In the simplest design, the virtual constituents are indivisible, and inalienable from the whole.[3] Yet no analysis can uncover a simple and direct relation of an element to the work as a whole. Its relations are with other elements, and are internal to these and to itself, i.e., constitutive of the elements which are related. One may trace the relations of an element in many directions, drawing in more and more elements, but never to a limit, i.e., to a point where the relationships form a set defining the elements which beyond that definition have no character, as in a coherent logical system; at some point, suddenly, its initial character changes, or some other element has taken its place, not dislodged it but swallowed it so that the old element is somehow "in" the new but is transformed by figuring in another functional pattern altogether.[4]

[3] Actual constituents—parts—may be dispensable, especially in large works; the virtual elements are restored by compensation, as will be shown.

[4] Cf. Joseph Goddard, *The Deeper Sources of Beauty and Expression in Music* (n.d.), p. 95: "That it is in the nature of a single musical effect to be versatile in

An excellent example of this is the transformation of a small, recurrent bass figure into a great motif, which Ernst Toch quotes from Brahms's Second Symphony (opus. 73).[5] In connection with this audible shift of meaning, he says: "Every combination of a few tones is apt to become a motif and, as such, to pervade and feed the cellular tissue of a composition, emerging and submerging alternately, giving and receiving support and significance by turns. . . . It lives on repetition and yet on constant metamorphosis; metamorphic, polymorphic, opalescent in itself, it takes on the hue, the flavor, the very mood of the environment in which it is embedded. But above all it creates and feeds movement, movement, movement, *the very essence of life.* . . ."[6] The material he refers to is a little figure, essentially:

It serves to create one element after another—motion, suspension of motion, recall, expectation, opening, sustainment, closure—what you will. In the adduced movement, after some sixty measures of "emerging and submerging alternately," it makes a final emergence as the theme; "*It, the little motif,*" says Toch, "*becomes the motive, the motive power, the* MOTOR."[7]

A figure may be transposed, inverted, enlarged and, if it is at all elaborate, even fragmented; but the elements created by means of it cannot. They can only reflect each other, enhance or otherwise modify each other, even seem to spring one from another. Their indivisible and inexhaustible character, due to the internality of all their relations, is what makes them expressive of life, even more than the fact that they assure movement; movement is not the essence of life, but a necessity of the primary illusion of music, time; it may be replaced in other arts by something nearer to vital essence, something which Joseph Segond has called "mobility."[8] But all elements of a work seem to arise out of that whole in which they may also disappear when they are submerged. All elements may grow into themes, into paramount forms. In painting quite elementary decorative figures may serve to organize gripping representa-

---

poetic expression is shown by the fact that in vocal music we are continually meeting the same turn of effect associated with a different idea or sentiment."

[5] *The Shaping Forces in Music* (1948), pp. 201–4. For all extensive musical examples, see the Appendix to this chapter; for the above, see Example 1. The clearest illustrations of almost every artistic principle are to be found in music, so a few are quoted in that postscript for the benefit of readers who can readily peruse and judge them.

[6] *Ibid.*, pp. 200–1.   [7] *Ibid.*

[8] "Esthétique de la mobilité" (1948).

tions, as the simple shape of the Cross the fusion of God and man, the crucified Christ; or in dance, a small technical motif may create gesture, light, emotion, space and time.[9] And once an element is created, it influences the entire work. A transcendent moment effects a rarification of the whole piece and lifts it as a whole above the level of sensuous tensities. Every abstraction, intensification, coloration, acts on it in this way; consequently even the most direct influence of one element on another seems to go via the implicit core of the art symbol.

Elements in art have not the character of things, but of acts. They are "active," act-like, even where they are not "acts" in the dramatic sense nor in the special sense which is sometimes given to the word by people who let it mean only moral acts. In a broad sense, which I find far more useful for philosophical purposes, any unit of activity is an act. Taken in this way, the term has an instrumental value for building up a coherent and adequate concept of mind, and on that pragmatic basis I use it in the broad sense here.[10]

People who pursue the general problems of art have to be steeped in their subject matter, so they tend to speak the language of artists, which is largely metaphorical and consequently less responsible and consistent than that of philosophers, but often more authoritative. Just because they are not aware of any metaphysical commitment they choose the most telling metaphors; and their turns of phrase support the finding that the dynamic structure of elements in art is cognate to that of acts in physical organisms. Especially musicians are ready simply to identify

[9] A striking example may be seen in Martha Graham's *Seraphic Dialogue*, based on Jeanne d'Arc's last vision of the Archangel, where one and the same gestic motif —a shimmering shake of the fingers and hands—is first the sun in the window (without any actual play of light), then the tremor of the girl's fear, then the intensity of her compulsion, responding to the same action of the compelling angel figure so that it makes the effect of a tense sympathetic vibration and the impression of a magnetic force; at last it occurs in slower tempo as life awaking in the angelic form that has stepped out of the window, then spreads to his draperies, flaring up in a shimmering shake as it becomes fire. All these phenomena, incorporating the same motif, are symbolic of each other—physical and emotional elements alike.

[10] Robert Vischer, in his little known, brief, but excellent *Über das optische Formgefühl*, clearly adopted the same broad sense, and for the same reason, when he wrote: "Die Empfindung ist die primitivste Lebensregung und aus ihr entwirkt sich erst der deutlichere Akt der Vorstellung, des Willens, der Erkenntnis und wir besitzen hiermit an derselben die ursprünglichste Form des Weltzusammenhangs. Mit und durch diesen allgemeinen Fortschritt wird auch die Empfindung etwas Anderes. Sie wird zum Gefühle. Das Gefühl ist objektiver als die Empfindung. . . . Indem ich abstrakt denken und mich als untergeordneten Theil eines untrennbaren Ganzen begreifen lerne, expandiert sich mein Gefühl zum Gemüth" (pp. 28–29). "*Gefühl*" seems to mean "emotional feeling," "*Empfindung*" being "sensibility." (His attribution of greater objectivity to "*Gefühl*" is puzzling to me.)

them. Heinrich Schenker, for instance, remarked in an essay on the organic structure of the fugue that Bach's fugal themes present "a strictly self-contained action" (*"eine streng geschlossene Handlung"*).[11] Jeanne Vial, in a book ostensibly metaphysical but more interesting as musical analysis, says explicitly: "Music is act," and justifies that dictum on grounds that come very close to some of those I have adduced above in drawing my own parallel.[12] The most corroborative statement, however, is in Hermann Nohl's *Die aesthetische Wirklichkeit, eine Einführung* (1935), for this author, schooled in philosophical thought and literal expression, makes clear that he sees the character of "acts" as such reflected in the structure of the art symbol, and is not finding certain acts, or kinds of acts, represented by such-and-such elements. His studies, moreover, embraced all the arts, so it was on a broad basis that he wrote: "The way the sensory elements in a work of art are combined, tones, colors, images, this stream of music, this cast of a poem, is the expression of an inward act in the agent; Schiller says, 'an inner movement.' In every work there is such an inner movement or act [*Handlung*], an internal delineation [*eine innere Linienführung*], which begets its style" (pp. 119–20).[13]

The peculiar dynamic structure of "acts" exemplified in all living things has baffled some of the most serious philosophers of science whose preoccupation is with biology and psychology—Klages, Buytendijk, even Weizsäcker—to the point of making them despair of defining the concept of "act" in any way, so they claim it is not a natural process—indeed, "not a process" at all.[14] One author, writing in a highly respected series of *Abhandlungen*, even says that all organic reactions to stimuli are genuine acts of will, that organic movements are in essence nonphysical, and that "psychophysical causality" (which, he holds, does exist) is truly and literally magical.[15] Such surrenders of reason bespeak

---

[11] *Das Meisterwerk in der Musik, ein Jahrbuch* (1925–30), Vol. II, No. 3, "Das Organische der Fuge," p. 60.

[12] *De l'être musical* (1952), pp. 10–11: "La musique est acte: elle se déroule dans une indivisible durée, souverainement libre et merveilleusement nécessaire: il est trop évident, en effet, que chaque note ne s'évanouit pas quand chante la suivante; la musique n'existe que parce que le passé ne meurt pas et que le futur ne nous semble pas un risque de rupture: son unité transcende l'opposition du présent au passé, du passé à l'avenir. L'expérience musicale est donc bien de l'ordre de l'être."

[13] If "movement" is taken in this sense, one might call it, with Toch, "the essence of life." But I do not think Toch meant it in the sense of an "inward act."

[14] See the discussion of this school of thought in Chapter 1 above, pp. 10–15.

[15] See J. A. Loeser, *Die psychologische Autonomie des organischen Handelns* (1931): "Alle Reizreaktionen sind echte Willenshandlungen. Nur graduell unterscheidet sich die einfachste von der kompliziertesten" (p. 10).

"Die Wirkung des Strebens oder Wollens auf die Physis ist uns hierbei ein

a failure of scientific imagination. An impasse, such as the inadequacy of classical mechanics to describe the biological phenomenon of variable responses to similar stimuli, should lead a philosophical mind to some intellectual pioneering, to a reconsideration of the physical laws which are supposed to be demonstrated by any and every "scientific" fact, and above all to a searching study of what really is the dynamism of an act, wherein it deviates from the standard of classical dynamics.

The logical form of acts is projected in the art symbol, though not necessarily or even primarily in images of acts. Virtual acts are the basic elements of one poetic mode, namely drama,[16] and may occur in other modes and other arts, too; but I am not referring to virtual acts. All artistic elements whatever—all distinguishable aspects of the created work—have formal properties which, in nature, characterize acts. Inviolability, fusability and the revivable retention of past phases in succeeding ones are some of those properties. Another very important one, already mentioned in passing, is the relationship of elements to the whole, which is very complex, so that it is ordinarily not possible to designate it as a single relation. Every element seems to emanate from the context in which it exists. This appearance shifts from one created element—tension, gestalt, contrast, accent, rhythm or any that one may select—to another; whichever one attends to carves out a context that is indispensable to its existence. This is, of course, a manifestation of the internality of relations among created forms, which is a principle of art, not of life. But it parallels a biological condition: in life, every act is motivated by a complex of past and/or concomitant acts, and motivates a great number of other acts, among which there are usually some that belong to its own motivating context. It is in this way that an organism is made up of its own acts, and at the same time is the source of all its acts.

The unity of a work of art stems primarily from the interdependence of its elements, and is further secured by this dialectical pattern of their relations. The principle of dialectic is a phase principle; the consummation of one phase is the preparation for another, which in its own consummation prepares its successor, often a replica of the predecessor. Dialectic is the basis of rhythm, which consequently is more than sheer

---

Grundphänomen. . . . In der direkten Einwirkung der Psyche auf die Physis kommt also *das Grundgesetz der organischen Bewegung* zum Ausdruck. Es ist grundsätzlich nicht physikalischer Natur. Es ist naturwissenschaftlich ein Wunder. Man könnte die *psychophysische Kausalität* daher *magisch* nennen, ein Ausdruck, dessen innerste Berechtigung in dieser Arbeit noch nicht erwiesen werden kann" (p. 12).

[16] Cf. F.F., chap. xvii, "The Dramatic Illusion."

periodicity, or evenly spaced repetition of any occurrence.[17] A rhythmic phenomenon may even involve no exact repetition, but is always a dialectical pattern in which the resolution of tensions sets up new tensions; the recession of one color brings its complementary to the fore, our close attention to the latter exhausts its domination and lets the former advance again; in a good composition of volumes, every boundary of a form is also a conjunction of forms, the surrounding spaces taking their gestalt from the volumes they limit. This mutual conditioning of forms is forcibly apparent in a rhythmic design, where tension and release, upswing and drop, or centripetal and centrifugal impulse do not succeed one another temporally, but are projected spatially (see, for example, Fig. 7–1). One of the far-reaching problems of mind, which we shall encounter in due course, is the organic basis of this visual projection; it seems to be differently developed in different creatures, highly specialized where it exists below the human level (for instance in bees), but in many animals entirely absent. Our clearest exhibit of it is in the arts, which consequently put one on the track of this as of other elementary traits of human perception.

Figure 7–1. Rhythmic Design
every boundary of a form is also a conjunction of forms
(After Alois Riegl, *Stilfragen.*)

The dialectical structure that pervades the virtual object is the main source of its unity, though not the only one. But that object, which is the work as a whole, has more than unity. It has a substantive character, which is, of course, wholly illusory; it is created by the artist, by technical devices, most of which he probably uses so intuitively that he could give no account of them if he were asked how the semblance of substantiality is achieved.[18] There are, moreover, so many ways of establishing, sustaining, momentarily increasing and again etherealizing yet never losing this basic presence to which all the elements of the work seem to owe their existence, that his answer in reference to any particular work would have to be long and circumstantial. But a few practices which normally give body to a composition may be found almost everywhere.

[17] For a fuller discussion of this topic see F.F., chap. viii, "The Musical Matrix," especially pp. 126–30.

[18] "Dionysius himself tells us, that the sound of Homer's verses sometimes exhibits the idea of corporal bulk . . ." (Samuel Johnson, quoted by William Empson in *Seven Types of Ambiguity* [1930], p. 12).

The most essential one is the interaction of the primary illusion with the highly variable secondary illusions that arise and dissolve again, while it remains steady, complete and all but imperceptible because of its ubiquity. The fact that secondary illusions never present completely developed realms of virtual time, space, etc., makes their manifestations appear against the plenum of the entirely developed primary illusion, which consequently seems like a negative background, supplying their complementary forms; and since secondary illusions may be of many kinds, that background has to have a protean character, which gives it an air of indefinite potentiality.

In nature, such indefinite potentiality is the essence of bodily existence, which feeds the continuous burgeoning of life. Life is the progressive realization of potential acts; and as every realized act changes the pattern and range of what is possible, the living body is an ever-new constellation of possibilities. In art the elusiveness of secondary illusions serves to give the work as a whole something of that same character; it seems to have a core from which all its elements emerge—figurations and rhythms and all the qualities to which these give rise. Many elements are but slightly articulated, as their individual developments limit each other; this creates an effect of constant becoming, because partially developed forms seem to be still unfolding. Also, very similar or even identical gestalten may make entirely different sense in their respective contexts, so they seem distinct and yet the same; the gestalt element then appears ambivalent, as though its nature were not wholly determined, but still open to modification by internal or external conditions.[19]

The interplay of identity and diversity of forms is a major factor in the dialectical structure of the art symbol. Especially the use of exactly the same gestalt in expressions of diametrically opposite feelings or impulses is a ready means of projecting the total qualitative dimension which separates yet links its extreme degrees. The explicit image is of two particulars, but the whole gamut is implicitly presented, so that they seem like selective realizations springing out of a matrix or body of potentialities. It is primarily this semblance of qualitative continua that establishes the virtual substance, and the appearance of partial realization that makes elements emerge and submerge like transient aspects

[19] Cf. the passage quoted from G. Anschütz, "Zur Frage der musikalischen Symbolik," above, p. 105. There is also an interesting article by Giulio Zambiasi, "Intorno alla misura degli intervalli melodici" (1901), 157–77, in which the author checks the results of older investigations, especially by Helmholtz and by Cornu and Mercadier, to the effect that melodic (i.e., successive) intervals are played on the Pythagorean scale, harmonic (simultaneous) ones in various systems; but the third, sixth and the perfect intervals practically follow the pure physical scale.

of its inward being. The illusion of bodily existence is so strong that writers on art sometimes designate purely visible, imaginable or audible forms as "tangible."[20]

Yet the "tangible" object exists only for perception, and any imperceptible properties that seem to underlie this sensuous object are made by intimation. Light caught on a sculptured surface makes its "texture"; it is not just the physical texture of the material that we see there, but something made of that texture by the chisel, the molding hand or even the electric buffer—whatever the tool may have been; the result is a "texture" made of light and shade and color. The actual surface yields to a virtual one which exhibits tensions that seem to come from an unseen interior, and are what creates this interior of the figure.[21] A stone statue is no longer a stone; its center is not a geological reality, but a vital one. And yet a statue of a man does not look like flesh; although the weight of the stone is many times that of a man, there is no attempt to make the figure appear as though it weighed no more than its model. It looks like stone, as a good bronze looks like bronze. The idea of specific gravity disappears altogether; where "great weight" is expressed, as in Rodin's massive "Balzac," it is still not the weight of the material. It is the sense of weight—weightiness, greatness, power—that is made visible in such a work.

Another condition that gives the art symbol its semblance of bodily existence is its complexity. Its preconceived structure—the "plot," as Aristotle called it, of a story, a painting or a building—may be the simplest of compositions; but in the course of its realization it entails more and more elements, all involved with others as reinforcements and foils and echoes. There are virtual tensions spanning tensions, resolutions that are the poles of new tensions; and on all levels, even that of purely

[20] Rudolf Arnheim, for instance, writes: "The artist's privilege is the capacity to apprehend the nature and meaning of an experience in terms of a given medium and thus make it tangible" (*Art and Visual Perception: A Psychology of the Creative Eye* [1954], p. 133), and Andreas Feininger, in his beautiful book of photographs, *The Anatomy of Nature* (1956), says of one picture that it makes the vast number of stars in a galaxy "somewhat tangible" (p. 7), obviously meaning "imaginable."

[21] Cf. Friedrich Theodor Vischer, *Kritische Gänge*, Vol. II, p. 55: "Wenn . . . vom Stoffe . . . abgesehen wird, so wird doch darum keineswegs abgesehen von der Kraft die diesen Stoff beherrscht. . . . Sie eben ist es, die auf der Oberfläche, in der Gesamtwirkung erscheint, in der Vielheit, welche diese darstellt, ist sie die Einheit."

Cf. also Paul Stern, "Über das Problem der künstlerischen Form" (1914–15), p. 168: "Wesentlich . . . für die ästhetische Anschauung ist, dass alles Aeussere in ihr ganz unmittelbar als Verkörperung eines Inneren, als Ausdruck einer bestimmten Verfassung inneren Lebens in Zustand oder Aktion, in Ruhe oder Bewegung empfunden wird."

THE IMPORT OF ART

technical construction, there may be mixed projections, such as Francastel remarked in Renaissance painting[22]—mixed for the sake of expressing visual concepts charged with important feeling. All levels of feeling are reflected, explicitly or implicitly, in art. This holds for any successful work, and constitutes its "depth," which therefore varies according to the degree of success with which the work is organized. A very light subject, a bagatelle, may have this sort of artistic depth.[23] It rests on a constructive process that is mainly or even wholly unconscious, and seems to the artist like "the happy hand of chance." But "depth" in this sense is a matter of logical structure, and as such can be understood instead of accepted as a mystery—like many other aspects of artistic expression, which become amenable to serious study only through the ways they are achieved.

It is often surprising to find that the most essential and obvious features in art are created semblances; not only substantiality and depth, but even the unity which is generally considered a *sine qua non* of all good art work, and the individuality or "uniqueness" characteristic of it, are virtual, and usually are achieved together, by the same intuitive moves. The inviolable unity of a total form, especially a highly developed one, is due in large measure to the play of several projections, which cannot be conceptually received at the same time, so that when one is in evidence the others recede, much as one visual gestalt disappears when another emerges, in the well-known ambiguous images of the psychology laboratory; and, for another part, to the "overdetermination" of elements, which makes their cathexes do the same thing. Those possible forms which are eclipsed by the realization of rival ones are still present as the potentialities of an implicit totality created by their dialectic.[24] But they do more than establish that inclusive whole: they constitute the

22 See Chapter 4 above, p. 87n.
23 For instance, some of Blake's *Songs of Innocence*, De La Mare's *Very Old Are the Woods*, or a little song like Mendelssohn's "Leise zieht durch mein Gemüt" (where the strange beauty is made chiefly by the ambiguity of almost every chord and the complete absence of dissonance, so the substitute for the usual harmonic tension and resolution is harmonic ambivalence and reassurance). Consider also some of Picasso's or Klee's or Rembrandt's sketches of casual and ordinary sights.
24 Roger Fry, in *Transformations* (1926), said of the Pointillists: "The desire for pure colour and luminosity in colour led them to conceive the possibility of even abolishing altogether from their palette the neutral and tertiary tints. Where these were inevitable they were made by placing bright complementary colours side by side in such small strokes or dots that they would mix to produce a grey on the retina by the overlapping of the complementary colour sensations, and yet leave in the mind a faint suggestion of the brightness of the colours of which they were composed" (p. 218).

"body" or "organism" out of which its realized elements seem to arise. That is why a good work of art presents itself as a matrix, from which all its sensuously given articulations are derived, while others which do not appear are nonetheless felt to lie somehow in limbo.[25]

Finally, there is a problem of which many people are quite unaware, concerning the individuality of every expressive form. The phenomenon is universally admitted, but what constitutes it, what makes each piece unique, is often a matter of debate. A beautiful object may impress one instantly as unique without being startlingly different from others; it may even exemplify a standard form, as many highly artistic products of traditional design do. Its uniqueness is popularly explained on the ground that every handmade thing shows some deviations, tiny perhaps, but many, from the conceptual standard. There may be some truth in that consideration, but it does not account for the force of the impression and the importance it has in one's experience of art. Many things (e.g., pipe cleaners, dishmops, cartons), though made on one pattern, are far from precisely identical, yet they do not seem unique. A statue, however, may be reproduced—in some media, such as metal casting, there is no "original," but a perfect democracy of "copies" as long as the mold is intact—yet every "copy" seems entirely individual. There are, furthermore, artifacts of glass, spun metal, etc., which have undeniable aesthetic value depending chiefly on their machined perfection; and they, too, with actually no individual differences at all, seem to be unique in their beauty.

The upshot of these paradoxical findings is that the individuality of a work of art is not a factual condition but a quality, as virtual as all other artistic qualities. It is the semblance of organism that creates the apparent uniqueness of a piece. The work seems unique when it is "alive," i.e., expressive.[26] Consequently this all-important character may grow but slowly under the artist's hand. Instead of being one of the elementary ingredients in the "Idea" with which he starts, it is the reward of his work. Sometimes its promise is evident from the beginning; but again sometimes it hinges on revelations of implicit form and even discoveries

---

[25] H. Mersmann observed this function of underdeveloped elements in his "Versuch einer musikalischen Wertästhetik" (1935), where he said: "Die Substanz eines Kunstwerks umfasst alle Bindungen durch die Elemente, welche unterhalb der Gestalt eines Motivs oder eines Themas liegen. Sie ist die unmittelbare Erscheinungsform des Individuellen" (p. 44).

[26] Donald Francis Tovey, in his article, "Music," in the *Encyclopaedia Britannica*, said: "Many movements by Mozart are as alike as peas. But, being alive, they are not as alike as buttons."

of possible feeling which come to him only as he concentrates on his idea and sees its image develop.

Most, if not all, artistic elements perform more than one function, and as a rule any artistic device can serve more than one purpose. The mutual limitation which possible subordinate forms set on each other's full realization, whereby some are automatically curtailed in favor of a few dominating ones and consequently remain half-realized, has already been mentioned as a source of the apparent fecundity and growth of the work. But it has a further and more elaborate use, in that the competing potential forms may be developed to varying degrees, so as to present a gradient of development. In a Cambodian Buddha statue, for instance, there is usually a perfect elaboration of the head, and a flowing line to the hands, which are given slightly less articulation; the torso and crossed legs are very simply treated as large surfaces and opposed curves. There is a gradient of development toward the head, culminating in the face, and a lesser one toward the hands, that leads up to their delicate form and gesture. Such a figure has the living stillness of a plant; its "inward action" is concentrated in its apex, the head, which consequently predominates without being given any other emphasis by way of extraordinary proportion, posture or features. Its expressiveness suffuses the figure and makes the columnar body seem to subserve the development of the meditative head and the reflection of its poise in the hands; the traditional lotus pedestal repeats the theme of slow and gradual efflorescence.

Another, less obvious example of gradients constructed by the use of crowding and partly developed forms may be found in the poetry of Robert Browning. Browning's most original poetic resource is the sound of conversation, which he renders largely by unfinished sentences, curtailed phrases, contractions and elisions of words, so that numberless incipient utterances seem never to reach their completion. But his monologues have an artful structure; the element that emerges from the many beginnings and short-lived passages of thought is a state of mind, a rhythm of reasoning that builds up an emotive tension. The average length of the separate, casually spoken thoughts produces the pace of the entire soliloquy. Consider the haphazard beginning of *Caliban upon Setebos* (a soliloquy kept in the third person, since the speaker is not a fully human being, so there is no "I" in the mental experience); the image of Caliban's world, his human masters and his God builds up from one chance perception and child-like rumination to another, until his scraps of thinking frighten him, and a clap of thunder accuses him of heresy. Without any real coherence the poem has a gradient of ideation and feeling up to the moment of crisis, where philosophizing turns to abject superstitious terror.

Gradients of all sorts—of relative clarity, complexity, tempo, intensity of feeling, interest, not to mention geometric gradations (the concept of "gradient" is a generalization from relations of height)—permeate all artistic structure. They also permeate most of animate nature as basic patterns of change and especially of growth; and gradients of growth make the intricate forms of plants and animals. This is the central theme of D'Arcy Thompson's famous morphological study, *On Growth and Form*, which I have already cited several times. There is an interesting and lucid passage in that book, wherein the author touches directly on three closely related aspects under which gradients figure in art, and proffers some incidental hints of their relatedness. Concluding a long and circumspect discussion of animal forms, he says: "In short, a large part of the morphology of the organism depends on the fact that there is not only an average, or aggregate rate of growth common to the whole, but also a *gradation* of rate from one part to another, tending towards a specific rate characteristic of each part or organ. The least change in the ratio, one to another, of these partial or localized rates of growth will soon be manifested in more and more striking differences of form; and this is as much as to say that the time-element, which is implicit in the idea of *growth*, can never (or very seldom) be wholly neglected in our consideration of form.

"A flowering spray of Montbretia or lily-of-the-valley exemplifies a growth-gradient. . . . Along the stalk the growth-gradient falls away; the florets are of descending age, from flower to bud; their graded differences of age lead to an exquisite gradation of size and form; the time-interval between one and another, or the space-time relation between them all, gives a peculiar quality—we may call it phase-beauty—to the whole. A clump of reeds or rushes shows this same phase-beauty, and so do the waves on a cornfield or on the sea."[27]

The significant points in these remarks, with respect to our present subject, are (1) the recognition of organic structure as a record of the processes of growth, a logical projection, in terms of physical form, of their various rates, directions, mutual interference or merging; (2) the consequent involvement of time in the spatial character of living forms; and (3) the appearance of the same logical pattern, which is statically projected as bodily shape, in the dynamic projection of waves running over fields and waters. All three of these conditions throw light on the relation of artistic form to vital form.

Taking these points in their order, we find that the first one, the eventuation of growth processes in articulations of shape, is reflected in

[27] Vol. I, pp. 193–94.

the semblance of constant becoming which belongs in some measure to all good art work and in a high degree to its really great products. This has, I think, been sufficiently discussed above; the only remark to be added is that in the plastic arts this semblance has to be created without any actual temporal ingredient. The fact that it takes some time to contemplate a picture or a statue in order to experience its subtler values, which is sometimes brought up in this connection, is really irrelevant, as is also the depiction, as such, of aged persons and objects which bespeak age or decay.[28] The sense of becoming, i.e., of process, is symbolically rendered very largely by gradients of apparent completeness, among elements so related to each other that they possess a visual unity and make a forcible impression of evolution from the slightest to the richest articulation of volume, line, surface and implied inner tensities. It may be reinforced by many other elements, including motifs, which often determine centers of interest and axes of symmetry. But gradients of all sorts run through every artistic structure and make its rhythmic quality.[29]

The second significant concept in the passage borrowed from Thompson is that of "phase beauty," the simultaneous expression of successive phases in a single form. This is, of course, just a special case of visually projected temporal process; but it is so familiar and almost ubiquitous an example that it explains in some measure the readiness with which we interpret the projection, and "see" the advance of nature in typical forms of life. It illumines the interchangeability of spatial and temporal concepts on an elementary level of human intuition, the level of "natural symbols" or spontaneous interpretation of visual data. Despite its basic role in human perception such intuition is probably peculiar, or very nearly so, to humanity, and absent (with a few possible exceptions) in animals. That problem and its significance will come under discussion

[28] See Paul Klee, *The Thinking Eye* (1961; translated from *Das bildnerische Denken*, 1956), p. 59: "The work grows in its own way. . . . As projection, as phenomenon, it is 'for ever starting' and 'for ever limited.' Art is a transmission of phenomena, projection from the hyper-dimensional, a metaphor for procreation, divination, mystery."

[29] Le Corbusier, in *Toward a New Architecture* (n.d.), provides an excellent example by his description of the interior of the Green Mosque at Brusa in Turkey: "You are in a great white marble space filled with light. Beyond you can see a second similar space of the same dimensions, but in half-light and raised on several steps (repetition in a minor key); on each side a still smaller space in subdued light; turning round, you have two very small spaces in shade. From full light to shade, a rhythm. Tiny doors and enormous bays. . . . You are enthralled by a sensorial rhythm (light and volume) and by able use of scale and measure into a world of its own . . ." (pp. 181–82).

later, with the origins and specialization of mind. What I would stress at present is the naturalness of the symbolic projection of vitality, especially growth and rhythmical activities, in essentially spatial as well as essentially temporal arts.

The third of Thompson's observations—the likeness of phase beauty in growing things to the form of unbroken waves—throws some light on the oft-remarked fact that in art all motion is growth, although the lines and volumes and tensions that seem to grow never reach any increased dimensions.[30] Wherever those dynamic patterns are projected in any medium, the phenomena which incarnate them assume the appearance of life; for the phases of growth and decay, rise and crisis and cadence, constitute the all-inclusive "greatest rhythm" of life, reflected in every completed subordinate rhythm, and tolerant of numberless variants and deviants among the lesser forms which it spans. The occurrence of such dynamic patterns in non-vital movements like waves, pendulums or spiraling storm winds, just by reason of being non-vital, uncomplicated by the lesser forms (and forms within forms) of life, makes the abstraction of the great overarching rhythm so impressively that it elicits our intuition of "living form" without any conscious judgment. The first such judgment, indeed, is likely to be the false one of taking the symbol for an actuality; but that is a natural stage in the evolution of symbolic thinking, a bridge from mythical to realistic reasoning, which need not be surpassed before the function of symbolic seeing and formal expression may be in full swing.[31] Once the abstraction is made by the eye or ear (the roar of a breaker has a similar cadential form), and spontaneously received as something charged with feeling, it tends to govern the impulse of every stroke or uttered sound or bodily motion. Then the course of art becomes an adventure in the growth and precision of feeling by virtue of its expressibility.

In the previous chapter, surveying the many principles of abstraction which operate naturally and, as a rule, unobserved in the human mind, a little attention to the history of some ordinary words which designate or suggest sensations revealed two such hidden abstractive processes: the distillation of feeling-value by the use of sensuous metaphors, and the unconscious relating of opposite values to each other as the two extremes of one continuum. In the present discussion of gradients in nature and their projection in art, the second of those two functions is relevant; for

---

[30] For a more detailed discussion of this topic, see F.F., pp. 63–67.

[31] On this general subject see especially Ernst Cassirer's *Philosophy of Symbolic Forms* ([1953–57], Vol. II; original German edition, 1924), and *Phil. N.K.*, chaps. vi and vii.

the tacit recognition of such qualitative continua, which is inherent in human perception itself, is the intuitive basis of our concepts of degree. The deployment of sensory materials by degrees is the chief device of visual and audial articulation. But degree as such, and therewith all increase and decrease, becomes thinkable and appreciable only along an ideal graduated scale.

The immediate relation, sometimes appearing as identity, of sensation and emotion (objective and subjective modes, respectively, of feeling) explains the appearance of sense data as possessing specific degrees in various respects: brightness, loudness, pungency, etc., and implying the entire range of every such quality from one extreme to the other; it is the natural form of internal action that is reflected in the primary, receptive formulation of sensory impacts. The fundamental dynamic pattern of rise and decline, *crescendo* and *diminuendo*, build-up and dissolution, is paralleled by the frame of our perceptual experience, and governs the world presented to us through our senses. Sensations, like emotions, like living bodies, like articulated forms, have gradients of growth and development.[32]

The rhythm of acts which characterizes organic forms pervades even the world of color and light, sheer sound, warmth, odor and taste. The implicit existence of gradients in all sensation reinforces our appreciation of living form by giving it an echo, or reiteration, in sense, which is always charged with feeling and consequently tends to subjectify the form, to make its import felt yet hold that import to the projective medium. This is probably the greatest single means artists have of "animating" their work; its importance has made many aestheticians believe direct sensory pleasure to be the prime purpose of art.

Every element in art has many connections with other elements, and every creative technique has many uses. There is no principle of abstraction or organization that is not exemplified in more than one procedure. So the apprehension of gamuts in the realms of sense, which permits an artist to create qualitative gradients that support and substantiate the "phase beauty" of rhythmic lines or motions, also serves to enhance the fundamental dialectic of elements in the work. When a quality reaches a great height of intensity it seems ready to induce its opposite. Nothing appears more intimately related in art than a value and its contrary.[33]

---

[32] Rudolf Arnheim defines a gradient as an increase or decrease in quality (*op. cit.*, p. 223).

[33] The tacit understanding of this relation is not peculiar to the artist's intuition. School children often express superlatives of quality by saying: "It's so hot, it's cold!" "It's so wet that it's dry!" And so forth.

This implicit interplay of extremes is, again, the sensory reinforcement of the many formal elements of dialectical structure which create the substance, unity and livingness of a piece.

The reinforcement of creative functions by what one might term "auxiliary" means is one of the internal bonds of artistic form, and reflects a trait of actual organic form, namely, the fact that in most multicellular organisms there are, besides special organs for the major functions, also some tissues which can take over those functions—perhaps parceling out their several aspects as separate actions—in case the special organ is impaired. Removal of that organ often induces growth of an equivalent one, as in the case of auxiliary buds. In art, the fact that a desired quality may be achieved by more than one means allows a good artist to reinforce his created elements by using one chosen device to establish a particular impression but letting many others incidentally serve the same purpose. This practice, which is intuitive and far-reaching, makes the created work seem like a direct exhibit of life rather than a symbolic presentation, because it obscures the technique in distributing it over many unapparent devices. It also has another, more remarkable, effect: it makes many subordinate features of a work reflect each other, so that each one of them connotes the whole, carrying the traditional and personal style like an "individuality factor" even in separation (within variable limits).[34] That is why a sculptural fragment may still be beautiful, and a mutilated work especially of sculpture or architecture, tends to close in on itself and restore its organic semblance, sometimes to perfection.

The versatility of almost all materials and technical practices in the arts makes possible the endless originality of expressive forms. All efforts to prescribe the means of creating this or that semblance run the risk of complete miscarriage. There are general studio adages, gained by experience, about the usual effects of colors, musical intervals, or in prosody tenses and word orders and the like, but a survey of the actual creative uses which have been made of the most familiar media and techniques soon proves that their potentialities go far beyond their normal services. Ordinarily, adjacent tones of a scale produce dissonant intervals, espe-

---

[34] Arnold Schoenberg, in *Style and Idea* (1950), declared that "the work of art is like every other complete organism. It is so homogeneous in its composition that in every little detail it reveals its truest, inmost essence. . . . When one hears a verse of a poem, a measure of a composition, one is in a position to comprehend the whole" (p. 4).

Cf. Frank Thiess, *Der Tanz als Kunstwerk* (1923): "Die einzelne rhythmische Körpergebärde . . . hat die Bedeutung des Ganzen und trägt die Dynamik des gesamten Tanzes in sich gesammelt" (p. 54).

cially when they form minor seconds, but sometimes they sound strangely consonant, or at least not dissonant. Tones that belong to differently directed progressions or different "sound communities" may coincide without setting up any interval for the ear to identify at all.[35] Any musical combination may, in fact, be either exciting or depressing;[36] and likewise, in the visual realm, although some colors are known as "approaching" and others as "receding" colors, any color really may approach or recede.[37] Abnormal functions are a powerful resource for artistic purposes.

As there are many uses for one and the same device, so there are many possible means to the same end. Where the usual method of meeting a structural necessity or attaining a desired effect cannot be used, either because of physical limitations of the medium or because of internal conflicts, a technical equivalent must be found instead; sometimes it is far removed and indirect, but the created impression is essentially the same.[38] The harpsichord, for instance, is unable to sustain a tone; the composers who wrote for that favorite eighteenth-century instrument contrived, therefore, to make a virtue out of a mannerism that verged on musical vice in their day: ornamentation. In ornamenting a tone with a mordent or a trill, they managed to strike that same tone several times without seeming to reiterate it, for the effect of such a

[35] For examples see the Appendix to this chapter, Examples 2 and 3.

[36] Cf. J. Vial, op. cit.: "Tout intervalle dissonant, déprimant lorsqu'il est entendu seul, peut devenir excitant dans certains conditions, et même s'il vient à la suite d'intervalles consonants" (pp. 74–75).

[37] Cf. E. von Sydow, Form und Symbol (1928), pp. 40–41, in which the author, whose interest is in art, comments on the psychological (laboratory) findings of G. J. von Allesch, to the effect "dass fast alle Farben die Eigenschaft des Heraus-springens annehmen können, . . . und ebenso wird die Richtung des Versinkens in die Tiefe den gleichen Farben zugeschrieben." (His reference is to reports in Die aesthetische Erscheinungsweise der Farben [1925].)

[38] An example of internal prevention of normal methods is given in Klee's discussion of the third dimension in painting: "The spatial character of the plane is imaginary. Often it represents a conflict for the painter. He does not wish to treat the third dimension illusionistically. Today a flat effect is often sought in painting. But if different parts of the plane are given different values, it is hard to avoid a certain effect of depth. . . . If it does not remain flat, we come to the formal problem of the third dimension. One of the artist's basic problems is how to enlarge space. We do it by means of overlapping planes" (The Thinking Eye, p. 49).

In another passage he calls attention to the versatility of all devices. "What is treated exotopically [shaded outside of the contour] on the picture plane tends to stand out. What is treated endotopically [shaded inside] tends to recede. . . . But there are exceptions to all general principles. Sometimes, it becomes impossible to say that something really goes back or comes forward. The third dimension asserts itself, but still operates with flat values that stand a little farther back or forward" (p. 54).

decorative figure is of adventitious tones playing around the harmony note which steadily holds its place among them.[39] The virtual element is a duration, while in fact the tone is repeated; and as each repetition is prepared by a tension which resolves into it, the semblance of rest on a single sustained pitch is created. But the creation is indirect and involves several factors. It is made by the unity of the "grace" as a gestalt, the harmonic identity of the central note, and the directional tendency of non-harmonic tones, which is to come to rest on the harmony note. Unity, identity and repose, deriving from three separate sources, are brought to bear in what seems like a stylistic convention, but is actually a transformation of superficial ornaments into a structural device.

A similar replacement of unavailable means through some technique that normally serves a different purpose is attested by some remarks which Pablo Casals made about the organ and its lack of nuance: "The organ is of itself a very mechanical instrument. The expressive value of string instruments can only be produced through a kind of phrasing that is done by the advancing and retarding of notes, by the pauses and the length of the sound . . . its tone cannot vibrate with the rhythm of the organist's hand—and yet, we can recognize immediately if the player is a real musician."[40] The little metric irregularities here referred to are usually employed to produce pathetic accent, and therefore belong pre-eminently to romantic music; the organist, however, at least in the context in which Casals spoke of him, uses them to replace the stress on particular notes which normally produces dynamic accent, i.e., a formal element.

The upshot of all these considerations is that the materials and processes which may be employed in artistic expression are practically limitless, and give an indefinite scope to new conception and invention. As Casals said in another conversation, "Composers can use any material as long as they make a coherent whole of it, a whole which will express true life in music."[41] This freedom of artistic creation has a peculiar in-

[39] Émile Bernard, in his *Essai de la philosophie musicale* (1917), remarked on this solution of the harpsichordist's prolongation problem.

[40] Reported by J. M. Corredor, in *Conversations with Casals* (1957), p. 126.

[41] *Ibid.*, p. 175. In a valuable, but apparently little-known, book by W. Harburger, *Form und Ausdrucksmittel in der Musik* (1926), there is a comparable passage, interesting particularly in that he makes two essential demands on any work of art, namely, that it shall have "substance" and "physiognomy." So he says: "Im Prinzip soll uns hier jedes Mittel recht sein, wenn nur das, was dabei heraus kommt, etwas taugt, ein 'Gesicht hat'" (p. 8). On the same page he inquires "worin . . . die individuelle wie auch die musikalische Substanz irgendeiner Komposition liegt." And on the next: "Aber auch manche Dinge, die uns heute abgetan erscheinen und gegen den Strich gehen, können trotzdem musikalisch gekonnt sein und Physiognomie haben."

fluence on the created form itself. Even the beholder who is not at all versed in the technique that produced the piece he is contemplating knows at sight that the artist could have done differently; not because that thought occurs to him, but because the work itself expresses intent. With its expression of the artistic idea it conveys a sense of reasonableness, a tacit explanation of its growth. In a good work, every subordinate form appears not only to emerge from the inner substance, but also to be prepared by something else, perhaps by many factors. In a picture, every color or line or even lacuna that sets up a virtual tension does so because other factors are there to meet it. If we start with any feature, its place appears to have been ready to receive it, and its occurrence, in turn, to support the creation of qualities that one would attribute primarily to other constituents of the picture. It is as though a series of preparations were simultaneously taking place. This means, of course, that the sense of preparation is symbolically projected as an aspect of the artistic import.

It is here that Immanuel Kant made his major contribution to the philosophy of art, in the concepts of telic form without purpose and perceptible rationality without discursive logic. That such insights could stem from a thinker generally as devoid of artistic leanings as Kant is really astounding, and attests the strength and candor of his mind to meet paradoxical conditions that fitted none of his intellectual habits. The *Critique of Judgment* bespeaks a recognition of the most baffling problems of art—the objective validity of judgments made without reasoning, the "free lawfulness" of beauty, the telic directedness of creation without practical purpose. That these problems are not really solved in the *Critique* is due mainly to the lack, in Kant's day, of the semantic concepts which have grown up in the wake of his own "Copernican Revolution" over a period of a century and a half. Perception, rather than significance, was the central epistemological theme of his time. But his third definition of beauty as the form of purposiveness without any idea of purpose[42] clearly denotes a presentational abstraction; and his fourth definition of the beautiful as that which is "necessarily" recognized, without the aid of a concept, as an object of aesthetic approval[43] certainly expresses the notion of formal intuition. That taste was the arbiter of beauty and pleasure its criterion were obvious assumptions in default of

[42] *Kritik der Urteilskraft* (1790 and later eds.), sec. 16: "*Schönheit* ist Form der *Zweckmässigkeit* eines Gegenstandes, sofern sie *ohne Vorstellung eines Zwecks* an ihm wahrgenommen wird."

[43] *Ibid.*, sec. 22: "*Schön* ist, was ohne Bergriff als Gegenstand eines *notwendigen* Wohlgefallens erkannt wird."

the concepts of artistic import and intuition; yet Kant's definition of taste as the power of judgment of the aesthetic pleasure-value of an object without any other interest,[44] and his imputation of universal validity to the deliverances of such judgment, bring him very close to the idea of the objectification of subjective realities; his "common aesthetic sense" is essentially the appreciation of projected living form.

Actually, the purposiveness of artistic forms is not so purely illusory as Kant thought, but the purpose is one he did not know—expressiveness. And the freedom is not absolute, but is bound to a range of options. In fashioning a work of art, its maker is constantly faced with choices of possible moves, every realized possibility precluding others that might otherwise have materialized, and with them many of their potential consequences; but even as every realization dooms its alternatives, it gives rise to new options. An option is a situation in which two or more incompatible acts are prepared so the performance of any one of them would appear reasonable. The artist's work proceeds from one option to another. In every instance, his choice is based on a great many considerations, ranging from stylistic habits, predictability, safety, etc., to the lure of novelty and the promise of creative functions in the various possible devices. Some combination of motives must, of course, determine him; but every decision rests on so many personal factors—tendencies, blocks, special powers and weaknesses, biographical twists—as well as technical circumstances like the nature of tools, or some unique advantage offered by the material (grain, color, hardness, etc.), or the foreseeable environment (especially for architecture, and sometimes sculpture), that for him the long series of decisions which make up the course of his progress may take on every kind of appearance; it may seem to develop automatically, or to be controlled entirely by the work, or to be imposed by his will on stubborn materials resisting his intent.

All these personal experiences and motivations, decisive as they may be for the success or failure of his work, do not belong to its import. Only one aspect of his creative adventure enters into the artistic projection itself: the sense of movement from option to option, the recurrent progression from potentiality to realization, every decision producing new possibilities and offering new choices. This dynamic pattern belongs to art itself, because it is an inescapable pattern of life. That is why a really "living" work always seems reasonable in every respect, yet not predictable, as though it could, nevertheless, have been different. All its

---

[44] *Ibid.*, sec. 5. "*Geschmack* ist das Beurteilungsvermögen eines Gegenstandes oder einer Vorstellungsart durch ein Wohlgefallen oder Missfallen *ohne alles Interesse*. Der Gegenstand eines solchen Wohlgefallens heisst *schön*."

internal articulations are prepared, yet not caused, by their environment; the resulting impression is of motivation instead of causation. Aristotle stated the case simply, when he said of the virtual events in tragedy: "Such incidents have the very greatest effect on the mind when they occur unexpectedly and at the same time in consequence of one another: there is more of the marvellous in them then than if they happened of themselves or by mere chance."[45]

The semblance of motivation is another powerful factor in making artistic elements similar to acts rather than to things. Where motivation appears as a fundamental and pervasive relation, its domain—that is, the work—exhibits the form of the great fabric of acts we call a "life." Motivation entails the concept of acts. But when one speaks of a "motive" in art, the word does not usually refer to what motivated the artist's procedure, but to relations of forms within the piece itself; the variant, "motif," is therefore a justified term, as it removes the ambiguity.[46] A motif serves to create a virtual motivation. It prepares a context for a number of developments, some of which are realized, their alternatives thereby ruled out.

The image of life as motivated activity reflects an aspect of animate nature that has baffled philosophers ever since physics rose to its supreme place among the sciences, because inanimate nature—by far the greatest concern of physics—has no such aspect: the telic phenomenon, the functional relation of needs and satisfactions, ends and their attainment, effort and success or failure. There are no failures among the stars. Rocks have no interests. The oceans roar for nothing. But earthworms eat that they may live, and draw themselves into the earth to escape robins, and seek other worms to mate and procreate. They need not know why they eat, contract or mate. Their acts are telic without being purposive.

Teleology is peculiar to vital process, and since it appears in psychical phase as the pattern of aims and voluntary acts, realization, frustration, questioning, solution or bafflement, it is a central theme in the study of mind; but it has had the least successful philosophical treatment of all our major problems. The main reason for this failure is, of course, that teleology is confusedly and yet intimately associated with so-called "free

---

[45] *De Poetica,* 1452ª.

[46] I was once invited to watch José Limon teach a class of young dancers. He demonstrated a rising movement from sitting on the floor to standing on the balls of his feet with uplifted arms, and asked: "What motivates the raising of the arms?" The students offered various answers, such as: "an expression of longing," "aspiration," "awakening." "No," he replied; "the movement of the hips motivates the raising of the arms." The students were searching for a motive, but the master had asked for a motif.

will," with vague notions of purpose carried into cosmology, and indeed with religious and moral issues generally, over which the protagonists of scientific thought and the defenders of the faith fight lusty battles. Let us, at this point, bypass those issues, and look at telic patterns as they appear in the artist's image of life, where no fear of doctrinal words besets their exhibition.

In art, as in life, and nowhere else in the universe as we know it, we find the conditions of necessity and freedom. Freedom is having an option; necessity is the lack of any option. But options belong only to living things, and where the concept is not relevant, as for instance in astronomy, the notion of necessity does not occur either. Only a mythical conception of physical nature leads moral philosophers to bemoan or admire its "unalterable law" and the "necessity" of causal sequences. Physicists do not think of the determinate quantities in their calculations as more "necessary" than indeterminate ones, because indeterminateness does not involve options. Neither has probability or randomness anything to do with choices.

If, therefore, a work of art makes an impression of free movement, whimsy or—in a serious vein—decision and sacrifice, it projects the basic character of life; and so it does if it seems to embody a necessity, an inevitable conclusion or completion. Necessity is as much a vital phenomenon as option; it is the final elimination of all options. Dwindling options and growing necessity, widening options and receding necessities make the "tide in the affairs of men." The semblance of necessity or "inevitability" in art, which has been remarked with wonder since the beginnings of aesthetic philosophizing, is of one piece with that of internal freedom, which is even more widely attested and praised. These two pervasive elements are counterparts, and can enter into all sorts of dialectical relations with each other, even intersect—which they do, indeed, in every optional act: for the agent, if he may choose, also must choose. That is the real "inexorable law" of nature: the perpetual advance of life, from one situation to another, unbroken from birth to death. But the sense of inevitability which all great art conveys is most perfectly made without reference to that actuality. It is created by the fittingness of forms, the build-up of tensions and the logic of their resolutions, the exact degrees to which the elements are articulated, etc.; the idea is completely abstracted from actual life, transformed into quality, projected in sensuous terms. Yet we know it for what it is.[47]

[47] There is a truly classical example of preparation making for a growing sense of inevitability, although the conclusion is not predictable, in Schubert's "Der

All the traits of "livingness" in art which have been discussed so far are prerequisites for the expression of the most consequential aspect of life, that which ties together its lowest and its highest forms: individuation.

Individuality has been mentioned before, in connection with the apparent uniqueness of every good art object even though it be actually one of many copies. But there is little one can say about "individuality," in the sense of "uniqueness," except that in art the semblance of it is what counts rather than the actuality, and that this semblance is part of the illusion of organic form. An object is unique or it is not, and a virtual object seems unique or it does not. With regard to the conception of actual living form, this quality of the art symbol cannot, in itself, give us much help.

But the conceptual difficulty that besets us here is the same one that we have already met in the handling of "the subject-object relation" broached in Chapter 1 (p. 31); it may be met by a different turn of phrase which simply replaces the concept of an attribute by that of a process. Individuality either may or may not be predicated of an object, but individuation is a process, capable of degree and direction, of being either steady or intermittent or transient, or incipient; it may be reversed, because it has a converse, which will, in fact, play a great part when we come to problems of actual life: involvement.

In artistic organization the mutual influence of forms is patent; they are often involved with each other to the point of complete interdependence. But at any point a subordinate form may take on an active, even ruling function, a motif may develop and detach itself from the fabric of the whole to some degree; it launches on a process of individuation. This is prepared by the semblance of becoming, of something constantly emerging, which is part of all "livingness" in art; but special internal individuation occurs in every piece that may properly be called a "work" —that is, for instance, a picture as against a decorative design, or a lyric

---

Wegweiser" (opus 98, *Winterreise*, No. 20). The occurrence that is presaged from the first measure is a monotone reiteration of the tonic in unvarying beats:

But the figure is shifted from one tone of the scale to another—*b*-flat, *e*-flat, *c*, through the major and back to the minor key; until, in measure 54, it becomes fixed on the tonic. Even then it alternates with *f*, *e*-flat, *a*, *d*; finally, however, the *g* is repeated steadily through measures 67–74, to the crisis in 75–76 and the cadence following close upon it. The impression of necessity is absolute. See Appendix, Example 4.

as against a nursery rhyme—to make it express not only uniqueness, but an internal individuation. This is not to belittle the artistic character of pure ornaments or Mother Goose rhymes; but their excellence lies in their expression of the more elementary vital rhythms, without any superstructure of higher developments, and this makes a categorial difference between them and "works," properly so called, in the arts.

The higher development of forms may be actually carried out without producing any impression of individuation; it may produce an appearance of great, even enormous, proliferation, teeming life, and yet maintain the character of universal rhythms in their endless regeneration. Usually such ambitious design animates a surface in a pictorial, sculptural or architectural work, and so enters into a higher form (see, for example, Fig. 7–2). As a rule, however, pure design, springing from a direct visual interest, is folk art; led on primarily by a love of color and form, it develops motifs and traditional decorative styles that belong to the people and embellish their homes, their weapons, tools, utensils and dress.

Where, on the other hand, the mainspring of an artistic tradition is the desire to depict objects, plastic art becomes pictorial at an early stage. But this does not mean, in itself, that it rises to the level of "works," as paleolithic art did; for it may be entirely subordinated to communication and become a cursory symbolism, even script, which has artistic aspects and uses (like Egyptian hieroglyphics or Chinese calligraphy), but only elementary expressive powers.

True works of art arise out of the standard artistic practices of a society, when someone who is prone to see the expressiveness of plastic forms in a way of his own modifies the familiar motifs to project that personal mode of feeling. In a gifted people this may happen frequently and freely, especially where the motifs are largely or wholly representational, so that new forms constantly present themselves to the reverberating sense and spontaneously abstractive eye. Then, among the countless animal images crowding walls and rock faces, there appears the splendid Altamira bison or the reindeer of Font-de-Gaume. But pure design may also, at almost any point, bring forth an autonomous form, and be transmuted into an individual expression, with the beauty of an emotional idea; or it may be invaded by a new element which changes its decorative motifs and rhythms into supports for a different composition altogether, and makes them seem to emerge from a matrix that did not exist before, but which now they serve to constitute and also derive from.

Wherever such a projection of inward vision takes place in the making of a piece, its maker is approaching it not in the spirit of a schooled craftsman, but as an artist, and the piece is a work of art, whatever the

Figure 7–2. Dome of the Mosque of the Shāh of Isfahān
an appearance of great, even enormous, proliferation, teeming life
(André Godard, *L'Art de l'Iran* [Paris: Arthaud, 1962]. Photo by Roger-Viollet.)

success or shortcoming of its realization; its measure is more than vital expression, it is the expression of his idea, his personal conception of the ways of feeling. From the very inception of the work it is not a familiar rhythm, the typical expression of a style, that guides the artist, but his personal idea; and the central feeling on this level is always specifically human feeling.

All that has been said here with reference to plastic art, especially painting, holds, of course, for all other arts as well. The same phenomenon occurs in them all: sooner or later, great minds depart from their popular inheritance, and seek to express new and personal conceptions of inward life, especially the rise and course of emotions, and the coloration that each artist's particular "habit of emotion" (to borrow Mr. Murry's term once more) lends to all the sights and sounds and events of the outward world. In art—any art—the effect of this departure is a new created element: as the artist's work proceeds, the development of forms seems like more than elaboration, it seems like a gradual individuation. The actual procedures—adding, filling, simplifying, clarifying, etc.—are not new, but the semblance of growth is new, for it seems to be an individuating process instead of a typical organic completion.

Since the new element appears as a process in the virtual life of the piece, it also has the appearance of reaching a degree beyond which it still could go save that something arrests it—the rivalry of other processes perhaps, or the exhaustion of its own impetus. All such effects, of course, are technically achieved, though perhaps the device is sought and found intuitively.[48] In Egyptian sculpture, for instance, every contour that is

[48] Cf. Schoenberg, *Style and Idea*, pp. 62–63: "When Brahms, toward the end of the last movement of his Fourth Symphony, carries out some of the variations by a succession of thirds,

he unveils the relationship of the theme of the Passacaglia to the first movement. Transposed a fifth up,

it is identical with the first eight notes of the main theme,

articulated at all is carried to completion; its growth impulse is spent. The represented being may be young, but the statue is always mature. This, more than any conventional pose, makes it hieratic, in spite of the frequent realism of physique and features. It is grown form, not growing; its limits of individuation are predetermined by those of its initial force. Greek sculpture, by contrast, seems in the process of individuation. Forms interplay and suggest each other, edges disappear in volumes, planes merge their movements. If Greek art (this goes for architecture, ceramics, drama, or what you will) makes an impression of perfection, it is different from the perfection of great Egyptian works; it reaches a perfect moment, like an eternalized act of coming into its own.[49]

Often it is not possible to say wherein the semblance of individuation consists, and what makes a piece really a "work"; but the personal idea is there. Subordinate forms in a design may be developed so their values arrest one's vision separately, and the rhythm of the whole is imbued with an apparent readiness to give way to new and growing tensions. Subtle asymmetries of placement or contours, accentuations, enhance this air of fecundity, which puts the piece on a higher level than traditional taste and craftsmanship. The bowl pictured in Fig. 7–3 is a case in point. Similarly, the luminosities in an essentially symmetrical rose

---

and the theme of the Passacaglia in its first half admits the contrapuntal combination with the descending thirds."

"It would look like a high accomplishment of intellectual gymnastics if all this had been 'constructed' prior to inspired composing. But men who know the power of inspiration, and how it can produce combinations no one can foresee, also know that Wagner's application of the Leitmotiv was, in the great majority of cases, of an inspired spontaneity."

The passage under discussion is given in the Appendix, Example 5.

[49] Helmuth Duve, in "Das Bewegungsprinzip in der Skulptur" (1928), says of Greek sculpture: "Die Form wird schliesslich in ihrer Erscheinung so eigenwertig, so durch sich selbst geltend, dass sie den Raum von sich abweist und dieser weder positiven noch negativen Anteil an ihr nimmt. . . . Die Form sättigt sich mit Bewegung, die sie veranschaulicht, so dass der Eindruck der gestalteten Masse gerade noch dem Ausdruck der aufgefassten Bewegung die Wage hält. . . . Die Begrenzung der skulpturalen Form in den Profilen ihrer vielfachen Ansichten—wobei die Linienzüge, die sie begrenzen, ineinander übergehen und aufeinander hinweisen— vollzieht sich in kontinuierlicher Rhythmik . . ." (pp. 439–40).

Figure 7–3. Greek Bowl with Fishes

*subordinate forms in a design may be developed so . . . the rhythm
of the whole is imbued with an apparent readiness to give way
to new and growing tensions*

(Collection of the Museo Nazionale, Taranto.)

window, modified by the deployment of colors, and constantly, properly,
subject to changes of light,[50] may serve to give the whole pattern an air
of incipient individuation which is finally realized in the vertical
windows.[51]

The essence of design is repetition, and almost any repetition of elements at regular intervals (i.e., intervals which also repeat each other)

[50] The tolerance of works of art to fortuitous conditions is highly variable, ranging
from almost zero in pieces which require very strictly prescribed performance, or
specialized environment for display, so that they are readily ruined by any outward
change, to a high degree of self-sustainment under various conditions, and even—
as in the case of stained glass designs—to their assimilating of chance factors and
transforming them into organic elements, making their very accidence an asset. This
is a trait which has a parallel in actual organisms.

[51] André Malraux has remarked that the great pictorial art of medieval Europe is
in the cathedral windows. See *Le Musée Imaginaire* (1949), p. 37: ". . . le sommet
de la peinture au XIIᵉ, au XIIIᵉ et au début du XIVᵉ siècle hors d'Italie, ce n'est ni
telle fresque, ni telle miniature, c'est telle morceau des verrières de Chartres."

or in immediate juxtaposition will engender a design, as the kaleidoscope readily shows. Bilateral symmetry is repetition, and so is every sort of periodicity. The principle of repetition goes all through nature, and this fact is reflected by its structural role in art. But some aspects of the units deployed in a pattern may be repeated while others are not, and those which are so keep up the rhythm of the larger whole despite the few that are breaking it; this produces the appearance of "irregular regularity"[52] which seems like an incipient individuation at the level of lowly, even vegetative, life.

Where the semblance of growing individuation is paramount, it makes a radical shift in the image of living form. It effects the humanization of artistic import; hence, whenever it gains ascendency in one of the great orders of art, that art begins its Golden Age. Even then, there are degrees of individuation, which are apparent in the qualities of the average achievements in the work of an age as against the qualities of its greatest masterpieces, where it eventuates in the dynamics of fully developed human emotion and even of thought and understanding.

The process of realization of a work of art is usually the development of various elements that are implicit in the "commanding form"—the artist's pristine "idea"—to unequal extents, giving rise to subordinate internal forms, centers or lines of highest activity. These are distinct within the whole, set off by boundaries generated from within. Their boundaries, however, articulate with other forms, being each other's intaglios, which makes the continuity of the whole.[53] This dialectic of separation

---

[52] Wladimir Weidlé, in an article entitled "Biologie de l'art" (1957), attributes this phrase to Buytendijk, who remarked on the pervasiveness of such quasi-regularity in nature, but, according to Weidlé, was not greatly interested in its exemplification in art, although he was aware of it.

[53] Such elements may be single tones or phrases, words or poetic ideas, virtual volumes positive or negative (i.e., empty), virtual motions, architectural spaces. So Casals said, when his interlocutor read him a comment on his playing, namely, that "when he plays, each note is a prophecy or a recollection; . . . the presentiment of what is to come, as well as the remembrance of what has been" (Divan Alexanian, in his *Traité théorique et pratique du violoncelle* [1922]): "That is what I imagine form should be . . . nothing should be isolated, each note is like a link in a chain—important in itself and also as a connection between what has been and what is to be" (Corredor, *op. cit.*, p. 194). Helmuth Duve (*op. cit.*, p. 440) says something similar about a moment of arrest as a phase division in virtual movement (which he calls "*extensive Bewegung*," meaning by this term what Segond calls "*mobilité*" as distinct from "*mouvement*"): "Das Wesen jener extensiven Bewegung, die aller wahren Plastik eigen ist, besteht in . . . der Veranschaulichung jenes kurzen Ruhemomentes, das zwei Bewegungsphasen trennt und zugleich verbindet." And subsequently, even more clearly illustrating the mutual articulation of elements, he

and connection is typical of organic structure, and has often been remarked by biologists. Its image in art is the image of individuating force, unequal growth, which underlies all morphology and is the fundamental mechanism of evolution; hence its power to raise artistic expression to a level of complexity that reflects not only universal vital rhythms, but particularly human ones.

Human mentality, however, is not set apart from that of other animals only by a higher emotional or sensory development; its unique characteristic, which makes it something different in kind from animal mentality, lies in the constant stream of cerebral activities which are essentially subjective, having no perceptible overt phases, but terminate as images, ideas, thoughts, recollections, often elaborate figments, entirely within the organism in which they take rise. Those phenomena have a pattern of their own, unlike any other large and well-known order of events; they are elusive, occur without visible source and usually end without visible trace; placeless, yet highly individual, sometimes systematically continuous, sometimes repetitious, at other times madly mixed and rapid in their passage, and usually tinged, if not saturated, with emotional feeling. The most persistent impression we have of such events is that they "are and are not really there." Yet they are generally estimated as the highest values in life, lifting human existence out of the animal world into a different realm altogether. Most people, therefore, ascribe to such purely subjective phenomena a different metaphysical status from that of material objects and physical events, and postulate a separate sort of reality, "consciousness," intrinsically psychical and not a phase of physiological functioning, which embraces them as its own, essentially non-physical, contents. How little any scientist can do with "contents of consciousness" the earnest efforts of great men have long demonstrated; how little can be done while ignoring the intraorganic climaxes of human mental acts psychologists great and small are still demonstrating. The fact is that people operating with familiar physicalist models do not see that the wraith-like character and mysterious coming and going of images and thoughts are a peculiarly interesting aspect of cerebral activity, which has never been studied or even precisely envisaged.

The art symbol, however, reflects the nature of mind as a culmination

points out how in Giovanni da Bologna's "Sabine Women" space is sculptured with the solid forms: "Die Raumabschnitte zwischen den Beinen und Armen gehören als negative plastische Werte zur Komposition und Kennzeichnen den Anteil des Raumes, der da, wo die Plastik endet, beginnt und an ihren Konturen allseitig sich abgrenzt, sein Dasein anschaulich geltend macht" (p. 442).

of life, and what it directly exhibits, first of all, is the mysterious quality of intangible elements which arise from the growth and activity of the organism, yet do not seem entirely of its substance. The most powerful means to this end is a practice which has already been discussed in other respects: the creation of secondary illusions. It figured in the previous chapter as an abstractive device, and earlier in the present one as a way of giving the work just that virtual substance which the image of psychical occurrences must appear to transcend (see above, pp. 206–7). At the same time, it serves most readily to produce that image, with its appearance of disembodied being.

The semblance of substantiality may arise from many conditions in a work, one of which is the contrast between the permanent, ubiquitous character of the primary illusion and the transient, incomplete character of all secondary ones in its domain. This same contrast stresses and enhances the insubstantial nature of secondary illusions. These result from special uses of the material, which are closely allied in principle with "sensuous metaphor," the symbolic equivalence of sensations. In the emergence of a secondary illusion, such as the sudden impression of color in music, or of eloquence in the lines of a statue, an element is created that seems to belong to a different symbolic projection altogether from the substance of the work. The effect is a sublimation of the expressive form.

In all advanced artistic creations there is some such play of secondary illusions over the unfailing, all-supporting primary illusion, with various effects, ranging from the expression of elementary feeling as a transient phase emerging from the organic matrix (as in rhythmic designs) to the appearance of acts that seem to depart from their somatic source, and form a separate, essentially mental pattern. A secondary illusion may be so elusive that it seems like a purely personal impression, and one is surprised to find that someone else has experienced it, too; or it may be so powerful that no one can fail to receive it. Of the latter sort is "harmonic space" in music, which musicians refer to as familiarly as they do to "tempo," "movement" or "sequence."[54] Similarly, in the plastic arts all movement is a secondary illusion, but its appearance as "spatial rhythm" is so normal that few people realize it is secondary. Yet space in

[54] Heinrich Schenker, for instance, assumes it in the very definition of music: "Musik ist lebendige Bewegung von Tönen im naturgegebenen Raum" (*Das Meisterwerk in der Musik*, Vol. I, p. 12). He does not even recognize his tonal space as virtual, but regards it as "*naturgegeben*." A discussion of spatial illusions in music, as well as corroborations from the literature, may be found in F.F., chap. vii, especially pp. 117–18; the significance of secondary illusions, however, was not clear to me at the time of writing that book.

music is not homogeneous and complete, nor is time or motion in visual works; both are secondary, but genuine artistic elements—that is, neither is made by extraneous suggestion, as motion, especially, is sometimes thought to be.

The readiest illustration of the difference between the representational suggestion of motions and the secondary illusion of movement may be seen by comparing a documentary photograph of persons dancing with a rendering of same theme in painting or sculpture. The camera records a momentary attitude which is obviously untenable beyond the captured moment, and therefore suggests a subsequent posture, the conclusion of the movement, to the percipient's common sense (Fig. 7–4). If the photographer uses his instrument primarily to create a picture, presenting movement instead of rationally implying it, the dance becomes material, and the presented movement may very well be something quite different from dance—a balletic leap, for instance, may appear as soaring flight (Fig. 7–7). But a freely created plastic image of a dancing figure holds its pose without inviting one to imagine a change, and yet it dances (Fig. 7–5). Its motion seems perfected, not suspended, in the image (Fig. 7–6); it is abstracted without being developed, like the "color" of music or speech, unanalyzable in any systematic terms (in this case, i.e., in vector terms), for it is a secondary illusion striking into the primary one of fully organized virtual space.

The role of secondary illusions, however, may be a major one, and the techniques that beget them sometimes lift artistic conception to entirely new heights. Then, naturally enough, the astonishing illusion seems even to expert critics to be the very substrate of their art.[55] The phenomenon of "harmonic space" is an interesting example of the creative power which such a principle of construction (as against a principle of art as such) may exercise; for the spatial illusion made by chordal structure has lifted European music to a level unrivaled by any other musical

[55] George Borodin holds that dance is the creation of virtual events in the dream mode, and as he recognizes this to be the poetic mode of cinematic art, he treats ballet and film as two variants of essentially the same art. In *This Thing Called Ballet* (n.d.; *ca.* 1944) he says: "The basic materials of both the ballet and the film are similar. . . . It is, in fact, only that the idiom, the turn of phrase, is different. The difference between ballet and film is very similar to that between two languages having a common origin—as, for example, Italian and Spanish, or Dutch and English" (p. 56).

"The ballet . . . is a projection of the dream into the rational life of man. . . . When the spectator looks at a ballet and listens to the music, he is, in effect, seeing a living dream" (p. 59). But the illusion of dream events in dance is a secondary illusion which arises in various degrees from the unbroken interaction of virtual powers.

Figure 7–4. Dancer's Leap (Merce Cunningham)
*the photograph arrests a momentary attitude which . . . suggests
a subsequent posture . . . to the percipient's common sense*
(Photo by Jack Mitchell.)

Figure 7–5. Greek Cup with Leap Dancers

but a freely created plastic image of a dancing figure holds its pose without
inviting one to imagine a change, and yet it dances

(Collection of the Staatliche Museen zu Berlin, East Berlin.)

tradition. In Victor Zuckerkandl's *Sound and Symbol*[56] there is a lucid exposition of the way in which simultaneous harmony, based on the complexities of tone itself, the overtones that yield the "cycle of fifths" and the natural relations of keys,[57] conjures up the inescapable secondary illusion of space. In this tonal framework, the intervals between tones become powerfully apparent virtual elements which take on every sort of significance according to their places in the new pattern of musical tensions. "The interval actually heard," he says, "does not extend be-

[56] 1956. An interesting book, unfortunately somewhat impaired by polemics against alleged beliefs in "science" and "philosophy" which are long out of date, though most of them were in vogue at some time, and by metaphysical concepts that betray the amateur philosopher. Its author's notion of the philosophy of physics fits the thought expressed in Büchner's *Kraft und Stoff* in 1855 (cf. especially *Sound and Symbol*, p. 55: "Without forces, no bodies; but equally, without bodies, no force"). He appears to be so convinced of the folly of "the philosophers" that he simply will not read them, and consequently offers as new discoveries such ideas as the

Figure 7–6. Shiva Nataraja, South India

its motion seems perfected, not suspended, in the image

(Collection of the Cleveland Museum of Art, Purchase from the J. H. Wade Fund.)

distinction between discursive and non-discursive symbols (in a chapter entitled "The Dynamic Symbol") without reference to Cassirer, and practically paraphrases the theory of "virtual time" and "virtual space" in F.F., which he appears not to have read (in his chapter xiii, "Tone as the Image of Time"). Yet the ideas he presents which are really his own outweigh all these failings.

[57] The structural aspects have been treated by many musicologists, such as Heinrich Schenker (see p. 158n) and W. Harburger (see p. 163). But Mr. Zuckerkandl is concerned with a more general problem of musical creation.

tween two different pitches; it extends between two different dynamic states" (p. 92). Yet the music is not static; the harmony itself is directed "forward," i.e., toward cadence. Most people will agree with his assertion: "When we hear music, what we hear is above all motions" (p. 76). It has been said countless times. But the usual meaning given to "motions" is "melodic motions," which are essentially motions up or down from one tone to another within the gamut of pitches. A progression of

Figure 7–7. Dancer's Leap (Jacques d'Amboise)
a balletic leap, for instance, may appear as soaring flight
(Photo by Fred Fehl.)

chords, however, is not a motion up or down; yet neither is it simply a succession, like a single tone sustained or repeated. In a chordal sequence, "the step from the root e to the root a is, if you like, a step from all the e's in tonal space to all the a's in tonal space—from what is common to all the e's to what is common to all the a's. Is this step in tonal space directed upward or downward? . . . We see that the question has no meaning . . . no chord can be 'higher' or 'lower' than another. The chord is, so to speak, a sound-state of all tonal space. Chordal steps lead from one sound-state of all tonal space to another sound-state of all tonal space. Such steps . . . seem steps *of* tonal space rather than *in* tonal space.

"The chordal step V–I is called the *cadence*, the 'fall.' It is so called in all languages. Does this not, after all, contain an explicit statement . . . that it is a step directed downward? We understand to what this statement refers: to the dynamic meaning of the chordal succession, . . . the arrival at the center of gravity."[58] The tones forming the chords may move upward or downward toward I (in fact, a standard cadence of our hymn tunes is

wherein the root of V moves up to the root of I, and the leading-tone in V moves up to the tonic above); the harmonic phenomenon is still cadence—"the arrival at the center of gravity." This is release of tension, a passage of feeling, which figures in our perceptual experience as a cadential quality.

Melodic motion abstracts the feeling of locomotion; it is a sound image of locomotion where nothing is transferred from one place to another. Even if we have a strong impression of musical space, and tones of different pitch appear to occupy different places in it, the movement of a melody from tone to tone is not a displacement of anything. But harmonic progression makes a further abstraction, for it does not even create any semblance of locomotion. It makes a direct and very pure abstraction of a feeling of pure temporal change, which is its outstanding quality.

Here, then, we have the "disembodiment" of an element emerging from organic structures. What occurs in harmonically structured music is a double abstraction, namely, of pure movement and furthermore of change without movement, pure transmutation. It is this double abstraction, made by a single new device—simultaneous harmony—that gives music the power to symbolize the passage and constant transformation

[58] *Op. cit.*, p. 114.

of acts which are finally consummated in purely inward events, in the emergence of fantasies, memories or verbally formed thoughts. The "intellectual" quality of such music has often been remarked, yet nothing is more emotional than the harmonic advance of chords.[59] Notice, however, that this advance does not create any semblance of "mental contents," but only of their psychical nature and the dynamism of their fluent shaping.

The development of a secondary illusion to such a degree as that attained by "musical space" with the advent of simultaneous harmony is exceptional; the escape of artistic import from somatic feeling is usually a matter of high moments, in which a work of art attains what seems like a sublimation of its perceptual form. In every humanly significant work there is a wealth of secondary illusions; they intersect with the primary illusion and with each other, and are often the appearances most keenly noticed by good judges of art. When Le Corbusier said: "The Cathedral is not a plastic work; it is a drama,"[60] he felt the dramatic illusion suddenly overriding the primary one of space, which in architecture is an ethnic domain.[61] But the building is not a drama; it is "dramatic." In drama, conversely, every sort of secondary illusion may occur, and may even seem to be its substance. Robert Edmond Jones says, "Every play is a living dream."[62] Jones usually knows very well what is a secondary illusion, even to the point of knowing his own work—stage design, the creation of virtual place—as secondary, destined to arise and

---

[59] Schopenhauer regarded music as a direct presentation of the Will, without embodiment, whereas all the other arts had to present its embodied forms (*Die Welt als Wille und Vorstellung* [1818], Bk. III, last section). For later observations on the transcendence of somatic feeling in music see, *inter alia*, Émile Bernard, *Essai de la philosophie musicale* (1917); and William A. Fisher, "What Is Music?" (1927). The last-named author writes: "Superficially it may seem to be a purely emotional experience, but the musician knows better, for the composer is a *thinker* in terms of music, and the great composer is a great thinker and a great artist in one. The substance of absolute music is therefore emotionalized, wordless thought . . . ; the expression of that which transcends the limitations of inelastic speech . . . ;" etc. (p. 14). Similarly, Richard Wallaschek, in an article "On the Origin of Music" (1891), wrote: "The power exerted over us by any rhythmical movement lies in its being adjusted to the form in which ideas and feelings succeed each other in our minds. . . . The effect produced in us . . . is due to our recognizing, in the intensity, strength, velocity, increase and decrease of the [musical] movements, forms, corresponding to the flow of our ideas and feelings, though the nature of that flow depends entirely on each individual psychical organism" (p. 378).

[60] *Op. cit.*, p. 30.

[61] Cf. F.F., pp. 95 ff.

[62] *The Dramatic Imagination* (1941), p. 28. In that same book he deplores the confusion of narrative fiction, drama and film (of which last virtual dream is the poetic mode), yet himself considers film as a new sort of theater art.

collapse again.[63] He says of a beggar's rags on stage, "These rags have somehow ceased to be rags. They have been transformed into moving sculpture."[64]

In most instances, secondary illusions present themselves as intangible elements deriving from some other order of existence than the virtual substance of the work in which they occur. The possibility of their occurrence makes the art symbol capable of reflecting the many-dimensional and incalculable character of experience. This widened scope of its expression may even be attained without going outside of its own primary illusion for a "secondary" effect, because various modes of the same primary illusion arise very readily—sculptural qualities in painting or architecture, pure lyric moments in drama or dramatic ones in prose narrative or lyric poetry. Heinrich Wölfflin once pointed out that when architecture falls into ruin its appearance becomes "picturesque," because the lines of walls and windows lose their space-dividing functions and are assimilated to the shifting and merging contours of solid volumes.[65] Also, it has often been noted how some monumental buildings and many Greek temples, viewed from an optimum distance, appear more like sculpture than like architecture. In the poetic arts the lines between literature, drama and film are sometimes ignored or even defied by serious scholars. Both T. S. Eliot and Francis Fergusson class Dante's *Divine Comedy* as a drama, and cite it in illustration of their dramatic theory.[66] With more precise evaluation of the modal relationships, Georg Simmel observed that in Michelangelo's painting the figures, though genuinely pictorial, had the loneliness which is intrinsic to sculpture, and that this essentially sculptural isolation gave them their "titanic" character, which is usually attributed to their exaggerated physique, but really is due to their isolation, the semblance of a superhuman self-sufficiency, complete individuation. In his painted human forms as in his statues all internal elements are completely interdependent so that the life of the figure seems to be consummated in its own confines.[67]

---

[63] *Ibid.*, p. 27: "The designer creates an environment in which all noble emotions are possible. Then he retires. The actor enters. If the designer's work has been good, it disappears from our consciousness at that moment. We do not notice it any more. It has apparently ceased to exist . . . and the designer's only reward lies in the praise bestowed on the actor."

[64] *Ibid.*, p. 33.

[65] "Über den Begriff des Malerischen" (1913).

[66] Cf. F.F., p. 334n.

[67] In his "Michelangelo" (1911) he says of the paintings in the Sistine Chapel: "Es sind keineswegs etwa 'gemalte Skulpturen,' . . . sie sind durchaus nur als

The interplay of modes in the selfsame primary illusion (while the work as a whole, nevertheless, keeps its particular modal character) rarely has the intense effect of a genuine secondary illusion, which seems to present to one sense what normally reaches us via another; yet both are of the same category. Their functional importance is the same, only at different points in the development of the life image. Any secondary illusion, whether it is striking or quite subdued, is perceived as an act of transition from one level of feeling to another. It is transition made perceptible. Such a presentation of pure passage always carries with it an effect of "strangeness," as Owen Barfield observed,[68] which more extravagant writers often call "magic." Where it is produced by a secondary illusion of the commoner sort, i.e., incursion of another mode, it tends to enhance the semblance of potentiality to the point where the work seems ready at any minute to take on a different organization. This is what makes an expressive form "exciting" without any daring novelty or representational appeal. It gives it a fullness of life that brings it to the brink of a transformation.

A powerful secondary illusion usually carries the virtual life beyond that brink and makes the impression of a real shift from one order of existence to another. Herein lies its intense emotional power: the transition occurs, the higher phase emerges.[69] Many thoughtful aestheticians

---

Gemälde gedacht, aber als solche haben sie von vornherein das eigentümliche Lebensgefühl der Skulptur; sie sind vielleicht die einzigen Erscheinungen in der Geschichte der Kunst, die völlig im Stil und den Formgesetzen ihrer Kunst bleiben und doch völlig aus dem Geiste einer anderen Kunst heraus empfunden sind" (p. 165). And, more generally; "Aus dieser Aufhebung aller gegenseitigen Fremdheit und Zufälligkeit der Wesenselemente entsteht das Gefühl einer in sich vollkommenen Existenz dieser Gestalten. Was man an ihnen von je als das Titanische . . . empfunden hat, ist nicht nur die Übergewalt ihrer Kräfte, sondern jene Geschlossenheit des innerlich-äusseren Wesens, deren Mangel das spezifisch Fragmentarische unserer Existenz ausmacht" (p. 163).

[68] *Poetic Diction, a Study in Meaning* (1928), p. 189: ". . . almost any kind of 'strangeness' may produce an aesthetic effect, that is to say, an effect which, however slight, is qualitatively the same as that of serious poetry." This strangeness, furthermore, "must be felt as arising from a different plane or mode of consciousness, and not merely as eccentricity of expression. It must be a strangeness of meaning" (pp. 189–90).

[69] Cf. Hermann Friedmann, *Die Welt der Formen* (1930; 1st ed., 1925), p. 129: "Wir unterscheiden die Funktionalität, bei welcher die beiden Variablen demselben sinnlichen Bereiche angehören, von derjenigen, die einen Übergang aus einem sinnlichen Bereich in einen anderen ausdrückt. Die erste bezeichnet eine Intensivierung der unabhängigen Reizvariablen, die zweite aber ihre Transformation. Das Intensitätserlebnis ist es, das charakteristisch ist für den haptischen Bereich; wird seine Grenze durch Transformation überschritten, so tritt ein neues Erlebnis auf: das Qualitätserlebnis."

have puzzled about the fact that so often the quintessence of art seems to lie in a division between two states, in which some mysterious passage or transmutation takes place. Kandinsky has remarked it, though one has to know his language and his peculiar mystique to gather from a brief excerpt what he means.[70] Professor Fergusson speaks of a "timeless moment" in drama, in which a destiny is realized.[71] A "timeless" moment is one in which action, the mode of drama, gives way to an inward vision, still on the part of the agent in the play (not the spectator), but apparent to the audience as a plastic image would be, though it is not a visible image. It is poetically created, a mental act of the dramatic personage that presents itself as though it were fixed in a sort of mental space.[72]

All secondary illusions, whether they serve primarily to intensify the expressiveness of a piece or whether they create a quintessential moment, have the same character of suddenly coming into existence from nowhere, apart from the virtual substance of the work (which is anchored in the primary illusion according to its proper mode), and fading again into nothing. In their very nature, therefore, they project the outstanding attribute of human mentality, the termination of autonomous acts in psychical phases that resemble those of perceptual acts in many respects; that is to say, the occurrence of images. Like fantasies, secondary illusions seem to have no somatic being; they are disembodied, yet they come out of the created form and heighten its livingness, even to a degree where the form in its entirety seems to be changed.

It has been said that the high moment in any work which achieves such a sublimation is "an unreality between two realities."[73] It is the

---

[70] *Point and Line to Plane* (1947; original German edition, 1926), p. 30: "Only by feeling, are we able to determine when the point is approaching its extreme limit. . . . The attainment of that moment when the point, as such, begins to disappear and the plane in its stead embarks upon its embryonic existence—this instant of transition is the means to the end.

"In this case, the end is the *veiling* of the absolute sound of the form: the emphasis on its dissolution, . . . the flickering, the tension, the abnormality in the abstraction . . . that is, the creation of a double sound by a *single* form."

[71] *The Idea of a Theater* (1949), p. 193.

[72] D. G. James, in *Skepticism and Poetry: An Essay on the Poetic Imagination* (1937), observes that this "inward vision" is always marked by a lyric utterance which makes it apparent to the audience (see p. 118).

[73] Some people prefer to describe it as "a reality between two unrealities." That phrasing, of course, expresses a value judgment (and certainly a proper one) on the element of sublimated feeling, the vision or inward illumination or whatever it seems to present; "reality" is a word of better "metaphysical pathos" than "unreality." But, rhetorical preferences apart, the intended phenomenon is the same.

element in which the transition from somatic feeling, which is our ordinary reality, to imagination, the autonomous act of envisagement, comes to formal expression. Here is the dynamic pattern of the conceptual act, the strangeness and "otherness" and bodilessness of symbolic imagery, projected in the structure of great art.

Owen Barfield, whose recognition of this phenomenon of "disembodiment" I have already cited, also realized the relation of its highest forms to the masterpieces of poetry. "There is a kind of strangeness," he said, in the work I have already quoted above, "which should be distinguished, as the truly *poetic*, from the other, merely *aesthetic*, varieties of strangeness. This kind depends, not so much upon the difference between two kinds of consciousness or outlook, as on the act of becoming conscious itself. It is the momentary apprehension of the poetic[74] by the rational into which the former is forever transmuting itself—which it is itself forever in process of becoming. This is what I would call pure poetry. This is the very moonlight of our experience, . . . the pure idea of strangeness, to which all the others are but imperfect approximations, tainted with personal accidents. It is this which gives to great poetry its 'inevitableness.'. . ."[75]

There is an interesting concept expressed in this passage, which really sums up the new function that characterizes the human mental process: "the rational into which the former [the primitive feeling] is forever transmuting itself." That process which symbolic projection brings with it is the objectification of feeling, which continues into the building up of a whole objective world of perceptible things and verifiable facts. But as soon as it begins to build the "world"—and that is probably very soon, almost *ab initio*—it also presents abstractable forms, such as the external world provides us with, and the process of objectification engenders its counterpart, the symbolic use of natural forms to envisage feeling, i.e., the endowment of such forms with emotional import, mystical and mythical and moral. That is the subjectification of nature. The dialectic of these two functions is, I think, the process of human experience. Its image is "the poetic," more generally "the artistic," and it can appear in art from either pole of the dialectical tension—the artist sees it constantly in the quality which he finds in his world, and which emerges in his successful work, and in his best work may even go beyond anything he has ever known.

There is nothing in art that must always be achieved by the same particular means, as there is no definable limit to what a technical device

[74] This term denotes the archaic, the pristine form of feeling.
[75] *Op. cit.*, p. 199.

can do. But the creation of secondary illusions is a powerful, not always feasible and controllable process; and the influence it exerts where it occurs involves the whole work. Those artists and art lovers who find the quintessence of art in that sublimation of the primary illusion where the piece seems to be given to more than one sense (taking poetic perception as one "sense," though that is a difficult and deep subject), are prone to attribute that power to art per se. It is, indeed, a key to the mystery of art as an image of mind; and as such it appears in Friedrich Hebbel's statement, preserved in a letter to a friend: "Every true work of art is *infinite* and *effects* the *infinite*; it stands, like a deed, as an isolated phenomenon, on which a dual light falls, between two irrationals; and as with a deed, one asks in vain what cause went before and what must follow. . . . That is why fragments have the same value as finished works; indeed, I wish some things had remained unfinished."[76]

It is hard to prove the achievement of a true secondary illusion, because many appreciative percipients of a work receive its emotive impact without realizing what makes it; also it may fail to occur, as artistic experience of any sort very easily does under unfavorable circumstances. The impression may be brief and seem personal, but nevertheless be made on a good many persons, so one feels that those who do not receive it have "missed" something that is—or was—there. In the British Museum there is a seated figure of Ceres, alone in its alcove on the stair landing, opposite a window from which daylight falls on it in full but diffuse brightness; I remember vividly the impression it creates as one comes up or down the stairs, which is of a sudden great silence. In a temple this effect might be even stronger; but it was acute when I received it in the museum, which was not particularly quiet at the time. If one approached the statue in company of twenty trudging people, with a guide reciting handbook history and calling attention to stylistic and material features, the silence of the piece would probably be destroyed before it could be felt by any beholder.

There is one step beyond that play of secondary illusions in which different perceptual realms seem to intersect, and make a semblance of rising and fading presences that in actuality belongs only to mental images—memory, fantasy and shifting perceptions. This further step I can only call "transcendence," because it seems to transcend the sensory vehicle altogether and make an almost pure presentation of the "Idea." Such a rarified projection cannot stand alone; it arises from a constellation of devices, making a manifold abstraction. Effects of this sort are

[76] From a letter to Elise Lensing, dated 1837. See Bernhard Münz, *Hebbel als Denker*, p. 367.

often designated by awed recipients as "magical." There is, by way of example, one line in Keats's *Ode to a Nightingale* in which he captures the quintessence of the nightingale's song—not the ordinary experience of bird song, but the nightingale's typical crescendo and break, and nothing more—without any imitation:

> Now more than ever seems it rich to die,
> To cease upon the midnight with no pain,
> *While thou art pouring forth thy soul abroad*
> *In such an extasy!*
> Still wouldst thou sing, and I have ears in vain—
> To thy high requiem become a sod.

The "o" sound, accented four times in succession, creates the essence of a crescendo, which is intensification, to be followed by the short line of three trivial words with short vowels that throw the only accent of the line on "extasy"—a word with an explosive consonant and two quick cadential syllables, on which the line breaks. The real stroke of genius here is the use of a vowel that has the least possible likeness to a bird's utterance to render the crescendo and forestall any onomatopoetic imitation. There is no bird voice, but only the form of its feeling.[77]

One more remark may be in order here, before we go on to the study of mind guided by a new image. Generally, when one receives an impression of "sublimation," it carries with it a sense of quite sudden and unaccountable simplification. When life reaches a limit of complexity and intensity it breaks over into a larger pattern that swallows its former elements and exhibits greater ones of its own. The sense of simplification attends that change.

There are many more things to say about art, but they will have to await the time when they become directly relevant. At the present juncture we turn to nature for the reality conveyed in the image, the actual phenomena it brings to light. For an image may—indeed, must—be dropped when it has done its work. To enlist its work at all is a new venture. The use of biological concepts and phenomena to illuminate problems of art is far from new, and has, indeed, a considerable vogue at present; witness all the beautiful books of nature photographs, from

---

[77] This instance may remind one of Schiller's comment in his essay "Über naive und sentimentalische Dichtung" (1795–96): "Eine solche Art des Ausdrucks, wo das Zeichen ganz in dem Bezeichneten verschwindet, und wo die Sprache den Gedanken, den sie ausdrückt, noch gleichsam nackend lässt, da ihn die andere [gewöhnliche] nie darstellen kann, ohne ihn zugleich zu verhüllen, ist es, was man in der Schreibart vorzugsweise genialisch und geistreich nennt."

Haeckel's *Kunstformen der Natur* (1899) to Postma's *Plant Marvels in Miniature* (1961).[78] Several writers have also made very serious analytic studies of art based on the analogies to artistic form that are provided by natural phenomena—cell division and replication figuring as formal repetition, organic rhythms as prototypes of artistic rhythms, etc.—much as they appear from the opposite angle which I am taking here.[79] Our difference is not one of findings or opinions, but of purpose. Their purpose is to penetrate the mysteries of art with the help of their biological knowledge; mine is to gain some biological and psychological insights through the suggestiveness of artistic forms. A symbol always presents its import in simplified form, which is exactly what makes that import accessible for us. No matter how complex, profound and fecund a work of art—or even the whole realm of art—may be, it is incomparably simpler than life. So the theory of art is really a prolegomenon to the much greater undertaking of constructing a concept of mind adequate to the living actuality.

[78] See also Wolf Strache, *Forms and Patterns in Nature* (1956); Andreas Feininger, *The Anatomy of Nature* (1956); Maurice Déribéré, *Images étranges de la Nature* (1951); Oskar Prochnow, *Formenkunst der Natur* (1934); and Karl Blossfeldt, *Urformen der Kunst* (1st ed., 1928).

[79] W. Harburger's *Form und Ausdrucksmittel in der Musik*, already mentioned several times, is particularly rich in such parallels. Even more so is Weidlé's article, "Biologie de l'art," also cited above.

Appendix to Chapter 7

Example 1: Brahms, Second Symphony, opus 73, first movement (see p. 201).

Example 1 (*cont'd*)

Example 2: Debussy, Preludes, No. 10. The sustained chords set up many dissonances with the moving quarter-note sequences, but the effect is vague and wandering rather than tense and directed toward resolution. Even the simultaneous *c* and *b* of the prevailing chord in measure 14, being very widely spaced, do not create a real dissonance for the ear (see p. 216, n. 35).

Example 3: J. S. Bach, "Nun lob, mein Seel, den Herren." At the point marked by a cross, the five tones sounding together are f, g, a, b-flat, c; yet the movements and suspensions of the voices are such that no startling dissonance results by that coincidence.

Example 4: Schubert, "Der Wegweiser," measure 40 to end (see p. 221n).

Example 4 (*cont'd*)

Appendix to Chapter 7

Example 5: Brahms, Fourth Symphony, last movement (see p. 225n).

Part III

Natura Naturans

8

The Act Concept and Its Principal Derivatives

ONCE the image of life is recognized in its artistic projection, where it is seen to include all mental life, we have some measure of adequacy for the terms of a conceptual structure which can support biological thinking of a sort that, in due course, will pose and resolve psychological issues. We are in the realm of nature; actual life, not virtual, is in view from the new angle which the central position of feeling imposes on the artist's vision. That vision remains in the philosophical background, to hold the work of logical construction to its course, from the choice of basic concepts to the most advanced formulation of scientific facts. Such elaborate, advanced factual thinking should arise in due course by the use of concepts which, in the beginning, may be roughly sketched and imperfectly defined; use has to establish their definition, and will occasionally even force their redefinition. At the outset it is better not to worry too much about scientific form, or even about unassailable proof of every assumption, as long as the assumption is known for what it is. But a steady increase in precision, a cumulative growth of decidable issues, a certain ease of technical formulations and a visible tendency toward systematic connections among those formulations may well be demanded of a philosophical theory that claims to provide a conceptual framework for the empirical study of mind. That challenge is to be met.

Turning from the symbolic presentation of life to the phenomena of its actual occurrence on nearly all of the earth's surface, one is most immediately struck by the difference between living and non-living entities, animate and inanimate nature. At first glance these two categories seem entirely distinct: the non-living things exhibiting a relatively simple system of interrelated events, the system of mechanical causes and effects, in which past and future conditions are mathematically calculable and predictable, and the living things defying the laws of

[257]

mechanics as by some inward power, so that their histories within the framework of inanimate nature are incalculable and essentially unpredictable. But upon closer inspection the boundaries between those two categories appear less and less sharp; there are borderline cases, such as viruses, which are hard to assign to one or the other,[1] and there are physicochemical particles which exhibit no life of their own in isolation, yet have vital functions within organisms, or assume them when introduced into living structures.[2] "Life" is obviously not easy to define. Yet there is little doubt that every attempt to produce life from lifeless substances, without any vital germ to animate them, has so far met with failure;[3] and as long as the results leave us in no serious doubt even in the minimal test case, we must have some fair idea of what we

[1] See, for instance, the discussion of this point by R. C. Williams at a symposium of the National Academy of Sciences, published in the Academy's *Proceedings* for 1956 under the title, "Relations between Structure and Biological Activity of Certain Viruses": "Prior to the purification of tobacco mosaic virus by Stanley twenty-one years ago, viruses were considered to be small-sized microbes, obviously living and multiplying within infected cells. Now that . . . certain virus-like manifestations have been found in other, simpler molecules, it is by no means clear that any criteria of life can be established that will allow all types of organic systems to be unequivocally listed among either the quick or the dead . . . whether or not viruses as they exist in a test tube are alive is a moot and fuzzy question. It is more relevant for us to recall that the interesting phase of a virus' existence is the intracellular one, where it is certainly a part of the stream of life. . . . An essential aspect of that which is living seems to reside in its high degree of organization, leading to functional activity not predictable from a consideration of its separate parts" (p. 811).

Applying this last criterion, F. M. Burnet wrote his interesting *Virus as Organism* (1945), which explores the possibility that viruses might be degenerate forms of one-time parasitic organisms, in which even the procreative functions have become vestigial and almost entirely relegated to the host cell.

[2] Ludwig von Bertalanffy, in *Das Gefüge des Lebens* (1937), p. 51, calls attention to the frequent inclusion of non-living matter in organisms where, however, such matter forms part of the living whole. Such tissues as hair, nails, tooth enamel, the outer bark of trees, or deposits like mollusks' shells or hard eggshells, furnish plenty of instances.

[3] Paul A. Weiss opened an address to the National Academy of Sciences with the words: "Preceding speakers at this symposium have convincingly traced the stepwise synthesis that leads . . . from molecules through macromolecules to ordered macromolecular systems. Extrapolating this trend, one might feel tempted to expect that it would take but another upward step of the same nature to carry us to the living cell and that this last step—the synthesis of the cell—would, like the preceding ones, turn out to be not just a logical construct but a demonstrable physical reality. Let us stress right at the start that there is nothing in our experience to justify this bold expectation" ("The Compounding of Complex Macromolecular and Cellular Units into Tissue Fabrics" [1956], p. 819).

would quite certainly accept as "living," though there might be dispute about the exact criterion.[4]

The difficulty of drawing a sharp line between animate and inanimate things reflects a principle which runs through the whole domain of biology; namely, that all categories tend to have imperfect boundaries.[5] Not only do genera or species merge into each other, but classifications made by one criterion do not cover the cases grouped together by another, so that almost all general attributions have exceptions, some of which are really mystifying.[6] And finally, even series of events, such as

[4] Even at a deeper level of biochemical research, Jean Brachet found reason to declare: "Il n'y a pas de 'loi du tout ou rien' qui sépare l'être vivant de la matière" ("Les Acides Nucléiques et l'origine des Protéines" [1959], p. 361). N. W. Pirie, in an article entitled "Chemical Diversity and the Origins of Life" (1959), goes so far as to say, "It seems better to recognize that life is not a definable quality but a statement of our attitude of mind towards a system" (p. 78). On the other hand, so justly famous a biochemist as N. H. Horowitz writes in a contrary vein: "Some biologists and biochemists tend to regard the question of the definition of life as essentially meaningless. They view living and non-living matter as forming a continuum, and the drawing of a line between them as arbitrary. . . . I do not accept this point of view, because I have not been convinced that the postulated continuum actually exists" ("On Defining Life" [1959], p. 106). The fact that we are at present unable to draw a dividing line does not prove, of course, that no discontinuity exists; but even Horowitz's criteria, offered in the same little paper, do not decide the borderline problems.

[5] Thus Johannes Holtfreter, after observing that induction, unlike hormone action, requires contact between the inducing and the reacting structures, adds that there are certain contrary observations, so that "if the necessity of cell contact is used as a criterion to distinguish between inductive and hormonal processes this may not hold true for every instance" ("Some Aspects of Embryonic Induction" [1951], p. 118). Likewise, E. Broda, in "Die Entstehung des dynamischen Zustandes" (1959), says: "Eine qualitative Beantwortung der Frage nach dem dynamischen Zustand eines Körperbestandteils mit einem absoluten 'Ja' oder 'Nein' scheint im Allgemeinen wenig sinnvoll. Man wird die Bestandteile besser quantitativ durch die Halbwertzeiten der Erneuerung kennzeichnen. . . . Dabei mögen sich allerdings in manchen Fällen so lange Halbwertzeiten ergeben, dass die Erneuerung sogar qualitativ nur mehr mit Schwierigkeiten nachzuweisen ist" (p. 337). Many years earlier Charles Manning Child, writing his *Individuality in Organisms* (1915), found almost at the outset that "it is sometimes difficult to determine whether a particular aggregation of protoplasmic substances is an individual or not" (p. 5).

[6] So, for instance, one finds the following statements in the literature: "Positive inotropic and chronotropic action of alendrine on isolated heart of cat or frog is 10 times stronger than that of adrenaline; no difference in their action on rat heart was observed" (from an English abstract [1950] of a Russian article by A. And'yan, K. Lissak and I. Martin, "New Data on the Mechanism of Action of Sympathomimetic Substances, in Particular Alendrine," [1948]). "Temporal polar stimulation gave a rise in B.P. [blood pressure] in cat and a fall in monkey. . . . Respiration was inhibited in the majority of temporal lobe stimulations and accelerated in the ma-

the successive stages of plant or animal growth and development, permit no sharp conceptual wedge to be driven between them. No one knows this as surely as the biologist in the laboratory, and the more exactly he experiments and observes, the better he knows it. Davenport Hooker, watching many hundred times the spectacular sequence of stages through which embryos pass in the course of gestation, was led by that experience to say: "It must be borne in mind that the development of activity is always a continuous process and that any division into phases or stages is entirely artificial. The whole process is a smoothly progressive affair and not a series of isolated posturings."[7]

But the distinction of phases is not really something entirely artificial, and need not lead to the division of a natural continuum into "a series of isolated posturings." It is, indeed, entirely pragmatic; but in order to be so it has to operate with some form which can be empirically found in the actual event and selected for conceptual manipulation. The continuous process is not composed of discrete episodes, but it has peaks of activity which are centers of recognizable phases, though these have no precise start or finish lines. What we need, then, by way of analytic terms are units with definite centers and labile limits. To identify such units in nature is relatively easy; the great structural problems lie in their complex and sometimes indeterminate contours. Where does an accent, a peak, a concentration begin or end? From what does it arise? Just what does it embrace?

A phenomenon viewed from its center has to be treated as indivisible, or its center as such would be lost. For the same reason these units cannot be homogeneous, but must have internal structure. A homogeneous quantity is always theoretically divisible; if it is taken as a unit, it is so by fiat, and then the analytic procedure has an arbitrary basis and is to that extent "artificial."[8] Units by virtue of inviolable structure, on

---

jority of frontal lobe stimulations, but opposite effects in some animals were also obtained" (B. K. Anand and S. Sua, "Circulatory and Respiratory Changes Induced by Electrical Stimulation of Limbic System (Visceral Brain)" [1956], p. 399). "Uric Acid. . . . Main excreted product of breakdown of proteins and nucleic acids in uricotelic animals. Excreted also by primates and Dalmatian dogs but not by other mammals . . ." (M. Abercrombie, C. J. Hickman and M. L. Johnson, A *Dictionary of Biology* [1954], p. 238). "The cardiovascular system . . . precedes all other organ systems in attaining initial functional capacity except, apparently, in some sharks" (Davenport Hooker, *Evidence of Prenatal Function of the Central Nervous System in Man* [1958], p. 1).

[7] *The Prenatal Origin of Behavior* (1952), p. 15.

[8] Cf. an observation in point made by Selig Hecht in "Energy and Vision" (1945), p. 75: ". . . at sea level it takes about 1,000 ergs to lift one gram one centimeter

the other hand, are not unanalyzable, though they cannot be divided without losing their identity. They may be great or small, permanent or transient, their limits may be sharp and clear, or obscure, untraceable beyond some vague and variable point. But a structural center determines and locates each unit.

Obviously we are not dealing here with material parts of a living thing, but with elements in the continuum of a life. Those elements may be termed "acts." It is with the concept of the act that I am approaching living form in nature, only to find it exemplified there at all levels of simplicity or complexity, in concatenations and in hierarchies, presenting many aspects and relationships that permit analysis and construction and special investigation. The act concept is a fecund and elastic concept. It applies to natural events, of a special form which is very widely represented on the surface of the earth (and probably on other heavenly bodies too); a form characteristic of living things, though not absolutely peculiar to them. Such events arise where there is already some fairly constant movement going on. They normally show a phase of acceleration, or intensification of a distinguishable dynamic pattern, then reach a point at which the pattern changes, whereupon the movement subsides. That point of general change is the consummation of the act. The subsequent phase, the conclusion or cadence, is the most variable aspect of the total process. It may be gradual or abrupt, run a clearly identifiable course or merge almost at once into other acts, or sink smoothly, imperceptibly back into the minutely structured general flow of events from which the act took rise.

An act may subsume another act, or even many other acts. It may also span other acts which go on during its rise and consummation and cadence without becoming part of it. Two acts of separate inception may merge so that they jointly engender a subsequent act. These and many other relations among acts form the intricate dynamism of life which becomes more and more articulated, more and more concentrated and intense, until some of its elements attain the phase of being felt, which I have termed "psychical," and the domain of psychology develops within the wider realm of biology, especially zoology. The work of tracing and understanding that ever-progressive, self-weaving web of life in

---

against gravity. One erg is therefore one thousandth of this energy, and in order to see we need only three-billionths of an erg.

"A datum of this kind is impressive by virtue of its size as compared to man. However, an erg is an arbitrary unit, and by choosing some other unit, whose value for example would be a micro-micro-erg, the amount of energy might be given as 3,000 such units and would surely not be nearly so impressive."

terms of acts and their interdependent functions will be the substance of what follows in this book. First, however, it is important to have a clearer idea of the act concept and a number of conditions which follow directly from it.

In the literature of biology one meets, practically at every turn, some technical or semitechnical words that occur rarely, if at all, in physics and chemistry: especially "induction," "norm," "preparation," "inhibition," "realization" and the adjectives "presumptive," "potential"[9] and "actual." These are terms which do not arise by logical or pragmatic necessity in the physicist's classical universe of discourse. Yet the biological sciences, ranging from biochemistry to psychoneurology, are too well advanced to be considered unscientific or prescientific in their procedure. Biologists are making free use of physical concepts and methods and penetrating deeply into chemistry, electrochemistry, radiology and other strict mathematical sciences, especially in pursuing genetics to the very beginnings of life; and in so doing, they are not using any borrowed concepts, but following out their own, in a direction where these ultimately become equivalent to physical terms. That is as it should be, and promises better for the "unity of science" than any conscious adoption of language or method. But in the opposite direction, where the study of life broadens out into the investigation of higher organisms, evolutionary processes, animal behavior and mental phenomena, the subject requires its own vocabulary. Its terminology has grown up naturally for its purposes, and reflects the peculiarity of its material.

So far, however, neither the professional philosophers who concern themselves with the nature of life and mind nor philosophizing scientists themselves have looked to that vocabulary which is actually used in the laboratory for the basic concepts with which the biological sciences at all levels tacitly and happily operate. Both are prone to generalize, almost

---

[9] "Potential" is, of course, familiar enough in physics, but with the specialized meaning of electrical potential or of "potential energy" (as against "kinetic energy"). In embryology it is related to "potency" (as, e.g., "multipotential") and opposed not to "kinetic," but to "actual" or "realized." See, for instance, S. Spiegelman, "Physiological Competition as a Regulatory Mechanism in Morphogenesis" (1945), p. 121: ". . . if the early blastomeres of some eggs are separated, each is capable of giving rise to a complete and perfectly proportioned embryo. Yet in the normal course of events this potentiality is not realized, each blastomere giving rise to but a part of an embryo."

C. H. Waddington, in *The Nature of Life* (1961), p. 29, wrote: "The first step in the understanding of heredity is to realize that what a pair of parents donate to their offspring is a set of potentialities, not a set of already formed characteristics."

Examples could be gathered from almost any biological treatise.

at once, either the condition at the molecular end of the ladder of life, where vital processes emerge from non-vital ones and their analysis becomes more and more feasible in physicochemical terms, or the condition which develops toward the other extreme, where such terms are clearly inadequate. The result is a war of attrition between the two venerable factions of physicalists (by various names) and vitalists (with equally various labels, sometimes accompanied by a special protestation that this brand is not vitalism).

Yet no insistence that all process is "transposition of matter in space" and must be viewed as such can make an ecologist really view the changes of plant and animal populations in a burned-over or newly inundated area as a "transposition of matter."[10] Neither can the earnest asseveration that, in the face of certain anomalies of animal behavior, "It is clear that for the interpretation of these remarkable facts we require the concepts of need, normal end-state or goal, and directive activity towards attainment of the goal"[11] make any scientist content with zoological explanations in terms of need and fulfillment. Not even the proponent of the teleological principle whom I have just quoted attempts to enter into any details of anatomy or physiology under its aegis, but resorts for such details to causal relations, with which he seems admirably well acquainted.[12] Details and framework of biological thought thus tend to fall apart; and the methodological platform known as "holism," like "telism," is an ineffectual protest against letting that happen. No matter how much we are adjured to "relate everything always to the whole,"[13] no direct relationship between a whole and any or all of its parts presents itself. Similarly, the watchword, "The whole is more than the sum of its parts," leaves us unable to measure how much more it is, and in what its excess over the sum of the parts consists. All too aptly, Paul Weiss has written: "As soon as one raises the

[10] On the resort to "physicalism" and especially this "extremist" terminology, see Chapter 1, p. 15, above.

[11] E. S. Russell, *The Directiveness of Organic Activities* (1945), p. 25. The particular anomaly in question is the acquisition of defensive nematocysts by the little flatworm *Microstomum* from *Hydra* which it eats.

[12] At the same time he says (pp. 94–95) of the cell divisions in the growing egg that "these forms of cleavage are directive towards future goals integrally related to the general process of development, and comprehensible only on this basis, *whatever their causal explanation, if any, may be*" (italics mine).

[13] See Jan C. Smuts, *Holism and Evolution* (1926); Adolf Meyer-Abich, "Organismen als Holismen" (1955); also R. H. Wheeler, *The Laws of Human Nature* (1931), which is avowedly based on the programmatic principle: "Begin with the whole and always refer everything back to it."

eye from the unit to the whole system, the subject becomes fuzzy, the problems ill-descript, and the prospect of fruitful attack discouraging in its indefiniteness. . . . the 'whole' gets a large share of one's thought and talk, but the elements get all the benefit of one's actual work; here the problems seem to be so infinitely more tangible."[14]

It is with elements that I propose to start, but with elements which show as much tendency to become expanded and elaborated into wholes as to yield further and further subordinate elements, to the limits of distinguishability. Acts are such elements; for their functional sub-units, separately considered, close in on themselves to present in minia-ture the typical act form,[15] and in contrary perspective acts merge and grow into whole lives, still maintaining that same essential structure.[16] The best support for this contention comes from the working vocabulary of our research biologists, geneticists, physiologists, evolutionists and the field zoologists who observe the behavior of higher animals, birds and mammals, in the wild. Through all their widely separated areas of work there runs the fundamental recognition of the act as the basic phenomenon.

In the study of behavior this is, of course, obvious enough; in watch-ing a creature's behavior we quite indubitably see one act after another begun, carried out or miscarried, and ended, sometimes by completion and return to quiescence, sometimes by the rise of another act. We see acts that seem to be barely begun and then "inhibited," and others which become crowded and confused, "disorganized" in panic. Incipient acts, finished and unfinished acts, predictable and unpredictable acts, are the ethologist's factual and theoretical units.

In witnessing and recording behavior, the line between observation and conceptual supplementation is not always easy to draw. No one can see an act being inhibited, eschewed, deflected or incompletely per-formed; the "act" in any such case is a conceptual unit partly exempli-

---

[14] "Self-Differentiation of the Basic Patterns of Coordination" (1941), p. 4.

[15] This phenomenon will be discussed and exemplified below.

[16] Existentialist philosophers speak of life as an act of being, or existence. See, for instance, Benoit Pruche, *Existentialisme et acte d'être* (1947). In a more immediately practical vein a psychiatrist, Stephan Krauss, describes the "Korsakoff syndrome" as marked by a fragmentation of acts due to the loss of the single central act matrix, which he calls *"die innere Haltung,"* the totality underlying and unifying the acts of a normal human being. So he says: "Man sollte überhaupt 'die Handlung' ansehen als *Figur* auf dem Grunde der gesamten Handlungszusammenhänge." And further: "Die innere Handlung ist . . . die Repräsentanz der Gesamtpersönlichkeit" ("Unter-suchungen über Aufbau und Störung der menschlichen Handlung" [1930], pp. 682, 684).

fied. But that is exactly where its scientific virtue lies; the act concept is a formal concept, and as such it can be manipulated to yield further, larger or smaller, simpler or more highly structured conceptual forms, for which one may seek exemplifications, perfect or partial, in observable organic processes. That is why acts are properly called "elements" rather than "factors" of biological analysis.

The word "act," like all words in common use, has different special connotations for different people. It is often understood to mean "conscious act," "intended act," or, even more particularly, "rational," "responsible" or "moral act." In adapting any familiar word to a technical use one has to divest it of all accrued shadings and apply it rather where its definition (or, failing that, its explicit characterization) permits than where popular associations do so. Restrictions may then be explicitly placed on its meaning for systematic purposes. In the present instance, this liberation from tacit value connotations brings the term "act" into very close accord with the sense in which scientists use it, not only for intentional acts but also for unconscious, involuntary or even purely somatic processes, of long or short duration, in plants as well as animals, even down to microscopic protozoa and germs.

Physiologists, psychologists and other scientific observers who deal with acts on the level of behavior have rarely given attention to the general form of those events, but have usually noted their complexity, the fact that a distinct behavioral act characteristically subsumes many movements of different muscles. They have sometimes made an explicit distinction between movements and acts, as for instance J. M. Nielsen in his studies of aphasia and related psychopathic conditions. Dr. Nielsen is an ardent believer in localized cerebral controls of somatic functions, and has surveyed in the greatest possible detail the "centers," especially cortical ones, where muscular movements as well as some perceptual responses are said to arise. After determining, item for item, what dysfunctions are involved in motor apraxia, and where they seem to be located, he concludes: "These various centers of performance of *movements* are in turn governed for the performance of *acts*. An arbitrary differentiation must be made between movements and acts, the latter consisting of a series of the former coordinated to the carrying out of a plan of action. There is no 'center' for carrying out an act."[17]

Here the word "act" is applied (somewhat arbitrarily, as Dr. Nielsen said) to preconceived, intended acts as against the incidental acts in-

[17] "Epitome of Agnosia, Apraxia, and Aphasia with Proposed Physiologic-Anatomic Nomenclature" (1942), p. 113.

volved in them; e.g., making a telephone call as against grasping the receiver, flexing the body toward the instrument, etc. Those incidental motions are distinguished as "movements." Extensive experiments with a considerable variety of animals have shown that particular movements can be elicited by electrical stimulation of the cortex at particular points, and that their voluntary performance can be frustrated in similar fashion or by lesions in circumscribed areas of the brain. The responses of the animal to the stimulating electrode are called "movements" because they are stereotyped elements which usually belong to behavioral acts rather than figure as such acts by themselves. If they are inhibited, some acts to which they are normally incidental may be rendered impossible, as in human beings immobilization of the tongue makes speaking impossible; but within wide limits a complex behavioral act, once initiated, tends to find some alternative course to its completion if its normal course is blocked by the arrest of one or more of its habitual constituent movements.[18]

If, however, we use "act" in the way I have proposed, the distinction drawn by Dr. Nielsen between acts and movements proves to be needless as well as artificial. All animal and human movements take the form of acts. Dr. Penfield has described the completely involuntary motions of persons undergoing cortical stimulation in neurological examinations to determine the exact foci of brain lesions; his description of their directly elicited behavior is very interesting, because it clearly shows that the response to even such abnormal peripheral stimuli as an electrical current applied to the bared brain always takes the form of an act.

"When any of the four cortical vocalization areas is stimulated," Dr. Penfield wrote, ". . . the conscious patient carries out the following complicated set of movements, quite against his own will.

"His mouth opens, the throat and larynx are held fixed, while diaphragm and abdominal muscles contract and force a stream of air through the larynx, making a vowel sound. If the stimulating electrode is held in place, this continues until it is interrupted by his taking a breath, after which he vocalizes again. (The cortical control over respiration seems to have been lost until the inspiration is completed.)"[19]

Obviously this "complicated set of movements" is a single act; the patient "vocalizes." No less telling is the parenthetical observation that

[18] Cf. K. S. Lashley, "The Problem of Cerebral Organization in Vision" (1942), p. 317: "When the rat learns the maze, he learns, not a stereotyped sequence of movements, but a series of distances and turns, which can be performed by unpracticed muscle groups, as when a cerebellar lesion forces him to roll through the maze."
[19] Wilder Penfield, "Mechanisms of Voluntary Movement" (1954), p. 11.

what interferes with this artificially induced, involuntary act is another act, "taking a breath," which ends when "the inspiration is completed," i.e., the act is consummated, its impulse spent.

Every movement, whether it be a whole behavioral act (like the twitch of a horse's skin to throw off a fly) or a subordinate element (like the tensing of a muscle in vocalization), exemplifies the basic act form, and so do its further and further constituent elements, down to the smallest units of vital action. This may sound like a sweeping assertion, but it is well supported by the careful, literal language (not the fashionable jargon of industrial metaphors) used by clinicians as well as field and laboratory scientists. Even botanists see the life they are dealing with in terms of concatenated acts; growth, in most plants, is one lifelong activity, with rapid phases and slow phases, a continuum of rhythmically engendered elements which are acts in form and in characteristic relations; growth itself is one great act, going through orderly successive phases.[20] In plants as in animals, the vital activities subsume smaller and smaller events, yet biologists still recognize the basic indivisible act form, even where they are delving down to the biochemical factors involved in those complex microscopic processes; as H. C. Dalton, speaking to cytologists at a symposium, did when he said: "Any attempt to understand the mechanisms of cellular differentiation must be based on the accurate description of molecular events associated with this crucial act of developing systems."[21]

Among those people who work with the human nervous system, too, the act is regarded as a basic phenomenon, potential or realized, com-

[20] Cf. C. W. Wardlaw, "The Floral Meristem as a Reaction System" (1955–57), pp. 401–2: "Some of the controversial matters, which figure so largely in the morphological literature, either become unimportant, or can be harmonized, if the concepts of organogenesis are transferred from the morphological to the physiological plane. Thus, such problems as the homology of the several floral organs and the vegetative leaves . . . disappear if we bear in mind that all the appendages of the axis are the products of the same reaction system. . . . This reaction system, located in a particular region of the apical meristem, retains its property of giving rise to regularly spaced growth centres throughout the whole development of the plant—from cotyledons to carpels. These, however, may vary in their chemical composition and potentiality for growth. . . . It is to differences in the metabolism of growth centres and young primordia that the visible morphological differences must be referred." And finally (p. 406): "The task, as it now presents itself, is to explain how an apical reaction system is constituted and how it becomes modified, phase by phase, in such a way that it both maintains a general pattern of morphogenetic activity and yet gives rise to successive groups of distinctive floral organs."

[21] Introduction to "Molecular Events in Differentiation Related to Specificity of Cell Type" (1955), p. 967.

plete or incomplete, overt or covert,[22] equally exemplified in elaborate performances and in practically momentary events such as catching sight of something. Thus Stephan Krauss, who speaks of human life as one inward act (*"die innere Handlung"*),[23] observes that the single act of grasping an object subsumes an act of vision and one of hand movement (*"Sehakt und Handbewegung"*).[24] The famous neurologist Sir Henry Head spoke of "a complete sensory act."[25] Henri Wallon, discussing the influence of language on human mental activity, describes the utterance of a single word as "an act which takes the imprint of things and gives its own to them."[26]

What gives every act its indivisible wholeness is that its initial phase is the building up of a tension, a store of energy which has to be spent; all subsequent phases are modes of meting out that charge, and the end of the act is the complete resolution of the tension. Sometimes an act is complicated in its build-up, that is, a number of more or less independently originating charges summate to create a synthetic high tension; its inception, then, is widely based in the organism, and despite the apparent singleness of the pool, each contributive charge may require its own release; the act, therefore, has to be correspondingly complex. Also, a tension in process of being spent may be reinforced by a new

---

[22] Not "behavioral." All psychical phases of acts are covert; so are many entire acts. The terms "overt" and "covert" for acts or elements of acts which are, respectively, accessible and inaccessible to observation from outside the organism wherein they occur, are used by an increasing number of writers, viz.: Mapheus Smith, in "An Approach to the Study of the Social Act" (1942); Justus Buchler, in *Toward a General Theory of Human Judgment* (1951); D. B. Lindsley, at the Nebraska Symposium on Motivation held in 1957, in commenting on C. T. Morgan's paper (1958); Talcott Parsons, in "Psychological Theory in Terms of Theory of Action" (1959); Davenport Hooker, in *Evidence of Prenatal Function of the Central Nervous System in Man*, where the close interrelations of covert and overt acts, especially the normal motivation of the latter by the former, are discussed. Paul Weiss has called "conditioning" a covert act (see R. L. Watterson [ed.], *Endocrines in Development* [1959], p. 76). So it is regrettable that Claire and W. M. S. Russell, in "An Approach to Human Ethology" (1957), p. 178, try to introduce this same pair of correlative terms for an entirely different purpose, and with a completely different meaning, namely, for emotionally admitted and unadmitted psychological motives.

[23] *Op. cit.* Cf. above, p. 264, n. 16.

[24] *Ibid.*, p. 674.

[25] "The Physiological Basis of the Spatial and Temporal Aspects of Sensation" (1919), p. 79.

[26] *De l'acte à la pensée* (1942), p. 248: "Le mot . . . est un acte qui prend l'impreinte des choses et leur donne la sienne, à l'égal du geste qui modifie les choses et se modifie à leur contacte."

charge which enters its path and heightens its potential again. The fact that locally developed energies have particular paths of release seems fairly certain, but only a very few such courses are known at all—the reflex arcs, which exhibit complete acts in what appears to be a genetically simplified, schematized form.[27] Even in low animals the acts of feeding and of withdrawal from injury—in many species, the only behavioral acts—are elaborate events with many variable elements. They are usually, perhaps always, responses to specific stimuli, but by no means stereotyped reflexes.[28]

---

[27] Cf. G. E. Coghill, *Anatomy and the Problem of Behaviour* (1929), pp. 19–20: in the salamander larva, "the first limb movement is an integral part of the total reaction of the animal and . . . it is only later that the limb acquires an individuality of its own in behaviour. The local reflex of the arm is not a primary or elementary behaviour pattern of the limb. It is secondary, and derived from the total pattern by a process of individuation. . . . The limb arises in absolute subjugation to the trunk. . . . The freedom which it ultimately attains, particularly under certain experimental conditions, has the appearance of being practically absolute, and the experimental reflex of this nature has come to be accepted as the elementary unit of behaviour."

Davenport Hooker, in a study of human embryos, has found the development of reflexes to bear out Coghill's observations, but quotes several dissenting opinions as deserving consideration. He found, moreover, that the stereotyped reflex was often a transient phenomenon, yielding in later stages to more adaptive forms. See *Evidence of Prenatal Function of the Central Nervous System in Man*, especially pp. 35–36.

[28] There is an interesting and detailed report on the feeding acts of sea anemones in C. F. A. Pantin's "Behavior Patterns in Lower Invertebrates" (1950). The movement pattern of *Metridium senile*, according to Pantin, "shows parallels with complex coordinated activity in the higher animals. It occupies the first stage in the behaviour pattern of food capture. The subsequent links vary in character so remarkably that the manner of achievement of the goal, food ingestion, has an air of improvisation." And further: "The preliminary food responses resemble the preliminary taxes of an instinctive behaviour pattern. At the end, the pattern is brought to a close by the achievement of the goal, and the pattern of activity then changes" (p. 190).

The event thus described certainly exemplifies the dynamic form here called an "act." It has its inception, development, consummation and close; "the presence of dissolved food substances in the surrounding medium brings about characteristic preparatory activity." The disk opens, the animal elongates its column and sways its head. But, as the experimenter remarks, individual behavior is extremely variable. Some specimens do not respond at all, and of those which do, some do not open their disk at the mere taste of foodstuff in the water, some do not sway, etc. "Each element involves a coordinated sequence of activities of several muscular systems, and this co-ordinated activity may continue for a long time after the application of the food-juice stimulus, even if this be removed" (p. 183). Finally, after this preliminary rousing to the presence of food, "a slow 'circular' contraction begins in the region of the sphincter of *Metridium*, and from this point a wave of circular tone passes slowly and peristaltically down the column of the animal. This in turn may be followed by other activities which finally merge into a general base-line of inherent activity" (p. 189).

[269]

The discovery of reflex action, however, worried the antimechanists—vitalists, teleologists, entelechists or whatever their denomination—because it seemed to be the discovery of a mechanism set in motion by a physical cause, and immediately conjured up the specter of man as a simple edifice of reflex arcs. It worries them to this day, and elicits many protestations that "the causal principle" is inadequate to explain our higher mental functions, or that the scientific treatment of life is but one way of understanding phenomena which are really comprehensible only in non-physical terms. But these philosophers and ideologists do the reflex too much honor; and, moreover, their conception of "causality" is all too often the conception of forces directly applied to masses, displacing them, according to the elementary laws of dynamics. Even a practicing psychiatrist, briefly venturing into pure theory, has declared: "The sensory-motor reflex arc is the most basic principle in the science of neurology. . . . In the reflex theory the causal principle has found a wide and fertile field of application."[29] But above the level of the spinal reflex arc, he claims, "we need auxiliary hypotheses, quite different from mechanical principles. Certain peculiarities of reflexes cannot be understood if they were [sic] purely mechanical in nature and therefore, reflex facilitation, reflex inhibition, and other biological, nonmechanistic concepts are added to the reflex mechanism."[30] Why these concepts (which are added to a mechanism) are non-mechanistic is not demonstrated; the author's retreat before the advancing front of science is step by step, clinging to every unexplained phenomenon for support.

The unexplained phenomena, meanwhile, are the physiologist's lure to further experimentation and new hypotheses that contain no auxiliary concepts whatever to supplement the principles of science. Though he is dealing primarily with acts, and employs the concept of "act" in his formulations of problems and findings, he is quite clearly bent on the discovery and understanding of mechanisms, and expects to trace acts to purely physical origins; that is, he regards acts as natural phenomena, and the realm of life as continuous with the vastly greater realm of inanimate nature. In the laboratory scientist's thinking, "vital act" and "mechanism" are not incompatible terms. It is true that he does not follow their theoretical implications down to their metaphysical pre-

---

[29] Adrian C. Moulyn, "Mechanisms and Mental Phenomena" (1947), p. 243. In the preceding paragraph he said that he would not here consider such elaborations of the system of mechanics as "the aspect of the transformations of energy" which may be found in it; but "with these restrictions one may say that . . . masses play a passive role and that forces are the active agents in this system."

[30] *Ibid.*

suppositions, as philosophers of science are called upon to do, and that this might be why the paradox does not trouble him; but it is somewhat strange that two principles which are philosophically conflicting should be so systematically conjoined in all scientific studies of life and mind. Yet it is the notion of the "true act" that is most often advanced to defy scientific explanation; think only of what Klages, Buytendijk and Palagyi made of it.[31]

As there is prescientific knowledge of feeling expressed in the work of artists, so there is prephilosophical abstract conception involved in the thinking of empirical scientists; and concepts which serve without stint or hazard in a limited field, where they are freely mingled and manipu- lated, should be capable of some consistent logical formulation in wider systematic thought, such as is the philosopher's business. If the concepts of mechanism and of vital act seem antithetic in philosophy, one or the other of them is probably misconceived.

The act concept has been so exhaustively discussed, especially by the above-mentioned writers, that one would hardly expect to find un- recognized presuppositions still lurking in the term. But the same can- not be said about the concept of mechanism. It has not received any such critical treatment or careful definition from the philosophers who declare it inadequate to the task of describing vital phenomena. Con- sequently their idea of the mechanism is still the traditional one of a machine made of prefabricated inert parts, powered from a single source and designed to perform a predetermined set of interlocking movements, producing a result intended by the makers and users of the contrivance.

Physicists have, of course, progressively modified their concepts, al- though they may become explicitly aware of such changes only retro- spectively. A thoughtful stock-taking article, with special reference to the demands of biological conception, is John R. Platt's "Properties of Large Molecules That Go Beyond the Properties of Their Chemical Sub-Groups" (1961). The first sub-section is entitled, "Changes of Physics with Changes of Scale," and begins with the observation: "We usually say, of course, that the fundamental principles of physics and chemistry do not change with a change in molecular dimensions. But this is only a theoretical dictum and . . . we generally use a different set of rules or equations for problems of different ranges of size. The basic laws may not alter, but they are combined in different proportions, so to speak; . . . consequently the dominant physical forces and the distinctive properties of matter change, with every alteration in scale.

[31] See Chapter 1, pp. 10–15, and Chapter 7, p. 203.

"Small water waves, for example, are governed principally by surface tension, while large ones are governed by gravity, although both forces are always present. Atoms and small molecules must be described by quantum mechanics, but this . . . goes over more and more exactly into classical mechanics in describing the macroscopic properties of large molecules.

"In biology . . . the complete change of structure and mechanism with every change of scale is particularly striking" (pp. 342–43). After discussing the surprising phenomena that arise with molecules of 500 atoms, 5,000 atoms or "in the 50,000 atom range" in formations of linear chains, helices and double helices, etc.—phenomena such as self-replication, electron transfer, muscle action—he concludes: "Evidently many strange phenomena of biology actually follow quite naturally from physics and chemistry, once we have adjusted our concepts to the changed conditions and the changed scale and the changed complexity. What is most important is that this makes it less necessary for us to be dependent on vitalist or epigenetic or other peculiarly biological explanations that have been proposed for these phenomena" (p. 358).

But the philosopher seeking the essence of mind, or even of life, usually has no such elastic and potent concepts of biophysics and biochemistry. Every discovery of physical laws exemplied in animal functions threatens to impose the image of "*l'homme machine*" on his mind, and drives him, in protest, to some form of antimechanism, because on the basis of his entire experience of vital feeling and intrinsic value, his intuitive judgment—perhaps even his sense of humor—tells him that the image is spurious.[32] Yet he has no other image that invites scientific minds to explore its object. So he can only resort to the logically inconceivable, but religiously familiar, animating soul, guiding entelechy, or numberless incarnations of an *élan vital*.

It is here that the artist's symbolic projection provides a principle of analysis applicable to the actual living form his work reflects: the principle of distinguishing, within a dynamic whole (i.e., a whole held to-

---

[32] The scariness of "reflexes" lies in their superficial resemblance to machine operations. Today there is a new flurry over "mechanical brains," due not only to the unprecedented efficiency and flexibility of electronic computers, but also to a new intellectual foible, the use of gross behavioral acts as the criterion of life and the guidance of such acts as the criterion of mentality. To any one who has ever worked with living matter *in vitro* or under the microscope, the synthetic production of a chemical particle that metabolized for a brief period would be a much more impressive approach to the creation of a brain than the invention of Eniac and all its successors. A machine, however powerful and versatile, is an entirely different mechanism from a cell, a multicellular organ or a complex organism controlled by its own brain.

gether only by activity) articulated elements, which nonetheless are indivisible in themselves, and inalienable from the whole, if they are not to give up their identity. In the preceding chapter I pointed out that artistic elements are "act-like." Their biological analogues in the world of nature are acts.

The analysis of acts leads one, not to inert permanent bits of matter being rearranged by impinging forces, but to further and further acts subsumed under almost any act with which one chooses empirically to begin. At length, some present-day researchers even reach a level of proto-acts—events which belong to chemistry or electrochemistry as much as to biology, because they exemplify the character of acts in many respects, yet not in all, so that viewed in a reverse perspective they would seem to foreshadow rather than present processes of genuine life. In this borderland it becomes relevant and justifiable to ask, where does life begin?—which is equivalent to the question, what are the most primitive full-fledged acts?

Clearly, at that point we are in the realm of science, and seeking the mechanisms of molecular integration in the evolution of organic compounds. By "mechanism" I mean any process we understand in terms of physics and chemistry.[33] The complexity of such processes is beyond the imagination of anyone who does not know some samples of them rather intimately; they grow up into self-sustaining rhythms and dialectical exchanges of energy, forms and qualities evolving and resolving, submicroscopic elements—already highly structured—merging and great dynamisms emerging. The common-sense tenet that such products of nature cannot attain feeling, awareness and thought loses its cogency when one is confronted by the actual intricacies of chemical and electrochemical organization.[34] The bridge to organism arises of itself, and the

---

[33] W. E. T. Trendelenburg, in *Einheitlichkeit und Anpassung in den Lebenserscheinungen* (1929), p. 18, defines "mechanism" as "alles das, was sich unter Anwendung physikalischer, chemischer, physikalisch-chemischer, kolloid-chemischer usw. Methoden, Begriffe und Vorstellungen an den Lebenserscheinungen erfassen lässt." This is exactly what I mean by "mechanism" (or "mechanical," etc.).

[34] One of the very few religious philosophers who realize the vast complexity of living matter is Teilhard de Chardin. In *La place de l'homme dans la nature* (1956) he attributes the vital character of organisms to their physicochemical "complexification." As a natural consequence he is not afraid of finding mechanisms in organic nature, but holds that their elaboration is precisely what makes mental advances possible. Unfortunately for the philosophy of science, however, he imposes his biological model on the entire universe, so as to make all nature culminate in the realm of the Spirit and the fulfillment of a divine plan. The demonstration of that metaphysical thesis not only leads him to adopt an orthogenetic doctrine of evolution which is far from demonstrable, but also to abandon his strictly biological reflections all too soon when they have served their purpose.

conviction that "extended substance" cannot think and "thinking substance" cannot have material properties appears as a medieval doctrine handed down to modern philosophy in Descartes' famous dictum, and with no firmer foundation than his word.

It is, in fact, noteworthy that there are rather few elaborate lifeless mechanisms in nature.[35] Geysers and blowholes, volcanoes, tides and whirlpools, the variable yet generally patterned dynamisms that effect cyclonic storms, are spectacular mechanisms, but they are simple; all very intricate natural mechanisms that we know are alive. Now, such vital working systems have no inanimate solid parts assembled and adapted. They show rhythms within rhythms, interlocking timed sequences of chemical changes, electrical fields and currents that induce the chemical actions or, conversely, are generated by them, the most elaborate physical processes under a network of homeostatic controls; the sum total is a matrix of acts within acts, organizing previously unrelated material units at the molecular and atomic levels in an unbroken continuum of activity that builds itself up into the incredible dynamism called a living body, which disintegrates if the creative activity stops.[36]

The study of living functions as acts thus leads backward into the physical sciences without coming to any dividing line that has to be crossed by a *saltus naturae*. We need not worry about its connection with more established systems of knowledge. Looking forward, however, to the fields of less well-known nature—psychology, sociology, ethics— we find ourselves in possession of a basic concept that can be developed to serve far more difficult ends than workers in those fields have ever dared to set themselves.

The semblance of living form, the "organic" character in any good work of art, contains more suggestions for the logical treatment of acts than just the image of their tensive internal structures which makes them indivisible elements. This character is, after all, not strictly peculiar to them; it belongs also to many chemical reactions and is, indeed, what makes the most complex ones that still belong to organic chemistry so "act-like" that they may well be called "proto-acts." The further condi-

[35] More accurately, in sublunar nature; solar systems are, of course, vast natural mechanisms, and their complexity may well be little known to us.

[36] Cf. the last paragraph of a paper by Marko Zalokar, "Ribonucleic Acid and the Control of Cellular Processes," presented at a symposium held at Woods Hole, Massachusetts, in 1960: "This Symposium has shown that control of cellular function can occur at many levels. . . . The quality and quantity of enzymes and other proteins formed in the cell starts an intricate interplay of many factors, each controlling another, and the concerted action of them all produces what we see as a living cell or organism" (p. 130).

tions distinguishing genuine acts are also reflected in the art symbol; they are relational, and their nature leads to the immense involution and evolution pervading the realm of life.

Perhaps the first and most arresting feature of that realm is that the causal order exemplified in nature as a whole seems to be somehow absent here. A little further reflection, however, makes it apparent that all mechanisms exhibit causal relationships, all movements have causal antecedents; the analysis of acts in terms of such sequences is not impossible, but for some reason it is generally irrelevant. A more important relation among acts is something called "motivation." The term is vaguely used and little understood. "Motivation," and especially its cognate "motive," like the word "act," is used with all sorts of gratuitous connotations of goodness and badness in popular speech, and usually likewise in ethical writings, which rarely leave the common-sense level of so-called "ordinary language." So far, even professional scholars who have taken account of the difference between causation and motivation have not transcended that level. As a result their reasoning has generally brought them to one or another of three basic positions, none of which, however, provides access to more advanced problems or leads to further conceptual developments and connections. The three positions are, in the main, (1) that motivation is something metaphysically distinct from causation, and to treat motives as causes is a basic error, for they are not physical events, but reasons, and as such they influence but do not produce their relata, which are acts of "free will";[37] (2) that motivation is a special kind of causation, other than physical causation, controlled by ends instead of antecedent conditions;[38] or (3) that causes and motives, respectively, belong to different "logical languages" and consequently obtain in different "universes of discourse."[39] Against all these views there is the insistence of many philosophers of science that motives

[37] See R. S. Peters, *The Concept of Motivation* (1958), especially p. 28: "My thesis is that the concept of 'motivation' has developed from that of 'motive' by attempting a causal interpretation of the logical force of the term." Also pp. 149–50: "The most obvious and usual answer to the question 'Why' about human actions is to find the goal or end towards which an action is directed or the rule to which the action is made to conform." And he praises William McDougall for the virtue that "he did agree with the main thesis of this monograph—that explanations of the purposive, rule-following type . . . cannot be *deduced* from more general postulates of a mechanical type as envisaged by Hobbes, Hull, and perhaps Freud in his more speculative physiological moments." In an earlier passage (p. 11) he said of Freud that "his account of the working of the primary processes creaks with causality."

[38] Adolf Meyer-Abich, *op. cit.*, *passim*. The original proposal of this notion by J. R. Mayer in 1876 is discussed below, p. 284.

[39] Cf. Chapter 1, pp. 16–17.

[275]

are ordinary causes, operating in systems which are essentially self-servicing computers provided with mechanisms to effect their fueling, repair and replacement.[40] There are also theorists who make the more moderate claim that motives are natural causes of the scientifically familiar sort, but that every act in a living system is due to a convergence of so many causal factors that no isolable previous event (or events) can be called the cause (or causes) of the act. A motive, therefore, must always be regarded as a contributing but not definitive cause.[41]

If, now, one examines the nature of acts more closely, the concept of motivation also takes on a more precise form. But in pursuing an analytic line of thought with constructive intent one cannot abide by the casually established, overlapping, ethically tinged meanings of words in common use. Despite the warnings sounded by champions of "ordinary language,"[42] any progressive thought soon breaks out of those loose fetters. There is no need of declaring war on popular usage; sometimes it does express important conceptual modalities, and then one may well follow its lead in choosing terms to be formally assigned to various distinct though related concepts that are but vaguely and inconsistently distinguished in colloquial speech. But to begin, as Mr. Peters does in *The Concept of Motivation*, by considering the usual overtones and cathexes of the word "motive" and then saying they are "implied" by its proper use[43] can only lead to sparring encounters with thinkers who have seen the word differently used and attribute different limitations and powers to it.

For further-reaching intellectual purposes it is best to take only suggestions from "ordinary language," and then use the key words as they

---

[40] The literature of this computer psychology is too extensive and—in my estimation—too ephemeral to require documentation here; but for a typical presentation see W. R. Ashby, "Principles of Self-Organizing Dynamic Systems" (1947), which offers a machine model whereby "all variables in the C.N.S." should be computable; and for an extreme case of "computerism" see K. W. Deutsch, "Mechanism, Organism and Society; Some Models in Natural and Social Science" (1951), where machines, people, societies, brains and camera films exchange places, and where values, consciousness, will and autonomy are defined so that computers may have them.

[41] So, for instance, A. H. Maslow, "A Theory of Human Motivations" (1943), p. 371: "While behavior is almost always motivated, it is also almost always biologically, culturally and situationally determined as well." Expressions of the same idea may be found in many other sources.

[42] One of these is R. S. Peters, in the work cited above, p. 154; despite the fact that in his Aristotelian Society symposium paper, "Motives and Causes," published in 1952, i.e., several years before the appearance of his little book, he rejected "the vagaries of 'ordinary language'" (p. 145).

[43] See p. 29 ff.

are needed to carry quasi-technical meanings. In the present case, "motive" is not even the first candidate; the relational concept of "motivation" is my immediate concern. This word, too, has already acquired some special meanings, not always the same for different authors. Some psychologists limit it to the operation of "drives," and claim that this excludes acts springing from emotion; some hold that it should be applied only to the element of goal-directedness in acts, whether they arise from "drives" or not, so that "random" acts are classed as "unmotivated." I do not know whether the intraorganic antecedents of behavioral acts are really distinct dynamic patterns which can be properly conceived as so many separate "drives"; and as for goal-directedness, though it is an important and obvious phenomenon, easily demonstrated in countless experiments, the theoretical results achieved by making it a central concept are not impressive. In view of the conflicting uses of the word "motivation" that are already current, I would prefer not to add another to the list, but have been unable so far to find any other term that seems at all suitable to denote the process which mainly distinguishes living from non-living systems. Perhaps, however, this is not really a matter of proposing one more sense for an overworked term, but of developing the general trend toward its broader and more serviceable definition, which is already well under way in some quarters. Physiologists have begun to employ it in preference to "causation," as for instance V. C. Twitty and M. C. Niu in an article entitled "The Motivation of Cell Migration, Studied by Isolation of Embryonic Pigment Cells Singly and in Small Groups in Vitro" (1954), or J. M. R. Delgado, W. W. Roberts and N. E. Miller in "Learning Motivated by Electrical Stimulation of the Brain" (1954). These uses bespeak the broad sense of "motivation" which I am accepting here.

Several psychologists in recent years, and indeed during the last two decades or more, have questioned the practice of restricting "motivation" to the causal patterns underlying particular kinds of behavior, and consequently distinguishing between motivated and unmotivated kinds. There seems to be some strong feeling that a special process, peculiar to living systems, is expressed in the origination of all behavior, animal and human, goal-directed or not, "tension-reducing" or not, born of "drives," or of other conditions; and that "motivation" should denote this distinctive process. In an excellent article, "Motivation Reconsidered: The Concept of Competence" (1959), R. W. White has written: "Something important is left out when we make drives the operating forces in animal and human behavior. . . . In this paper I shall attempt a conceptualization which gathers up some of the important things left out by drive theory. To give the concept a name I have chosen the word *competence,* which is intended in a broad biological sense rather than in its narrow everyday meaning. As used here, competence will refer to an organism's capacity to interact effectively with its environment. . . . My central argument will be that the motivation needed to attain competence cannot be wholly derived from sources of energy currently conceptualized as drives or instincts" (p. 297).

After discussing the work of other psychologists—Lashley, Hendrick, Head and some of the contributors to the Nebraska Symposia on Motivation—which led them to serious criticism of the drive principle and to some interesting counter-suggestions, he concludes: "Twenty years of research have thus pretty much destroyed the orthodox drive model. . . . Instead we have a complex picture in which humoral factors and neural centers occupy a prominent position; in which, moreover, the concept of neurogenic motives without consummatory ends ["goals"] appears to be entirely legitimate.

". . . I prefer in this essay to speak of the urge that makes for competence simply as motivation rather than as drive" (p. 305).

This "urge that makes for competence" expresses itself in exploratory behavior whereby a creature finds out the nature of its surroundings, and in nosing, mouthing or manipulating objects, whereby it learns its own powers to effect changes, to make things happen in the external world. The aim of such "monkeying" behavior, acting just for the sake of causing and controlling events, Dr. White calls "effectance"; and he pertinently points out that the conventional stimulus-response formula, the rat-runners' "S → R," is not really applicable to this kind of activity, in which the sequence is better described as R → S; the creature finds out what will happen in consequence of an act. The motivation of that act lies in his own previous behavior of exploring the environment, and the changes that occur in his situation are due to his own postural changes and shifts of focus or attention.

Other psychologists are moving in the same direction. R. S. Woodworth has called the traditional distinction between motivated and emotional acts in question, pointing out that emotion under some conditions provides additional power to clearly telic motivated acts.[44] P. T.

---

[44] See *Dynamics of Behavior* (1958), pp. 60–61: "High activation is likely to have an emotional character, especially when a drive is prevented from reaching its goal quickly. Intense hunger or thirst is not usually called an emotion, but it undeniably has an emotional character. Intense sex drive is admittedly emotional, and so is an intense drive to avoid or escape from danger. . . . Intense emotion is beneficial in situations demanding . . . a great output of muscular energy. . . ." Cf. Edgard

The Act Concept and Its Derivatives

Young treats feelings (in the traditional sense of pleasure and displeasure, which he calls—more forcibly—"delight" and "distress") as the primary motives of behavior.[45] Yet he regards emotion as undirected, "unmotivated" activity.[46] Several writers have paralleled White's "competence motivation" and "exploratory drive" with a still less goal-directed, sheer "activity drive," arising from a basic need to have something to do.[47]

In all these and many other writings on motivation, wherein the extension of the term is gradually widened from the restricted category of "drives" engendered by basic physiological needs to that of any and all purposive activations, there still remains the reservation that not all the processes which constitute a life can be traced to identifiable motives; so that even A. H. Maslow, one of the most liberal theorists, felt called upon to say: "Not all behavior is determined by the basic needs. We might even say that not all behavior is motivated. There are many determinants of behavior other than motives. For instance, one other important class of determinants is the so-called 'field' determinants. Theoretically, at least, behavior may be determined completely by the field, or even by specific isolated external stimuli, as in association of ideas, or certain conditioned reflexes.

Forti, "La nature de l'émotion" (1936), who remarks that behavioristic psychology can view emotion only as "un phénomène perturbateur qui désorganise l'action régulière," but that in fact "l'émotion comprend des *mouvements positifs* qui ont un sense et une efficacité . . . malgré le désordre et l'excès . . . nous trouverons des gestes qui ont une intention définie, des actions qui tendent vers un but" (p. 361). "L'émotion se modèle, se règle sur des circonstances parfois délicates que notre intelligence apprécie; ce que nous fait toucher du doigt qu'elle est en rapport avec l'activité psychique supérieure et n'a point le caractère simple d'un réflexe" (p. 366).

[45] In his Nebraska Symposium paper, "The Role of Hedonic Processes in Motivation" (1955), he wrote (p. 194): "Affective processes as primary motives arouse behavior; they sustain or terminate an activity in progress; they regulate and organize behavior according to the hedonic principle; and they lead to the acquisition of motives, stable dispositions to act, and value systems." (The motives acquired under the pressure of primary motives are presumably secondary ones.)

[46] See especially *Emotion in Man and Animal* (1943), and, most recently, *Motivation and Emotion: A Survey of the Determinants of Human and Animal Activity* (1961).

[47] Cf. R. W. White, *op. cit.*: "Murray and Kluckhohn . . . have made a case for pleasure in activity for its own sake, reviving the *Funktionslust* proposed many years ago by Karl Bühler and recently developed in some detail by French. They also argue for intrinsic mental needs: the infant's mind is not acting most of the time as the instrument of some urgent animal drive, but is preoccupied with *gratifying itself*" (pp. 312–13).
See also W. F. Hill, "Activity as an Autonomous Drive" (1956).

"Some behavior is highly motivated, other behavior is only weakly motivated. Some is not motivated at all (but all behavior is determined)." And in a cautionary footnote, he adds: "I am aware that many psychologists and psychoanalysts use the terms 'motivated' and 'determined' synonymously, e.g., Freud. But I consider this an obfuscating usage."[48]

Yet it is interesting to find that the generalization of "motivation," when it is systematically pursued, leads to a basic concept that applies impartially to the generation of all sorts of acts, so that "motivation psychology" as the study of a particular psychological phenomenon becomes more and more tenuous, until its very name seems redundant, indistinguishable from "psychology" pure and simple. Dr. White recognized this as a danger to the present organization of research in a conveniently restricted field, and chose to stop short of such a radical move.[49] But another pioneering thinker on the same topic (and, incidentally, on the basis of almost exactly the same literature) has no compunction at drawing the disconcerting inference; Klaus Foppa, at a symposium on motivation held in Tübingen in 1963, declared roundly that "motivation" really means nothing else than "causation of psychical events," that every act is motivated, and that consequently it is misleading to speak of "motivation psychology" as a special line of research; motivation psychology means nothing else than psychology *per se*.[50]

---

[48] *Op. cit.*, pp. 390–91.

[49] See "Motivation Reconsidered," p. 318. After gathering exploration, effectance and activity for its own sake under the rubric of "competence," and postulating a general "competence motivation," he says: "No doubt it will at first seem arbitrary to propose a single motivational conception in connection with so many and such diverse kinds of behavior. . . . We could go further and say that each item of behavior has its intrinsic motive—but this makes the concept of motivation redundant."

[50] "Motivation und Bekräftigungswirkungen im Lernen" (1963), pp. 656–57: "Die Steuerung des Verhaltens erfolgt [im Experiment] durch die Situation (nämlich das Furchtsignal) einerseits und die Antizipation der Signalfolgen durch das Versuchstier. Analoge Steuerungen und entsprechende Vorwegnahmen künftiger Ereignisse kann man aber in jeder einfachen Konditionierungs-Situation beobachten. Selbstverständlich könnte man auch dort von 'motiviertem,' d. h. gesteuertem Verhalten sprechen, nur glaube ich, dass man dadurch den Begriff ad absurdum führte, weil schliesslich nichts anderes mehr damit gesagt wäre, als 'Ursache von Psychischem.' Vom Lernvorgang her sehe ich keine Möglichkeit, eine sinnvolle Abgrenzung zwischen motiviertem und unmotiviertem Verhalten zu treffen. . . . Solange der Nachweis nicht geführt ist, wodurch sich diese Funktionsgefüge von anderen Funktionszusammenhängen unterscheiden, halte ich es für irreführend, von einer eigenen Motivationspsychologie zu sprechen. Denn Motivationspsychologie in diesem Sinn bedeutet nichts anderes als Psychologie überhaupt."

It is not only as "causation of the psychical," but as this and more, that I would use the term "motivation": as "causation of acts." Since acts are natural events of a highly variable, yet fundamentally typical form, it is not surprising that their causation by other acts as well as by non-vital, circumstantial events shows a comparably typical, complex pattern. The entire motivation of an act can never be summed up in a "motive," nor in several motives, although motives are important definable elements among mental acts, to be discussed in due course. For the moment let us consider the peculiar process of motivation, in contrast to causal relationships which do not fall under that heading.

Every act arises from a situation. The situation is a constellation of other acts in progress, often including some which develop with the acute initial phase of peripherally originating acts, such as we feel as impact if they are intense enough to develop a psychical phase. But the substance of a situation is always the stream of advancing acts which have already arisen from previous situations punctuated by previous impacts; centripetally proceeding acts (usually but not always of brief, quickly completed build-up) impinge on that continuous integral process, differentiating and developing its myriad possible forms. All distinguishable acts arise from this matrix, which is their situation.

The process whereby they arise is a basic causal relation obtaining among acts, which may be termed "induction." One act, or a complex of acts, may be said to induce a new act; ultimately the entire situation, whatever its stage at the time in question, induces any and every act. There may, however, be a range of immediate relevance to a given act within the situation. Induction is a very complicated process that is but little understood; this does not mean, however, that we should dismiss it for some less difficult "conceptualization" which will give us law-like formulas with simple predictive uses but no instrumental promise in the face of true scientific problems.[51]

[51] The predictions which test such formulas rarely go beyond what common sense would lead one to predict, and this has even been taken as their justification; so, for instance Werner Traxel, summing up his system of postulates for predicting behavior, says with evident satisfaction: "Das System wird in einigen Details erklärt, und es wird am Beispiel der Verbindung von je zwei Motivierungen zu einer neuen Resultante gezeigt, dass es zu Aussagen führt, die psychologisch plausibel erscheinen" ("Über Dimensionen und Dynamik der Motivierung" [1961], p. 427).
A similar passion for prediction as the sole purpose of science is evinced in J. B. Rotter's criticism of P. T. Young's paper, "The Role of Hedonic Processes in Motivation," proffered at the Nebraska Symposium on Motivation (1961) and printed in conjunction with the paper: "There seems to be little doubt that there are differences in the effects of first presentation of the same reinforcement to different

An organism normally exists in a non-living physical environment,[52] which is constantly undergoing changes that affect the active being, the organism, at its center. This environment at any particular time is generally referred to as "the environmental situation," and is the psychologist's chief object of consideration in judging the acts of the organism; the "inward situation," being unsurveyable, is usually dismissed with a phrase of recognition, except where special pharmacological experiments are the issue. But the way in which any event in the "environmental situation" affects the life is deeply influenced by the fact that it is a life, not a non-living physical complex, on which the event impinges. That impingement, again, presents us with a causal relation of specialized form, and fortunately one that has been more intimately studied from a psychological angle than induction, though by no means sufficiently. Many years ago, Jakob von Uexküll called attention to the fact that two different organisms in the same environment were likely to exist in widely differing environmental situations, or, as he called them, different "ambient worlds" ("*Umwelten*"), due to the selective powers with which their respective peripheral organs (their integuments, as well as specialized sense organs) could filter out noxious or even merely useless influences. The world of extraorganic events is not the same for a mosquito, a snake and a cat—not even for a mosquito and a honeybee.[53] Besides these differences in the reception of outside influences, there is an immense variation in the value an influence, once received, has for various creatures. This is because the external event can keep its formal self-identity only to the point of making peripheral contact with the system of acts; if it invades that system (as, for instance, a chemical

species and to different individuals within a species. . . . To be able to predict these differences may be important for a true psychophysiology or for a neurophysiology but they are not of *necessary* significance for the problem of the psychologist. To directly observe these initial effects behaviorally allows us to predict what we need to predict" (p. 241).

[52] Normally—not always; some parasitic organisms, such as *Trichinella spiralis*, have no life episodes outside of a living host.

[53] In his *Umwelt und Innenwelt der Tiere* (1909), Von Uexküll said of an amoeba's *Umwelt*: "Um dieser Welt näher zu kommen, müssen wir vergessen, welchen Eindruck die Umgebung der Amöbe auf unser Auge macht. Von all den bunten vielgestaltigen Gegenständen, wie Oszillarien, Infusorien, Rotatorien, Steinchen und Detritus ist nicht die Rede. Schwache und starke Reize gibt es, die nur der Intensität nach unterschieden werden, mögen sie mechanisch oder chemisch oder durch das Licht ausgelöst sein. Dazu kommen die spezifischen Reize der Nahrungsmittel, die das Ektoplasma klebrig machen und erweichen" (p. 39). A less systematic and more humanistic use of the idea of relative *Umwelten* goes through his later book, *Nie geschaute Welten* (1936).

action may enter it as food or poison, a rise in temperature may be propagated inward from the contact surface), that importation falls at once under the sway of the vital processes, and becomes an element in a new phase of the organism; that is, it engenders a new situation. This sort of influence is continually taking place. Energy is fed into living systems all the time,[54] as oxygen, nutriment, warmth or light, from their peculiarly specific *Umwelt*, which is not "the environment" in a usual sense; Uexküll's word, "*Umwelt*," has been translated "ambient," which is the term I shall use.[55]

Such impingement of outside events on an organism is, of course, a causal relation; but it is not a simple causal relation between a non-vital force—the push of an object, or even a current or a ray—and an act produced by it. The only way an external influence can produce an act is to alter the organic situation that induces acts; and to do this it must strike into a matrix of ongoing activity, in which it is immediately lost, replaced by a change of phase in the activity. The new phase induces new distinguishable acts.[56] This indirect causation of acts via the prevailing dynamic situation is "motivation." It may arise from intraorganic sources, too, which is to say from autogenous acts that radically alter the general activity, inducing a new phase (i.e., a new situation); in that case, overt acts of the organism may be motivated by entirely covert events.[57]

[54] The recognition of organisms as "open systems," receiving energy from outside sources to make up for the inevitable entropy that accompanies patterned activity, has been ardently discussed in the meetings and writings of the Society for General Systems Theory headed by Ludwig von Bertalanffy, who was the first to stress, and perhaps to note, "open systems."

[55] It is interesting to compare, with this use of "ambient," the definition of the cognate word "ambiance" in *Webster's New World Dictionary*: "1. Surroundings; milieu. 2. In art, especially painting, the configuration of secondary designs, themes etc. enhancing and extending the central design or theme."

[56] Cf. K. S. Lashley, *op. cit.*, p. 320: ". . . in cortical activity there must be postulated a persistent substratum of tonic innervation upon which are superimposed the fluctuating patterns resulting from current stimulation, in the same way that the innervation of voluntary movement is superimposed upon the spinal pattern of postural tonus." Also J. Z. Young, "The Evolution of the Nervous System and of the Relationship of Organism and Environment" (1938), p. 198: "In general any set of afferent impulses arriving at a centre may be supposed to produce an effect which will depend on the pattern into which it fires. . . . The nervous system thus provides a matrix of neurons, interconnected with each other in most various ways, and the patterns build up and maintain themselves within the matrix."

[57] *Ibid.*, p. 200: "The most important factor in determining the course of a reaction in such an animal as man is probably the intrinsic activity of the nervous system itself, and it is therefore convenient in such cases to speak of the behavior as controlled by *neural motivation* or a *neural drive*."

An external event that initiates a behavioral act in an organism is a "stimulus."[58] It is quite obviously a cause of the act (the latter being known as the "reaction" or "response"), yet the sequence of cause and effect is just as obviously not a simple transmission of motion, heat, or other physical quantity from one body to another. The disproportion between the impinging force and the action produced in the organism is so spectacular that some thinkers have questioned the applicability of ordinary causal laws to behavioral responses; the first to do so appears to have been one of the discoverers of the conservation of energy in physical transformations, Julius Robert Mayer. Having formulated the law of conservation for classical mechanics, he admitted the "triggered" reaction as an exception, and proposed a distinction between "true" causality and another sort "wherein the cause is not equivalent to the effect." But the philosophically-minded German chemist, Wilhelm Ostwald, who left an unpublished lecture on Mayer's doctrine in the form of a dictaphone record made in 1914,[59] saw that the action "triggered" by a stimulus drew its energy from the organism, not from the stimulus, and that the physical effect, elaborate and sometimes violent as it might be, indicated the existence of complex and labile structures within the organism in which great energies were chemically bound and could be released by a small catalytic contact. Mayer, and Ostwald in his earlier works, had spoken of a special kind of energy, "Nervenenergie," which obeyed the unknown laws of the special kind of causality, "Auslösungskausalität."[60] But Ostwald soon realized that the resort to such concepts was a counsel of despair—an admission of scientific defeat, with the decision to fall back on incomputable quantities and functions instead of the terms and relations Mayer had happily established for the transformation of measurable energies and true causality," Umsetzungskausalität." For such counsel, surely, the time

[58] This definition seems to be generally (though sometimes tacitly) accepted, indeed assumed, by psychologists. Physiologists do not limit the term to causes of circumscribed, overt behavior, but regard many drugs and natural endocrine products as "stimuli" starting or furthering entirely intraorganic processes.

[59] Posthumously published as *Julius Robert Mayer über Auslösung* (1953), and prefaced by a complete reprint of Mayer's short treatise of 1876.

[60] Mayer wrote "*Auflösung*," but even in his lifetime "*Auslösung*" came into general use for the action of a stimulus. See Alwin Mittasch, *Wilhelm Ostwald's Auslösungslehre* (1951), p. 21.

The assumption of a special kind of causality is a device used again by Richard Semon in *Die Mneme als erhaltendes Prinzip im Wechsel des organischen Geschehens* (1911) and, following him, by Bertrand Russell in *The Analysis of Mind* (1921), to account for the discontinuous memory of past occurrences.

was not ripe; there was the whole field of neurology and neurochemistry to explore for possible scientific explanations, before scientific principles themselves might have to be given up.[61]

Few psychologists, however, and few philosophers concerned with the nature of mind, have had the audacity to make the understanding of those systematic, mainly intraorganic processes their aim. Religiously oriented thinkers who held a brief for the radical difference between mind and matter welcomed the special "nervous and mental energy," which could be vaguely imagined, but not measured by any parameter, so that the mind was safe from "materialist reduction." For psychological theory this cleared the way to all sorts of assumptions—libido, soma-psyche, psychical factor, will to live.[62] Such entities may be justified as aids to moral orientation or to practical, psychiatric conception, but for scientific purposes they are useless. The laboratory psychologists therefore had to find some other framework in which to develop their young "-ology," sprung like Athena from the head of Modern Science, full-armed with method, apparatus and technical terms. Their original intent was, of course, to tackle the study of mental phenomena with the same directness and the same systematic exactitude with which investigations of physical phenomena had been carried to their astounding results. But several untoward circumstances conspired to deflect the project from this ambitious course. In the first place, mental phenomena as such are not publicly available for direct inspection; in the second place the universal belief that there were mental entities, metaphysically distinct

[61] Ostwald, *op. cit.*, p. 31, editor's note: "Ostwald verfolt nun den Gedanken, dass alle psychischen Vorgänge 'in Umwandlung chemischer Energie in andere Formen bestehen, deren Gesetze wir nur sehr unvollkommen überschauen, deren Wirkung wir aber als Gedanken, Gefühle, Willensbetätigungen usw. sehr genau kennen.' "

[62] The most notable and influential protagonist of the doctrine of a special energy with laws of its own is C. G. Jung. In his article "Über psychische Energetik" (1928), in which he credited Groot and Lipps with the idea of "psychic energy," he wrote: "Da wir nun leider wissenschaftlich nicht nachweisen können, dass ein Äquivalenzverhältnis zwischen physischer und psychischer Energie besteht, so bleibt uns nichts anderes übrig, als entweder die energetische Betrachtungsweise fallen zu lassen, oder eine besondere psychische Energie zu postulieren. . . ." He elected the second alternative, and called the postulated psychical energy "libido." (The "soma-psyche" made its appearance in an article by G. R. Heyer, "The Psycho-Somatic Unity—A Few Practical Inferences" [1938], which recommends yoga breathing exercises to activate and readjust "the organs and organic systems of that part of our essential soul that we call the somapsyche.") For the methodological conse-quences of assuming a non-physical sort of energy, see Jung's subsequent book, *Naturerklärung und Psyche, Synchronizität als ein Prinzip akausaler Zusammenhänge* (1952) (published in a somewhat inexact English translation in 1955).

from material entities, to be dealt with defeated any hope of naturalistic access. Then, too, mental data were bound up in living organisms and could not be investigated at autopsy; even in life, any experimental procedure might be expected to interfere more or less with their pre-sentation, for introspection as well as for objective diagnosis. Altogether, it soon appeared that the principles and methods of physics were not applicable to *res cogitans*, and the hope of revealing the structure of Mind by such well-established means fell very low.

It was then that John B. Watson proposed to abandon the pursuit of what psychologists really wanted to know, and study, instead, what could be scientifically studied—the overt behavior of animals and men.[63] Here was something related to mental phenomena but, unlike them, observable, and capable of being recorded, repeated and influenced by controllable conditions. It showed some inherent regularities, familiar to common sense as "habits," that might well be following natural laws. Above all behavior could often be predicted, and the prospects for a technique of controlled variation and prediction were certainly good. Almost at once, behaviorism swept the psychological laboratories, lecture rooms and literature, especially in America, but also in large measure throughout the rest of the civilized world.

The wave of pristine enthusiasm has abated, but the effect it has had on psychological research is radical. The original project to study mental phenomena has been swept away; instead, we have a busy host of "psychologists engaged in their paradoxical flight from the psychologi-cal," as one member of their profession has put it.[64] Directly observable motions of organisms, sequentially related to observable changes in the environment, were the physical phenomena to which the first rush of that flight took them. These typical relata in a sequence that certainly looks like direct causation are the "stimulus and response"[65] of today's "behavioral" (as distinct from "behaviorist") psychology. They looked so beautifully physical, accessible and simple that an empirical science of psychology was quickly proclaimed as a *fait accompli*.

No one saw, from the outset, that they were too simple. Most labo-ratory psychologists were not trained in any more venerable science to

---

[63] See *Psychology from the Standpoint of a Behaviorist* (1924). He had published *Behavior; an Introduction to Comparative Psychology* ten years earlier, but the great impact seems to have been made only by the second book.

[64] C. O. Weber, "Homeostasis and Servo-Mechanisms for What?" (1949), p. 238.

[65] Probably derived from the "stimulus and reflex" of the great physiologist I. P. Pavlov. The "conditioned reflex" promised to extend the principle beyond the physiological reaction to higher levels, and was hopefully regarded, for a while, as the key to all mentality.

the length of having done advanced research work; their ideal of empirical science was derived from Bacon and Mill. The collation and comparison of data and occasional generalization of findings in the form of "laws," and especially the predictability of much stimulated behavior, satisfied the canons of those eminent methodologists, and seemed to follow Bacon's admonition to students of nature to "lay their notions by, and attend solely to the facts." Concern for the purity of the facts tended to rule out all theoretical constructs and risky hypotheses; "the facts" came to mean directly observable facts. The peculiar result of this stern "scientism" was that psychology, instead of broadening out by conceptual growth to reveal more and more unsuspected facts and penetrate deeper and deeper into detail, shrank up into ever closer confines, eschewing all researches that required any extensive assumptions and especially any departure from the stimulus-response pattern. The pattern has lent itself to quasi-mathematical formulations, as in the ambitious systems of E. C. Tolman or C. L. Hull; but it has engendered no powerful concepts. The main reason is that the ideas whereon the systems are built are categories, not abstractions. Tolman, after his long and intensive work with them, finally wrote: "I still feel that 'response' is one of the most slippery and unanalyzed of our current concepts. We all gaily use the term to mean anything from a secretion of 10 drops of saliva to entering a given alley, to running an entire maze, to the slope of a Skinner box curve, to achieving a Ph.D., or to the symbolic act of hostility against one's father by attacking some authority figure. Now, I ask you!" And further: "As for perception, I have always been about as sleazy about it as I have been about responses. The 'stimulus' as we use it today seems to be just as slippery a term as that of response."[66] To build a life work on such a foundation is indeed a heroic response to the stimulus word "Science."

A less fiery but more diagnostic criticism is proffered by D. B. Lindsley in his Nebraska symposium paper, "Psychophysiology and Motivation," where he has written: "The separation of motivational forces into drives and incentives is arbitrary and tends to delude us when we think only about end results, or behavior. Presumably it is done to facilitate and simplify the design of experiments. But do we necessarily have to simplify at this stage? Sometimes it is better to look at a problem in its broadest context in order to map or conceive a plan before attacking it. . . . Have we settled too soon upon our parameters of measurement? Have we become overly preoccupied with the need for finding a simple

[66] "Principles of Purposive Behavior" (Koch II, 1959), p. 95.

method of measuring and evaluating drive and motivation? Do we have to find in a specific technique of behavioral measurement the one and only approach to this problem? Obviously not, for the problem is a broad and complex one" (p. 49).

Woodworth, P. T. Young, R. W. White and others have given motivation theory a wider basis and larger scope than it could aspire to in the days of orthodox behaviorism; yet their development of it is still handicapped by their heritage from that school, namely the identification of "activity" with "behavior," which limits their data to behavioral acts, so that only such acts may be considered as motivated, directed, consummated or themselves effective.[67] The meaning I have given to "motivation" is dictated by the much more abstract act concept with which I am operating; it is not really at odds with the broadest current usage, but makes a more radical break with behaviorism, freeing the study of mind from its supposed restriction to molar observations and descriptions. No systematic study—let alone a genuine science—can remain on the molar level where all knowledge has to begin. Whether there can be sciences of behavior comparable to physics in their abstract structure I am not sure; but there can very probably be an understanding of mental phenomena that would permit us to trace them in a downward direction, at least, to their roots in biological events we have hopes of knowing systematically and precisely.

To this end I have introduced the notion of the act as a formal unit, or modulus, of living processes. A response, such as aggression against a father-surrogate, is an act, arising from a situation that reflects a general, accumulated condition of the organism. The condition stems from acts long past that still affect all emerging situations through their (now completely integrated) traces in the matrix of activities, the organism itself; they have continued themselves in its special form, its disposition. A response such as a sudden increase in saliva flow is also an act, arising from a situation that may be just as complex but less individual, so that analogous acts with analogous motivations may be induced under many circumstances, in countless organisms. A heartbeat is an act, and so are all distinguishable physiological events that compose it.

Returning, at last, to the analysis of the act form itself, with its typical phases of incipience, acceleration (with or without growing articulation),

---

[67] The restriction of psychology to behavioral acts—that is, to overt acts of the organism as a whole—is attested by a passage in Woodworth's *Dynamics of Behavior*: "Tolman . . . in 1932 . . . introduced a distinction between *molar* and *molecular* behavior which is acceptable to many and perhaps all psychologists . . . it is a distinction between two kinds of description, a psychological description being always molar and a physiological description being often though not always molecular" (p. 22).

consummation and cadence, we are arrested at the very outset by the challenging complexity and importance of the first phase. Most of the motivation psychologists, who conceive motivation as the evocation (or, sometimes, the releasing) of a "drive" by a stimulus, and the behavioral act, therefore, as the overt termination of covert processes in the organism (a concept to which I certainly can take no exception), have observed that "drive" is not simply a general impetus to motion, but is directed to and through particular effectors, and subject to guidance along the course of its swift or gradual expenditure. Woodworth, for instance, has remarked this specialization in "drives," which distinguishes them from inorganic forces; for machine power will drive one machine as well as another to which it may be coupled, but "drives" are discriminate. As he words it, "In general, a drive has direction as well as intensity; it is selective as well as activating."[68] Its direction is toward a particular object or away from it; but the specificity of the drive goes further, according to the particular nature of the object to be approached or avoided. He cites, by way of example, the difference between the danger of falling and that of being run over, which are met by quite different movements implementing the "avoidance drive."

Motivation, effected by a stimulus, is thus conceived to be the activation of a "drive." Just what constitutes the "drive," however, is hard to tell. The different drives directed toward goals or away from dangers certainly appear as "intervening variables." They are events in the organism, started by stimulation and terminating in overt acts that have obvious reference to specific things, beings or events (such as electric shocks occurring in particular places). Yet some psychologists hold that "drive" designates no actual process, but a concept in terms of which actual phenomena are describable and measurable. Woodworth, who introduced the term, originally used it in this abstract way;[69] even in his much later *Dynamics of Behavior* from which I have just quoted, in

---

[68] *Ibid.*, p. 63.

[69] See P. T. Young, *Motivation of Behavior* (1936): "The term 'drive,' . . . in its modern sense, was first used by Woodworth in 1918, in his *Dynamic Psychology*, and from this source was taken up by animal psychologists" (p. 70).

"Woodworth spoke of *drive*, not *drives*; using the word in a general, not a specific, sense. He derived the term from mechanics; a machine must be driven if it is to move . . ." (p. 71).

"When Moss used the term 'drive' in 1924 he wrote of specific concrete drives such as hunger, thirst, and sex" (p. 72).

"We have assumed only one kind of energy—physical energy; differentiation among drives is not possible from this point of view. We may say that individuals differ in the degree of drive or energy known in their activity and that the drive of an organism varies with conditions, but when the word is used in this sense we cannot speak of different drives" (p. 79).

which he has certainly shifted from "drive" as a conceptual measure of motivation to "drives" as actual "intervening variables" between stimuli and responses, he quotes C. P. Stone in support of the old abstract sense: " 'Drive, like force in the realm of physical phenomena, has conceptual as opposed to phenomenal reality.' "[70] But one cannot have it both ways; if "drive" is a systematic concept like "force," then "drives" must be manifestations of "drive" as "forces" are of "force," capable of transformation or redirection (as Lorenz and Tinbergen consider them in their theory of "displaced" consummations); but if, as Woodworth holds, drives are goal-specific and have inherent direction toward or away from certain kinds of objects related to the organism's welfare,[71] then drives are intraorganic events with "phenomenal reality," not scientific magnitudes with "conceptual reality."

That is not a good state of affairs. Obviously, since "motives" elicit responses by activating "drives," "drives" are operative conditions which build up in the organism and terminate in behavior; they are events, or aspects of events, and may occur or not occur. The conception of behavioral acts as terminations of "drives" has the great virtue of emphasizing the intraorganic dynamism that underlies overt responses, and thus breaking away from the simple "billiard ball" model of causality. But as a principle for the theoretical analysis of life in terms of acts it is not suitable, because it does not characterize acts as such. This is immediately apparent from the fact that not every act is supposed to stem from a drive; some elementary responses, such as withdrawal of the hand from contact with fire or a hot object, are said to be caused directly by the stimulus.[72] Furthermore, the same drive may be thought to terminate in any one of several possible acts, which are formally distinct, though pragmatically equivalent. Above all, one can speak of "drives" only in relation to behavioral acts; but the understanding of such acts requires conceptual terms that can be carried to the level Tolman has dubbed "molecular," as against "molar." (Here I would rather speak of "microscopic" versus "macroscopic"; for if the study of mind is ever to be merged with the advancing sciences of life, in which many phenomena are actually molecular, the use of that term in a sense not referring to molecules will make for quite needless ambiguity.)

[70] Woodworth, *op. cit.*, p. 58.

[71] See *Dynamics of Behavior*, pp. 63–66. On p. 64, the author raises two questions: "1. Is there such a thing as *non-directive* drive? 2. Is it possible to *redirect* a drive so as to make the same drive activate different mechanisms, or so as to make different drives activate the same mechanism?" He answers both in the negative.

[72] *Ibid.*, p. 52: "If a response is sufficiently accounted for by a stimulus, there is no room for a motive—no sense in assuming a motive."

The Act Concept and Its Derivatives

The dynamism of life lies in the nature of acts as such; it is incorporated in their structure and gives them their typical form. Every act, as I said early in the present chapter,[73] has an initial phase, a phase of acceleration and sometimes increasing complexity, a turning point, or consummation, and a closing phase, or cadence. The initial phase is its impulse. This is an integral part of the act itself; it is the incipient act.

An impulse is usually conceived to be a homogeneous discharge of energy, the equivalent in animate nature of a force, or impetus, in the inorganic realm. But no vital process is exactly like anything outside the context of life; it is always more complex. An impulse, or nascent act, is an offshoot of a fluid situation which, because of its unstable character from one moment to another, is probably never altogether determinable. Within the active organism, the very matter which is implementing an act may undergo dissolution in the process and be transformed, so that it enters into another mechanism that functions in another situation. The first really identifiable element is the impulse; and this is already an articulated process.

Even in acts which are complete, all-or-none discharges of energy, like the "firing" of a single neuron in the brain,[74] the impulse must be supposed, usually if not always, to be composite. The complexity of the cell makes it unlikely that a simple stimulus normally evokes the act of discharge,[75] although the latter, when it does occur, seems to be always exhaustive of the entire impulse. In such minute acts, therefore, the building up of the impulse may be essentially a summation of converging

[73] See p. 261.

[74] See particularly D. O. Hebb, *The Organization of Behavior*, which contains his brilliant hypothesis of the functional "cell-assembly."

[75] In a study by W. H. Marshall and S. A. Talbot, "Recent Evidence for Neural Mechanisms in Vision Leading to a General Theory of Sensory Acuity" (1942), p. 122, the investigators report: "In the cat, optic tract endings in the geniculate divide into several branches and as many as 40 ring-shaped boutons have been seen on single radiation cells which may come from as many as 10 optic tract fibers. Each fiber also divides to form synapses with several radiation cells. In addition to bouton contacts, the radiation cells have numerous dendritic processes, with which the optic tract endings make apparently more numerous synapses (of the free axon type), than with the radiation cells themselves."

To gain still further insight into this vast complexity the reader may be referred to the extensive and detailed neurological researches by Rafael Lorente de Nó, especially his "Anatomy of the Eighth Nerve" (1933) and his "Cerebral Cortex: Architecture, Intracortical Connections, Motor Projections" (1943). Lorente de Nó maintained that it always requires more than one electrical impulse to "fire" a brain cell—a statement with which Hebb takes issue (*The Organization of Behavior*, a long footnote to p. 66).

[291]

charges raising the electrical potential of the neuron to the point of release, whereupon the cell, having "fired," is briefly "refractory," unable to act, and perhaps unable to receive a charge;[76] the activity runs through a series of cells as a course of successive discharges, each motivating the next.

In more complex acts, however, various elements of the total event often seem to occur as by a "pre-established harmony," rather than in the manner of a chain where each movement induces the next. Subordinate acts appear to converge on the achievement of an integral larger whole.[77] The result is that a dynamic pattern is realized in the occurrence of the total act. Every constituent act has its particular impulse, with its own intensity and easiest path, its possible alternatives if that path is obstructed and its own rate of progress; and every impulse, when it issues in action, resolves the particular organic tension it represents, which is an accumulation (perhaps infinitesimal) of potential energy ready for transformation into some other phase. In living systems such charges tend to form integrated patterns, i.e., unitary but organized impulses, and spend themselves, under the influence of one largest, unifying impulse, in a flow of events that takes the characteristic form of a single act.

The only really decisive way of justifying a new philosophical departure such as we are making in treating the act as the unit of living form is to apply the proposed concepts to some field of empirical research to which they properly relate, and show that they are adequate, or at least more nearly adequate than any older ones, to the logical demands of the material. Fortunately—as it often happens along the broad fronts of intellectual advance—such a test has already been made, with gratifying results, in the biological science of ethology. In a pioneer-

---

[76] The latter condition is assumed by Hebb as a ground for the widening of electrical circuits within the cell assembly; but discussion of that point belongs to a subsequent chapter, and will be taken up there.

[77] A striking and convincing demonstration of such mutually independent timing of subordinate acts in a single larger act is presented with meticulous care and precision by R. W. Doty and J. F. Bosma in "An Electromyographic Analysis of Reflex Deglutition" (1956). The experimenters obtained concurrent electromyographic records of ten muscles involved in the swallowing reflex, plus the diaphragm. The contractions of these muscles occur in a functionally ranged series, but apparently without successive inducement of each act by its predecessor. "Participating muscles," the report says, "could be excised or procainized without affecting the sequence of events in the others; nor did concurrent stimulation of hypoglossal or lingual nerves change the response in any but the muscles which they innervate. Cocainization of the mucosal surface of the pharynx did not alter swallowing evoked by SL [superior laryngeal nerve] stimulation" (p. 53).

ing theoretical article, entitled "A Basis for the Quantitative Structure of Behaviour" (1954), the authors, W. M. S. Russell, A. P. Mead and J. S. Hayes, have taken the act concept as the systematic unit for the understanding of behavior.[78] They limit their theoretical discussion to the treatment of such phenomena as ethologists ordinarily deal with, i.e., the more or less stereotyped, often species-specific responses of animals that have been scientifically observed in field or laboratory: courtship behavior, parental behavior, hoarding, fighting and other distinct reactions to identifiable stimuli. They begin with the assumption that there are basic whole acts not further analyzable into subsumed lesser acts. This postulate may invite objection at first, but it is really only a conceptual device to achieve a tentative definition; in the course of the article the composite act is restored, superacts and subacts defined and even the possibility recognized that any act may subsume other acts.[79] But what might be called the "basic behavioral act," in their language simply "Act," is defined as follows:

"An Act = a set of observable activities in different effectors (*e.g.* muscle tensions), *regularly observed in combination* (thus not analysable into separate components . . .) hence recognisably different from other such acts observed in the same species."[80] Thereupon, almost immediately they go on to expound another basic idea, saying:

"From similar considerations to those advanced by HESS (1948)[81] and TINBERGEN (1951),[82] we take the existence of an Act so defined as sufficient and necessary evidence for the existence of a central nervous unitary mechanism of co-ordination, to be called: An Act Centre.

"Evidentally an Act Centre is a special case of Tinbergen's more general concept of Centre, referring to a low level. Nothing whatever is implied about the localisation or spatial discreteness of the Act Centre, or about its nature, with the single exception that it must act as a functional unit."[83]

---

[78] This paper, and another one ("The Mechanism of an Instinctive Control System" [1954]) which carries out the application of the theory to more special problems, came to my notice when my own ideas on the subject were already written down. The convergence of our independent lines of thought is a most encouraging case of corroboration.

[79] *Ibid.*, p. 183: "It might conceivably be found that all reactions previously supposed to have been Acts in their own right . . . were analysable into components and were, therefore, Superacts themselves."

[80] *Ibid.*, p. 163. (I am not altogether satisfied by the use of regular conjunction in all observed cases as a criterion of *unanalyzable* unity.)

[81] W. R. Hess, *Die funktionelle Organisation des vegetativen Nervensystems.*

[82] Nikko Tinbergen, *The Study of Instinct.*

[83] Russell, Mead and Hayes, *op. cit.*, pp. 163–64.

Another concept required from the outset for their analysis of behavior in terms of acts is that of *act tendency*, which is measured by the frequency of occurrence of an act in a certain number of trials under experimentally varied conditions. Fluctuations in the tendency to a particular act are traceable to changes in those conditions, factor by factor; the factors being all controllable items of organic state (hunger, hormone levels, blood pressure, fatigue) and environment (light, temperature, etc., and acute stimuli); all these are "variables" which may or may not affect the "threshold" for the act, or the "height" of the act tendency.

It is, unfortunately, impossible to present in due order even the main points of the conceptual structure built up in that very systematically written article, but certain similarities between its treatment of acts and mine are readily visible: first, the unity of the act, though Russell, Mead and Hayes base it on an empirically observed, regular coincidence of elements, and I find it in the singleness of an over-all tension; second, the development of the act from a functional "center," which is not said to be a local mechanism or "brain center," but some sort of intraorganic starting point where the act is formed as a unit charge, ready to be triggered by an external or even internal stimulus—clearly an analogue to my notion of "impulse" (though the difference between our respective treatments of that starting point is a matter to be further elucidated); and, finally, the tenet that extraneous influences on the act reach it indirectly, through the act tendency, which is conceived as "a *continuous variable* associated with the Act Centre, where any change in conditions producing an increase in Act Tendency increases the number of combinations of states of other variables associated with the Act's occurrence."[84] This leads directly to the concept of the situation, or current phase of the life in which—and from which—the act arises.

It is here that I find the contribution of Dr. Russell and his collaborators most helpful; for they succeed in constructing the concept of "situation" by a simple yet astute logical device, which completely eludes the confusing, prejudicate language of "environmental" versus "internal" events. They have, it is true, borrowed a terminology which threatens to introduce another kind of misguiding figure of thought, the currently fashionable language of small-current electrical engineering; and although they assure us that it is harmless and helpful, because "it is becoming well recognized that the nervous system does not work in detail exactly like a radar outfit or computing machine," they do accept "the general point made by Wiener—that the C.N.S. is a communica-

84 *Ibid.*, pp. 167–68.

tion system."[85] That such systems furnish models of some highly important neural mechanisms is demonstrated, I think, by the advances they have implemented in the field of brain physiology and neurology; especially the basic recognition that nervous activity involves electrical potential and current. The insidious influence of the model, however, is the apparent implication "that the C.N.S. is a communication system." The central nervous system effects communication in the course of its total operation, which may be not just somewhat, but radically different from that of a machine dedicated to communication as its primary function. Wiener remarked that the development of the human brain may already have passed its optimum stage, being so complex that communication paths are becoming longer instead of ever shorter and readier.[86] That means, of course, judging its "optimum" as that of a communication system. The central nervous system may have more to do than messenger service, and may be quite efficiently reducing its capacity for one function in the course of augmenting that for another; developing new forms of memory, symbolic transformation, even synaptic slowing and stowing that makes for more complex psychical phases—that is, for lower thresholds of emotive and sensory feeling.

After recording all these misgivings, however, I want to return to the constructive treatment of the situation from which an act arises, as given in the article by Russell, Mead and Hayes. They begin with the notion of an indefinite number of variables which might have causal effects on the "Act Centre" and thus alter the degree of the "Act Tendency," to the extremes of precipitating or inhibiting the act. Not all of these variables of circumstance, however, really exercise their potential influences, because their effectiveness depends in large part on the receptivity of the "Act Centre." So, the authors say, "We can express the situation at any

---

[85] *Ibid.*, p. 169. They come back to this point a few paragraphs later (p. 170), saying: "It is further to be stressed that in using communication language we are not committed to the application in behaviour of the mathematical theory of communication or information, developed mainly in the context of communication engineering, nor to the strict usage of any of the terms in that subject—on whose definitions there is in any case considerable disagreement at present." To subject already ambiguous terms to further loose usage is a dangerous practice. This criticism, however, must not be taken to be leveled at all biological and psychological studies employing the language of physics, probability theory or other mathematical science; there are works, such as J. W. S. Pringle's paper "On the Parallel Between Learning and Evolution" (1951), in which the terms of scientific discourse are strictly literal in their application to vital phenomena.

[86] See Norbert Wiener, *Cybernetics: Or Control and Communication in the Animal and the Machine* (1948), p. 180.

Act Centre, in a general way by means of the following equation for its Act Tendency . . . :

"$T_x = f(a, b, c \ldots)$

In this equation $T_x =$ Act Tendency$_x$.

a, b, c, *etc.* represent physical variables.

The symbol f denotes a *single-valued, undetermined, unique function*."

By "undetermined" is meant "that the function can be of any kind"; by "unique," that only one value of T_x can correspond to f, though several f's may have the same value, T_x. [These f's would, then, be equivalent for the level of the act tendency.]

Since the value of T_x varies with that of $f(a, b, c \ldots)$, the righthand term of the equation can have any number and any combination of elements *a, b, c,* etc. We may put everything in the universe into the argument to f if we choose. "Obviously, however, these variables can in fact be divided into two sets. Those in one of these sets can vary throughout their range without causing any variation in Tendency; those in the other cannot vary without producing such a variation. . . . We propose to call the former *Ineffective* and the latter *Effective* Variables. (These terms seem to us intuitively clearer than Non-Causal and Causal.)"

Omissions or additions of elements in the class of ineffective variables make no difference in T_x; but any change in the membership of the effective class does. "Many Variables can be confidently regarded as Ineffective for a particular Act *a priori*; others can be determined by experiment; in this way some variables can progressively be eliminated from the right-hand side. But until and unless this is explicitly done the 'doubtfuls' must be left in."[87]

This formulation has the great intellectual advantage (as the authors point out) that it lets one operate with a causal function, $f(a, b, c,$ etc.), without knowing specifically what elements it actually contains, and which of its members are effective, i.e., which ones determine T_x and which are "vacuous" conditions for it.

"Any single combination of the states of the right-hand variables," the text of the paper continues, "may be regarded as potentially one individual message. The set of all such messages we may call M.

"Evidentially one sub-set of these messages . . . the set of messages consisting of combinations of states of the Effective Variables for the

---

[87] *Op. cit.*, pp. 171–73. The possibility of experimental investigation of what variables are effective "for a particular Act," and of progressively eliminating ineffective ones, shows that the "Act" is not a specific occurrence but a class of acts, or (defined by intension) an act form.

Act. We shall call this m. . . . it amounts to the set of messages reaching, or *available* to the Act Centre."[88]

There may be several combinations, permutations, etc., of the effective variables "available" to a particular act center; these may determine different values of T_x, or some may determine the same value (though each can only correspond to one value of T_x). All these potential m's together constitute the "Act-Available Message Set."[89]

A particular m is what I have called a "situation." Since the authors I am citing are dealing with act forms whereas I am dealing with the individual act (which may or may not have recognizable repetitions, or be a repetition), T_x, the act tendency, has to be replaced by a real occurrence; this is, in fact, the impulse. The maximal value of T_x, which in their system would be the occurrence of the act, would to my thinking be the carrying out or actualization of the impulse. We are concerned with the same phenomena. In their terms as in mine, any motivating influence upon an act works through the situation that induces the impulse or tendency, wherein the tension pattern (form, direction and cathexis) of the act is figured, to be projected—perhaps under many modifying, deflecting, even distorting accessory impulses—in the course of its actualization, if such occurs.

There is one further step in the same direction, taken by these three authors, leading to another construction which demonstrates the aptness of the act concept: that construction is the set of all variables (with reference to any one animal) each of which is effective for at least one act, and which may, of course, be variously combined to yield the set of all possible m's for all potential acts of the animal, the "Animal Available Message Set," briefly designated M.[90] The set M is what I call the organism's ambient. The great advantage I see in the proffered definitions is that the effective variables underlying a situation (m) may be either environmental or internal, present or past, so that we have no longer the awkward dichotomy of the "environmental situation" (the modifier often taken for granted), and the "internal condition," sometimes conjoined as "the external and internal environment."[91] The wider concept, M, still consists of the same elements, which make situations that engender impulses and motivate acts.

In view of all our agreements on the use of the act as a basic unit, the indirectness of the causal influences of other events on it, the practically

---

[88] *Ibid.*, p. 173.   [89] *Ibid.*, p. 177.   [90] *Ibid.*, p. 179.

[91] "*Milieu interne*," a term introduced by Claude Bernard, referred strictly to the body fluids surrounding or filling the more solid structures of the body. "Environment" is a most interesting but difficult concept, to which we shall return.

inexhaustible complexity of conditions relevant to it, the conceptions of "situation" and "ambient" and a number of other developments which I cannot adduce here,[92] and especially in view of the priority of that remarkably rich theoretical article, it might well be asked why I could not simply accept its conceptual frame and terminology, and follow the line of work it indicates. The chief reason is that, as it stands, the only acts which fit into its frame are behavioral acts, whereas acts as I propose to treat them are the lowest terms in the study of all life, plant and animal and human. That is why I have not begun with more or less fixed forms, such as instinctive "consummatory acts," but with the general act form exemplified in all instances of life, and foreshadowed even below the most primitive level of unambiguously vital existence. Such a wide range of acts makes one of the key concepts of Messrs. Russell, Mead and Hayes—namely, the concept of an "Act Centre"—problematical in some contexts. Even though they assure us that "nothing whatever is implied about the localisation or spatial discreteness of the Act Centre, or about its nature, with the exception that it must act as a functional unit," the assumption of different act centers does carry with it the connotation of some permanent mechanisms, distinctively geared for different acts. In the higher animals, especially vertebrates, with their developed central nervous systems, such mechanisms very probably exist. But in an amoeba, where the same protoplasm serves now as the cell's gelatinous structure and the next minute as its fluid content, and may belong, in random succession, to pseudopodium, dividing membrane (if the cell splits), absorptive, digestive and eliminative endoplasm, there can be only one center of all acts, the creature. If we carry the act concept into plant life, where the predominant acts are vital, and only a few (such as the explosive opening of some seed capsules or the closing of the Mimosa's leaflets) are behavioral, and there is no nervous apparatus to our knowledge, messages available to an act center become a difficult assumption.

The difficulty is obviated by assuming that the impulse arises from the situation. The trouble with "Messages" is that they connote a receiving center where they are combined and processed; this is one of the insidious by-products of "a useful form of language." Someone or something must get the messages. I consider the effective variables as events (strictly speaking, peripherally started acts, chiefly acts of perception) and conditions (many of which may be traces of former acts or im-

---

[92] The article goes on to a logical construction of "superacts," which parallels perfectly my own treatment of acts subsuming other acts; also to conjunctions, incompatibility and inhibition, and further derivative concepts.

pulses) that compose the situation of the organism. The situation motivates the impulse. The presumptive form of the act is established in
the formation of the impulse. This conception seems to me more negotiable in the very large field of philosophy of mind than that which
is rather specially fitted to the demands of ethology,[93] and much more
so than the motivation theory that attributes the properties of directedness and selectivity to the "drives" which are presumed to lie behind
animal behavior. The notion of "drive" is not amenable to manipulation
and generalization, as basic analytic concepts have to be. "Impulse" is
a far better unit; one can face a hypothesis postulating thousands of
impulses, but thousands of drives would be formidable. The greatest
advantage I see, however, in regarding impulses as the starting points
of acts is that this conception takes one smoothly from the determining
conditions to the organic act, because the impulse is the first phase of
the act itself, even while it expresses the entire relevant situation (the
"effective" aspects of $f_x(a, b, c \ldots)$).

The supposition that the act takes its presumptive shape and scope
in the intial phase, the impulse, has several further advantages, too, for
it allows one to define potential acts, actualization (and at a later stage,
realization), inhibition, and finally intention, opportunity, choice and the
effects of decision. The concept of the potential act is of special importance here, for it has sometimes been said that such a concept,
necessary as it is for understanding choice, intent, foresight and any
effect of negative conditions (as in "sins of omission"), cannot be construed in terms of natural science, which can admit only actual occurrences to its realm of causes and effects.[94]

A potential act, is, then, an impulse. It need not reach a stage of
muscular contractions, nor of cerebration that records itself in galvanic
skin reflexes. Its overt development may be wholly suppressed by the
actualization of another, incompatible impulse; then it remains poten-

[93] Russell, Mead and Hayes, *op. cit.*, p. 164, say: "Evidently an Act Centre is a
special case of Tinbergen's more general concept of Centre, referring to a low
level." Tinbergen's concept is based on observations of the phases of instinctive
acts—the very stock in trade of ethology. To the term "Act Centre" I would also
bring an objection similar to that which I raised against Tolman's use of "molecular"
for "minute" or "microscopic": the term "center" has such an established sense in
neurology that the application of the term to something else than a localized brain
area supposed to control specific functions is confusing, and if biological science is
ever to bring neurology and ethology together in one system of thought (as it soon
may), such half-literal, half-metaphorical usages—really just loose usages—will stand
in the way of the merger.

[94] See, for an example from a notable source, William McDougall, *Body and
Mind: A History and Defense of Animism* (1911).

tial, though it may be so for a long time, continued by repetitive pulses, or held unspent in its nascent state. The impulse, or potential act, is something that really occurs; even without the normal development here called "actualization," an impulse is an event.

That the essential form of a complete act is prefigured in its impulse has been experimentally established in recent years by records obtained with microelectrodes inserted directly into individual neurons. This method, combined with macroelectrode measurements of changes in tissues and with behavioral observations, has yielded some surprisingly detailed knowledge of the incipience of acts. The all-important finding in these empirical studies is that the so-called "firing" of a neuron is not a simple discharge, such as we see in an electric spark, but has a typical and often elaborate spatiotemporal pattern. The cell is completely depolarized, its polarization being briefly reversed, so the "all-or-none" principle which has often been asserted is evidently valid; but that does not mean—as the first speculative theorists assumed, on the analogy of electrical conduction and discharge—that the depolarization of a nerve cell is a formless process. In the light of the new researches, T. H. Bullock, who is one of the pioneers in that field, states as a well-grounded conclusion: "The output of single neurons and of groups of neurons is normally probably always patterned, i.e., temporally and spatially distributed in a meaningful, non-random way."[95] Several investigators have analyzed the heartbeats of lobsters and crabs, which lend themselves particularly well to this purpose, and their findings all converge to corroborate the hypothesis that a prepattern of the act (the heartbeat) is exhibited in the impulse. As Bullock reports: "In each heartbeat of a lobster a complex burst of several dozen impulses arises in a pacemaker neuron of the cardiac ganglion. . . . It fires many times during a burst, and the other neurons—followers—fire repeatedly, each in a different pattern, for example starting at high frequency and declining. The actual frequencies, time courses, durations and numbers of impulses are individually characteristic and recur consistently for hundreds of heartbeats.

"This has been analyzed in some detail and the conclusion reached that *a primary patterned burst arises in a single cell*, the pacemaker, not dependent on feedback of spikes from other cells to formulate it. Intracellular mechanisms, presumably two or more interacting processes, must be postulated."[96]

"In the lobster cardiac ganglion and in some other cases, we now have . . . an intimate view of the truly spontaneous activity of single neurons.

[95] "The Origins of Patterned Nervous Discharge" (1961), p. 48.
[96] *Ibid.*, p. 53.

We even have evidence of the confinement of the pacemaker to a limited part of a neuron for there can be two separate rhythms in one cell at one time. . . ."[97] Other neurons in the ganglion contribute their own specialized parts of the impulse initiating the single act called a "heartbeat."[98]

Such elementary somatic acts, moreover, are not the only ones that seem to be generally and tentatively, at least, preformed in the impulse, to be modified by situational pressures in the course of their actualization; and preoccupation with lobster hearts does not necessarily blind a laboratory biologist to the presence and significance of analogous instances at higher levels. Dr. Bullock, for his part, observes such wider connections, saying: "In predisposed humans, Penfield has found that certain regions of the cortex can be crudely stimulated electrically and complex, vivid audio-visual experiences triggered. . . . Less complex but normally coordinated movements occur in lower forms. Although many regions of the brain are 'silent' or yield simple jerks or twitches, particular loci call up entirely normal-appearing sequences such as chirping or antenna-cleaning in crickets, compulsive drinking in a goat, stuffing imaginary seeds into the mouth in a pocket mouse, crowing and many other actions in chickens.

"The category of patterned movements represented by most of the responses just mentioned ('fixed action patterns') is of special interest for our problem. Other familiar examples are sneezing, jumping, spitting, stinging, swallowing, seizing, displaying, and the like. . . . It seems at present likely that for many relatively complex behavioral actions the nervous system contains not only genetically determined circuits but also genetically determined physiological properties of their components so that the complete act is represented in coded form and awaits only an adequate trigger, either internal or external."[99]

[97] *Ibid.*, p. 56.
[98] A detailed account of the experimental work here referred to, and done in large part by Dr. Bullock, is given in an earlier article by that author and C. A. Terzuolo, "Diverse Forms of Activity in the Somata of Spontaneous and Integrative Ganglion Cells" (1957).
[99] "The Origins of Patterned Nervous Discharge," p. 55. The present chapter was complete before any of this neurological material came to my notice. In that first draft I had written: "In microscopic acts, such as vital acts at the cellular level, the impulse must be too small to be known by any but stochastic means; its existence is really hypothetical; yet the assumption of it is, I think, theoretically justified, in that it offers a way of conceiving the intricate dynamism of life from the lowest, biochemical, to the highest mental and moral level." The finding of a considerable body of experimental work corroborating that purely theoretical guess and hazardous assumption came as sheer intellectual treasure trove.

In another detailed study of essentially the same sort,[100] M. G. F. Fuortes noted the presence of the background activity from which every individually identifiable act arises. "If identical shocks are applied at regular intervals to a skin nerve," he observed, "while activity is recorded from a flexor motoneurone, it is often seen that the first shocks produce each a single reflex impulse, but the later shocks evoke a progressively increasing repetitive firing. This is an example of temporal summation. . . . The change responsible for temporal summation was called by Eccles and Sherrington the *central excitatory state*" (pp. 214–15).[101] And further: "When the firing of a motoneurone is elicited by stimulation of sensory fibres, one sees that the postsynaptic spike[102] which corresponds to the reflex impulse is preceded by a slow depolarization which has been called the excitatory postsynaptic potential. Even when it is elicited by a synchronous afferent volley, this potential has a relatively long duration and it has a graded character, its height being controlled by the intensity of the presynaptic volley. A reflex spike is produced by a motoneurone only if the synaptic potential exceeds a certain amplitude, which can be reached not only by combination of simultaneously arriving impulses but also by summation of the potentials evoked by impulses arriving in succession" (p. 217).

The use of microelectrodes has revealed unexpected functional specialization even within single cells; it appears that, far from being just storage batteries periodically emptied and recharged, nerve cells have physiologically distinct parts, and their charges and discharges involve those parts in rhythmic sequences for the inception of apparently simple, integral acts. The mechanisms are still in the realm of conjecture, but the fact that the act form is established in the impulse is fairly certain. So Fuortes feels safe to say: "The interpretation of these results is that

[100] "Integrative Mechanisms in the Nervous System" (1959).

[101] Almost two decades before microelectrodes made intracellular observations possible, Paul Weiss wrote: ". . . if we accord to the centers [of muscular action] an unbiased investigation, we might some day discover that the 'excitatory state' of *Sherrington*, for instance, may not be after all such a trivial and colorless condition as electrotonus, or the concentration of a chemical substance, but that it may assume a variety of differentiated states in which *specific biochemical configurations* or specific *electric parameters* would appear as keys selectively unlocking for the discharge of excitation peripheral channels of correspondingly specific configuration" ("Self-Differentiation of the Basic Patterns of Coordination," p. 68).

[102] The frequent mention of "spikes" in neurological literature refers to the sudden deflections of the recording stylus that make sharp spikes appear on the rotating drum. To illustrate the graphic projection of cellular activity, see Fig. 10–3 below, from Bullock and Terzuoio, "Diverse Forms of Activity in the Somata of Spontaneous and Integrative Ganglion Cells," in the text.

impulse activity originates in an area which is separated from the record-ing microelectrodes by an appreciable resistance, and . . . that this area is located in the axon. . . . the function of the motoneurone soma will then be to . . . integrate the effects induced by all impingements present during a certain time interval, and to transmit this integrated result to the pacemaker area where impulses are originated" (p. 220).

The "integrated result" is the microsituation that motivates the act by begetting its impulse, as one further passage seems to me to make quite clear: "These conclusions have been reached . . . by experiments in which transmission was evoked by single electric shocks giving rise to large synchronous volleys . . . but it is now appropriate to reconstruct from these results the events occurring in the more normal conditions in which afferent signals are . . . more or less prolonged trains of impulses proceeding asynchronously in different sensory fibres. The receptive activity of each fiber carrying excitatory impulses will elicit in moto-neurones a series of small excitatory synaptic potentials and, as long as no reflex firing is produced, the combination of the effects exerted by several of these fibers will be a larger, smooth and sustained depolariza-tion, the intensity of which will be a function of the number of impulses reaching the cell in a given time. With sufficiently intense stimulation an impulse will be discharged as soon as the depolarization has reached a sufficient level by virtue of temporal summation" (p. 221).

The build-up of tension in a single cell can terminate only in a minute impulse, but cell assemblies can form impulses to larger acts. The lobster's cardiac ganglion contains only nine cells, but these form the integral, patterned impulse to a heartbeat, which is no microscopic act.[103] There are billions of cells in a mammalian nervous system; cerebral cells, especially, seem to work in large assemblies, initiating behavioral acts of wide range and duration that go through many phases to their over-all consummation. And for every actualized impulse there are densely crowded others which reach only a momentary state of incipience be-cause the on-going act abrogates their expression, so they terminate as infinitesimal modifications in the matrix of life. Those are the potential acts eschewed by the individual, and they are countless.

A potential act, then, does have physical existence, though this may be far below any threshold of observability by feeling or (at present) by laboratory devices. If an act runs a normal course, steered, as it inevitably is, by the concomitant development of the surrounding situation, modi-fied by contingent further impulses and confluent acts in passage but nevertheless "carried out" (an uncommonly apt term in common use)

---

[103] Bullock and Terzuolo, *op. cit.*, p. 341.

to its consummation and the exhaustion of its original impulse, its potentiality has been actualized.

Impulses whose realizations would run counter to each other can occur in such rapid alternation that they practically coexist, springing from one and the same situation in the organism. If they originated by separate paths they might even actually coexist; the complexity of individual neurons found by Bullock and Terzuolo certainly attests the possibility. But their respective developments certainly cannot both be carried out; they are incompatible, and the occurrence of both impulses in one situation presents an option. The organism has to select its course. Thousands of options are presumably decided automatically all the time; in lowly animals this may even hold for larger behavioral acts. But whenever an act is induced by a change in the vital situation, such as the life process itself constantly engenders (thereby motivating an endless stream of acts), it is likely that not only the impulse of that act, but also one or more conflicting impulses or alternative potential acts are formed, which are doomed to speedy abrogation. This play of impulses forms the dynamic matrix of life, a plexus even more involuted and compounded than the metabolizing, differentiating, ever-changing structure that is the material organism, because the latter consists only of actualized events, but the life comprises also all the potential acts which exist only for milliseconds or less. Perhaps that is why we feel that life is "in" the body and pervades its actuality. Out of this matrix all mental and behavioral acts arise.

If the basic concept of the act is that of an event, a spatiotemporal occurrence, the Act in the ethologist's sense—strictly, the act form—has to be distinguished by a convenient term; for in any theoretical context it is this abstractable form, which two acts may have in common, that is of interest. Different act forms are the concern of physiology, psychology, ethics, jurisprudence, sociology; forms whereby acts may be classified, recognized, and treated as instances of "Acts," or as expressions of "Tendencies." This formal aspect of acts I shall call "action." In so doing one really just establishes explicitly what is rather generally accepted in common parlance, for such typical forms may be found in inorganic nature too, in which case they are usually referred to as the actions of machines, winds and waves, corrosives, etc. We speak of the "action" of pistons or gears rather than of their "acts." It is action that living and non-living mechanisms may have in common. The same action may be performed by an animal's heart and by a pump, or by a rodent's teeth and by a file in a man's hand, or possibly by some electrical circuits in the man's brain and those in a modern computer which

therefore is viewed as a "mechanical brain." The term "action" is sometimes extended to specific acts, usually in cases where it is the nature of
the action rather than the act that counts, as for instance in impersonal,
collective acts, like the conduct of a war.[104] But in the strict sense here
given to it, it means the causal pattern, or operative principle, according
to which an organic or inorganic mechanism works.

There is one further derivative concept which is of major importance
if acts are to be taken as the basic units of all biological thinking: the
concept of vital activity. For this we require the notion of action, as just
defined, for separate acts may be hard or even impossible to identify in
that department, let alone to predict. The situations inducing acts are
so unsurveyable that a situation can practically never be calculated with
enough certainty to assure the occurrence of any specific act. In the study
of life, therefore, any safely acceptable "laws" usually take the statistical
form of probability laws. Most of the "laws" which have been established with any degree of certainty in that domain concern acts which
are exemplified by thousands of instances, namely, repetitive vital acts,
of which the readiest examples at a molar level are ciliary or flagellate
movements, heartbeat, and gill or lung breathing. Such acts are very
commonly concatenated into series, wherein the same general form is
discernible over and over again, often with considerable deviations that
cancel out in a larger statistical pattern.[105] I shall refer to act sequences

[104] Ignace Meyerson, in *Les Fonctions psychologiques et les oeuvres* (1948), uses
"action" in this sense, but refers to subordinate acts incidental to the planned total
performance as *"actes."* See especially p. 21: "Les actes de l'homme présentent une
forme . . . ils sont systématiques, mais ils s'isolent en unités; les actes deviennent
des actions, des touts pourvus d'une sorte d'existence et de qualité propre . . . les actes
d'une guerre seront 'une expédition' ou 'une bataille': ce sont là des ensembles denses
et complexes dont on peut dire que par la nature ils ont un début et une termi-
naison."

Meyerson remarks (without a reference) on Pierre Janet's similar use of "action."

[105] A good example is given in G. E. Coghill's study of the differentiation and
functional developments in the spinal cord of the *Ambystoma* larva, where, in early
stages of growth, mitosis occurs, according to his observations, "in numerous, irregularly distributed, restricted foci. These foci are obviously transitory, for they are
found in various locations in different specimens. . . . They are rarely found to be
bilaterally symmetrical in their arrangement, and they hold no definite or constant
relation to the motor or sensory nerve roots. Accordingly, proliferation as indicated
by mitosis appears to be unorganized. . . . [But] where the number of cells in each
segment are charted on the longitudinal axis of the spinal cord, the graph so derived
conforms to a definite and constant pattern. . . . Individual parts may grow
differently from moment to moment, apparently haphazardly, but they are regulated
in such a way that they produce a definite and constant result" ("Growth of a
Localized Functional Center in a Relatively Equipotential Nervous Organ" [1933],
p. 1087).

of this sort as "activities," and regard that term as strictly proper to them except where it is explicitly extended beyond its strict meaning (in the literature, of course, it is often more freely applied). This practice will provide us with a distinctive name for a very important type of construct in the biological universe of discourse, and at the same time not violate common usage; the constant events of life, such as circulation and breathing, are generally called "activities."

The introduction of "activities" brings us very naturally to the all-important topics of rhythm, dialectic, self-continuation and finally to the possibility of defining "agent" in terms of acts. But concepts developed in the course of that undertaking really belong to the next chapter, and will have to wait their turn.

9

On Individuation and Involvement

TO SPEAK of acts without assuming the existence of agents may appear to be an odd philosophical vagary; and those philosophers who begin their conceptual analyses by deciding what is "implied" in the use of selected key words[1] are likely to take exception to the practice, on the ground that an act is necessarily performed by an agent. Their argument rests, of course, on their concept of "act," which is at first vaguely accepted, in order to be precisely established by analysis of its contents or implicit "sense." I have not taken this approach, but have assumed the notion of "natural event," and distinguished acts as events of a particular sort. The defining function of the class of acts established in this way does not entail the assumption of an agent.

Agents, in such a context, cannot figure as ultimate unanalyzable entities, like the "metaphysical Subject," the Mover of "his" arms and legs, completely self-identical, "implied" by the occurrence of acts, but not accessible to empirical or historical study. If acts are to be treated as natural events, so must agents be; and the act concept proves adequate for the construction of the much more elaborate concepts "agent" and "agency." Though it may seem somewhat Pickwickian to construe the agent in terms of acts rather than vice versa, any logician would find it natural enough (logicians are generally on the side of Mr. Pickwick); philosophically, what justifies the logical device is that it lets one inquire into the origin and development of life, the rise of psychical phenomena in the animalian branch and the evolution of the unparalleled human specialty, Mind, in a scientific way that the initial assumption of a physical, psychical, or "psychophysical" entity, the subject, agent or individual, does not lay open. Even "organism" as a basic concept has not the constructive power of the act concept for the understanding of Mind, because it tends to lead too insistently into morphology at the expense of physiology and psychology.

[1] Cf. J. O. Urmson's contribution to the Aristotelian Society's Symposium on Motives and Causes (1952), especially p. 181.

In the field of philosophy there is a large and important literature dealing with the individual—understood to mean the human individual —in its relation to the world. The problem has a long history, going back to Plato's doctrine of the intellectual but passion-blinded soul and to the subjectivism of the Sophists and Cyrenaics, and attaining its full meta-physical status in the Renaissance,[2] from which we inherit it with all the added weight of its medieval religious cargo. So our own century has seen such works as N. S. Shaler's *The Individual* (1900), Josiah Royce's famous *The World and the Individual* (1900), Fawcett's *The Individual and Reality* (1909), and Bernard Bosanquet's *The Principle of Individuality and Value* (1912), to mention only a few obvious English titles. Here, as in the older post-Kantian literature, interest was centered largely on the metaphysical Subject and the difficulty of its liaison with phe-nomenal Nature—even including "its" own body—on the one hand, and on the other hand with the noumenal Object, i.e., the "real World," and God. Somewhat later, perhaps under the growing influence of Jungian and Adlerian psychology, the Individual gave way to the Self in moral and metaphysical writings, and we have *The Self, Its Body and Freedom,*[3] *The Self and Its World,*[4] *The Self and Nature,*[5] *Problems of the Self*[6] and most recently, *On Selfhood and Godhead.*[7]

All these are essentially metaphysical works; but there is also a wide literature dealing with empirical relations between the Individual, or Self, and its surrounding human World—the Community, the State, the Social Order, Society. Such works, which would fill a large catalogue, cover a great many topics and problems, reaching out from philosophical ethics toward politics and economics,[8] and in another direction toward dynamic psychology and personal religion.[9] In the former group the Individual is generally not defined, but is assumed to be the physical human being recognized by common sense, specified as adult or infant, male or female, legally responsible or irresponsible, and so forth. The problems revolve around his rights, duties, deserts, and his benefits from

[2] Cf. Ernst Cassirer, *Individuum und Kosmos in der Philosophie der Renaissance* (1927).

[3] W. E. Hocking (The Terry Lectures), 1928.

[4] G. A. Wilson, 1926.

[5] De Witt Parker, 1917.

[6] John Laird (based on the Shaw Lectures of 1914), 1917.

[7] C. A. Campbell (The Gifford Lectures), 1957.

[8] Here belong such books as G. H. Mead's *Mind, Self and Society from the Standpoint of a Social Behaviorist* (1934), R. B. Perry's *The Moral Economy* (1909), and P. E. Pfuetze's *The Social Self* (1954).

[9] The names of Reinhold Niebuhr, Erich Fromm and Paul Tillich naturally come to mind; but surely the list could be extended *ad libitum*.

being a member of society; in short, they concern the Individual as the unit of organized human life. In the latter group of writings, however, the social world is usually taken for granted, and in that frame the problems center on the human being's unity, growth, and self-realization as an Individual. The meaning of "individuality" is important here; for it is conceived not as a basic attribute, but as an ideal, or at least a standard, to be attained in the course of life. Individuality is something to be defined in terms of mental functions—emotion, conation, envisagement, perception and judgment—that lead to behavioral acts.

Viewed from the angle of a constructive theory of mind, this is a better conceptual basis than the absolute Self or Subject. But it does not take us far enough in the direction of infrahuman life to let us trace back the genesis of mind to the beginning of vital evolution. There have been some studies in dynamic psychology of animals, but of course one cannot apply human standards to simian or canine subjects; the "dynamisms" of behavioral response may be studied in monkeys, dogs and even rats, but the building up of the Individual and its coming to terms with Society are processes that commence only with human estate.

Yet biologists who are not concerned with mental phenomena of any sort—let alone an Ego, phenomenal or transcendental—use the term "individual" freely and, unfortunately, sometimes in questionable ways. Charles Manning Child called every specialized structure, such as a branch, or even a transient amoeboid pseudopodium, an individual.[10] He had his reasons, of course, the chief one of which was to emphasize the sameness of the functions that produce whole organisms and those which articulate appendages and organs. But the disadvantages of Child's usage seem to me to outweigh its virtues, for he himself soon slips into the more ordinary practice of equating "individual" with "organism."[11] I shall therefore call his subordinate "individuals," such as

---

[10] See *Individuality in Organisms* (1915), p. 1: "The individual is not necessarily a single whole organism; it may be part of a cell, a single cell, or a many-celled organ or complex part of the organism; or, as in most plants and some of the lower animals, a number of organisms possessing certain organs or parts in common, and therefore remaining in organic continuity with each other, may together constitute an individual." It follows from this use of the term that "many individuals are made up of other individuals and these in turn of others" (p. 2).

[11] *Ibid.*, p. 5: "Life in general consists of the life-histories of individuals. Individuals arise from other individuals, undergo a more or less definite and orderly series of changes known as development, usually reproduce new individuals in a more or less orderly way, and either undergo complete physiological distintegration into new individuals in the process of reproduction or else finally lose their unity by the cessation of their activity in death." Limbs and organs do not reproduce themselves; cells in a multicellular organism, on the other hand, still fit the modified concept.

organs, limbs, mouth parts and horns, "articulations" in organisms. This distinction brings us back to the normal use of the term "individual" in the writings of geneticists, physiologists and biochemists, despite some exceptions that are harder to justify than Child's departure.[12] Such use, moreover, is in agreement with that made by social, legal and political philosophers; both terminologies seem to meet on a level of common sense and tacit convention.

Yet it is in keeping with common sense, too, and not offensive to philosophical thinking, to say that one person is "more of an individual" than another, though no one would say he is more of an organism. The absurdity of that substitution reveals a difference of meanings, after all, made by the difference between the biological and psychological-ethical contexts. Oddly enough, however, it does not sound absurd in biology to say that one living thing is "more of an organism" than another—for instance, that a bacterium is more of an organism than a virus; and to say it is more of an individual means the same thing on that level. We have to ascend far up the evolutionary stair before "individual" takes on a different meaning from "organism."

To trace the development of mind from the earliest forms of life that we can determine, through primitive acts which may have vague psychical moments, to more certain mental acts and finally the human level of "mind," requires a more fertile concept than "individual," "self" or even "organism"; not a categorial concept, but a functional one, whereby entities of various categories may be defined and related. The most promising operational principle for this purpose is the principle of individuation. It is exemplified everywhere in animate nature, in processes that eventuate in the existence of self-identical organisms; it may work in different directions, and to different degrees; that is, an organism, proto-organism, or pseudo-organism may be individuated to a low or high degree, in some respects but not in others, and anomalies of individuality—double-headed monsters, parabiotic twins, as well as properly semi-individual plants and animals—may arise by imperfect or by normally only partial individuation. Under widely various conditions, this ubiquitous process may give rise to equally various kinds of individuality, from the physical self-identity of a metabolizing cell to the intangible but impressive individuality of an exceptional human being, a Beethoven or a Churchill, who consequently seems "more of an individual" than the common run of mankind.

---

[12] Ernest Baldwin, for instance, in his essay, "Rigidification in Phylogeny" (1937), calls enzymes "individuals," and refers to other "chemical individuals," meaning simply various chemical substances.

Individuation and Involvement

Individuation is a process consisting of acts; every act is motivated by a vital situation, a moment in the frontal advance of antecedent acts composed of more and more closely linked elements, ultimately a texture of activities. The situation, uniquely given for each act (and therefore not amenable to specific description), is a phase of the total life, the matrix, from which motivation constantly arises. The acts that effect the individuation of a being emerge from some larger matrix, such as a biotypic stock; or a generally only semi-individuated organism may undergo a higher individuation in one of its life episodes, as some colonial creatures, for instance corals, have a free-swimming, independent juvenile stage; or again, a species may develop a particular line of individual action, as the hairs of the sundew (*Drosera*), unlike any other parts, and quite unlike the hairs of most other hirsute plants, respond directly and independently to stimulation by contact.[13] Here it is the direction rather than the degree of individuation that is interesting. In some exhibits of the principle, for instance the individuation of tumors within the normally subordinate tissues of plants and animals, both the direction and the degree to which those special acts can go are of prime interest to the pathologist. But it is in the embryologist's laboratory that we find the functional concept of individuation explicitly recognized as more negotiable, and therefore more appropriate to research, than the older substantive concept of "individual." Ever since Roux's and Spemann's classical experiments, in which Roux, by killing one half of a frog's egg, induced the production of half a larva, whereas Spemann (and also Hans Driesch) removed the sacrificed half and saw the remaining part become a miniature complete larva,[14] the mechanisms of twinning and parabiosis have engaged the speculative thought as well as the further experimental work of embryologists. Naturally enough, it was in this sort of research that the limitations set by the concept of "individual" and the greater theoretical value of "individuation" became clearly apparent.[15] What was not immediately evident was the philo-

[13] Charles Darwin, in his *Insectivorous Plants* (1897), made a thorough study of *Drosera rotundifolia* and related species. He refers to the hairs as "tentacles" and to the secretory cells at their bases as "glands." See especially chaps. i and ii.

[14] For an authoritative account of these experiments see Hans Spemann, *Embryonic Development and Induction* (1938), chap. ii.

[15] The difficulty of identifying individuals becomes acute in the case of cells or relatively simple, though multicellular, organisms. Thus A. B. Dawson, in a symposium paper, "Cell Division in Relation to Differentiation" (1940), remarked: "Cell division carries the implication that there are always individual, discrete units which may be multiplied by this process. However, cell individuality cannot always be assessed by the criterion of morphological delimitation. Morphological discreteness

sophical constructive power of the operational term; for the process of individuation runs through the whole realm of life, and is the thread of Ariadne that guides our understanding from one level to another, where the shifts formerly seemed to be breaks: from inanimate nature to animate, from physiology to psychology, from animal life to human.

A very appreciable logical virtue of the concept of individuation is that it has a converse which is also a functional concept, namely, "involvement." The two principles are opposed, yet interdependent, in more intricate ways than simply balancing each other or alternating. In most vital phenomena both of them are in operation, and the processes that exemplify them are numberless. Individuation occurs in cytological differentiation, in parturition, in speciation, in the origination of mind and in the development of personality. Involvement proceeds from the earliest coazervation[16] (if such there was) of proto-organic droplets to the most extraordinary acts of integration—mutual control of cells in a tissue and of tissues in a body, physically connected populations (plants conjoined by rhizoids, stolons or rhizophores, colonial animals by a coenoecium or a coenosarc), bisexual reproduction, and beyond such physical levels, herds, communal nests and hives, processes of human communication, human society. These are but a few of the forms that involvement of acts and of whole lives can take. Most of them—as also the diverse forms of individuation—will require discussion below, because they have played their parts in the long history and vastly longer prehistory of "mind." The development of beings with minds is probably the highest individuation the world has ever known, and its prehistory is the history of life on earth.

---

is not constantly encountered. . . . Accordingly, any discussion of cell division cannot entirely ignore these limitations of the concept of cell individuality" (p. 95).

Similarly, Paul Weiss wrote, in "The Problem of Cell Individuality in Development" (1940), p. 98: ". . . cell discreteness is of a transitory character, which comes and goes according to circumstances. Repeatedly, cells merge into large plasmodial masses, thereby losing their outlines, and later emerge again as individualized, well-circumscribed units."

In a much older work, N. S. Shaler's *The Individual: A Study of Life and Death* (1900), the notion of individuation is already in force to resolve such difficulties, which are not limited to the cytological level; for Shaler, speaking of crinoids or "sea lilies," observed that "in the crinoids, the first of the animals of distinctly radiated structure . . . to attain a relatively high station as a solitary form, the initial step in the advancing series consists in the adoption of an external covering made up of very many polygonal plates. Each of these has its separate centre of growth, and . . . *each is as completely individualized as is consistent with the needs of its place in an organic whole and its relations to adjacent units of like nature*" (p. 80, italics mine).

16 This term will shortly be explained and come in for discussion below, p. 317 f.

It is hard to realize in scientific or historical imagination the actual backward continuity of every individual with whom one does business in office or church or shopping center. Yet rational reflection makes perfectly clear that not just some vague "species" to which we "belong" only as to a club, but the direct personal ancestry of each one of us goes back, one generation at a time, through periods in which the whole line of that person's particular progenitors had four feet, and further to their great-great-great-grandparents—but still real ones—that crawled, and further to great-greater ones that probably swam and beyond that to smaller and smaller great-greater ones, until self-identity becomes meaningless because their lives did not end in death but in fission and re-individuation. Even here, however, our personal origin goes back through some specific succession of episodic creatures whose activity has never broken. Where it began, suddenly or gradually, and how it established itself to last until today, we do not know.

To relate such a prehistory to one's own being is so difficult that most people's ideas of ancestry lead back from themselves only to earlier phases of whatever ethnic group or groups they consider as of their own "blood," and then jump to an impersonal relation between "man" and "his" possible forerunners, all ape-like; further back there is only a vague "evolution" that produced the ape, but their own "blood" is no longer specifically represented in that natural history.[17] Even in scientific works on evolution one rarely finds such appreciation of the length of individual ancestries as Robert B. Livingston expressed in a recent *Bulletin of the Atomic Scientists*, where he wrote: "The chains of life to which each of us constitutes a biological witness go back without interruption to the first life on earth, beyond mammals to slithering creatures who crawled along mud flats and ledges, blinking at the bright sky, bailing this chancey atmosphere in and out of their new-found lungs. Doubtless,

---

[17] This tendency may prompt the oft-expressed wonderment at the "recency" of the human race. Caryl P. Haskins, for instance, remarks: "A most striking feature marking the contrast between the evolution of man and that of other social creatures is the extraordinary youth of *Homo sapiens* as a biological species" (*Of Societies and Men* [1951], p. 4). In a similar vein, Alexander Petrunkevitch wrote, in "Environment as a Stabilizing Factor" (1924): "We can safely assume that man being the most recent of all animals is also the most recent host for parasites" (see p. 86). Where did this prodigy suddenly come from? Comparative studies of human and other large primate fossil remains permit some doubt whether any modern species of ape is older than man. A comparison of the known remains of *Homo sapiens* with those of past generations of modern apes shows the oldest specimens of *Pan* and *Gorilla* to be younger than *Homo*, and *Pongo* not much earlier. See the family tree presented by Elwyn L. Simons in "The Early Relatives of Man" (1964), p. 55.

our survivorship goes all the way back to what was no more than a fateful cluster of successful catalysts."[18]

If mentality, and more particularly the human version of it which is properly called "mind," is to be conceived as a natural outcome of the irregular, yet cumulatively progressive development of animalian life, we have to look for all explanations of its powers and limits, its mechanisms and vagaries, in the ruling tendencies of the great animal kingdom itself; and these appear, often in their purest modes, on the lowest levels of life that biological investigations have reached, even far below the separation of its main orders. Going still further backward, we can only extrapolate the directions and rates of presumable earlier events in the formation of life itself. Yet there, in those events, lies one of the key problems for the life sciences: the origin of agents.

In discussing the nature of acts I have already touched upon the "act-like" character of some non-vital chemical transformations, especially interactions of different substances. Why are these events not bona fide acts? Chiefly because they do not develop into a self-continuing system of actions proliferating and differentiating in more and more centralized and interdependent ways; that is, they do not enter into the constitution of an agent.[19] As soon as they do so, they become acts. An agent is a complex of actions, and all actions that belong to that complex are acts of that agent. All true acts, therefore, are to some extent involved with other acts; very intimately with their own subacts, and with superacts into which they themselves enter, and not much less so with acts to which they stand in some relation—active or passive—of induction. The wider relationships that compose the basic dynamism of life, and are all

[18] "Perception and Commitment" (1963), p. 14. One other similar statement occurs in Raymond Ruyer's *Éléments de psycho-biologie* (1946), on the opening page, which reads: "Le protoplasme de mon corps est vivant depuis des millions d'années, puisque chacun de ses fragments dérive directement, par dédoublement ou fusion, d'un fragment vivant. Cette cellule de mon épiderme qui va mourir et se détacher, elle n'avait connu que la vie depuis des siècles et des siècles. Mon corps est en continuité, par mes ancêtres humains et animaux, avec les vivants les plus primitifs. Il dure depuis les origines même de la vie. Et l'on peut en dire autant de tout ce qui vit aujourd'hui sur la terre: pas un brin d'herbe qui ne remonte au commencement de la vie."

[19] D'Arcy Thompson pointed out this crucial difference between chemical reactions and acts, especially the organic acts of growth, saying: "As the chemical reaction draws to a close, it is by the gradual attainment of chemical equilibrium; but when organic growth comes to an end, it is . . . due in the main to the gradual differentiation of the organism into parts, among whose peculiar properties or functions that of growth or multiplication falls into abeyance" (*On Growth and Form* [1942], Vol. I, p. 258).

subsumed under "motivation," probably hold in one direction or the other between any acts belonging to the same organism.

The question of the origin of organisms is, then, how some of the chemical actions on the surface of the earth or in its surrounding gaseous envelope ever became involved with each other so as to form centers of activity which maintained themselves for a while amid the changes of forming and dissolving compounds around them. They need not have been endlessly self-perpetuating from their beginning; in a chemically active cloud, or the "rich oceanic broth" of a lately cooled planet, where such compositions could occur, they would be likely to happen by the million, so the first proto-organisms might have been of short duration—unfixed, unrelated myriads of briefly viable forms—constantly replaced by means of the same causes that produced the earliest ones, and not yet by means of each other.

As a serious scientific problem, the beginning of life on earth has been mooted only since the latter half of the nineteenth century, when Pasteur's sterilization experiments were widely accepted as definitive disproof of any and all theories of spontaneous generation. Then, of course, the question became urgent among naturalists and philosophers: where has life come from? If it could not originate on earth, where did it arise, and how did it get here? Imaginative scientists as well as philosophers published some bold speculations on possible ways in which some germ of living substance might have come to a lifeless planet from outer space. Radiation, transport by meteorites and other improbable means were proposed, but of course all such solutions only pushed back the issue to the similar question of how the "spark of life" had appeared on other cool stars in the first place. The significance of Pasteur's negative findings was exaggerated until they were supposed to prove that spontaneous generation was impossible under any circumstances, present or past, that the cosmos could offer. Since life did exist, it must be co-eternal with matter. This doctrine, known as "panspermism," was popularized by Svante Arrhenius,[20] who held that life sperms must come to each star when it is in condition to harbor them, if there is to be life on it.

But scientific interest in the origin of life on earth was brief. For three or four decades, Ernst Haeckel's *Natürliche Schöpfungsgeschichte* (1868) and *Die Welträtsel* (1899), H. Bastian's *The Beginnings of Life* (1872), Arrhenius' book and several others enjoyed a vogue among scholars and general readers alike; then the conflict of theories, all as

[20] In his *Worlds in the Making: The Evolution of the Universe* (1908), especially chap. viii.

undemonstrable as they were irrefutable, together with the methodological dogmas of an impatient and self-conscious scientific age, killed academic as well as popular interest, and the problem of how life began or where it came from henceforth was considered insoluble and therefore "unscientific." It was "metaphysical."

The cause of the failure, curiously enough, really was metaphysical; not in the derogatory sense which scientists all too often give to that term, but in its perfectly respectable sense of dealing with the basic assumptions implicit in our formulation of "facts." The metaphysical fallacy lay in the conception of "life" as a special essence different from "matter," something that pervaded "living matter" and set it apart from "mere matter" which obeyed the laws of physics. This something was what had to come, supposedly, as a spark from outside the purely physical system of events on the virgin earth, and somehow make liaison with the lifeless matter there.

This approach to the problem of biogenesis presents an interesting parallel to the frustrating search for the "liaison" of mind with brain, which also springs from an assumption of two metaphysically distinct substrata of reality, whose manifestations are supposed to follow different laws while commanding the same locus in the spatiotemporal world. The escape from the dilemma is similar in both cases, too; namely, to abandon the metaphysical dualism of ultimate *substantiae,* and try to make the logically more amenable system of physical, chemical and electrical events yield a functional explanation of vital and—in due course—of mental phenomena.

The first person to undertake this task and carry it to considerable lengths was the Russian biochemist A. P. Oparin. His book, *Origin of Life* (1936), appeared at a time when its very title aroused skepticism and prejudice; yet it was almost immediately translated into other languages (English in 1938), because even a casual perusal sufficed to show his colleagues that this was an entirely new approach to the old problem. His hypothesis contained three novel features: (1) he thought of the origin of vital processes as a heightening of chemical actions rather than the incursion of an ontologically unique "living spark"; (2) he postulated past, not present, conditions of the earth's surface as the environment of such accelerating changes;[21] and (3) he anticipated an insight

---

[21] On this point, cf. H. C. Urey's paper "On the Early Chemical History of the Earth and the Origin of Life" (1952), which opens with the statements: "In the course of an extended study on the origin of the planets I have come to certain definite conclusions relative to the early chemical conditions on the earth and their bearing on the origin of life. Oparin has presented the arguments for the origin of

which is only now receiving general recognition—namely, that the phenomenon of "life" is a wide, varied and unbelievably complex functional pattern rather than a single attribute or essence which either is or is not possessed by any given physical object. Some aspects of that pattern may be found in a physical region where others are lacking; one could not say, perhaps, that there is life at that point, yet there is action that produces organic structure.

Briefly, Oparin's theory runs somewhat as follows: All action—mechanical, electrical, chemical, organic or inorganic—occurs at some time and in some place of the material universe; that is, it has a physical location and effects changes in the material condition of that place at that time. If any repetitive action (be it the beat of waves on a rock or the passage of an invisible ray through an invisible gas) is concentrated for an appreciable time in a particular location, the matter which occupies that location, provided it continues to do so for a long time, becomes organized in a spatial pattern reflecting the dynamics of the action. Once the imprint of the action is established it facilitates further repetitions, and at the same time standardizes their form by channeling the course of motions, meting out the substances or charges involved (as filters, pores, tubules and electrical conductors do), thus modifying the activity which is itself deepening and distorting its own frame and thereby establishing and modifying itself. The relation between structure and process is mutually adaptive.

Before such organization even of the simplest sort can occur, however, there has to be some stability of substance in a circumscribed locality. In a perfectly liquid or gaseous medium there will obviously be no chance of developing a permanent imprint. Lightning passing through vapors or striking into waters may cause a dramatic molecular reorientation and induce great chemical actions, but all effects are soon lost in the chaotic motions of the fluid media. The first requisite for complex structures, therefore, is an isolated and bounded region in which chemical changes can happen to a self-identical substance. Oparin suggested that colloidal gels, which (following their most notable investigator,

---

life under anaerobic conditions which seem to me to be very convincing. . . .

"During the past years a number of discussions on the spontaneous origin of life have appeared in addition to that by Oparin. One of the most extensive and also the most exact from the standpoint of physical chemistry is that by Blum [in *Time's Arrow and Evolution* (1951)]. It seems to me that his discussion meets its greatest difficulty in accounting for organic compounds from inorganic sources. This problem practically disappears if Oparin's assumptions in regard to the early reducing character of the atmosphere are adopted" (pp. 351–52).

Budenberg de Jong) Oparin calls "coazervates,"[22] are very likely to have formed in the welter of the earth's early hydrosphere, and possess many properties and especially potentialities to make them probable forerunners of primitive living things. Unlike other colloidal particles, these coazervates—which arise in all sorts of sols—effect a rigid orientation of the water molecules around their envelope of equilibrium liquid, so there is a real shell around the little two-phase particle.

In a solution of two colloids with opposite electrical charges, coazervation results in so-called "complex coazervates"; for, as Oparin explains, "the stability of hydrophil colloidal sols is determined partly by electrostatic and partly by hydration forces. . . . But when oppositely charged colloidal particles are present in the same system . . . the electrostatic and the hydration forces are antagonistic to each other, the hydration still promoting the stability of the sol, but the electrostatic forces . . . tending to bring the colloidal particles together. . . . [So] the stability of the system is determined by the cooperation of two opposed but balanced influences. This imparts to the systems the property of extreme lability, making it possible for them to shift easily in either direction from the equilibrium under the influence of the smallest change in external conditions."[23]

[22] *Origin of Life* (1936), pp. 149–50: ". . . in solutions of hydrophil colloids, besides coagulation, another phenomenon of separation may be observed. . . . Under this condition the colloidal solution separates into two layers, a fluid sediment, rich in colloidal substance, which is in equilibrium with a liquid layer, free from colloids. During the past several years this phenomenon has been subjected to careful investigation by B. de Jong, who designated it as *coazervation*. . . . He calls the fluid sediment . . . the *coazervate*, while the non-colloidal solution he calls the *equilibrium liquid*. . . . In a number of instances this coazervate . . . remains in the form of minute microscopic droplets floating in the equilibrium fluid. . . . The droplets of coazervate are sharply demarcated from the surrounding medium by a clearly discernible surface. These droplets may fuse with each other but they never mix with the equilibrium liquid."

[23] *Ibid.*, p. 153. The lability here postulated is suggestive, in view of the character of some protozoa, such as *Paramecium*, in which G. H. Beale remarked the great tendency to variation, which could not be put down simply to "natural selection," since "some alleles can rarely if ever be expressed under natural conditions, e.g. certain types are only formed under conditions of high salinity, where the species is not usually to be found, or at temperatures never reached in the locality where the stock is found. . . . What may, however, be an advantage is the ability to form a series of readily interchangeable types under diverse conditions. A free-living, unicellular organism like *Paramecium* is extremely exposed to the effect of environmental fluctuations, and it is to be presumed that the organism possesses some mechanisms of internal adjustment to cope with such fluctuations" (*Genetics of Paramecium Aurelia* [1954], p. 143).

Here, then, we have a fulfillment of the first requirement for an organization of matter by localized activity—a bounded, inwardly active particle. Oparin goes on to describe further properties, such as absorption and adsorption of molecules from the surrounding medium, and how these acts might lead to growth, differentiation, proliferation, growth competition and so-called "natural selection." Some of those further speculations raise difficult problems; but the problems themselves invite experimental tests and, where the findings do not bear out the scheme, alternative hypotheses. The main point is that all the later details are developed from his theory of how the most primitive life was formed, which shows clearly that the theory is conceived in genuinely biological terms.

At once, the proscribed topic of the nature and sources of life was brought out of its shadowy limbo into a new light, and scientists in many fields—colloid chemistry, electrochemistry, biochemistry, photodynamics, genetics, paleontology and even geology, astronomy and other tangential disciplines—felt the lure of its mysteries again.[24] All these sciences had made such rapid advances in the early decades of the twentieth century that there was a great wealth of factual knowledge on which speculation could now be based. Oparin was not reasoning in a vacuum. Even the concepts which directed his researches, i.e., the active nature of "living matter," the acceleration and heightening of chemical actions by increasing autocatalysis, and the development of "steady states" in centers of balanced action, were not unshared notions when he professed them. Bertalanffy said as much at just about the same time.[25]

---

[24] Just two decades after the original appearance of Oparin's book, the International Union of Biochemistry decided to hold a Symposium on the Origin of Life on the Earth, which, in special honor of "Academician Oparin," met at Moscow in 1957, and assembled an impressive group of scientists. The proceedings, published in Russian by the Academy of Sciences of the U.S.S.R. and in English, French and German by the International Union of Biochemistry, comprise sixty-one papers of technical character, besides discussions in which further participants joined (though it must be admitted that the Russians had the lion's share of the comments as of the papers, and awarded themselves the laurels).

Apart from this cooperative venture, there is a formidable array of studies in biogenetic theory. See, e.g., the collection of papers published by Penguin Books as No. 16 of *New Biology* (1954).

[25] Cf. Ludwig von Bertalanffy, *Das Gefüge des Lebens* (1937), p. 61: "Den dauernden Geschehensfluss, den Heraklit im Gleichnis des Stromes und des Menschen erblickte, nennen wir in exakter Sprache einen 'stationären Zustand.' Der Fluss—ein hydrodynamischer stationärer Zustand, Erhaltung einer bestimmten Strömung im Wechsel der Wasserteilchen. Das Lebendige—ein chemischer stationärer Zustand, erhalten im ständigen Wechsel, Auf- und Abbau seiner Materien." And further (p. 74): "Das, was in der Zelle vorhanden ist, sind nicht

So as soon as their relevance to the conception of vital origins was pointed out, the metaphysical chasm between the inanimate world and the world of life began to close, and biological research struck its natural connection with fundamental laws of physical science.

Today this functional character of organic existence is a commonplace, though philosophers and social scientists who speak of "mutual dependence of parts" and "certain interactions" usually have no idea of how deep those interactions go, actually to a constant exchange of atoms between molecules within the protoplasm of a cell. An admirable demonstration of the intimacy of such organic interactions is given in R. Schoenheimer's *The Dynamic State of Bodily Constituents* (1942), which records his tracing of substances through metabolic transformations by means of "marked" metabolites, "heavy water" (H^2) and "heavy nitrogen" (N^{15}). On the basis of these experiments, he says: "In mammals most of the fat is located under the skin, between the muscles, or around internal organs. This fat is usually called depot or storage fat and has generally been regarded as a biological energy store. According to this concept, the fat depot is outside the general metabolism and becomes active only in times of need" (p. 8). But the first experiments convinced him "that the depots are not inert storage materials but are constantly involved in metabolic reactions" (p. 12). "The fatty acids of the depot fat ... are constantly transported, in the form of fats or phosphatides, to and from the organs. ... When fat is absorbed, the acids of dietary origin merge with those from the depot, thereby forming a mixture indistinguishable as to origin. ... Some of this pool of acids is degraded and some of it re-enters ester linkages to regenerate fat, which is transported back to the depots" (p. 23). Proteins, too, undergo constant breaking down and reconstitution within the living organism, and their constituents pass from one chemical structure to another. "When leucine was fed [to rats], the isolated glycine contained the tracer, and the reverse was found when glycine was given. ... The amino acids continuously interchange nitrogen atoms" (p. 30). Schoenheimer's general conclusion is that "if the starting materials are available, all chemical reactions which the animal is capable of performing are carried out continually" (p. 39).[26]

---

chemische Substanzen im statischen Sinn, . . . sondern in der Zelle bestehen ungeheuer komplizierte Gleichgewichte zwischen den . . . Stoffen." The total substance of an organism is a dynamic equilibrium, "ein dynamisches Gleichgewicht" (p. 81).

[26] This intimacy of functional involvements within an organism makes the claim of mental activity for man-made machines absurd. The convergence of gross effects

Once the philosophical atmosphere was cleared of its worst fog, the resort to non-physical ingredients, so that scientific research could be aimed at the dim beginnings of life, various hypotheses were soon set up as to the actual processes of biogenesis. Oparin's theory was, after all, only a tentative proposal. H. F. Blum, for instance, found it more plausible to assume that life began not in a chemically rich ocean, but in slowly drying puddles, into which amino acids, formed in the atmosphere, had been precipitated.[27] N. H. Horowitz, whose brilliant reasonings, borne out by experiment, will engage us later in another connection, concluded that the first beings were probably not molecular aggregates, but single macromolecules.[28]

In the course of such speculations and debates, one fact became apparent to all the participants: namely, that the nature of the earliest living entities is as problematical, as hard for us to reconstruct on the basis of any life that exists today, as their environment. They were almost certainly nothing as complicated as a cell, with a nucleus harboring a chromosomal system of inestimable potentialities; nor is it likely that their metabolism was comparable to that which occurs in even the simplest living cell today. N. W. Pirie pointed out the difficulty of envisaging the conditions of those first "eobionts" and their possible ambient, when he wrote: "If *Corycium* is indeed the fossil of an organism it was enigmatic in more ways than the paleontologists have

---

achieved by animalian cerebral acts and by electronic computers, respectively, may impress the engineer whose imagination of the mental process only penetrates as far as the parallel really goes (and where, by the way, it has furnished some wonderfully useful models to neurology); but no psychologist with a modicum of insight into the physiological basis of mental phenomena could be beguiled by their astounding mechanical mimics to credit the latter with living powers.

[27] See his essay "On the Origin of Self-Replicating Systems" (1957), p. 161: "From the experiments of Miller we may believe that amino acids were formed in the earth's primitive atmosphere, and that they were carried down in water droplets to accumulate in the surface waters of the earth." Pools of amino acid solution may thus have been formed, and if with further changes of the crust these slowly dried up, "the formation of polypeptides should have been favored, and if . . . the residue was redissolved, there could have been present polypeptides in solution and suspension. Such polymerization might also have taken place in the nonaqueous phase of a coazervate. . . . Once formed, and lacking enzymatic forwarding of their hydrolysis, the long polypeptides could conceivably have remained for a long time, and have been ready at hand to play a role in the first replicating systems." Cf. also his *Time's Arrow and Evolution*, p. 528.

[28] "On Defining Life" (1959). This short paper ends with the statement: ". . . the question raised by Professor Oparin—namely, did life arise as individual molecules or in the form of complex polymolecular systems—can be answered in the following way: Life arose as individual molecules in a polymolecular environment" (p. 107).

thought of and probably organized its metabolism with the help of molecules that would not fit neatly into our systems of classification."[29] In the same vein, in weighing the hypothesis that all modern organisms have arisen from very few original eobionts, if not from a single one, he pressed home the point, saying: "This is logical enough but it is illogical to think that this ancestor was the original organism. It seems much more probable that this, the beginning of morphological complexity, was nearly the limit of biochemical complexity."[30] And finally: "The metabolic processes used by present-day organisms were probably present in some of the early forms of life but there is no reason to think that they were all present, that they were present in all forms, or that when present they were quantitatively important."[31]

The beginning of the life process, whether it stemmed from the chance occurrence of a single viable molecule or aggregate, or—as I find far more plausible—lay in a generative phase of the earth's history during which millions of eobionts arose,[32] certainly appears to all scientific inquirers as a formation of patterned activities and their more and more perfect integration until they constitute a matrix in which their own form becomes modified or even entirely blurred, so it can only be found again by analytic abstraction. Such living matrices may have various degrees of coherence and persistence; but they are systems, self-sustaining, and (as we know them) self-propagating, wherein every event is prepared by progressively changing conditions of the integral whole. Every distinguishable change, therefore, arises out of the matrix, and emerges as an act of an agent; for such a vital matrix is an agent.

In this earliest phase of life, however it may have proceeded in detail, the principle of involvement certainly appears as dominant. There must have been strong ruling tendencies toward organization, which led to increasing interdependence of actions and eventuated in the forma-

---

[29] "Chemical Diversity and the Origin of Life" (1959), p. 79.

[30] *Ibid.*

[31] *Ibid.*, pp. 81–82.

[32] J. L. Kavanau, in "A Theory of Causal Factors in the Origin of Life" (1945), supports my view that natural conditions which could give rise to such active structures would probably produce them over and over, as common chemical products. Of these he says: "Once such reactive associations were established the systems would become primordial living complexes" (p. 192). The alternative hypothesis, suggested by Horowitz, that the formation of life was a single, local phenomenon, has found its protagonists too, and is defended notably by G. W. Beadle (see "Genes and Biological Enigmas" [1949], especially p. 241). But, although I agree with Beadle that "it is logically unnecessary to assume that existing organisms trace back to more than one such origin," I cannot agree that this is likely, and that all life very probably stems from a unique event in geologic history.

tion of biological mechanisms. The most important factor in that process, the main source of all functional continuity, must have been the establishment of rhythms. Rhythmic concatenation is what really holds an organism together from moment to moment; it is a dynamic pattern, i.e., a pattern of events, into which acts and act-like phenomena very readily fall: a sequence wherein the subsiding phase, or cadence, of one act (or similar element) is the up-take for its successor. It occurs in non-vital as well as in vital processes, but in the latter it is paramount, and reaches degrees of differentiation and intensity unrivaled by anything in the inanimate realm.

Rhythm is not usually defined in terms of the concatenation of the elements which compose the rhythmic series. It is commonly assumed in scientific thinking that the essential character of rhythm is the repetition of any distinct, recognizable event at equal intervals of time; i.e., that rhythm is periodic repetition. But, in fact, not all rhythmic repetitions are strictly periodic.[33] There may even be rhythmic sequences of events which are not really repetitive; for instance, a performer of "modern dance" rarely repeats a movement, yet every least motion of the dance has to be rhythmic. Similarly, if a cat runs across a floor and suddenly leaps up on a table, the leap is as rhythmic as the repeated loping movements, although it is a unique element. It need not even terminate the series; if the animal goes on, leaping down again and bounding away, the unrepeated movements may have broken its course, but not the rhythm of its over-all act.[34]

[33] One physiologist, insisting on this definition because it was the only "scientific" one, but confronted in his laboratory with the irregularities of organic functions, found himself driven to the conclusion that heart action in most animals is not rhythmic. After remarking that the alternation of systole and diastole is the most strikingly rhythmic bodily process, he continues: "Wie in neuerer Zeit entwickelt wurde, kann man im *strengen* Sinne des Wortes von einem Rhythmus nur dann reden, wenn sich *ein gleiches Geschehen in gleichen Zeiten wiederholt.* Es steht nun von vornherein zu erwarten, dass unter diesem strengen Gesichtspunkt die Herztätigkeit niemals rhythmisch vor sich geht" (Helmut Wolf, "Über die Genauigkeit der Herztätigkeit in der Tierreihe" [1941], p. 182).

A definition which makes the defined concept never strictly applicable to the phenomena for which it is intended is not "scientifically exact."

[34] L. J. Cloudsley-Thompson is one of the few writers on biological rhythms who explicitly notes this fact. In *Rhythmic Activity in Animal Physiology and Behaviour* (1961), p. 24, he says: "Rhythmic or cyclic phenomena can be observed on all levels of biological organization, from cell, tissues and organs, to organisms and communities. . . .

"It is not essential that the repetitions should be identical in every detail or that comparable phases should follow each other in equal or even approximately equal intervals of time." It is somewhat surprising, therefore, that he does not recognize

The essence of rhythm is the alternation of tension building up to a crisis, and ebbing away in a graduated course of relaxation whereby a new build-up of tension is prepared and driven to the next crisis, which necessitates the next cadence. If the series of actions thus engendered consists of alternating contraries, such as rise and fall, push and pull, suction and expulsion, and each element in spending itself prepares and initiates its own converse, the resulting rhythm is a dialectic.

Dialectical rhythms, like rhythms *per se*, are not limited to the actions of vital systems (the swing of a pendulum, in which each fall builds up the potential energy for the opposite rise, or the wave returning down the slope of a beach and hastening the next breaker, are familiar examples), but they play such a major role in vital functions that their importance in the activity and even the physical existence of organisms makes them an essential mark of living form in nature, as their virtual image is of "living form" in art. The concatenation of minute acts—close to the molecular level—into continuous series, self-sustaining by virtue of the cyclic structure of their elements, each of which has a definite magnitude measurable from any phase of the cycle to its repetition, is the basic pattern of life. It is carried on by such diminutive and intricate mechanisms that every development of our optical enlarging instruments reveals new functioning structures in what theretofore looked like homogeneous substance. Most, if not all, of the processes we can distinguish as yet are summations of smaller ones; and the summations take the form of rhythmically concatenated acts, which either summate or differentially interact to produce larger rhythms. The parabolic curve which expresses the typical act form emerges again and again, at each level of integration, in the physiological rhythms of every organism; and this form, with its main phases of inception, acceleration, consummation and cadential finish, is what makes the rhythmic pattern, and is accordingly the basis not only of the distinguishable unit acts in a continuous activity, but also of their self-concatenation, and the consequent self-perpetuation of the continuum.

The events that go on in organic particles visible only under the

---

the distinction Kleitman made between rhythms of "cyclic" structure, in which "the sequence of occurrence rather than that of duration tends to be constant," and rhythms kept going by constant environmental changes, so they are lost as soon as those concomitant changes cease, "referred to by Kleitman as 'causal, synchronous, associated or coupled periodicities.'" Obviously Kleitman's "cyclic" rhythms are concatenations, the "periodicities" effects of extrinsic events. Yet Cloudsley-Thompson objects that "there seems to be no valid justification for not applying the term 'rhythm' to such phenomena" (p. 3).

highest magnification, and perhaps in still smaller ones too, defy the imagination even of biophysicists, who are used to fantastic extremes. Dr. Francis O. Schmitt, for instance, in reporting on the "macromolecular assemblies" that seem to bridge the gap between "biological chemistry" and "cellular biology" (see the remark by Paul Weiss, p. 258, n. 3), has written: "Recent work has shown that the major biochemical processes of protein synthesis in most cells are carried out in the ribosomes, a complex of RNA and protein in which a variety of ordered enzyme reactions may be presumed to occur. These clearly required a structural entity or MMA [Macromolecular Assembly] fundamental to the biosynthetic task. . . ."

"The primary sequence of the protein chain has the three-fold responsibility of providing (1) the enzymatic site, (2) the capability of combining with a dozen or more other species in forming the characteristically structured MMA, and (3) the ability to perform its catalytic task in the manner appropriate to the overall pattern of cyclic and coupled reactions characteristic of the MMA. Such specificity and complexity is too staggering to comprehend.

"It is highly probable that cellular MMA's exist whose functional parameters are wholly outside conventional concepts of present-day molecular biology. To penetrate such barriers requires approaching biological problems from fresh viewpoints."[35]

All these incredible processes are activities, unbroken rhythmic series of acts, largely of a dialectical sort, every cycle engendering its own repetition in completing itself.[36] Under ordinary conditions of chemical change, such intensely active complexes could not arise and maintain themselves. All the transformations found in animate nature take place in the inanimate realm, too, but at incomparably slower rates. Their tremendous acceleration—sometimes several hundredfold—is the work of catalysts; and one of the essential dialectical processes characteristic of organisms (though, even here, not strictly peculiar to them) is the act of autocatalysis, i.e., the synthesis of the catalyzer for a given function as a by-product of the function itself. So the activities of the living matrix become more and more concentrated and rhythmicized; they also be-

---

[35] "The Macromolecular Assembly—A Hierarchical Entity in Cellular Organization" (1963). See pp. 552–54.

[36] The exemplification of this principle in the simple mechanism of a "blinker" light has been noted by several authors, and its (apparent or true) biological analogues presented, with some attention to special cases, by Albrecht Bethe and Hans Schaefer in a paper entitled "Erregungsgesetze einer Blinkschaltung in Vergleich zu denen biologischer Objekte" (1948). See also H. Selbach, "Das Kippschwingungsprinzip in der Analyse der vegetativen Selbststeuerung" (1949).

come accelerated, by virtue of their autocatalysis, and would presumably reach excessive rates and break up under their own force, were it not for another circumstance, just as remarkable as their self-catalyzing functions, namely, that at some point they also produce checks on their own activities. So a complementary set of enzymes, synthesized by differentiating cells as they approach their adult stages, inhibits the action of their own morphogenic substances at their location and in their vicinity, and very frequently in the whole organism. A new balance of contraries is thus set up, a shift between self-induction and self-inhibition which reverses whenever the structure in question reaches completion or suffers loss. Paul Weiss summarized this controlling action with the statement: "Briefly, each organ or each cell type produces inhibitors of its own growth, and the total concentration of these in the circulation comes into equilibrium with the productive or generative mass to produce a steady state."[37]

This dialectic of induction and inhibition may exhibit very elaborate patterns, in which more than two processes become involved to effect the prevailing and continuous balance of life. Sometimes the activity of one functional unit, instead of directly initiating its antagonist and so limiting its own operation, induces the activity of another unit, which then provides the inhibition of the first. Sometimes they become each other's monitors, and sometimes the second unit—the inhibitor of the first—is limited by still another factor in the organic situation. Where many rapid, circumscribed rhythms are going on, some of these may summate and set up new, superimposed rhythms; or powerful periodicities extraneous to the organism may "entrain" its natural rhythms,[38] as the stronger intraorganic cycles always tend to entrain the weaker ones; so the sum total is still further complicated by the fact that activities deriving from widely different sources, from central and peripheral impulses, may intersect and interweave and modify each other in the system.

[37] Made at the Shelter Island symposium of 1956. See R. L. Watterson (ed.), *Endocrines in Development* (1959), p. 101.

[38] The term "entrain" is here borrowed from C. S. Pittendrigh and V. C. Bruce, who use it in "An Oscillator Model for Biological Clocks" (1957) as a transitive verb, in the sense of the French cognate *entrainer*; unsanctioned by Webster, yet justifiably used, I think, since we lack an equivalent single word. Their article is an interesting study in the "circadian rhythms" which, because of their current great practical interest for certain political and military projects, may be attracting rather exaggerated attention today (Janet Harker's "Diurnal Rhythms in the Animal Kingdom" [1958] presents a bibliography of about four hundred titles), but which also reveal some very important relations between the autonomous and heteronomous activities of different organisms.

[326]

Individuation and Involvement

The resultant steady state of actively maintained balance, which appears superficially as inactive existence, is the tonus of living tissues, the only constant sign of the basic vital process always going on, which makes the difference between the functioning leaf and the withered leaf, the man at rest and the man dead.[39] Without this unceasing, self-inducing and self-limiting dialectic ground base no stimulus can initiate any biological response from plant or animal. Only anatomical structures which are already active can be stimulated.[40] That is, of course, no more than a corollary to the proposition, set up in the previous chapter, that all acts are motivated by other acts, and any extraorganic events which elicit specific acts do so via the matrix of activities, the agent; or, as D. O. Hebb has put it, ". . . an afferent excitation does not arouse inactive tissue but feeds into an activity that is already going on."[41]

The organism is made entirely by processes which are vital acts; not necessarily all its own. Animals, with very few exceptions, are heterotrophic; they cannot synthesize their major nutrients from abiotic elements. They can only take over acts of transformation begun by autotrophs—that is, ultimately, by plants—and continue them in ways and patterns of their own. But from the time an agent initiates or assumes vital acts it performs them in systematic ways, making its more and more deeply involved and integrated matrix, its life.

[39] Henri Wallon, in De l'acte à la pensée: Essai de psychologie comparée (1942), p. 153, says of the motionless attitude: "C'est une contraction des muscles qui n'entraîne pas leur raccourcissement, mais qui les tient au contraire dans leur forme présente et leur donne un degré variable de consistance, de résistance au déplacements; c'est leur fonction tonique." The existence of muscle tonus, upheld by some unremitting internal activity even in bodily relaxation, e.g., in sleep, was actually discovered in the seventeenth century by Jan Swammerdam (see B. H. Pubols, Jr., "Jan Swammerdam and the History of Reflex Action" [1959]); but the mechanistic ideal of the next two centuries ruled out such apparently autonomous action, and demanded a passive body to which measurable outside forces could stand in predictable causal relations. A less sanguine and more sophisticated age has had to rediscover the physical fact.

J. H. Parsons interprets the inherent activity of living organisms in terms of stimulation and response, saying, ". . . the body, even in a state of repose, is bombarded with stimuli derived not only from extero-, but also from proprio- and intero-receptors; and it is upon this low-toned, but very complex background that the excitatory stimuli fall. This coenaesthesis is comparable to muscle tonus" ["Adequate Stimuli (with Special Reference to Cutaneous Nerves)" (1949), p. 308]. See also Emiroğlu and Winterstein, "Über den Ruhetonus des Nervensystems" (1956).

[40] Cf. the discussion of environment and Umwelt in Chapter 8 above, pp. 282–83; also p. 283, n. 56.

[41] The Organization of Behavior, p. 121.

All agents that we know have been engaged in this process at least since the time when life ceased to emerge from the ferment of electrochemical substances that probably covered the earth. When the phase of biogenesis ended, there must have been a good many persistent dynamisms already distinct in their synthetic powers and the forms of their operation; and well before the end of the originative period—perhaps from the very beginning, even before one could surely call them alive—those complexes must have shown the tendency to proliferate. They may have crept, by growth, over surfaces, and broken up into fragments that grew; they may have divided into drops by breaking of their surface films, rolled apart, again outgrown the unifying tension of their surfaces, etc.; some may have done things that we shall never know, because nothing in the present state of nature is doing them. Whatever did take place, the power of replication, by one means or another, must have been well established before spontaneous creation ceased.

The matrix of life processes in a primitive favorable location may have lasted for ages, dying in parts while growing or rejuvenating in others, spreading its patchy cultures from a center. Or a tiny—perhaps microscopic—metabolizing drop may have divided (in much simpler ways than a cell) and scattered itself through an area which soon was populated by similar metabolizing, growing and proliferating drops, a little cloud of life in a nutritive fluid medium. Such matrices, whether cohering or collective, are stocks; it is in them that the continuity of life has apparently been unbroken since the age of biosynthesis.[42]

At this point, the act form characteristic of life presents itself in an unexpectedly significant role: for in the dynamism of continuous stocks it exhibits not only the principle of progressive involvement, but also the converse principle of individuation. This principle comes to the fore as soon as one considers the sequences of growth, maturation, senescence and rejuvenescence or death that punctuate the history of a stock; then it becomes apparent that this history itself proceeds by cycles of successive individuations, which take the most diverse forms of pace and direction in their development, yet all exemplify the general pattern of acts; so that this pattern runs through the entire gamut of biological functioning, from its tiniest metabolic elements even to its largest ones, whole lives. Paul Weiss suggested just such a view of organic development as a whole, which is to say, of an entire individual life, as one

---

[42] The term "stock" is here used in a broader sense than it has in the geneticist's language, in which, according to a leading authority, "A *Stock* is a culture derived by the multiplication of a single individual, collected at a particular place" (Beale, *op. cit.*, p. 21).

widest superact, when he said: "Development may be considered in a broad sense as the large tidal wave of change that leads from conception to death, not broken up by the ripples of cyclic physiological activities."[43]

The materials metabolized by the basic life processes are, of course, minutely distributed and redistributed, held together by chemical linkages, oriented by ever-shifting polarizations and elaborated into a single, vastly intricate physical structure. That structure is the agent's body. The body is the perceptible record of the life which produced it, and still produces it as long as that life continues. Were it not for this cumulative recording of acts in their passage, we could not know most of them at all, because of their infinitesimal scope and often their deep involvement in greater acts, wherefore the constituent cycles are only statistically observable, not singly distinguishable; or, were we to devise means of watching them as specific events, we would surely not be able to keep track of their relations to the macroscopic movements that are the more spectacular exhibits of animal life. What makes the microscopic and submicroscopic events knowable at all is the fact that they inscribe themselves on the substances they utilize, and leave these organized into enduring structures by the perpetual round of maintaining acts. Other acts on a similar scale—acts of chemical transformation, transport, absorption and adsorption and changing polarization—augment the organism, usually up to a fairly fixed limit, but in some kinds (e.g., many trees, fish, perhaps some reptiles) indefinitely, though at a decremental rate that becomes very slow indeed at a great age.[44] The body, throughout life, is the "dynamic equilibrium" itself, growing and differentiating into articulate forms: leaves and rootlets and elaborate floral organs, or scales, feathers, membranes of many sorts, eyes and hearts, all unfathomably structured (Fig. 9–1). Underlying all its variegation is always the "substrate," the vast molecular structure that includes the cells, the intercellular fluids with their complicated cargo of alimentation products

---

[43] See R. L. Watterson (ed.), *op. cit.*, p. 1.

[44] The act form of growth processes is attested in the words of a biologist, N. C. Wetzel, who wrote: "Numerous studies on the part of antecedent investigators have established an almost universal recognition of the common outward characteristics of growth: the lagging start, the succeeding phase of acceleration, and lastly, the unavoidable decline in rate of gain as the stationary state is more nearly approached." And, interestingly enough, he continues: "But the corresponding trend and features of metabolism are certainly less familiar, and are accurately known, indeed, for very few examples" ("On the Motion of Growth" [1934], p. 187). The upshot of his work was a demonstration of the same dynamic form—inception, increase, consummation and decline—in metabolic processes generally; the act "writ large" and "writ small," even in a microscopic hand.

Figure 9–1. Hair Scales of the Silver Tree
leaves and rootlets and elaborate floral organs . . . all unfathomably structured

(C. Postma, *Plant Marvels in Miniature* [New York: The John Day Company Inc., 1961].
Reproduced by permission of H. J. W. Becht's Uitgeversmaatschappij N.V., Amsterdam.)

and metabolic by-products, either excreta or precious catalyzers, and the non-cellular deposits of calcium and other inorganic substances. This is the material matrix, the counterpart of the functional matrix of activities, and indeed the product, and therefore the exact reflection, of the latter.

What makes an organism look individual is not the possession of unique features, nor of many slight deviations from a standard, e.g., a known type specimen, but the fact that the ontogenetic processes of its individuation, which may be perfectly normal, are encoded in its bodily form. That is the basis of "living form" in nature; and this projection of dynamic pattern into relatively fixed material pattern holds good to the most elementary level of cellular and even molecular structure. The body consequently is the tangible record of past and passing activities, which have inscribed themselves, late ones over earlier ones, new ones constantly altering the latest summary, so that in a mature metazoan body it takes great acumen and patience to trace back the earlier phases of some lifelong continuities. As D'Arcy Thompson said, "In the marble columns and architraves of a Greek temple we still trace the timbers of its wooden prototype, and see beyond these the tree-trunks of a primeval sacred grove; roof and eaves of a pagoda recall the sagging mats which roofed an earlier edifice. . . . So we see enduring traces of the past in the living organism—landmarks which have lasted on through altered functions and altered needs; and yet at every stage new needs are met and new functions effectively performed."[45]

However difficult this record may be to uncover and unravel, it is the biologist's prime asset. No matter at what moment of an organism's development we view its body—perhaps even after its rhythms have been abrogated by death—every line and texture of that body expresses the directions, intensities and reaches of the acts that made it: growth and differentiation, intersections or fusions of separately originated processes,

---

[45] *Op. cit.*, Vol. II, pp. 1020–21. But while this great morphologist repeatedly pointed out the significance of gradients in living forms (cf. the quotation of his words in Chapter 7, p. 211), he also observed, near the beginning of his famous treatise, the restriction of that finding to macroscopic and chiefly multicellular plants and animals. In discussing forms that cannot readily be traced to growth processes, he remarked: ". . . the tiny medusoids of *Obelia*, for instance, are budded off with a rapidity and a complete perfection which suggests an automatic and all but instantaneous act of conformation, rather than a gradual process of growth" (p. 397; note that although this process seems automatic, he treats it, nonetheless, as an act). And further: "In the case of the smallest organisms . . . a certain species will not only change its shape from stage to stage of its little 'cycle' of life; but it will be remarkably different in outward form according to the circumstances under which we find it, or the histological treatment to which we subject it" (p. 405).

Figure 9–2. Growth Pattern of Mushroom (*Lenzites betulina* Fries)
fronts of arrest, waves of successive impulsion

(After M. C. Cooke, *Illustrations of British Fungi (Hymenomycetes)*,
To Serve as an Atlas to the "Handbook of the British Fungi,"
Vol. VII [London: Williams and Norgate, 1888–90].)

fronts of arrest, waves of successive impulsion (Fig. 9–2). Without the cumulative evidence of the physical organism at progressively ranged stages, we would have no clue to most physiological processes, and certainly no means of analyzing their complexity. It is by looking at the intricate, delicate details of bodily structure that one becomes aware of the tensions that hold them, or held them until they hardened in sudden or in gradual death.

The most illuminating exhibits, common though they may be, are the visible gradients of growth and other changes in organic forms, for by following a gradient one can see in spatial projection the beginning and end, the steady or changing rate and the direction, or radiating directions, of a process. Nothing could show the act form more clearly than do the gradients of ontogenesis, differentiation and growth to maturity; every change of proportions in a multicellular organism, normal or pathological, bespeaks a difference in rates of growth between comparable articulated parts; and most developments, whatever their rate, exhibit either an over-all gradient or an interrupted series of processes, each of which has a characteristic rise, acceleration, slowing and termination. Even the deployment of colors in flowers, insect wings or animal skins usually shows gradients of diffusion from a center of highest activity, arrested at boundaries of contact with other processes, so that lines of stopping and stowing mark their interpenetrating extremes.

A good demonstration of such pattern-making is to be found in the

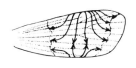

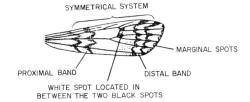

Figure 9–3. Wing of the Flour Moth (*Ephestia kühniella*)

(After Alfred Kühn, *Vorlesungen über Entwicklungsphysiologie* [Berlin-Göt-tingen-Heidelberg: Springer, 1955].)

literature on insect wings, which comprises many detailed studies of wing growth, veining, mutant abberations and experimental distortions of wings, at least in the genera *Drosophila* and *Ephestia*, and also some investigations of color distribution in butterflies' wings.[46] The final gradients of pigmentation are seen to arise, in the main, not from the points of origin of the wing, but from several peripheral areas, which induce "determination streams" of activity[47]—mitosis, cell migration and melanin synthesis—that seem to proceed by rhythmic waves of impulse. Such rhythms are not necessarily simple or single, since several areas of the wing are under separate controls, and have been experimentally found to express the actions of different genes. Henke distinguished two types of morphogenetic rhythms in the determination of wing patterns, "diffusion rhythms," which start from an area of earliest activity and gradually extend along distinct lines, motivating successive developments, and "simultaneous rhythms," which start from many centers at once, giving rise to scattered dots, that fall into more or less even spatial patterns by virtue of the competition among the simultaneous impulses[48] (Fig. 9–3). A similar visible projection of this type of rhythm may be

[46] The most notable pioneers in this field are R. Goldschmidt, K. Henke, A. Kühn, B. N. Schwanwitsch and F. Süffert. A fairly detailed survey and critical treatment of the many researches which have followed their leads is presented by K. C. Sondhi in "The Biological Foundations of Animal Patterns" (1963).

[47] "Determination stream" is R. Goldschmidt's term; see Sondhi, *op. cit*, p. 294: "Adopting a concept introduced by Spemann, he [Goldschmidt] called it a 'determination stream' and suggested that genes controlling the structure and the color of a wing may act by controlling a determination stream of definite quantity, speed of progress, pattern of flow, and of action upon different processes in morphogenesis." His analysis proved insufficient to explain the patterning of lepidopteran wings; according to Sondhi, "This was partly owing to the Goldschmidt's assumption that during development the whole wing acts as a unit. The situation was clarified when Kühn and von Engelhardt not only demonstrated the existence of a determination stream in the developing wing of *Ephestia* but also showed the presence of a definite arrangement of areas or fields forming specific parts in the wing pattern."

[48] Sondhi, *op. cit.*, p. 296 ff.

Figure 9–4. Jaguar's Fur
differential influences on the hair follicles
induce many similar groups of dots . . .
at fairly even distances from each other

(Wolf Strache, *Forms and Patterns in Nature* [New York: Pantheon
Books, Inc., 1956]. Copyright © 1956 by Pantheon Books, Inc.
Reproduced by permission of Random House, Inc.)

seen in the jaguar's fur, where differential influences on the hair follicles
induce many similar groups of dots deployed in an all-over pattern at
fairly even distances from each other (Fig. 9–4).

One of the most significant results of all these highly detailed studies
of pattern was Goldschmidt's discovery of the "prepattern," a purely
functional condition, primarily a differential rate of mitosis, in apparently similar cells which afterwards exhibit different qualities and degrees
of pigmentation. The distribution of colors on the wing is only a final
record of processes which have been going on over a protracted period,
and which started at different stages of pupal development (if not
earlier) and advanced at different rates, so they attain their irreversible,
final forms at different times. This means that those scales or other

[334]

tissues which reach the adult form soonest may block or change the course of others that are still labile in the "determination stream." The pattern is gradually decided, although under normal conditions the outcome is invariant; so the spatial design of typical colored forms really summarizes a complicated temporal process.

K. C. Sondhi, whose summary of this line of research I have largely followed here, observes: "The importance of Goldschmidt's findings lies in that they demonstrated for the first time that a primary pattern may be laid down before a pattern appears visibly."[49] The value he attaches to the prepattern is mainly that its formation, and the centers and gradients of differentiation involved in it, explain the deployment of pigments in insect wings, as well as some other, related phenomena. There is, however, another significant aspect to the functional design that precedes the visible phenomenon: it is in the active phase of laying down the prepattern that genes exert their influence.[50] No matter which or how many genes affect a particular process, the final bodily expression of it is still open to other than genetic determinants (for instance, the

[49] *Ibid.*, p. 293.

[50] J. T. Bonner, in *Morphogenesis: An Essay on Development* (1952), describes a comparable prepattern determining an entirely different mechanism of color distribution in the skins of two species of newt, *Triturus rivularis* and *Triturus torosus*, respectively, in which patterning is effected not by chromatin synthesis *in situ*, but by the migration of existing chromatophores to points of concentration or from points of dispersal. "The distribution of the chromatophores is characteristic of the species, for in *T. rivularis* they are quite evenly distributed over the body wall, and in *T. torosus* they are densely congregated in a streak along the region of the lateral line. In experiments on the dispersed type Twitty found that they spread in all directions to wherever there was an absence of chromatophores . . ." (p. 202). By means of tissue cultures (in which they maintained their tendencies), he was able to show "that the cells gave off some substance, a gradient of which would orient the cells away from the source of production or high concentration point. By this means their movement is controlled, for they attempted to spread as far away as possible from one another." Cells of *T. torosus*, on the other hand, migrated only short distances and then tended to clump, as they do *in vivo* along the lateral line. Somewhat later, Bonner speaks of the basic function of polarity, saying, ". . . in the beginning of development it is the first recognizable indication of the order that is to follow. Polarity is then a descriptive term referring to the geometry of the pattern . . ." (p. 205). It is, indeed, the prepattern of morphogenetic movements, determining the ultimate color pattern; and those morphogenetic movements are the site of genetic control, as he explicitly states, for he says apropos of Twitty's cultures: ". . . the tendency to migrate, which is the basis of the pattern, is consistently different for the two species. Even though we do not know the genes involved, we may call this an inherited and therefore a genetic difference, and so here is a good case of the control of morphogenetic movements by genes" (p. 203).

interference of scientists), so that the phenotype contains an element of accident which the underlying gene complex does not share at all.

Genetic inheritance makes the unity and continuity of a biotypic stock; it is every living being's indissoluble bond to its own ancestry, its commitment to the life of its kind and none other. The gene complement is the one permanent factor in its lifelong, progressive situation. Everything else may change; everything else that motivates the advancing flow of its impulses and its realized acts plays on this anciently established core. Heredity is the primary involvement of every organism with other organisms; not with a "kind" distinguished by characteristic traits, but with its stock defined by its own actual descent and its resultant common ancestry with some—possibly, all—others of its taxonomic "kind." The stock is the largest natural unit of life.[51] Its beginning goes back to the unknown days of biogenesis, and its end is unpredictable, though some forms of life that have ended may have been total stocks. If any entire plant or animal stock has ever really disappeared without progeny, a multimillennial generic process has finally vanished from the world; nothing less than that could be accounted the end of a phyletic history. The gene complement, with all its possibilities of combination and mutation, is what holds such a mighty unit together, and sets the limits of what may happen within its scope.[52]

But the stream of hereditary life moves with a characteristic pulse of its own, by a constant succession of individuations. The stock may be regarded as a unit of activity, but it is not an agent, an individual, or an organism; such entities arise only by processes of individuation.[53] And

[51] Compare J. Z. Young, *Doubt and Certainty in Science* (1951), p. 154: "[Individuals] are not the basic units of life. Each individual is part of a much larger system, which continues over millions of years, changing slowly by the process of evolution. This maintenance of continuity is the most fundamental feature. . . ."

[52] Cf. G. R. de Beer, "Embryology and Evolution" (1938), p. 63: ". . . it is not phylogeny or the series of adult ancestral forms, but the inherited and undeveloped germ-plasm which imparts to each organism the capacity to respond to certain stimuli in the manner recognized as development of the egg, ensuring that if development occurs at all, it conforms to the type of the species. Development has thus been shown to be an epigenesis of material of which the possible qualities are predetermined."

[53] Several philosophizing biologists to the contrary notwithstanding; e.g., T. W. Torrey, with whose views I am in general agreement, save that he follows our great inspirer, A. N. Whitehead, beyond the points where I have to depart from "organicism." See Torrey's "Organisms in Time" (1939): "The thing is, the individual as an ovum is inviolably linked with its progenitors and thus represents the initial stage in *a new cycle of the hyperspatial individual* rather than the beginning of an entirely new entity" (p. 283). What is meant here is certainly true; but to call a

here we are at the starting-line of today's most exciting and penetrating scientific researches, which are in genetics, embryology, cytology and related subjects that bear on the origin and development of organisms; for every such development is a course of individuation. The activities that compose a particular life, however, are so many and so variable that even in a single organism individuation may go to different lengths in separate directions, and even its greatest degree may be high or low, so that some living things *in toto* are more individuated than others. But one cannot range organisms in a simple hierarchy of progressive total individuation, because the many directions make a single criterion impossible. Sometimes, where a basic criterion obviously serves to class a particular living structure as an individual organism, although in some respects it is low in scale, it is better to analyze its peculiar involvements as such rather than as lacunae in its individuation. This applies particularly to symbiotic relationships and to mutual controls via the normal medium, for instance, polluting the water with chemicals that affect the growth or development of other individuals, as tadpoles and most fish do in close quarters,[54] and as many plants seem to do, by exudations directly (by roots) or indirectly (by leaves, via rain) into the soil.[55]

The most primitive act of individuation is the isolation of a protoplasmic unit by a completely surrounding membrane, selectively penetrable under osmotic pressure. Life probably began with such isolating acts or proto-acts of polarized drops, coazervates, highly structured macromolecules or other, now inconceivable eobionts. Today, the cell seems to be the smallest unit of life capable of functioning as an or-

---

vital stock "the individual" or (as he presently does) "the organism" obliterates all those distinctions which are the scientist's milestones on the way to insight. It only leads to the conclusion he expresses—"It is the universe, then, that is *the* organism, and all the nexūs prehended by it are organisms of a gradating [*sic*] subsidiary character." The subsidiary organisms "take the form progressively of 'electrons, molecules, rocks, and men'" (p. 287, quoting Whitehead); but this gives us no clue to the progression from rocks to men. See also John Phillips, "Succession, Development, the Climax, and the Complex Organism: An Analysis of Concepts" (1934–35), in which "the complex organism" is the ecological community of a region.

[54] S. M. Rose, in "Cellular Interaction during Differentiation" (1957), p. 382, cites C. M. Richards to the effect that "the agent of mutual inhibition in crowded tadpoles (*Rana pipiens*) is a cell found in the digestive tract and in the faeces. Other parts of tadpoles are not inhibitive."

[55] See J. T. Bonner, "The Role of Toxic Substances in the Interactions of Higher Plants" (1950).

ganism (the status of viruses is uncertain).[56] Within the cell wall—which may, in some cases, be of amazing complexity—the cell metabolizes and grows to the stage that marks the consummation of its life act, when mitosis begins; and when, at the end of cleavage, the cell membrane has met itself from all sides at the center of its constriction, two daughter cells have completed their respective individuations. The mother cell did not die, yet it is no more, for its individuation was only a brief episode in the life of the stock, and its two "daughters" have inherited every bit of its substance, without reservation, in equal shares. The same material exists, the same activities go on; the unicellular being does not.

The protozoa are undoubtedly the most adaptable organisms in the world, which is probably the reason for their continuance (assuming that they are also the most primitive) from distant beginnings to the present age. Their numberless "kinds" are really impossible to count or classify, because many of them change so radically from one form to another that the same animal may figure now as amoeba, now as flagellate infusorian with a different label.[57] Such changes seem to occur in direct response to environmental changes, which the animal cannot escape, but which—unlike higher and more permanently organized creatures—it can negotiate in this radical manner.[58]

Most cells, however, do not lead an independent life, as the free-swimming protozoans do. The protoplasmic cell is the elementary form of life;[59] it is represented by numberless strains, only some of which tend to individuate as self-sufficient organisms. The majority have special

[56] Some biologists treat the virus as a "naked gene," because it has some striking similarity to genes; see, e.g., G. W. Beadle, *op. cit.* Others regard it as a degenerate organism, for example, F. M. Burnet, in his *Virus as Organism.* Some have even advanced the opinion that it is not a living substance at all, but a purely chemical compound. A most interesting presentation of the baffling problems met in the study of viruses and our essential ignorance of virus action may be found in L. M. Kozloff's "Virus Reproduction and Replication of Protoplasmic Units" (1954).

[57] A good example is adduced by N. D. Newell in his article, "Biology of the Corals" (1959), p. 227: "The simple structure and globular form of the zooxanthellae [symbiotic algae in corals] make them difficult to classify, but . . . they may be dynoflagellates in disguise. Recently, scientists at the Haskins Laboratories extracted zooxanthellae from jellyfish, anemones and corals and placed them in suitable media. After a few days, the simple cells developed flagella and began to swim. In this form, they had already been described as the dynoflagellate, *Gymnodinium adriaticum.*"

[58] Cf. above, p. 318, n. 23; see also L. L. Woodruff, "The Protozoa and the Problem of Adaptation" (1924).

[59] R. H. Francé called the cell "die technische Form des Lebens" (*Die Pflanze als Erfinder* [1920], p. 13).

sensitivities to the presence of other cells, particularly others of closely related origin, so that the "daughter cells" created by an act of mitosis remain conjoined, and their "daughters" in turn augment the clump, which may grow to comprise millions. In such an aggregate, however, other potentialities naturally come into play, because each cell has to adapt itself to its position in the mass. A cell at the center has a different ambient from one on the surface. The central one has no access to metabolites from the outside, unless streams of fluids form passages, or some system of molecular transport develops whereby vitally necessary chemicals reach it. By the time they do so, they will not, of course, be in the same condition in which they reached the outermost cells. Also, the pressures and electrical tensions at the periphery and the interior, respectively, are apt to be quite unlike; so are the concentrations of waste products and enzymes manufactured by these microscopic agents. Considering, then, the inherent versatility of cells, one can understand—at least in principle—the processes of differentiation that arise spontaneously in a multicellular complex and progress with its increase. They are adaptations, just like the changes of protozoan forms, and the "work" each cell contributes is the result of its exercising all the functions it can in the straitjacket of limitation imposed by the activity of its fellows, and by the further environmental conditions that hedge the whole group.

It is in multicellular structures that the varying directions and degrees of individuation have astonishing effects. All living stocks are of generally similar constitution chemically and perhaps also in their electronic properties; they all perpetuate themselves by procreation, that is, by continually forming new organisms out of living matter developed in older ones; but the forms of their individuations are fantastically complicated by the tendency of cells to divide without casting loose from each other, and so remain closely involved with more and more others in all their acts. That means that the mass as a whole becomes the matrix from which all acts arise; the entire "culture" of genetically related cells constitutes a single organism. Intercellular substances arise with the proliferation of cells, intercellular electrical currents develop, groups of cells rather than separate ones become polarized in various patterns and receive influences of inductive or inhibitive substances, so that very soon, despite their genetic identity, they differentiate into diverse tissues, each with only partial possibilities of actualizing the impulses of life. Every fiber, ultimately every cell, has its own situation from moment to moment, somewhat as every plant in a colony has its own microclimate, though this may be practically indistinguishable from that of its nearest neighbor.

The internal microsituations, together with the original equipotentiality of cells, go far to account for the special articulations that tissues undergo in the development of a metazoan organism. But these changing and interplaying situations are not haphazard in their advance; they unfold in a perfectly logical over-all pattern, one total conformation preparing another in typical succession, and in relative independence of events that do not belong to the differentiating protoplasmic unit.[60] This ever-repeating pattern is peculiar to the stock, and is the basic framework of its characteristic impulses, actualized in the growth and activity of each individual. The course of actualization, however, is always modified to some extent by contingencies of the organism's situation as a whole as well as its internal transient conditions; and the degree to which its deviations from the generic norm can go without destroying the particular life—that is, the amount of leeway in its individuation—is itself part of its heritage. In the smallest, unicellular creatures, as already remarked, the leeway of form and function is bewildering to any biologist, and to the taxonomist, frustrating. In some higher forms it is so narrow that small irregularities, such as a malformation of an organ in embryonic development or, later, the lack of quite specialized forms of some metabolites, may prove fatal.

The vast complexity that results from all the possible relationships of active cells lets processes of individuation take many different directions and continue to different lengths. These processes begin in previously existing protoplasmic structures with which the new organism is all but indistinguishably involved at first, and from which its activities gradually free themselves; but any particular line of its individuation may end before that total superact is completed, so that in some vital stocks the new being retains its involvement with the parent, and through it, with other incomplete individuals. Some coelenterates, for instance, remain physically connected with one another through a common alimentary system, although each of the hydranths which grow from that conjoining organ performs its perceptive and conative acts—trapping and devouring food—in full independence of the others. Bonner gives a graphic description of such bodily involvement among some coelenter-

---

[60] J. W. Buchanan concluded from his experiments with a large variety of adult animals and embryos that "variations in external environmental conditions to which various parts of the animal are subjected are not material; . . . the internal environment and the channels of communication remain highly constant. . . . The whole picture of the physical unity and integration of an organism is one of an internal environment that is highly constant and uniform as compared with the environmental conditions which separate the individuals of a group" ("Intermediate Levels of Organismic Integration" [1942], pp. 57–58).

ates: "In the colonial forms a whole group of hydras are, to put the matter crudely, hooked together on a piping system, so that in *Obelia*, for instance, at the base of each hydra or hydranth (or polyp), the gastro-intestinal cavity narrows into a hollow tube, which is connected to all the other hydranths in the colony. It is a communal intestine, where the meal of some fortunate polyp may be shared by his relatives. . . . While the overall length of an individual remains approximately the same, it grows continuously at one end and degenerates continuously at the other."[61]

An interesting ambiguity is created here by Bonner's use of "individual" to designate the colony rather than one of its members. Except for the post-buccal stages of alimentation, the hydranths perform all animal functions, even procreation, for in most species (*Obelia* among them) the sexually produced colonies give rise by budding to an alternate, parthenogenetic, medusoid generation. Their entirely independent tentacular motions, moreover, make them seem far more like connected individuals than like articulations of one body. The underlying "piping system" engenders them, but then takes no part in their acts beyond receiving, digesting and passively (without peristalsis) redistributing their nourishment; it does not directly motivate any of their overt acts. One cannot really call either the colony or a polyp an individual; the colony is a complex of partial individuations.[62]

All that is necessary to preserve the life of the stock is that the rhythm of successive individuations be kept going without any break over hundreds of million of years. If once it is broken along the whole front of radiating family trees that represent the stock, then the great life that started when the earth was young is ended. But with regard to its continuity it does not matter whether its individuations are long or short, nor how many of their primitive involvements are broken or unbroken, just so long as the vital activity renews itself steadily.

It is for other reasons than their upholding the continuity of life that forms and degrees of individuation interest us; their importance for our understanding of human life lies largely in the kaleidoscopic combinations in which they occur. In a general way it might be said that the

---

[61] *Morphogenesis*, pp. 120–21.

[62] Cf. Bertalanffy's similar observation on a two-headed (solitary) hydra, produced by splitting its anterior end in embryo, whose two fully developed heads later fought each other ferociously for the possession of a morsel of food, though either one would send it into the selfsame "stomach": "Hier ist die Frage sinnlos, ob wir es mit 'einem' oder 'zwei' Individuen zu tun haben. Aber auch bei höheren Organismen ist, mindestens in frühen Entwicklungsstadien, die 'Individualität' eine prekäre Sache" (*op. cit.*, p. 56).

least individuated organisms are low in the evolutionary scale, all very high ones being at least spatially distinct and, in the animal kingdom, separately mobile. But to infer from this observation that there is anything like a simple gradient of increasing individuation with the advance of evolution would be a gross mistake. The internal involvement of acts with each other, known as "integration of functions," is the most important factor in individuation, i.e., in the establishment of self-contained, stabile, vitally active systems.[63] Yet structural stability does not always keep step with functional capacity; for instance, some creatures with elaborate methods of procuring food or of procreating, evading adverse conditions, etc., may be still divisible into major parts, which can all survive as individuals by rearranging their unbalanced organs and regenerating such as did not fall to their share, as starfish can re-form from a severed "leg." Surely Buchanan is right—speaking from experience with many kinds of animal life—that "when one compares these properties among organisms it is found impossible to arrange a sequence of increasing structural stability which in any close degree parallels the commonly accepted sequence of increasing complexity of structure."[64] Similarly, the physical connection of completely formed organisms with each other may occur in animals which perform more "advanced" procreative functions than some separate, free organisms—even such as may represent their own alternate generations. A single free-swimming *Salpa* procreates agamically, by budding; it produces a *Salpa* chain of many connected, but otherwise individuated, animals, which produce eggs and sperm in the manner of much higher organisms, to make the next structurally individuated generation. Here the same stock, in the same age of its evolution, exhibits both degrees of individuation.[65]

[63] Cf. Charles Manning Child, *op. cit.*, pp 2–3: ". . . the organic individual appears to be a unity of some sort, its individuality consists primarily in this unity, and the process of individuation is the process of integration of a mere aggregation into such a unity. . . ."

[64] *Op. cit.*, pp. 49–50.

[65] For a good descriptive account of the life history of this species see N. J. Berrill, "Salpa" (1961); for an appreciation of the individuality problem it presents, see W. K. Gregory's highly suggestive as well as technical article, "Polyisomerism and Anisomerism in Cranial and Dental Evolution among Vertebrates" (1934), dealing with the broader implications of relative individuation, especially its bearing on the articulation of anatomical structures. *Salpa* is adduced only by way of illustration; his general conclusion is stated in the words: "Polyisomerism and its opposite derivatives, anisomerism and hyperpolyisomerism, evidently make possible the almost infinitely varied structural and physiological combinations that are recorded in the millions of plants and animals. Adaptive radiation seems to have resulted from the summation along divergent lines of the results of secular polyisomerism, anisomerism and hyperpolyisomerism, while the production of allometrons, rectigradations and aristogenes appear to have been incidental to the same process" (p. 8).

The reason for this incommensurability of many aspects of evolution is that the entire process of individuating consists in so many lines of activity, moving at different rates and building up the organism in different respects—stabilizing its outward structure under the influence of an environment in which its ambient defines itself and changes with the creature, developing organs specially suited and also limited to the performance of particular functions—that a considerable degree of unity may be attained, or an impressive complex of interrelated organic structures articulated, even while some mechanisms remain primitive and below the evolutionary level of the organism as a whole. This sort of allometry runs all through the animal kingdom, and makes it well-nigh impossible to plot definite distances, on any taxonomic, physiological or psychological scale, between two members of different orders or even phyla.[66] In the study of human development it becomes of paramount importance, because it directs one's attention to analogues, if not precursors, of human traits in forms of life far removed from our own, or even from the primate stocks we are accustomed to compare with hominids.

First, however, let us turn once more to the elementary steps in the formation of organisms, because physical separation from a progenitor is not the only aspect of primary individuation. There are also degrees in the separateness of organisms from the inorganic world, as parts of the latter may become included in the former to serve physiological purposes. In some worms statocysts may occur, in the head segment or in others; these are sometimes open to the outside, in which case sand grains or other foreign bodies are picked up in the statocyst and serve as statoliths. In lobsters the statocyst always contains sand, which is eliminated at moulting and later replenished; the animal, consequently, is easily disoriented after ecdysis until its extraneous otoliths are replaced.[67] Seed-

[66] See, for example, E. C. Olson's *Origin of Mammals Based upon Cranial Morphology of the Therapsid Suborders* (1944), p. 125: "By the end of the Permian, a vast assemblage of animals, mammal-like in varying degree and with an almost infinite variety of association of mammalian and reptilian characters, had developed. . . . The very advanced ictidosaurs suggest that they possessed accelerated development of mammalian trends, a process which might be anticipated in the ancestry of the marsupials and placentals and their probable predecessors, the pantotheres. Certain of the very primitive characters of the monotremes suggest origin from a stock, which, though it passed the mammalian threshold, did not progress as rapidly as the others."

[67] Both examples are taken from Louis Guggenheim's *Phylogenesis of the Ear* (1948), pp. 54 and 66, respectively. This author is exceedingly credulous of fanciful evolutionary speculations, sometimes from doubtful sources, but that should not lessen his authority on facts of otology.

eating birds of many kinds carry gravel in their crops, which serves to grind up the hard-coated grains. A similar use of pebbles was apparently made by the dinosaurs, before the day of birds;[68] and it is interesting to note that this phenomenon of inorganic inclusions in animal bodies, which one would expect to find at low evolutionary stages rather than high ones, has been preserved in the rise of the avian phylum from the ancient *Reptilia*.[69]

There are other conditions, too, that produce ambiguities in the boundary between individuals and the external world, particularly the exact opposite of such prosthetic "organs," namely, the extrusion of organic products to form part of the animal's ambient. The most familiar instance is the spider's web. It is essentially an extension of the spider's body; the animal's reactions to the slightest touch of anything upon the web are like those of many other animals to a touch on their antennae, or of a person to a contact made with his groping hand, which explores and is ready to grasp. Yet the web can be abandoned; in this sense it is in the spider's ambient, rather than in the confines of the body as an antenna or a hand is. It may be regarded as a physical possession taken of the immediate environment, or as an extension of the organism that is not ended by a subsequent contraction, but by a casting off, like the shedding of hair, skin, milk teeth, and antlers, except that the abandonment of the web is under the immediate voluntary control of the animal, which true bodily castings are not. But the form of the web is due to an organic process, genetically determined, species-specific and subject to characteristic disturbances by various drugs or other physiological influences.[70] Its status with respect to the physical individual is really ambiguous.

There are few other examples in nature of organic extensions having such complex functions, serving for perception and food capture, and— in the case of the tunnel spider, for instance—even for refuge. The delicate aquatic nets of some caddis worms, such as *Hydropsyche*, may be classed in the same category. Caterpillars that make silken cocoons as they begin to pupate fabricate a functionally simpler product, since the cocoon is for protection only. The more familiar kinds of caddis case, made of little stones by some species and of straws and other plant

[68] See, e.g., William E. Swinton, *The Dinosaurs; a Short History of a Great Group of Extinct Reptiles* (1934): "It seems probable that the stomach of the Sauropoda was a powerful, muscular, and gizzard-like organ in which gastroliths, or stomach-stones, helped to pulverize the ingested matter" (p. 86).

[69] In a brief report contributed to *La Nature* in 1947, R. Legendre gives an account of seals eating stones with their meals and later regurgitating them.

[70] For a brief factual account see P. N. Witt, "Spider Webs and Drugs" (1954).

fragments by others, as well as sand collars, leaf cocoons, honeycombs and paper wasps' nests, though they all incorporate some substances produced by their respective makers, are not primarily extensions of the animal body; they are better described as exploitations of the environment, like borrowed otoliths or gastroliths, though they are not incorporated in the organism, but are what Sewall Wright has called "hereditary extraorganic structures."[71]

There are other ways in which living things, plants as well as animals, modify their environment by exuding chemical substances. Molds and fungi often emit enzymes into the substratum from which they draw their nourishment so that part of their digestion is carried out before they absorb their food; their immediate vicinity functions as a sort of external stomach.[72] A somewhat similar phenomenon in the animal kingdom is the feeding habit of the bug *Platymeris rhadamanthus*, that pierces its prey with a sharp, dagger-like proboscis, and pumps the victim's body full of a venom that not only kills it, but also digests all the fatty and other edible parts, which the bug then consumes in liquid form.[73] Here, again, digestion is performed through enzymes possessed by the eater, but outside of his body—this time in the body of the victim itself.

Such imprecise division between organisms and their surroundings may lead to curious forms of involvement with other creatures; for just as physiological products released into the circumambient medium—air, water, soil or a particular piece of flesh—may act as digestive enzymes on metabolites occurring there, so they may act very much like hormones on other members of the agent's stock that happen to be in the vicinity. Wiener characterized musk, or other sexually attractive substances, as "communal exterior hormones."[74] D'Arcy Thompson furnished a striking example of such action when he reported, concerning some protozoans of the genera *Enchelys* and *Colpodium*: "Two minute individuals kept in the same drop of water, so enhance each other's rate of asexual reproduction that it may be many times as great when two are together as

---

[71] "Genes as Physiological Agents; General Considerations" (1945), p. 292.

[72] According to J. A. V. Butler, "They liberate enzymes into the medium in which they are growing in order to solubilize substances and break them down into a state in which they can be absorbed. They use the liquid around them as a sort of auxiliary stomach. Some thirty or forty different enzymes have been distinguished in the liquid in which some fungi grow!" (*Man Is a Microcosm* [1951], p. 25).

[73] See John S. Edwards, "Insect Assassins" (1960), p. 72.

[74] Norbert Wiener, *Cybernetics: Or Control and Communication in the Animal and the Machine*, p. 182.

when one is alone; the phenomenon has been called allelocatalysis."[75] Still more astounding is the interaction of slipper limpets in the course of their development: "Young limpets clustered round an old female grow slower than others which live solitary and apart. The solitary forms become in turn male, hermaphrodite and at last female, but the gregarious or clustered forms develop into males, and so remain; development of male characters and duration of the male phase depend on the presence or absence of a female in the near neighborhood."[76]

In the cases adduced by Thompson the influence exerted by organisms on each other is certainly very much like hormone action. But this is not the only possible means of interaction between members of a stock, affecting ontogenesis; there seem to be also relations among individuating centers that may determine the polarity of their future organization. Studies of the "eggs" of *Fucus* have shown a "group effect" on the differentiation gradients of the individual eggs,[77] which looks like the work of an electric emanation, but under experimental conditions appears rather to be chemically engendered.[78] However, as the experimenter himself observed, "the distinction between physical and chemical largely

[75] *On Growth and Form*, Vol. I, p. 257, citing T. B. Robertson, *The Chemical Basis of Growth and Senescence* (1923).

[76] Thompson, *op. cit.*, p. 167.

[77] *Fucus* is an amazing zoophyte. It is certainly a plant, yet it propagates by eggs and sperm. D. M. Whitaker, in "Physical Factors of Growth" (1940), one of his many studies on this subject, gives the following report: "*Fucus* is a Marine Brown Alga which attaches to fixed rock in the upper part of the tidal zone. Like most animals, it is diploid except for a few cell generations leading up to the formation of gametes. Most species are hermaphroditic, but some are dioecious. The spermatozoa or antherizoids are motile and fertilization normally takes place in the sea water" (pp. 75–76).

"*Fucus* eggs exert mutual effects on the differentiation of neighboring cells which act at a surprising distance (e.g., through 0.3 mm. of sea water . . .)" (p. 77).

". . . *Fucus* eggs tend to form rhizoids in the . . . direction of neighboring eggs. [Above, he stated that "any region of the egg cytoplasm which is caused to form a rhizoid protuberance differentiates and grows in a special manner."] Hurd found that the effect was strong enough to overcome the response to light if the eggs were within several egg diameters of each other. Whitaker found that two eggs alone in a dish in normal sea water did not form rhizoids toward each other, but that the attraction ('group effect') depends on appreciable masses of eggs as an attractive body. The effect was also found to be non-specific, i.e. masses of unfertilized eggs of *F. vesiculosus* exerted a strong group effect on developing eggs of *F. evenescens*" (p. 78).

[78] On the basis of experiments, he concludes: "The egg responds to diffusion gradients from itself as well as to the products of another egg. It does not appear possible that mitogenic radiation, as ordinarily conceived, could be involved in this action of a single cell upon itself" (p. 79).

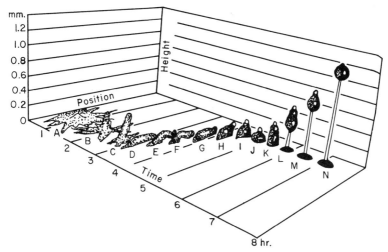

Figure 9–5. Slime Mold (*Dictyostelium discoideum*)
*probably the most complicated purely organic integration of
separate organisms into a true unit that we know*

(From John Tyler Bonner, *The Cellular Slime Molds* [Princeton: Princeton University
Press, 1959]. Reproduced by permission of Princeton University Press.
Copyright © 1959 by Princeton University Press.)

disappears at the molecular and sub-molecular levels at which we believe
the ultimate, even if mostly unidentified, processes of differentiation and
growth take place."[79]

All these relationships among members of identical or related stocks
seem to be residual links, not yet broken by processes of individuation.
But involvements may also be progressively established by organisms
with others of their own kind, or of entirely different kinds. The first
case is most perfectly illustrated by the well-known but still amazing
morphogenesis of the slime mold *Dictyostelium discoideum*, a multicellular fungus built up not by division of a parent cell, but out of free,
amoeboid cells which come together and find their places in relation to
others, whereupon each undergoes some particular modification according to the place it has taken, so that the self-arranging cells form a
plasmodium, stem and fruiting crown—a well-formed, multicellular,
individual plant (Fig. 9–5).[80] This is probably the most complicated

[79] *Ibid.*, p. 75.
[80] Excellent accounts of this process are given by J. T. Bonner, *Morphogenesis*,
p. 173 ff, and by Edmund W. Sinnott in *The Problem of Organic Form* (1963),
where he writes (pp. 18–21): "These organisms are technically plants, since they

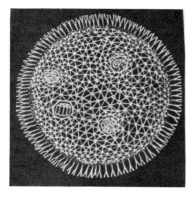

Figure 9–6. Volvox

*a spinning, minute ball of green
algal cells bearing flagella that
beat in unison*

(Illustration by Robert Barrell in C.
Brooke Worth and Robert K. Enders, *The
Nature of Living Things* [New York: New
American Library, 1955].)

purely organic integration of separate organisms into a true unit that we
know; what evolutionary course has led up to it no one (to my knowledge) has ever ventured to guess. But, phylogenetically, multicellular
individuals seem to have formed by aggregation of unicellular organisms
as well as by the process that is still reflected in metazoan growth, the
increase without complete separation of "daughter cells" produced by
mitosis.

Sometimes there is uncertainty or even flat disagreement about the
single or colonial status of an organic complex. In the case of *Volvox*,
for instance—a spinning, minute ball of green algal cells bearing flagella
that beat in unison (Fig. 9–6)—most textbooks, including some works
of distinguished authorship,[81] class it unhesitatingly as a protozoan

---

reproduce by spores, but their behavior is more like that of animals. The vegetative
unit is a single, minute, one-celled *myxamoeba* that lives chiefly on bacteria and
multiplies by vegetative division. After a time . . . multiplication ceases and the
myxamoebae begin to move toward a center through the chemotropic action of a
specific substance, *acrasin*." [Their movement stops when ten thousand to fifty thousand have gathered. These form the pseudoplasmodium, which may be about 2 mm.
long.] "This now begins to glide over the substratum with a motion apparently
produced by the rolling of the cells in the mass. The front end . . . seems to have
some directive function although the cells that compose it are continually changing."
[The mass seems like an organism, except that the cells remain free and move
singly, in amoeboid fashion, within the aggregate. Presently the migration stops, the
bottom cells anchor themselves and make a disk.] ". . . the cells in the center of
the mass, along its vertical axis, become thick-walled and adhere to each other and to
the disk to form a solid stalk." [Other cells slide upward over the fixed ones and
build the stalk higher, until the last ones gather at the top in a spherical body, where
each one develops into a spore, and blows off.]

[81] E.g., *Life: An Introduction to Biology* (1957) by George Gaylord Simpson,
Colin S. Pittendrigh and Lewis H. Tiffany, in which the resemblance to multicellular organisms is noted, but the structure is referred to as a colony (pp. 55–56).

colony. But D'Arcy Thompson,[82] and more recently Libbie Hyman,[83] speak of it as a multicellular individual. Nevertheless, Dr. Hyman refers to its constituent cells as "zooids" and then slips into the more accepted fashion of treating it as a "polarized colony." Obviously she regards it as a transitional form—although Moore declares flatly, "No one maintains that Volvox is an intermediate stage in the evolution of Metazoa from Protozoa."[84] Perhaps no one would subscribe to just those words;[85] but a number of biologists do seem to think that it represents such a stage, through which some existing metazoan stocks may have passed, even if none are doing it now (which we have no way of knowing).

The making of a metazoan organism by progressive involvement of previously separate cells (or other units) is known as integration, and most biologists prefer the latter term. My reason for using the former as the converse of individuation is that not all processes of involvement are

Also John A. Moore, *Principles of Zoology* (1957). One considerably older text does observe that the complexity of Volvox foreshadows several metazoan conditions, to wit, sexual dimorphism, gametic reproduction and natural death of the parental soma after procreation. See Michael F. Guyer, *Animal Biology* (1948; 1st ed., 1931), pp. 384–86.

Bonner, in *Morphogenesis*, p. 188, calls Volvox simply a colony.

[82] Thompson, *On Growth and Form*, Vol. I, p. 406, speaks of "the lovely green multicellular Volvox of our ponds, . . . whose spherical body is covered wholly and uniformly with minute cilia," and in Vol. II, p. 613, classes it as one of "the lower algae" without mention of its colonial character.

[83] "The Transition from the Unicellular to the Multicellular Individual" (1942), p. 38: "The difference between a spherical colony of protozoans and a spherical multicellular individual like Volvox is obviously a functional one. In the first case, the cells act in independence of each other, each is capable of performing all functions, and the colony rolls about aimlessly. In the second case, the cells act in coordination, the colony is polarized and swims always with one pole forward, as is actually the case in the colonial Volvocales, and each cell is not capable of all possible functions." And further, p. 39: ". . . in certain genera, notably Pleodorina and Volvox, the anterior pole zooids and even the whole anterior hemisphere loses the power of both asexual and sexual reproduction and becomes purely locomotory, and reproduction is limited to most or a few zooids of the posterior part of the colony. When such polar differences become established, we have in effect multicellular individuals with division of labor."

[84] *Op. cit.*, p. 109.

[85] Even that may be doubted; in another, still current, textbook, W. F. Pauli's *The World of Life: A General Biology* (1949), the author refers to "the spherical Volvox, a colonial organism on the verge of becoming multicellular and differentiated" (p. 110), and later, speaking of the green algae, he says: "Of these one order, the Volvocales, . . . allows us to reconstruct the whole process of evolution from single-celled to multicellular as well as the first important steps in the evolution of sexual reproduction. . . . [Of the Volvocales] some are single celled, some colonial, and some on the verge of being multicellular" (p. 176).

integrations; there are dovetailing functions and forms of life which never become integrated, but always remain distinct and even separable. Consider the numberless phenomena—some of them really bizarre—of parasitism, symbiosis and commensalism, which show all possible patterns of involvement and all possible degrees of histological and physiological integration. The incomplete individuation of social insects, especially honeybees, is the best-known example; but there is one less familiar case—that of the termites—which displays in each "queen's" life history the passage from a morphologically and functionally complete individual to the colonial condition in which that same individual loses its self-sufficiency, giving rise to a swarm of incomplete organisms that take over most of the functions of the mother, who finally can no longer move about, but is fed and tended by her sterile offspring, while she herself grows to enormous size and becomes, in effect, nothing but an organ embedded in the hill—the womb of the colony.[86] Yet she and her mate, who (unlike the fathering drone among the bees) lives with her in her chamber, and all their small specialized children, though collectively functioning, retain their physical separateness, and interact mainly if not wholly by behavior, and not by physiological integration (Fig. 9–7).

It is different with parasites and internal symbionts. Parasites become, ontogenetically or phylogenetically, more and more involved with the organic functions of a host, and quite generally lose the homologous functions from their own repertoire. Some exploit the host to a fatal extent. If the parasite has only one phase of its life to complete in that host, its parasitic acts will not lead to its own destruction, as it is through with the situation it has destroyed; otherwise it must have some way to shift to a new victim, if it is to carry on. Many parasites do not kill their hosts, and consequently can draw on them indefinitely for food and shelter, and in cases where locomotion to the next host, or to its likeliest whereabouts, is required by the parasite, this also is usually effected by the current host, either in defecation, or in stinging the next one, or in being eaten by it.[87] But as long as an organism is parasitic on another, its

---

[86] A detailed life history and ethological account of termites by a noted authority is given in Wilhelm Bölsche's *Der Termitenstaat: Schilderung eines geheimnisvollen Volkes* (1931), though unfortunately literary ambitions have beguiled the author into the usual anthropomorphic and moralistic reflections on insect "Society." The interpretation of trails made by droppings, to guide the blind foragers home again, as "a veritable sort of Braille writing" which they invented for that purpose (see p. 60, "Dazu hatten sie sich nun eine richtige Art *Blindenschrift* erfunden") is inexcusable even in a popular book by a scientist.

[87] For many examples of these various conditions, see Alexander Petrunkevitch, *op cit., passim.*

Figure 9–7. Termite Queen
nothing but an organ embedded in the hill—the womb of the colony
(Buffalo Museum of Science.)

activities remain entirely its own, and do not fall under the control of the individual it inhabits.

But there is also a looser sort of involvement of heterogeneous creatures with each other, in which neither seems to be entirely dependent on the other: the relation of commensalism. This condition may range from mere voluntary tolerance of the lesser partner by the greater one—with some quite unexploitative advantage to the former which makes it seek the partnership[88]—to an established and even hereditary association.[89]

[88] One of the most puzzling commensal relationships is that between the African army ants known as "mañans" and a horde of beetles, great and small, of many diverse but related species, which march with the fierce, insect-eating ants, living on the offal of the mañans' dinners, but themselves apparently immune to attack by the "army." No one has yet found an explanation of the tolerance they enjoy. A good

[351]

The phylogeny of such relationships is rarely determinable. Commensalism may sometimes, perhaps often, have resulted simply from the coincidence of suitable conditions for two very different vital stocks; each of the two might thus become so intimate a part of the other's normal ambient that their life patterns reflected each other's presence without real dependence on that anciently familiar factor. Under such circumstances one might easily come to make some use of the other, perhaps to prey on it, finally on it alone, and if the difference in size were enormous, as for instance between a small worm and a fish, the loss to the larger might be tolerable even as the worm became a true parasite. (Of course, a parasitical condition might also arise from an original purely predatory relation, in which the small fierce attacker attached itself to, or invaded, the larger organism, as ticks and burrowing mites do. There is generally more than one possible source of a phenomenon that becomes prevalent and many-faceted.)

But a third form of involvement that seems to be progressive rather than residual has probably always, or very nearly always, originated in commensalism: that is symbiosis. From an ecological and evolutionary standpoint, this is the most significant interorganismic relation, for it ranges from slight, probably dispensable or even casual mutual benefits, like those which ants and aphids derive from each other, to complete integration of the functions of both symbionts. Consider, for instance, the intimate relation that obtains between termites and the *Trichonympha* which inhabit their intestines and furnish the enzymes that digest cellulose, enabling the termites to eat our houses, furniture and musical instruments. As L. L. Woodruff put it, "the protozoans apparently act essentially as glandular cells within the digestive tract of termites; securing in return for their enzymal action an abode, somewhat

---

factual account of the phenomenon is presented by R. Paulian and F. Coric in "Les staphylins, commensaux des mañans" (1947).

[89] Cf. George H. F. Nuttall's old but still generally valid account "Symbiosis in Animals and Plants" (1923), where he says, in the concluding summary: "Symbiosis occurs frequently among animals and plants, the symbionts (Algae, Fungi, Bacteria) becoming in some cases permanent intracellular inhabitants of their hosts, and at times being transmitted from host to host hereditarily. Among parasites, non-pathogenic and pathogenic, we know of cases wherein hereditary transmission occurs from host to host" (pp. 474–75).

A more recent survey, very careful and highly technical, of advances in the study of hereditary symbioses and related conditions, is Joshua Lederberg's "Cell Genetics and Hereditary Symbiosis" (1952). The progress made in this field in the course of three decades, to which a comparison of the two articles bears witness, is not only gratifying, but astounding.

aloof from the general competition in nature."[90] And, as he goes on to say, "In some cases so intimate are the relations which are established that it is still debatable whether the intrusive 'symbionts' are actually distinct organisms or merely organelles of their bearer!"[91] A similar statement is made twenty-five years later by T. M. Sonneborn, who writes: "Animal and plant cells alike normally contain in their cytoplasm bodies called mitochondria. All of these structures have been interpreted in two ways. Some authors interpret them as parasites or symbionts; others, as native components of the cells. Even the nucleus with its genes has been held to live as a symbiont in the cytoplasm!"[92] The most far-reaching of such speculations, however, is Pirie's, in "Chemical Diversity and the Origins of Life," where he says, "The ability to handle the halogen-to-carbon bond may . . . have been useful to an eobiont. Sponges still make extensive use of bromine compounds, so do the various species of mollusc that make tyrian purple. . . . But the vertebrates have nearly given up this type of metabolic expertise. Not completely however; we still use iodine in the thyroid and normal life depends on this otherwise exceptional synthesis. There is no reason to think that this use of iodine is new; it is found in present-day amphibia and fish and may well be ancient. I suggest that it is a relic of an initially more catholic approach to metabolism. . . . The only logical alternative is to look on the thyroid as the remains of a commensal sponge that lodged in the gill arches of a primitive vertebrate giving rise ultimately to symbiosis as intimate and essential as that of the lichens" (pp. 80–81).

Some scientists, on the other hand, have suggested an opposite hypothesis, namely that present symbionts, such as *Trichonympha* and termites, or luminous bacteria and the fireflies, molluscs or fish they now invade and inhabit, were derived by individuation from anciently single matrices.[93] Both roads to what one might call integral symbiosis may have been taken in different phylogenies. Convergent phenomena are the rule rather than the exception in nature; though in most cases a process of assimilation of small commensals to the vital economy of the single, larger partner seems more plausible.

Just how a separate, heterogenic organism can fall so completely under the control of a larger one that its own relatively simple functions be-

---

[90] *Op. cit.*, p. 57.   [91] *Ibid.*, pp. 57–58.

[92] "Beyond the Gene" (1949), p. 51.

[93] See, e.g., S. J. Holmes, *Organic Form and Related Biological Problems* (1948), p. 5: "If we suppose that an organism should exist in which the animal cells and symbiotic algae, instead of being of independent origin, were both derived from undifferentiated cells with the potency of developing into one or the other of these two types, would not the same automatic regulation of relative numbers still occur?"

come part and parcel of its host's, subject to the latter's physiological regulations, is a difficult question; the actual processes are hard to envisage. My guess would be, however, that they were first of all entrainments of physiological rhythms that drew the lesser organism into the dynamism of the greater one, and tended to wipe out any of its formerly established activities which did not fit somehow into the stronger system —especially if the commensal situation gave the protozoan new and simpler means of life. The power of on-going rhythmic acts, and the entrainment of smaller cycles by larger ones, is the main principle of integration in organic structure; and as it makes the original matrix, more or less individuated in space, time and action, so it also might be supposed to subjugate lesser matrices that come within its immediate range, and modify their vital acts.

As life began with increasing concentrations and protractions of interactive ferments on the earth's surface, so it continues in an ever-mounting advance of devouring, integrating, self-maintaining activities. The progressive individuation of organisms, especially in the animal kingdom, at present culminating in human individuals, is so striking that it has led many people to regard individuation as the sole principle of evolution. But from a higher standpoint—an Olympian point of view— such an outlook might seem narrow. The gradients of individuation might appear as strong lines in a swirling flow of ecological involvement of species with species, life with life, wherein every impulse to individuation sets up its own course, that may be long and become spectacular on the way, or may come to a stop very soon. Individuation goes on all the time, but it can proceed only in a framework of active involvements with the generating stock and the nourishing substrate of an ambient that is a small detail in the whole vast biosphere. Often, indeed, the very means of individuation, being chiefly powers of aggression against other individuals, lead to new involvements which become paramount, as with organisms that exploit others to the point of becoming entirely dependent on them. Then the life of the parasite is strengthened and developed in a special direction, as, for instance, the tapeworm's incredible procreative power,[94] at the expense of its general self-sufficiency.

Yet with every acquired involvement, like that of the so-called "social

---

[94] Petrunkevitch (*op. cit.*, p. 83) recorded: "A *Taenia solium* . . . may live from five to fifteen years and during all this time from three to five gravid proglottids are broken off daily and passed with the stool. As each proglottid contains somewhere between 20,000 and 50,000 eggs, the yearly production of eggs amounts to between 7,300,000 and 18,250,00 and the total production during the lifetime of the worm to almost a quarter of a billion (maximum)."

insects," some increase of repertoire is gained even as individuation is blocked or lost. The over-all growth and elaboration of vital activity goes forward, sometimes by processes of individuation, sometimes by the opposite, the amalgamation of individual lives in a joint existence. Nothing in inanimate nature can serve as a model of this aspect of vital process. Life is so opportunistic that every possible avenue of implementation and continuation is exploited, and evolution seems to progress by movements that sometimes are dialectical dynamisms, individuation and involvement pushing each other to growth of activities. In human life, corporate acts are the most spectacular assertions of the species, extending its ambient even beyond the terrestrial surface; but they spring from the most individuating element in each brief life, the mind, and as soon as individuation is seriously frustrated, they also fall apart.

Though we have no physical model of this endless rhythm of individuation and involvement, we do have its image in the world of art, most purely in the dance; for this dialectic of vital continuity is the very essence of the classical ballet. Think only of that perfect example, *Les Sylphides*: individual figures emerge and submerge, *pas de deux* develop and melt back into the web of choric movement, divisions form only to close over what was, for a moment, the path of an advancing stream. And not only in dance but in all choric works of wide range this largest rhythm appears: the "tide in the affairs of men, that, taken at the full, leads on to fortune"; or, in the highest musical form that has yet been developed, the sonata, which is choric in structure whether scored for the keyboard or the full symphonic orchestra: a scarcely discernible new theme may begin a history, but even if it rises to apotheosis it can never transcend the stream, which may finally integrate it with another individual form or even simply engulf it. Individuation and involvement are the extremes of the great rhythm of evolution, which moves between them in a direction of its own, always toward more intense activity and gradually increasing ambients of the generic lines that survive. A degenerating activity is usually making way for the upsetting impetus of another kind of action; under such conditions the organism can persist only by being involved with others of its own kind or of alien kinds that vicariously perform its waning function. Thus the stock itself, which has evolved its own vital activities, may give up one or another of them in the course of its own expanding life.

> Like a child from the womb,
> Like a ghost from the tomb,
> I arise and unbuild it again.

10

The Evolution of Acts

*Nicht der Reiz, sondern die Veränderung die er
im inneren Zustand des Organismus hervorbringt
motiviert die Bewegung.*

—Ludwig von Bertalanffy

THE complexity of the numberless organic forms which have
evolved in the long course of life on the earth never fails to
impress the investigator who gives them his attention, though by practical necessity he has to restrict his special studies to a single genus, and
perhaps even to a relatively tiny part of that.[1] Our progressive insight
into physical, chemical and electronic mechanisms and the scope of
their possible functions has led to a general change of intellectual focus
from anatomy to physiology, that is, from the description of the myriad
forms as such to their description in terms of observed, or imputed, or
sometimes purely hypothetical functions. The conception of organisms
as self-fueling, self-maintaining and self-replacing mechanisms, and of
their internal structures as working or supporting parts, has opened
avenues of scientific advance in more than one direction, and made a
heroic onslaught against the mystical doctrines of non-physical agencies,
special kinds of energy ("nervous" or "spiritual"), final causes, and other
ineffable noumena supposed to invade the phenomenal realm to produce
physical effects.

The shift of interest from the great store of rather meaningless taxonomic details to functions that give some significance even to the
smallest of them has been the real revolution in philosophy of science in
our day. That conceptual change was, of course, not sudden, nor immediately recognized as a philosophical insight. In biology it was pre-

---

[1] Cf., for instance, A. H. Sturtevant's "The Evolution and Function of Genes"
(1949), p. 252, where the author says in passing: "There are about 500 known
species of *Drosophila*; of these, about eight have been reasonably well studied genetically, as many more are less well understood, and there are scattered data on still
others."

pared mainly by the classical Darwinian theory of evolution through random variation and "natural selection" of the most useful structures, which emphasized the functional aspects of organs, and consequently the importance of small changes and persistent tendencies in development as adaptations to environmental conditions.

In Darwin's own day, the hypothesis that random variation and mutual pressure for survival could account for living form in nature met with many difficulties, the main one of which was that the random element appeared to be unlimited, and the formation by chance—even unlimited chance—of any elaborate servosystems for the conduct of life, beyond all probability. This formidable problem was unexpectedly brought within view of solution by the rediscovery of Mendel's laws of heredity, which traced the phenotypical changes to the system of chromosomes with their specialized loci, and thus revealed a principle of general preservation of traits against which the spontaneous mutations, too, looked less disordered. Like the hereditary elements, they were traceable to bounded structures where their changes originated; though still unpredictable, they ceased to defy all hope of being found interconnected and ultimately understood. The convergence of the two epoch-making theories, Darwin's and Mendel's, in what Julian Huxley called "the modern synthesis," has provided at least an approach to the vast phenomenon of life on earth, ever self-renewing and producing more and more elaborate forms; for, as genetics can account with fair precision for the elaboration of intricate mechanisms by crossing and recrossing of mutant with conservative strains, so the classical doctrine of evolution contributes the concept of "survival value" of new traits for their possessor, to explain their differential continuity through further generations.

The essential rightness of Darwin's theory is strongly attested by the way it draws whole sciences into its orbit, and inspires new departures in their advances. The notion of "survival value" introduced a practical element into biological thinking that focused the attention of anatomists on physiological mechanisms and the functions whereby plants and animals conduct their lives, feed and defend themselves, keep from freezing, find mates or utilize wind, water and insect carriers to fertilize their seed, and perform all sorts of astounding acts, by virtue of which they survive, on the average, for a length of time characteristic of their species. Once the functional interpretation of all tissue structure became established, the discovery of more and more amazing mechanisms and further "servomechanisms" supporting them became the new line of advance in biology.

The Evolution of Acts

It was here that a still further conjunction of scientific theories occurred: this time, the importation of physical models from the advancing fronts of chemical and electrical engineering into biology. Despite the fact that some speculative thinkers are promptly beguiled, by the possibility of finding such models for vital processes, into seeing the human brain as a computer (a pathetically bad one by modern standards), and into speaking of cybernetic machines as "brains" or even "organisms" and of their parts as "organs," the suggestions derived from mechanics and electronics have led to revelations in physiology. The practice of such borrowing is, of course, very old; what gives it a special significance in the present case is that two fast-developing sciences are linked by it to progress together. The perfectly serious use of biological and even social terms for very complex machines shows that their operations are based on schematized biological acts.[2] As for the converse relationship, the influence of electrochemistry and other advanced physical sciences on the development of biological theory (not to mention their tangible contributions, such as high-powered microscopes, electrodes and dissecting instruments), it widens the physiologist's insight into living organs and functions with almost every new departure in purely mechanical principle or practice. Physiology, genetics, embryology, as well as paleontology and related studies, united by the basic concepts of evolutionary natural history, are building squarely on the older and riper sciences, especially by using the feats of modern engineering as guidelines and models.

There is, however, one serious pitfall in the use of technological models in conjunction with the classical concepts of evolution, i.e., survival value, adaptation and "natural selection" of the fittest. Man-made mechanisms are normally devised for some purpose; as a whole and in all their parts, which may be intricate servosystems, they are intended and designed to serve preconceived ends. To meet particular needs is what adaptation, fitness and value mean with respect to human contrivances. Consequently the notions of requirement, adequacy, conscious aim and choice of means are inherent in the thought and language of engineering, whether chemical, small-current or simply mechanical. When this language is carried over into biology, it naturally brings with

---

[2] Norbert Wiener, in *Cybernetics*, p. 55, declared: "the many automata of the present age are coupled to the outside world both for the reception of impressions and for the performance of actions. They contain sense-organs, effectors, and the equivalent of a nervous system to integrate the transfer of information from the one to the other. They lend themselves very well to description in physiological terms."

[359]

it the technician's view of mechanisms as inventions and of their functions as dictated by a presumptive user's needs. So all the considerations
of economy, time-saving, margins of safety, reserves of material, storage
and deployment of power, and the principles of coordination and communication governing our industry are read into the organic forms and
functions that have taken shape in the course of evolution.[3] In the
structural analysis of living matter, for which the technological models
are indispensable aids, this fallacy may be unimportant; but when we
come to the philosophical formulation of concepts for broad as well as
narrow and detailed biological thinking—concepts that can link all the
sciences which the "modern synthesis" has brought into the compass of
evolution theory, and can be further extended to cast psychological and
moral thinking in commensurable terms—the implicit misconception in
the formative principles becomes seriously misleading. Then it prompts
biologists to raise pseudo-problems, and to resort to teleological explanations that pass as "scientific" because they stem from a scientifically
grounded technology. So G. R. De Beer, for instance, speaks of the
"principle of double assurance," "of which the formation of the lens of
the eye in *Rana esculenta* may ... be taken as the best authenticated
case." After describing two different ways in which lens formation may
be induced in this frog (but not in *Rana fusca*), he observes: "The two
mechanisms thus involve different kinds of processes, and their existence
raises the problems as to how the 'additional' mechanism was evolved if
one would give the organism survival value against the rigour with which
natural selection would treat a lensless eye."[4] The "problems" arise only
with the assumption that biological mechanisms are servosystems developed to meet needs. Then, of course, one may wonder why *Rana fusca*
and many other species, with probably similar needs, should not have
received the same "additions."[5]

The obstacle which this economic misinterpretation of living structure
and function sets up against a truly biological conception of evolutionary
processes is increased by the fact that machines and programs are made
not only for a purpose, but by a designer and organizer. Since the assumption of a divine Creator, who might exercise the required foresight
and ingenuity, is proscribed in the scientific sphere, the analogy of the

---

[3] The effects of this practice have already been discussed in Chapter 8. For examples, see especially p. 276, n. 40. Our concern here is with its broad (and
easily comprehensible) motivation.

[4] "Embryology and Evolution" (1938), pp. 63–64.

[5] A. R. Moore, in *The Individual in Simpler Forms* (1945), pp. 71–72, cites the
"principle of multiple assurance" as a basic biological principle. Paul Weiss speaks
of "spare equipment" ("The Biological Basis of Adaptation" [1949], p. 9).

industrial plant can be carried out only with a replacement in the managerial and planning departments; and this is commonly made surreptitiously, by a literary trick of using what purports to be a mere figure of speech—the introduction of "Nature" or "Evolution" as the agent who supplies the blueprints and materials and guides the attainment of her (instead of His) purposes. This ready evasion of a difficulty which really shows up the weakness of the machine model has become the stock in trade not only of science writers, but of excellent, authoritative scientists writing on problems of adaptation, organic integration and evolutionary tendencies. The snare may be their conception of the public for which one writes on such subjects, whom they tend to regard as an utterly ignorant laity;[6] to such people one had better speak in terms of their habitual thinking, in metaphors, personifications, etc. But in fact, the serious lay reader of non-technical scientific essays, or of philosophical reflections on science itself, is trying to break away from the mythical conceptions which are still traditional in our everyday thought and speech. To be told, in the midst of a causal account of natural phenomena, that this or that course of events is what Nature has devised for the good of her children, or that such-and-such is Evolution's method of controlling them, does not help him to understand vital processes; only a superficial reader is pleased to meet those familiar, vague references to final causes, which require no envisagement of any connecting steps.

But a more peculiar and somewhat suspicious fact is that the mythical agents, ostensibly introduced to indulge the general public's literary taste, find their way into places where they will never meet that public. Readers unversed in science are not likely to choose, for instructive entertainment, a monograph on the polar rotation of molecules that constitute living matter, nor on the synthesis of new hormones, with ten-syllable names, by a mutant gene differing from its predecessor by one molecule, nor yet one on oxidative versus fermentative processes in the liberation of chemically bound energy. Still one finds, even in such literature, constant reference to nature's plans, goals and methods.[7]

[6] I have heard one of our most noted (and notable) biologists express precisely this view, and define "popular writing" as any discourse that is not at least in part mathematical; whereupon he designated Whitehead's *Process and Reality* as "popular."

[7] In A. von Szent-Györgyi's "Oxydation and Fermentation" (1939), one encounters such statements as that "Nature discovered oxydation by molecular oxygen," and that "we usually find that the way Nature reaches its purpose is the only possible way, and yet, in spite of its simplicity, the most admirably ingenious way" (pp. 172–73). Purpose and admirable ingenuity can only be imputed to an agent, which is none other than the "unscientific" and disavowed God, transvested into a goddess named "Nature" (with a capital N).

Another current device to evade theological implications yet keep the teleological outlook is to attribute purpose, foresight and even social conscience not only to primitive organisms (plants as well as animals), but also to organs, tissues or cells. So one reads of coelenterates "solving the problem" created by their increasing bulk, and slime molds "circumventing" other engineering problems by clever inventions;[8] of creatures that "succeed in evolving" new anatomical structures;[9] of flowers "learning to mimic" animals, and animals "knowing how to synthesize" chemicals;[10] of cells "inventing" their functions, and even "knowing how to count"! In a technical article, "Phyllotaxis, Anthotaxis and Semataxis" (1961), the author, E. E. Leppïk, attributes to the apical cells of a growing plant stem "a power of numerical choice" with regard to the number of buds it is to bear; "the number," he says, "is fairly constant, one numeral [sic; numerosity?] being selected by each apex.... But we do not know how the apex has the power to 'count' ...." (p. 7).

This rather recent fashion of imputing high mental functions to any and all sorts of living matter is not as insidiously confusing to the amateur philosopher (who may be a first-rate laboratory scientist) as attributing them to Nature, Evolution or other such Jove-like agent, because this sort of talk does not drag him back to his traditional way of simply referring all vast and general phenomena to divine aims instead of seeking causal explanations. The physical experiments dreamed up by jellyfish and the astonishing I.Q. of apical meristems are fairly sure to be taken as "poetic license." The interesting aspect of the new literary ruse is not its effect, but its source, which is the same as that of the old "Mother Nature" cliché; the ineradicable feeling that something must plan and direct, set goals and figure out ways and means to reach them; if not Nature, then cells, or even "Director DNA." Information is coded by one agent, decoded and acted upon by another, "input" is processed and "output" furnished "as desired" or "as needed," though we are never told who needs and desires it. Instructions go with it from spinal or cerebral centers. Only the factory manager is discreetly left nameless.

The fact that common sense is evidently unable to do without assuming some plan and method in the evolution of animate forms shows that there is, to date, no satisfactory alternative way of conceiving

[8] J. T. Bonner, *Morphogenesis: An Essay on Development* (1952), pp. 16 and 22–23. The use of such language is made a conscious mannerism in S. Granick's "Inventions in Iron Metabolism" (1953).

[9] A. Tumarkin, "On the Evolution of the Auditory Conducting Apparatus: A New Theory Based on Functional Considerations" (1955), p. 233.

C. H. Waddington entitled a book on genetics *The Strategy of the Genes* (1957).

[10] W. La Barre, *The Human Animal* (1954), *passim;* see especially p. 336.

life; no illuminating and systematizing conception of vital events, as natural (i.e., physically determined) progressions, in the course of which such phenomena as needs, values, purposes, means, successes and failures are generated.[11]

Darwin, for all his intellectual daring and reinterpretation of facts, and for all the challenge he proffered to the religious outlook of his contemporaries, did not directly challenge their common-sense image of natural law, which was that of a statute set up by God and obeyed by nature. As he viewed the vast array of plant and animal forms with which he was familiar, what aroused his scientific wonder was just what had always aroused people's religious wonder—the special fitness of each kind of organism to cope with the conditions of its existence. It was the scientific problem of how this evident pre-fitting came about in the turbulent course of nature that exercised his inventive mind; and when he found his ecological answer, he set it forth on the basis of well-known facts, which had never before tangled with any notions of divine law because they had never been seen as exemplifications of any large natural law: the results of selective breeding, long familiar in the kennel, the stable and the greenhouse. His famous phrase, "natural selection," was intended to generalize the principle on which the farmer's practices were based, and show that in nature a similar elimination of many incipient forms is constantly going on, leaving only some strains to perpetuate themselves, so that wild species were generated by differential mating just as cultivated species were produced by mating of selected stock. Darwin's aim was, obviously, to explain the screening out of some characteristics and the progressive development of others as an automatic consequence of competition among organisms for their common means of life; the fact that natural forces limited and channeled the formation of species without anyone to select the breeding stock for desired traits was the core of his theory. The phrase "natural selection" seemed a happy means to connect the genetic principles known in agriculture and animal husbandry with the new theory of evolution.

This purpose it probably served, but it also served another: after the

---

[11] A reflective psychologist, J. H. Parsons, whose explanatory theories are all in neurological terms, nevertheless finds it impossible to carry such thinking beyond his concrete investigations. In his article "Adequate Stimuli," quoted in Chapter 9 as a reliable source of scientific facts, he declares that "in the present state of knowledge it is not only undesirable, but unwarranted, to ignore teleological speculations. . . . In fact, it is almost impossible to do so; and those most averse from them frequently indulge in them unwittingly" (p. 303). Instead of "in the present state of knowledge," he might more accurately and aptly have said, "in the present state of conception."

first shock of Darwin's naturalism had worn off, the popular notion of "natural law" as a divine statute law obeyed by nature came to the rescue of religious traditions, because where there was a law, it was argued, there must be a lawgiver; and similarly, where there was any act of selection there was an agent who performed it. What Darwin had discovered was merely the mode of operation of a divinely given law to make divinely selected specimens. This piece of cultural history is well known, and needs no recounting; its important bearing on present-day evolutionary theory, however, is that it established Darwin's phrase, "natural selection," as descriptive of an act analogous not only in principle but in purpose ("improvement" or "survival" of the race) to the stockbreeder's practice.

Modern evolution theory has inherited the term; and even if no one takes it literally to denote an act of selecting, the word "selection" still has a strong tendency to make one act-like event out of the hazards and fortunes of life which constantly eliminate ill-adapted organisms and aid the better adapted to continue their lineage. Indeed, it is sometimes hard to believe that no one takes it literally, for instance when so important an evolutionist as C. D. Darlington speaks of plants that "anticipate the act of selection,"[12] and F. M. Burnet of "a blind selection for immediate survival only,"[13] as though selection were usually more wisely exercised. But the most serious influence of the phrase is a formal one; for where a choice is made—no matter how impersonally—there must be fixed forms from which to select. So, in various ways, the language of Darwin's heterodox *Origin of Species* tends to hold his successors to a traditional frame of thought. Every venture into functional thinking comes as a difficult "breakthrough" out of that frame.

One purely practical further circumstance which gave Darwin's brilliant formulation a somewhat rigid character was that all the data on which it was based were taxonomic. This condition, too, we have inherited. Adaptation is the key concept in the theory of evolution; but evolutionary history is long, so most of the materials at our disposal are fossil remains, or else distinct but anatomically comparable living stocks which have been ranged in hypothetical series of progressive development along various lines.[14] What we usually witness, consequently, is

[12] *The Evolution of Genetic Systems* (1939), p. 131.

[13] *Virus as Organism* (1945), p. 125.

[14] Compare the observations of Christian B. Anfinsen, in *The Molecular Basis of Evolution* (1959), on this state of affairs: "As long as crosses can be made between different family lines, phenotypic changes can generally be related to specific genes and the spread of these genes through a population can be fairly accurately mapped. Thus, the basic assumptions of evolutionary theory may be directly tested,

adaptedness; adaptation, the actual occurrence of changes to meet special conditions, must be inferred as having taken place in the past.[15] How it operated, and presumably still operates, has been speculatively but nonetheless hotly debated; at present, it is almost unanimously held that heredity, random mutation and "natural selection" suffice to account for the development of all plant and animal forms on the earth from their "eobiont" phases to the degrees of bodily, physiological and behavioral complexity they possess today.[16]

The patterns of heredity are certainly known in the large, at least for fairly high organisms such as flies or, in the plant kingdom, peas, four-o'clocks, corn and countless other spermatophytes; and few biologists doubt their determination essentially (if not absolutely) by nuclear genes.[17] That means that any novel trait which arises in a stock and is transmitted from parents to offspring must stem from a change in some gene or genes. In protozoa the mechanisms of heredity seem to be somewhat less simple and fixed; according to recent observations, the cortex of the cell in the little flagellate *Paramecium aurelia* has some morphogenetic powers over subsequent generations,[18] and particles in the

---

when the segment of time under consideration is small, and we are able to describe the process in terms of changes in *genotype*. When we deal with evolution on a large scale, however, the tools of the geneticist are no longer applicable. The evolutionist must now rely on the study of relative morphology and ecology as deduced from the fossil record, or on the comparative anatomy and physiology of living representatives of surviving species" (p. 212). And further: "The paleontologist, in estimating the rates and directions of evolution, must depend almost entirely on morphological evidence. . . . He is limited . . . to the *results* of evolution and can never hope to elucidate the underlying physiological changes that participate to produce new phyla" (p. 213).

[15] Cf. Paul Weiss, *op. cit.*, p. 3: "Adaptation . . . is the fitting, and adaptedness the fitness, through which a system is harmonized with the conditions of its existence."

[16] There are, of course, some dissenters, and even notable ones—for instance, William M. Wheeler. See the essay "On Instincts" in his *Essays in Philosophical Biology* (1939), especially pp. 49 ff. Also, on very different grounds, Lancelot Law Whyte's recent book, *Internal Factors in Evolution* (1965), which will come in for discussion a little later.

[17] Richard B. Goldschmidt, however, has challenged the theory of gene action with the counter-proposal that chromosomes act as a whole, though the mutation may have a minute locus. See especially his Cold Spring Harbor Symposium paper, "Chromosomes and Genes" (1951).

[18] T. M. Sonneborn, after many unsuccessful attempts at grafting cortex from one *Paramecium* into another, had the wonderful good luck to find two specimens which, in separating from their respective partners after conjugation, had taken a piece of the partner's cortex along and incorporated it; and in both cases the union

cytoplasm, which have been termed "plasmagenes," also appear to transmit hereditary traits in independence of the nuclear cistron.[19] The bulk of the inheritance, however—the real continuity of the stock—is transmitted by the chromosomes, in protozoa as in metazoa, and in all plants; though certainly the exceptions and additions, which bespeak a less restricted potentiality in the several parts of the cell, are greater in the simpler organisms. Alternative modes, entirely different expressions in form and function under different conditions, are characteristic of primitive lives,[20] and seem to yield gradually to fixed forms and strictly specialized functions with the development of larger, more elaborately organized individuations.

The occurrence of mutations, too, is a well-established fact, but beyond the patterns of their occurrence and their gross effects little is known; so little, indeed, about their natural causes that spontaneous mutations are generally termed "random." This really means no more than "unpredictable." Scientists are not likely to think that changes in gene constitution happen without causes. Only, since the causes are usually unknown (except in cases where exposure of organisms to chemical or radiant influences can be correlated with new heritable traits), mutations have to be treated statistically, and statistical methods require random occurrences. All we know of genotypes is deduced from the distribution and recurrence of phenotypes in breeding populations. But what is known in this way is astonishingly systematic and detailed, and thanks

---

had been between a singlet and a doublet animal. In the first case the singlet was the robber and survivor and in the second, the doublet. Each, under the influence of the acquired piece of cortex, gave rise by division to daughter cells of abnormal form, reflecting the character of the partner (ordinarily the mixing of singlets and doublets does not affect the "daughters," each partner breeding true to its own form), and the hybrid form became established in the descendant clone. See his paper, "Does Preformed Cell Structure Play an Essential Role in Cell Heredity?" (1963).

[19] Sonneborn, "Beyond the Gene" (1949). (In the studies of cortical influences discussed above, the action of "plasmagenes" was experimentally ruled out.)

[20] Cf. pp. 318, n. 23, 331n. As for still lower forms—bacteria, and viruses if they be admitted as organisms—we know something of their adaptability, but not the nature of their adaptive processes. As Julian Huxley has said, "The entire organism appears to function both as soma and germplasm, and evolution must be a matter of alteration in the reaction-system as a whole. That occasional 'mutations' occur we know, but there is no ground for supposing that they are similar in nature to those of higher organisms, nor . . . that they play the same part in evolution. We must, in fact, expect that the processes of variation, heredity, and evolution in bacteria are quite different from the corresponding processes in multicellular organisms" (*Evolution, the Modern Synthesis* [1943], pp. 131–32).

[366]

to the development of the ultra-highpowered microscopes the deductions can often be checked against anatomical findings. Whether "genes" are infinitesimal organs with individual functions, or (as suggested by Goldschmidt) focal parts each controlling a particular function of the harboring chromosome as a whole, and whether or not the visible bands on chromosomes are really the active "genes," the essential mechanism of heredity has certainly been identified as the "gene complex" that repeats itself in every cell of a multicellular individual. This pervasive and immense repetition is one aspect, often overlooked, of the intrinsic organismic unity which has no parallel at all in the gross functional unity of electronic machines that work faster than any brain.[21]

Mutations have been known to occur in somatic cells, but such mutations are not handed on to posterity. The whole inheritance of a new individual is packed into the nuclei of the parental gametes, and only what happens to the germ plasm before or during procreation matters to the stock.[22] Upon this minute bearer of heredity rests the initial fitness of the resultant individual to survive in a hard world, and the constitution of the stock depends on the nature of the family lines which in fact do survive. So, on each phenotype that comes to expression, "natural selection" is said to act, letting the new organism live to procreate ("selecting" it), or causing it to be extinguished without replacing itself (awkwardly termed "selecting against" it); indirectly, therefore, "natural selection" acts to shape the gene pool that is the potentiality of the stock, and establishes a statistically best adapted type.

The data for the differential ability of individuals to survive, and for its molding effects on what mutant genes can express themselves without upsetting their possessors' applecart and their own evolutionary continuity with it, are really too full and systematic to leave

[21] Leo Loeb, in *The Biological Basis of Individuality* (1945), p. 13, observes that the "individual differential" (more often called the "individuality factor") in the higher metazoa, which is the peculiar protein structure arising from the organism's individual gene complex, may be not only what sets the individual apart from others, but may also have a binding function within the organism and cause the tempi and rhythms of organic activities to be attuned to each other. The microstructures of its microscopic parts—the cells—all reflect each other.

[22] The fullest statement of "neo-Darwinian" evolution theory (with polemical insistence on its complete and sole adequacy) is presented in Huxley's famous *Evolution, the Modern Synthesis*. Early in the book (p. 16), discussing the struggle for survival and the differential transmissions of advantageous and disadvantageous genes, Sir Julian remarks: "The term Natural Selection is thus seen to have two rather different meanings. In a broad sense it covers all cases of differential survival: but from the evolutionary point of view it covers only the differential transmission of inheritable variations."

much question of their significance. Neither can the other factor in Darwinian theory—the heritable character of mutations, for good or evil, once they have occurred—be reasonably denied. Yet the theory of evolution by "natural selection" continues to be attacked as inadequate, oversimplified, "purely negative" and "too mechanical." Part of such criticism may be due to traditional attitudes and affronted feelings, and is to be discounted accordingly; but that fraction would not greatly trouble scientists writing for their colleagues, nor evoke such polemics as one finds in the works of "Selectionists" and their critics. There is something more important that is lacking or askew in our best and most systematic evolution theory, and that fault is philosophical; it lies deep in the conceptual structure which has been built up, by fortuitous, unrecognized steps, on the peculiar Darwinian vocabulary of "traits," "species," "adaptation" and "selection."

The gist of the philosophical failing is that in these terms the environment figures as the principal agent, while the organism is given an essentially passive role. Chemical processes (which Darwin did not know) endow it with shapes and colors, size, mobility and organic functions acceptable or unacceptable to the environment; it consequently is more or less adapted to an "ecological niche." In competition with other organisms, of other kinds as well as its own, the "more or less" of adaptedness is what matters; the less adapted will be reduced and gradually eliminated, leaving the better adapted to continue their lines. And, as heredity and mutation constantly produce new relative fits and misfits, there is a constant reviewing and culling of forms, which results in some really amazing adaptations; fishes that brood their offspring in their mouths, moths that sit on the bark of trees in such positions that their wing patterns line up exactly with the bark patterns and make them all but invisible, and countless other improbable but actual wiles.

Since Darwin's day, and especially with the growing exploration and understanding of genetic inheritance, the theory of evolution by "natural selection" of the best adapted forms among all those which appear in the history of self-continuing stocks has been highly developed and modified,[23] tested and established on many fronts. But the picture it

[23] The simple concept of hereditary traits, for instance, has been replaced by the more complicated but more accurate concept of inherited potentialities. Cf. the discussion of "potentiality" as a biological concept in Chapter 8, p. 262, especially the statements by S. Spiegelman and C. H. Waddington, respectively (n. 9), which illustrate the perfectly literal use of that concept and, indeed, its indispensability in their science. The naïvely assumed fixity and definiteness of species has given way to avowedly tentative and pragmatic definitions. See, e.g., Huxley, *Evolution*, p. 156: "It is . . . the existence of different kinds of species and of different

draws of the evolution of high forms of life from lower ones remains the same: environmental forces act in concert on the products of randomly varying procreation to select viable phenotypes, and so, indirectly, mold the genotypes by balancing the gene pools of the earth's plant and animal populations. I believe it is this picture—the image of life, not any working model—that is inadequate; and to amend that is a task in philosophy of science, for it requires no new factual data, but a reformulation of the basic conceptual scheme in somewhat different terms, to let the organism count for what it is, i.e., as the agent.

The conception of vital events as acts, or natural processes having a typical dynamic structure, automatically initiates just such a change. Indeed, it does something more than to improve appearances: it leads biological inquiry smoothly from the evolution of organisms to the development, in the zoological realm, of behavior, or spontaneous acts of an animal as a whole, probably with the first glimmering of a psychical phase, or perhaps only some momentary, protopsychical approach to such an intraorganic event. This transition is hard to make in the terms of classical evolution theory, where behavior has to be treated as part and parcel of the phenotype, "selected for" or "selected against" and, in the positive case, adapted by chance to some "ecological niche."[24] As one goes on to the higher forms of feeling, especially in human life, it becomes more and more difficult to extend the scientific frame without breaking it. But the act concept erects a different scaffolding for the same researches. The basic terms it logically provides are act and situation, motivation and actualization; in these terms potentiality, impulse, activity, rhythm, dialectic, entrainment and other essentially biological notions may be defined, as some already have been in the foregoing chapters. And just as the beginnings of life on earth fall within the

---

degrees of speciation within each kind, which makes it difficult to give a satisfactory definition of a species, and makes us sometimes wonder whether the term itself should not be abandoned in favor of several new terms, each with a more precise connotation. However, . . . the term species has a practical as well as a theoretical aspect," and so forth. The greatest improvements and refinements have been made on the concept of the gene, which has developed from the simple conception of "one gene—one characteristic" to the present ever-growing conceptual edifice that no one can sum up in a brief passage. The rest of this chapter will have to furnish evidence for it.

[24] Cf. H. Kacser, "The Kinetic Structure of Organisms" (1963), p. 39: "The essence of organisms is process. The measurement of static properties conceals their dynamic nature. However, when we carry out the analysis of the processes themselves, relations are revealed which are the foundations of behavior which we call biological."

compass of that conceptual structure, so does the whole evolutionary pattern of generation and differential survival, which underlies the phenomena of mind for which the framework is constructed.

At this point it is unavoidable, I think, to introduce a relational concept that lends itself to more precise uses than "selection" in tracing and describing what happens during the growth, life and procreative success or failure of individuals; that is, a general concept from which many special relations and logical constructions can be derived, and such phenomena as "natural selection" can be analyzed in more literal terms. Although I am reluctant to coin words (a favorite technique of pseudo-science-making), I have resorted to the new word, "pression." The semantic virtues of that term are that in the first place it is not a current word with some previously fixed connotation, yet is an obvious substantive, with a significant "root"; and second, it is capable of all sorts of specifications that do yield familiar words, most of which are definable and usable in the biological realm: impression, expression, compression, repression, oppression, suppression, all capable of grammatical transformation into active and passive verbs or adjectival forms; besides the quantitative term "pressure." "Pression" is a general designation for a class of relations which obtain between situations and acts: those relations that determine the form of an act in the course of its development, i.e., beyond its determination in the generating impulse, and conversely, such as shape a situation for subsequent or sometimes concurrent acts. The advance of life is a fabric of burgeoning acts, in literally billions of pressive relations which automatically adjust the elements of that incredibly complex dynamism to each other, so that it exhibits itself as an inscrutable matrix of "living matter."

Before taking up any more detailed discussion of pression, however, another, even more essential relationship between acts and the situations in which they develop should be recognized: that is implementation. Every act requires some support from its environment, be it intraorganic or extraorganic. Breathing is, and must be, implemented by a constant availability of oxygen. The metabolic processes have to be fed continually[25] with their special and essential metabolites. The consummation of acts—indeed, every moment of their proceeding—depends on opportunity given by steadily moving situations that keep pace with the evolving acts. The degree to which an act requires implementation is variable; the very high activity of cell division seems to require little material aid, once the cell has reached a stage of internal imbalance that motivates its

---

[25] Some, perhaps, continuously; but many appear to be rhythmically intermittent.

division[26]—unless an electrical charge or graded release of some stored energy, unknown to us, is its "supporting cause";[27] whereas normally acts of hearing, vision, tactual perception or of manipulation require a relatively great and constant implementation. Since an organism is built up by its own acts, its whole structure naturally reflects this need, which is inherent in the act form itself. As for those intraorganic events which seem quite self-contained, acts of differentiation going on in single cells, their implementing situation has been prepared before they begin; a cell may well be in a state of awaiting only one outside act, such as the synthesis of a sufficiency of one enzyme, to trigger its mitosis.

Motivation (at the organic level often appearing as "induction"), implementation and the many forms of pression are the influences that produce and shape the evolving system of acts which we see as a life, and in larger extent as the life of a stock. Evolution is primarily a development of acts, and secondarily of taxons; and all the principles of evolution spring from the nature of acts. Each separately distinguishable act evolves from its patterned impulse, through all the modifying exigencies of its course, to whatever consummation it achieves. Each situation evolves, with greater and greater complexity, to give more scope to subsequent acts than previous ones had. Every individuation, or ontogenesis, is an evolution. The evolutionary pattern is inherent in acts, and in all the complexes they form: lives, populations, stocks, and finally the whole history of life on earth that we usually mean by "evolution."

This being the case, the so-called "mechanisms" of evolution have their prototypes in the processes that beget and regulate all acts. The continuity of the vital matrix rests on the rhythmic self-renewal of its activities. In the larger advance of life, however—that is, the stock—this continuity, which is a result of the basic motivation pattern, is complicated by the ebb and flow of successive generations; and here the self-renewal is not a relatively simple repetition of acts, but breaks up, at the

[26] See Norman Anderson, "Cell Division. Part One: A Theoretical Approach to the Primeval Mechanism, the Initiation of Cell Division, and Chromosomal Condensation" (1956), p. 180: "In the most general case, cell division may be considered as resulting from an imbalance between the non-growing nucleus and the growing cytoplasm. If a substance X is produced in proportion to the mass of the cytoplasm, but is utilized in proportion to the mass of the nucleus, then it will vary rhythmically in quantity, reaching a peak before each cell division."

[27] It is interesting to note that Aristotle's concept of "supporting cause," which proved to be inapplicable to molar physical movement, for which he had postulated it, may have a different meaning in biophysical (i.e., mainly electrochemical) terms. Aristotle was by nature a biologist, and intuitively construed the whole world in terms of life, far beyond any explicit knowledge he could have had of vital phenomena.

starting point of each procreation, into a new pattern of impulses, each one of which stems, nevertheless, from the parent matrix of ongoing or potential activities. This extremely complex and varied phenomenon of rejuvenescence brings us to another characteristic of acts, which is of their essence, though it is seldom remarked except in rare and striking exhibits.

Total acts, whether relatively simple or unfathomably involved and synthetic, may expand or contract, without losing the unity of their over-all form that was determined in the primary impulse. Although observations of this general fact are few, there is at least one very searching, experimentally grounded study of acts as persistent gestalten throughout their growth from impulse to consummation and decline, no matter what subacts they may comprise or what superacts they may enter. That study, entitled "Über Handlungsganzheiten und ihre Bedeutung für die Rückfälligkeit," was published in 1933 by Georg Schwarz, then working with Kurt Lewin.[28] To my knowledge, it has not been superseded. Schwarz's findings lend strong empirical support to the act concept I have developed from an entirely different source, namely, the image of life that appears as the "livingness" in good works of art. Schwarz deals only with behavioral acts on a macroscopic level. He presents a careful analysis of such acts into elements which are themselves complexes, engendered either by articulation of subacts in the course of the evolving behavioral unit, or by entrainment of weaker acts of separate provenance by the greater dynamism. Such a superact, which swallows lesser acts already in process, Schwarz calls *"eine Mantelhandlung,"* a "mantling," i.e., covering or enfolding, act.[29] He observes,

---

[28] The article was written in the course of participating in a somewhat differently oriented research, conducted by Lewin, on the facts and implications of backsliding in the process of changing acquired habits. That experimental and analytic work revealed the interesting forms and relations of acts in general, reported by Schwarz. The experiments on which Schwarz's findings are based were made with a rather simple apparatus, a slot machine into which the subject dropped a colored marble, which he had selected for color from a varied collection, then pressed a lever that delivered the marble from a slot at the other end of the machine, where the subject had to catch it and lay it on a formboard to build up a simple pattern. Directions could be varied to demand working with one hand or both, the dexterous hand or the other, etc., and acts could be repeated to the point of becoming automatic, concatenated, rhythmic, and then varied in part or in entirety.

[29] "Über Handlungsganzheiten und ihre Bedeutung für die Rückfälligkeit," p. 149: "In der Regel folgen einander [anfangs] die Entschlüsse, Greifhandlungen, Augenblicke des Abwartens, Überraschungserlebnisse als ziemlich deutlich getrennte Phasen des Gesamtprozesses. Nach den ersten 2 oder 3 Würfen . . . pflegen bei allen Vpn. [Versuchspersonen] folgende Handlungsganzheiten vorhanden zu sein: Das Wählen

furthermore, in an early part of his paper that the unity of acts is not determined by a goal, but by the impulse, in the case of behavioral acts the intention, which he calls the "quasi-need" for their performance, and that subacts, though they be only contributory elements in larger wholes, yet have the typical structure of rise, consummation and close.[30]

This article will command our attention again, for it corroborates not only the validity of the act concept, but also the theory of rhythm, and a principle still awaiting discussion, the segregation of subacts in the accomplishment of phasic shifts and other progressive functions. For the present, it is the unity and persistent self-identity of the act that are of immediate importance, because they underlie its power of expansion in opportune circumstances, and of contraction, sometimes to no more than a complex of balanced tensions, under many forms of pression.

The growth and progressive integration of acts is generally recognized; but for the converse phenomenon—their reduction, which may amount to suspension of all movement for indefinite time—one has to turn to

---

der Kugel, das Einwerfen, die Hebelbedienung, das Fangen und das Legen der Kugel. . . ." After some interesting protocol statements on the shift of gestalten in the course of repeated trials, the account continues (p. 153): "Wenn jetzt mit Wurf 28 durch eine speziellere Instruktion über die Verwendung der Kugeln Wählen und Legen entscheidende Bedeutung gewinnen, bahnt sich eine neue Aufbauform, die *Mantelhandlung* an, die schliesslich mit Wurf 53 f. eindeutig ausgeprägt ist. An dieser Stelle liegt der Höhepunkt des Gesamthandlungsverlaufs überhaupt."

[30] *Ibid.*, pp. 144–46: "Die seelischen Energien, die zu Fehlhandlungen beim Umgewöhnen führen, stammen . . . aus dem als Wirkung des Vornahmeaktes entstandenen Quasibedürfnis zur (richtigen) Handlung. . . . Auf dem Gebiete der Trieb- und Willensvorgänge z. B. kommt den Handlungsganzheiten—ihrem Ausmaas und Aufbau, ihrer Entstehungs- und Veränderungsweise—eine grundlegende Bedeutung zu. . . . Dass die Einheit der 'Leistung' oder des "Werkes' kein hinreichendes Kriterium der natürlichen Handlungsganzheit ist, beginnt immer mehr Anerkennung zu finden. . . . In unseren Untersuchungen wird der Begriff der Handlung nicht von vornherein auf den der "motorischen Bewegung' beschränkt. Vielfach . . . bildet die motorische Bewegung nur ein sehr unwesentliches Moment; auch ist der Aufbau der Handlung, etwa ihre Gliederung in Unterganze, im wesentlichen durch Entschlüsse, Erwartungen, Hoffnungen, Schwierigkeiten und ähnliches bedingt. Handlungsganzheiten geben sich typisch als Geschehnisse, für die derartige 'innere' Vorgänge neben der blossen Motorik entscheidende Konstituentien sind. . . . (Dabei erweisen sich letzthin nicht motorische Handlungsganzheiten, sondern die zugrunde liegenden dynamischen Spannungssysteme als entscheidend.) Ovsiankina erwähnt auch die Gliederung der Handlungsganzheiten in Unterganze und zeigt, wie diese natürlichen Unterganzen sich dynamisch, z. B. in der Tendenz zum Abschluss, funktional ähnlich verhalten wie die umfassenden Ganzheiten. Lewin hebt die Bedeutung von 'Einleitung' und "Schluss' als typischen Handlungsphasen hervor."

biological, psychological and pathological observations. Most of the vital functions never stop entirely during the course of a life. They may undergo temporary or permanent transmutation into what looks like an entirely new and original act from an unrelated source, though it actually has arisen by functional metamorphosis, a change of dynamic pattern just as permutational and continuous as the structural changes literally denoted by "metamorphosis," in which we have learned to recognize orderly rearrangements of matter. Those dynamic changes are mileposts, as it were, in a continuum of acts, each act in the process of self-completion, each pressing for expansion, engulfing whatever substance will serve to implement it and meeting either entrainment or repression by other acts, or exercising such pressions on others. A completed structural metamorphosis is a consummated superact. But it does not truly mark the end of the activities that have recorded themselves in it; they were noticeable as a transient articulation in the matrix of acts while the new instar took shape, and at its completion they sink to low intensities, continuing as minute cycles, and perhaps being dispersed to other functional units in the new phase they have prepared.[31]

The degrees to which acts may be reduced are astounding. They may even fall below the threshold of impulse, to the level of potentiality, which is a biochemical level. For purposes of observation and understanding, it is fortunate that "form follows function" in nature as in art, because it is only in the relatively stable projection of material forms and properties that we can deal with very small, very slow or otherwise elusive events. But cytologists today are quite aware that the "properties" of biochemical structures are essentially functional characteristics, potentialities, which (to fall in with today's jargon) are "coded" in their molecular constitutions. Consider, for instance, E. N. Willmer's summary of protozoan chemistry: "The picture which emerges . . . is that at the evolutionary level of the protozoa, the foundations of the biochemistry of living matter are largely complete. Biochemically speaking all subsequent evolution has been in the nature of variations on a theme or themes already developed. Contractility, the mechanism for conductivity, excitability, ciliary movement, pinocytosis, amoeboid movement, photosensitivity, are all there in addition to the basic qualities of

---

[31] Cf. Davenport Hooker, *The Prenatal Origin of Behavior* (1952), p. 1: "It must be borne in mind that the development of activity is always a continuous process and that any division into phases or stages is entirely artificial." Not entirely, perhaps; the cadence that closes an act is a natural termination. The artificial division is only produced by regarding it as absolute, ignoring its motivating force that holds the finished act to the life pattern—not to mention its countless contemporaneous involvements with other acts.

living matter, growth, reproduction, respiration, metabolism, excretion, etc."[32]

In a slightly earlier passage, Willmer also made a remark which is directly relevant to our present topic: "The environment in which the activity of such an organism [i.e., a protozoan] is possible may have to be rather special and the organism may have already developed some resistant form into which it can retire when the environment becomes unsuitable."[33] Such a form is, of course, an expression of what the organism is doing: reducing its vital activities to such a degree that its periphery shrinks to the minimum of surface outwardly exposed and having to be inwardly maintained, i.e., a sphere, and its inward changes may even come to perfect rest, preserved only as a biochemical complex, a suspended internal situation. In such a waiting state, some microscopic lives appear to have remained unbroken for millions of years, and resumed their rhythms of vital impulses when a new ambient, fit to implement their actualization, was given again.[34]

The special relevance of this reducibility or contractility of acts to the conception of life as a progression of activities, rather than a series of anatomical structures made according to a given "blueprint," varied by chance and "selected" mainly by mischance, is that it suggests a process whereby vital activities themselves may be inherited via genetic transmission. Activities as here defined, i.e., dialectically concatenated series of acts (such acts being aptly pictured as "cycles"), even in countless conjunctions and mutual modifications, may be reduced so they are nearly or wholly suspended, if they have only a minimum ambient to implement their progress; they may contract to no more than a complex of electrochemical tensions or sustained patterns. In this state a concentrate of the uniquely structured, enzyme-controlling processes that characterize a particular organism may be physically separated from it to start a new life, alone or with a conjugate in which a corresponding stream of activities is temporarily stowed, chemically encoded, ready to

---

[32] *Cytology and Evolution* (1960), pp. 398–99. In a similar vein C. H. Waddington raises the question "how such-and-such particular biological properties operate." See his Foreword to B. C. Goodwin's *Temporal Organization in Cells* (1963).

[33] *Ibid.*, p. 397.

[34] Such bacteria have recently been found, first in Europe and then in America, in the deepest strata of salt mines, completely insulated by rock salt. Some of these deposits were laid down as long as half a billion years ago; the recovered bacteria, therefore, are not "modern" organisms, but probably very nearly "eobionts" of the primeval ocean, pools, or mud. For an excellent account of their discovery and characteristics see H. J. Dombrowski, "Lebende Bakterien aus dem Paläozoicum" (1963).

expand into impulse and action. In this way a germ cell carries a "genetic code," not as a "blueprint" to be followed or a set of "instructions" to be obeyed, but as an organically engendered crowd of suspended activities ready to resume their advance whenever possible, in any subsequently possible ways.

The most essential and ubiquitous kind of pression in the whole realm of life is the self-expression of impulses, i.e., acts. Every impulse presses to actualization. Every act, in the course of development, expands as far as its initial impulse and the buffering impulses of acts entrained by it or implementing it can press, and as its situation permits. Since the most immediate part of its situation is formed by other acts in process or impulses pushing toward actualization, it gathers their impressions as it moves to its own consummation. Its form, prepared in the impulse, evolves under impression and compression in conflict with contemporary acts, and in the confines of the situation established by its entire past of former acts—back to the beginning of life.[35] If the conditions it encounters do not let it come to consummation as a whole, it is repressed; it may be consummated belatedly, or some of its elements may find their own expressions or be dispersed to other superacts.[36] If it never, or

[35] There are not many discussions of unspent impulse in the scientific literature of genetics or physiology, presumably because such impulse lends itself to no exact tracking or measuring technique. In somewhat freer, behavioral studies, however, its existence and influence are sometimes noted; and such observational material may help one to evaluate conceptual formulations designed for its interpretation. See, for instance, E. von Holst and Ursula Saint Paul, "Vom Wirkungsgefüge der Triebe" (1960), p. 417: after a discussion of experimentally (electrically) induced conflicting impulses and their mutual alterations, these authors say: "Wir sind damit beim nächsten Typ angelangt, dem *Unterdrücken*, das weitverbreitet ist. Wir verwenden diesen Ausdruck dann, wenn ein Verhalten ein zweites unsichtbar macht und sich doch zeigen lässt, dass der latente Drang nicht annulliert werde.

"Das unterdrückte Tun wird als 'Nachexplosion' doch noch sichtbar; sein Drang muss also latent bestanden haben, aber irgendwie auf dem Wege zur Motorik blockiert worden sein.

". . . Im ganzen ergibt sich aus solchen Versuchen, dass bei den einzelnen Verhaltensweisen—ähnlich wie bei den Stimmungen—eine Aufstellung antagonistischer Paare meist nicht gelingt; es gibt gegenseitiges Unterdrücken (Verhindern) in sehr verschiedenen Abstufungen. . . ."

[36] Konrad Lorenz is obviously dealing with the same phenomenon of repressed impulse in his symposium paper, "The Comparative Method in Studying Innate Behaviour Patterns" (1950), when he sets up the hypothesis *"that some sort of energy, specific to one definite activity, is stored up while this activity remains quiescent, and is consumed in its discharge"* (p. 248). I doubt the existence of different kinds of energy; but the patterned impulse seems to be a biophysical reality which answers to his finding of a charge that will spend itself only in a particular sort of act. Cf. Chapter 8 above, pp. 300 ff.

scarcely, gets under way from its central impulse at all, it is suppressed, and its abortive dynamism adds itself to the unanalyzable matrix of the agent.

The pressure of billions of impulses, ever pushing to actualization in every single organism, entering or failing to enter the moving stream of acts that constitutes the life of the agent, and beyond the agent, the stock, and enfolding the stock, the whole teeming life process on earth, is a force that fairly defies imagination. Perhaps that is why Evolutionists do not imagine it. But their failure to do so has resulted in the "mechanical," "negative" and peculiarly passive conception of organisms which are, *ab ovo*, determinate future entities being fitted into "ecological niches" by an outside agent, the environment, that performs the acts of "natural selection." The use of the word "niche" for an ecological ambient strengthens the classical image. If, however, one regards the hereditary complex of vital rhythms as the initial datum, which consists essentially of dynamic elements—potential impulses,[37] coded in chromosome structures in ways which are not yet understood—the fact that the actual evolution of life proceeds by pressions becomes self-evident, for life is itself a central pattern of expression, in which each impulse determines its own obstacles, i.e., its own counter-pressures, intraorganic or external. In this way, each agent defines its own ambient, which is not the environment it shares with other beings, but is a unique tangle of pressive forces that encircles its acts, through which they wind or push their way, and in midst of which they also find implementing and promoting conditions which yield to their advance and take their imprint.

As a result of starting with the very static traditional projection of the evolutionary pattern, biologists today are but gradually finding out the dynamic, elusive and intrinsically directed character of native impulse, the extent to which early activities prepare not only the vital mechanisms but even the habitat for their own growth and completion and continuance, the processes of adaptation to opportunities offered by the ambient as well as to its dangers. The concept of a genetically predetermined zygote that is like clay in the potter's hands—the potter being "the Environment" or "Nature" or even "Natural Selection"—is giving way, inch by inch, to a much more difficult but also more adequate view of ontogenetic progressions, which certainly presents the phylogenetic picture, too, in a different perspective. So T. M. Sonneborn, working with the

---

[37] An impulse is a potential act; a potential impulse is a "competence," in the sense generally used by geneticists and embryologists. As impulses take shape, competence is lost.

relationships between nuclei and other parts of developing cells—especially nucleus and cortex, through the complex intervening cytoplasm—discovered that the immediate potentialities of cellular actions change from minute to minute with the state of the cytoplasm, especially the precise distribution of its inclusions, and that as cellular acts constitute the formation and growth of tissues, organic structure is far from preformed in the gene complex, though its central, lifelong impulses and widest limits are established there.[38] A. H. Sturtevant, in an older study of gene action, had already broached the problem of the variety of effects that could result from homologous genes in different and, indeed, highly differentiated cells of one organism, and proposed that differences in the substrate (which might be due to neighborhood, or other systemic or environmental influences) could radically alter the expression of genetic activities.[39] These are but a couple of randomly chosen examples. The evidence of such complexities grows constantly with the progress of physiological, and especially genetic, research: on pseudo-alleles, position effects, coincident mutations that may modify each other's influence, multipotential genes, plasmagenes, pleitropy, multigenic controls of development. Every discovery makes the living organism look less like a predesigned object and more like an embodied drama of evolving acts, intricately prepared by the past, yet all improvising their moves to consummation.

A closer study of the nature of acts suggests systematic explanations of many puzzling phenomena, such as the elaborate timing of organic processes, the apparent pre-fitting of organs for future functions, the fre-

[38] In his essay, "Does Preformed Cell Structure Play an Essential Role in Cell Heredity?," p. 217, Sonneborn writes: "Preexisting structure determines processes that lead to different structures and different processes in sequences that are self-determined at every step and that lead back cyclically to the starting point. This dynamic interplay of structure and process contrasts with the static view of an unchanging, persistent, fundamental ground substance or organization of the cytoplasm which always underlies the developmental and regenerative capacities of the cell. . . . Present knowledge rather indicates a cyclically changing scene dependent at every level on preexisting structure, from the preformed enzymes and ribosomes that are indispensable for genic replication and action to the organized cortical fields, gradients, and localized inductor response systems required for morphogenesis and hereditary transmission of cortical structure and action systems."

[39] *Op cit.*, p. 262: "Every cell of an organism has, in general, the same set of genes as the other cells in the same individual. The problem that arises at once is: how is differentiation possible? . . . the various parts of one organism are under gene control, yet the same genes are present in all these parts. This may mean . . . that in any given part of the organism all the keys are present but only some of the locks; in other words, that the outcome is determined by which substrates are present rather than which gene-controlled enzymes."

quent convergences of form in entirely different kinds of animals or plants, the semblance of orthogenesis in some evolutionary lines which is, nevertheless, not consistent or general enough to convince most biologists of a special tendency in all living things toward the perfection of some particular pattern.[40] Finally, the great central fact of development itself derives from the structure of acts and the radiating progressions engendered by that structure.

The fact that each act arises from an impulse in a highly organized protoplasmic complex gives it a primitive impetus which often carries it through sudden adverse conditions to a normal expression. This is well illustrated, on the level of somatic activities, in cases where elaborate but integral acts such as mitosis or procreation have been observed during radical changes in the implementing situation, and found to run safely to completion, apparently by the gathered energy, or impetus, behind their original motivation (induction).[41] What cytologists thus witness in miniature may happen on a great, super-individual scale to acts formed eons ago, no one knows at what juncture in the history of the race, and perpetuated as hereditary tendencies; so that many plant or animal stocks living under identical conditions of moisture, temperature and atmosphere, having essentially the same natural enemies and

---

[40] Again, Julian Huxley is the most categorical objector; quite early in *Evolution, The Modern Synthesis*, he says: "Quite bluntly and simply, all such assertions are unjustified" (p. 40). But subsequent arguments against the notion of orthogenesis show at best that it is not necessitated by the facts which are adduced to support it.

[41] See Dorothy Price (ed.), *Dynamics of Proliferating Tissues* (1958), p. 33 (citing Dr. C. P. Leblond): "When an individual dies, the mitoses that are started go to completion, but no new ones begin. If the tissue is not taken soon after death . . . there will be no mitoses." Also Allison L. Burnett, "The Maintenance of Form in Hydra" (1962), p. 48 (reporting an experiment in selective cell destruction by nitrogen mustard solution): "Animals that have begun to bud prior to mustard treatment will complete the budding process after removal from the mustard. If budding has not yet begun, however, this process is inhibited after treatment with mustard." In other experiments, where the active "growth region," which inhibits growth in its vicinity, was grafted to the "budding region" of a hydra not in process of budding, it was found that "budding will not occur until the head has, through growth processes, become removed some distance away from the budding region." But "if a growth region of a hydra is transplanted adjacent to a budding region which has already begun bud formation there is no inhibitory effect whatever" (p. 32).

Another indication that an act, once launched, has an inherent impetus may be found even on the genetic level; E. N. Willmer, in *Cytology and Evolution*, p. 126, remarks that "the activities of different sets of genes are called out by particular events in the cytoplasm, and once they have been called out, it becomes more difficult to suppress them and to evoke the activity of an alternate set of genes."

perhaps even drawing on almost the same food supply, are still enacting very different early impulses that have become more and more rhythmicized, and hence dialectically self-renewing; impulses that override minor obstacles and rivalries, and that shrink, in the procreative crisis, to molecular tensive patterns within single cells and even infinitesimal parts of cells—parts of chromosomes—and expand again into effective activities with each new individuation. It seems that in the reduced condition before a rejuvenation impulses are intensively built up but unspent, so that a young organism has a vast competence to be gradually determined, chiefly under the leadership of a few persistent or unequivocally encoded actual rhythms.

The impetus of an act is certainly the main source of its persistent advance as a recognizable unit even when complications largely mask its form or, sometimes, when its normal course is distorted by abnormal circumstances. The "displacements" of instinctive activities studied by ethologists and the symbolic substitutes for suppressed acts which psychoanalysts try to trace back to their central impulses are cases in point. At lower levels one finds examples of distorted consummation, still traceable to obvious impulsive beginnings, in the cytologist's and the neurologist's laboratories.[42] But the primary impulse alone does not account for the whole history of a far-expanded or long-sustained act. Certainly no impulse could spend itself through millions of years in the continuous vital activities that maintain a stock. Large public acts, too, such as

---

[42] See, for instance, P. Vogel, "Beiträge zur Physiologie des vestibulären Systems beim Menschen" (1932), p. 21, describing alternative reactions to galvanic stimulation of the vestibular nerve, which were either the normal responses of bodily displacement (leaning, falling) or, where these were inhibited, experiences of vertigo, i.e., visual illusions of motion. "Unsere Vpn. bekamen den Auftrag ganz steif zu stehen, ihre Haltung willkürlich zu fixieren. Applizierte man dann der Vp. einen Reiz, der in gewöhnlicher Rombergstellung eine deutliche Fallreaktion hervorrief, so trat statt dieser das Erlebnis einer Scheinbewegung auf." A footnote (p. 22, n) adds: "Die beiden Phänomene haben eine Tendenz sich gegenseitig zu verdrängen, sich allein durchzusetzen."

On the still lower level of somatic impulses, consider the report of H. S. Burr, "Some Experiments on the Transplantation of the Olfactory Placode in Amblystoma, I: An Experimentally Produced Aberrant Cranial Nerve" (1924), in which he describes the result of transplanting an extra nasal placode to the side of the head of Amblystoma embryos. Nerve fibers from the transplant, it appeared, might follow any one of four courses: (1) fuse with the olfactory nerve and make one larger unit —this happened less than half of the time; (2) in somewhat more than half, an aberrant nerve coursed right through the brain case to the diencephalon; (3) fibers from the transplant joined the ophthalmic division of the fifth cranial instead of the olfactory; (4) some may get lost in mesenchyme around the eye (near the place of the transplant).

making war, cannot be derived from one central impulse without involving further aspects of the basic act pattern; and although the consideration of such developments belongs to a later part of this essay, the principles of their formation lie in the nature of acts, and are operative in much earlier phases of life.

Dialectical concatenation into rhythmic series is, of course, the first higher complication that comes to mind: the cyclic repetition which makes "activities" in the sense adopted here. This has already been discussed. Each act, or "cycle," in such a series motivates[43] its successor by completing itself. But in the beginning of a life, as activities emerge again from their extreme reduction in the germ, something more than recurrence takes place in the course of such a rhythmic progression of acts: their passage tends to accelerate, and the cycles to become more precisely alike. Both tendencies stem from the fact that every act makes some impression on the system in which it occurs; that is to say, it sets the stage in some special way for subsequent events. As a rule it clears a path for its own passage, and thereby for that of any other act very much like itself. This pathfinding process is generally called "facilitation." Naturally enough, an act facilitated by a precursor is apt to run its course somewhat more rapidly than one that finds no obvious track to follow. Since cycles of organic activity—metabolism, respiration, circulation, etc. —are facilitated as well as motivated by their immediate predecessors, they are inclined to quicken unless some limiting condition in the system holds them to a maximum speed (as is sure to happen in an organism).

The same influence—the impression of past acts in the maturing system—accounts for the increasingly standardized form of the recurrences, because the readiest way to their consummation is to fit exactly, move by move, into the prepared form. It may also account for the fact that developments once fairly started sometimes continue, and undergo quite excessive elaborations which embarrass the devout Selectionist who can find no survival value for them, and cheer the Orthogeneticist because they so obviously express an inherited unadaptive tendency.[44] Such phe-

[43] Strictly speaking, every act is motivated by a whole organic situation; but often one act or circumstance (e.g., the presence of a catalyst substance) is paramount in making that situation, so in a somewhat looser but convenient sense one may well speak of a single event as "motivating" an act or act complex.

[44] The most extreme manifesto for the orthogenetic doctrine that has ever come to my notice is J. H. Schaffner's "Orthogenetic Series Involving a Diversity of Morphological Systems" (1929), in which he shows, by beautiful drawings, six examples of plant families that can be arranged (or better, of which selected species can be arranged) in a series with respect to some trait that seems to develop without any particular survival value; and he does not hesitate to say that anyone who

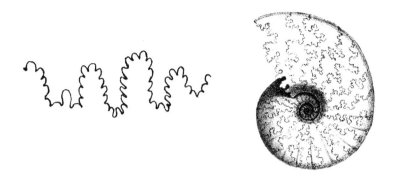

Figure 10–1. Ammonites

the spectacular evolution and devolution of ammonite sutures

(*Top,* from F. J. Pictet, *Traité élémentaire de Paléontologie* [Geneva: Jules-Guillaume Fick, 1845], Vol. II; *Bottom,* from Franz Ritter von Hauer, *Beiträge zur Palaeontographie von Österreich* [Vienna: Eduard Hölzel, 1858].)

nomena as the spectacular evolution and devolution of ammonite sutures (Fig. 10–1), and the well-known progressive exaggeration of saurian armor beyond any possible benefit to its bearers, certainly bespeak a tendency of impulses dominantly established in the gene pool of a stock to express themselves more and more characteristically as long as their

---

even admits the possibility of survival value is indulging in "the same kind of credulity as is necessary for a belief in fairy tales" (p. 53). He traces the development of the "alabaster box," *Euchlaena mexicana* Schrad., in which, he declares, "the point of perfection of workmanship has been attained. . . . There is one further possibility. The surface, instead of being smooth, might be ornamented. . . . In the higher plants, ornamentation is commonly an accompaniment of the evolutionary movement of induration" (p. 50).

actualization is ecologically possible. But it is rarely possible for long—so rarely that ammonites and ankylosaurs are famous. If the champions of orthogenesis build their evolutionary theory on such cases, it rests on slender supports, being founded on exceptional rather than normal facts.

Yet the much-decried principle does exist; only, it is not peculiar to phylogenesis. It is the principle of continuity characterizing all acts, the effect of their original impetus, on which internal and external pressions operate. As a rule it is heavily masked by limiting conditions, especially the processes of adaptation and ecological competition. What the spectacular overdevelopments teach us is that in the fight for survival the tendency to continuous elaboration of started activities is always there; and it is unfortunate that serious study of such implications is often discouraged by fear of an epithet—"orthogeneticist," "teleologist," "Lamarckist," or any so-and-soist that nobody wants to be.

The acceleration engendered by long repetition may also have some important consequences at a deeper evolutionary level, namely, for the building up of the first organisms, because (if one may trust physical models) rhythms with a tendency to accelerate appear to be "pacemakers," i.e., to entrain others of slower frequencies.[45] The earliest established activities in pristine "living matter," whatever was their nature, would then be likely to have assumed a guiding function, and on this simple physical basis provided a mechanism of integration from the very start of life. Similarly, in the course of the earliest individuation and growth, older rhythms would normally entrain younger ones, and keep the harmony of the organism as a whole, in that each new activity would have to reflect, *ab initio*, the inherited dynamism of the oldest complex. That is, indeed, a highly simplified skeleton concept of vital structure; even a bacterium is not only "a population of oscillators." But this skeleton concept does set forth what is perhaps the simplest and most essential principle underlying the original dynamic unity of each vital stock that has continued itself from generation to generation. So much

---

[45] Cf. J. W. S. Pringle, "On the Parallel between Learning and Evolution" (1951), p. 189: "Once the steady-state distribution of frequencies of oscillation has become established in a population of coupled oscillators, the asymmetry of mutual interaction ensures that an increase in the natural frequency of one oscillator has a greater effect on the state of the population than a decrease." See also—in reference to the possibility of primordial effects of established rhythms—C. S. Pittendrigh and V. C. Bruce, "An Oscillator Model for Biological Clocks" (1957), p. 103: ". . . we emphasize our view that the basic mechanism has only been elaborated, not created, in the course of metazoan evolution—that it is present, posing the fundamental problem, in the organization of a single cell."

of ancient determination, at least, is encoded in every daughter cell or gamete that passes on to a new life.

There are, of course, other factors that enable some rhythms to modify others: more powerful impulses, perhaps built up by previous confluences, as of nerve potentials in some dominant region of a nervous system; or special implementing conditions like optimum temperature; or facilitation by other activities, especially the catalyzing work of enzymes. The slight advantage bestowed by the effects of long repetition is certainly not the most commonly decisive factor in determining the pacemaker among the countless competing rhythms in a higher organism, but it may have been the most important in much earlier and simpler phases, before the formation of elaborate enzyme systems, nervous systems or special organs, when conservation of the established dynamism was paramount; so that the dialectical repetitiousness of the earliest supermolecular events, i.e., of the most primitive acts, may be one of the reasons why life persisted and expanded.[46]

But microscopic, cyclic acts are not the only ones which facilitate and often motivate their own repetition; it is a tendency of acts in general to "recur"—by which is meant, of course, to be followed by acts very similar to themselves—in similar situations, which are often only slightly or partially similar, much less so than the acts. The same principle that makes biochemical activity become established in what we sometimes refer to as a "groove" also leads to the gradual perfection of the much less steady, more varied functions of special organs. In a young infant's alimentary tract, peristalsis is still readily reversed. The heart action, on the other hand, having begun early in prenatal life, is already well regulated and fitted in with the enzyme patterns that seem mainly to control the situations in which it slows or hastens, or maintains its own pace against other specialized organ functions, e.g., of the lungs or the sporadically very active liver.

---

[46] Even at much later stages and higher evolutionary levels, the conservative functions are assured first. Joseph Pick, for instance, in "The Evolution of Homeostasis" (1954), says: "Functionally, the available evidence indicates that in the beginning all autonomic nerve activity is directed towards accumulation and storage of reserves. Only later, some autonomic nerves become gradually concerned with spending of energies" (p. 300). V. E. Negus, in The Comparative Anatomy and Physiology of the Larynx (1949), p. 208, states that if the laryngeal nerves are pressed or in any way obstructed, the subject will be able to close but not open the larynx; and he attributes this to the fact that the closure muscle is of earlier origin and was probably for a long time the simple sphincter that merely relaxed to open the larynx, and that a specific muscle to pull it open was a later development. At a still higher level, that of primitive learning, A. R. Moore remarks: "Learned inhibition is perhaps the most elementary type of learning to be found" (op. cit., p. 126).

[384]

The vertebrate heart offers a paradigmatic case of entrainment, in that it has really not only a single rhythm, but several, one of which—the "pacemaker," in most creatures the rhythm of the sinoatrial node—entrains the others, so the "beat" presents as a unit act; but upon removal of the "pacemaker," the several entrained activities on which it had imposed its own tempo resume their original rhythms.[47] This reversion is interesting in that it exhibits a persistence of the original impulse patterns in the older elements, which yield to the faster rhythm, but were not formed under its influence. They are not elaborations, but of independent origin, secondarily entrained by a more vigorous activity. Obviously, in the ontogeny of a highly advanced type of individual, we are not dealing with the simple case of elementary rhythms dominating by virtue of prior development. Yet it is suggestive to find that in so integral an act as a heartbeat subacts are still capable of segregating and running their own courses. The dissociation devised in the physiologist's laboratory shows in reverse the evolution of functional complexity.

Because acts are events of distinct form, held together by the impetus given in an original impulse, they can be partially altered in passage and

---

[47] See J. D. Ebert, *Interacting Systems in Development* (1965), p. 66 (summarizing researches on the origin of the heartbeat in vertebrate embryos): "The slow but rhythmical beat begins along the right side of the ventricle and gradually involves the whole ventricular wall. Soon the entire muscle of the ventricle is contracting synchronously. . . . Meanwhile, the atrium has been forming. As it takes shape, it too begins to contract but at a more rapid rate, which governs the rate of the heart as a whole, the ventricular rate being increased. . . .

"Finally, the *pacemaker* develops. When this region, which controls the contractions of the fully formed heart, starts contracting, the whole heart accelerates. . . . If the regions of the heart are cut apart and isolated, each tends to revert to its characteristic rhythm. If they are combined, again the slower is increased to keep pace with the faster."

A comparable relationship between the brain and the spinal cord is reported by Victor Hamburger and Martin Balaban in "Observations and Experiments on Spontaneous Rhythmical Behavior in the Chick Embryo" (1963), where they write: "The normal healthy embryo performs continuously, in a rather regular sequence of cycles, each of which is composed of what we call an 'activity' and an 'inactivity' phase" (p. 537). "It is obvious that isolated parts can generate periodic motility autonomously, not requiring a pacemaker at cranial levels. . . . The activity phases of isolated parts are more regular, and the relative share of longer [spinal cord] phases is greater than in intact embryos" (pp. 539–40). "In summary: the total activity time per unit in normal embryos is 25% higher than in isolated parts. . . . one is led to the hypothesis that these short bursts signalize an intrinsic pattern of the brain, differing from the spinal cord pattern essentially in the shorter duration of activity phases" (p. 542). "The hypothesis is advanced that the periodicity of the motility of normal embryos results from a superimposition of a shorter cycle originating in the brain over a longer cycle intrinsic to the spinal cord" (p. 544).

still achieve a relatively normal consummation. "Pacemaker" action, whereby one process controls the frequency of other processes with which it interacts, is only one of many integrating factors in organic activity. Slow rhythms of great amplitude may entrain, wholly or partially, much faster ones of smaller compass; the variable yet dialectical rhythm of breathing produces some alteration in the much less variable pace of the heart, through a vast intermediary complex of biochemical mechanisms. This whole subject of physiological rhythms is far too great to be pursued here, though the scientific literature on it is almost always relevant to the study of life with a view to the nature of mind. But some notion of the complexity of rhythms, the constant and countless oscillations that go on all the time in a metazoan organism from the first cleavage of the fertile egg to the death of the individual, may be gathered from the following records, which are but a few random samples, taken, of necessity, wherever the literature makes them available. The first is a standard electrocardiogram that shows the beat of a normal human heart. Clearly, the heartbeat is a complex act; but the graph shows nothing of the nervous oscillations that underlie its impulses. The next tracing reflects such a pattern, recorded by a microelectrode placed in a single cell of a nine-cell ganglion that controls a crab's heart. The beat, determined by a "pacemaker" cell which the recording cell is following, is imposed on the latter's spontaneous activity, and is projected as a "spike" among the smaller discharges. The blood pressure fluctuates minutely with pulse and respiration (Fig. 10–4), each breath inscribing

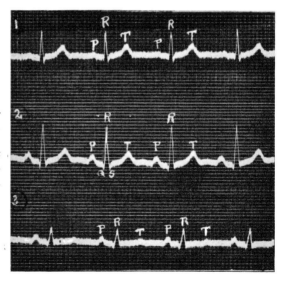

Figure 10–2. Normal Electrocardiogram, Leads I, II, and III

"Note that the height of any upward deflection or depth of downward deflection in II equals the sum of the corresponding deflections in I and III."

(From J. J. R. Macleod, *Physiology in Modern Medicine* [St. Louis: C. V. Mosby Company, 1935].)

Figure 10–3. Action Pattern of Cardiac Ganglion Cell

One cell of the nine-cell ganglion providing the impulses of the heartbeat of the crab. The tracing

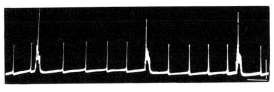

shows three bursts, in which this cell is a follower of the pacemaker. Note that its spontaneous rhythm is reset if a burst interrupts it. 20 mv., 200 msec.

(From T. H. Bulleck and C. A. Terzuolo, "Diverse Forms of Activity in the Somata of Spontaneous and Integrative Ganglion Cells," *Journal of Physiology*, CXXXVIII [1957].)

Figure 10–4. Arterial Blood Pressure

"Normal tracing, somewhat magnified, of arterial blood pressure in the rabbit. . . . The small undulations correspond with the heartbeats, the larger curves with the respiratory movements.

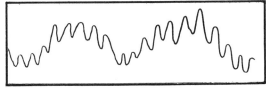

(From W. D. Halliburton, *Handbook of Physiology* [Philadelphia: P. Blakiston's Sons & Co., 1917]; after Burdon-Sanderson.)

itself as a curve on the series of small steep curves that represent the heartbeats. Every activity reflects the immediate dynamic situation that motivates it.

Many muscle fibers respond to electrical stimulation, and the successive oscillations of any fiber in its response to a single application of current vary widely (Fig. 10–5).

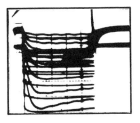

Figure 10–5. Muscle Twitch

"Response of a distal muscle fiber to directly applied depolarizing current (*Left*), and hyperpolarizing (*Right*). 10 mv., 20 mS."

(From B. S. Dorai Raj, "Diversity of Crab Muscle Fibers Innervated by a Single Motor Axon," *Journal of Cellular and Comparative Physiology*, LXIV [1964], 47, Fig. 5.)

Besides these and numberless other motor reactions to impinging or internal stimuli, the reactions in their turn motivate further acts, as stretching or contracting striated muscle fibers, for instance, activates sensory nerves. The neurogram from a sensory nerve in the leg of a frog records the reaction of the nerve (with or without sensation) at the stretching of a leg muscle (Fig. 10–6).

Some little-known sense organs lie in skin, muscles and joints; such are the Pacinian corpuscles, long known to anatomists, but only rather recently

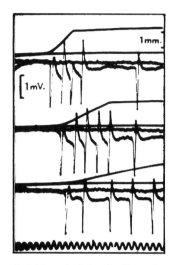

Figure 10–6 (*Above*). Muscle Stretch

"Electric responses (*lower trace of each pair*) during three mechanical stretches at different rates (*upper traces*). The response consists of spikes and of a slow local depolarization which depends upon rate and amplitude of stretching. Time signal (*lowest trace*): 500 cyc./sec."

(From A. L. Hodgkin and B. Katz, "The Electric Responses at a Sensory Nerve Ending," *Journal of Physiology*, CIX [1949], 9P.)

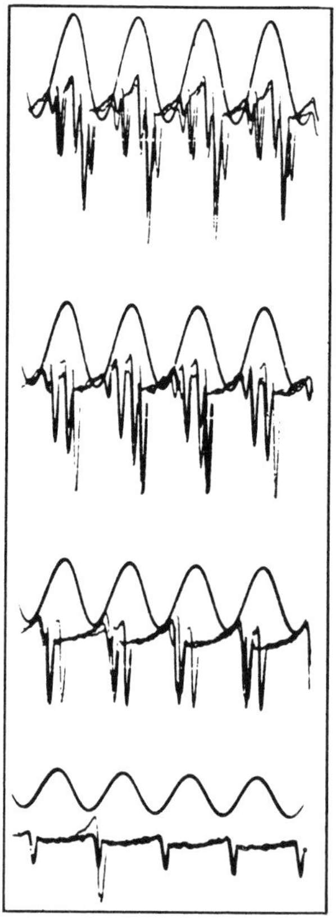

Figure 10–7A (*Left*). Electroneurogram

"Recording from nerve branch to several Pacinian corpuscles. Amplitude of vibration at 200 cyc./sec. increased from above downward, indicated on lower trace. Note recruitment of units."

(From C. C. Hunt, "On the Nature of Vibration Receptors in the Hind Limb of the Cat," *Journal of Physiology*, CLV [1955].)

identified as organs of the vibratory sense (Fig. 10–7A). The better-known receptors, above all the organs of the "higher" senses—the eye and the ear—have been studied in great detail; they respond to minute patterns of impinging vibrations, as the accompanying records—the retinogram and audiogram, respectively (Fig. 10–7B and C)—testify.

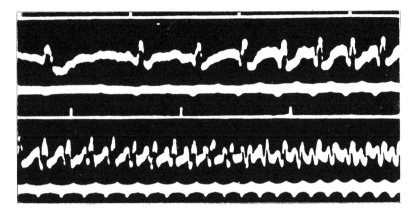

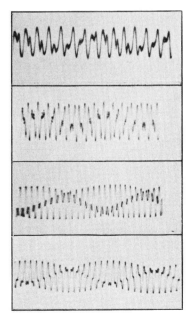

Figure 10–7B (*Above*). Electroretinogram

Demonstration of off-effect in response to flicker. . . . Lower record continuous with upper. Tracings (*top to bottom*): time (2 sec.), retinal response, stimulus (light on, downward; off, upward).

(From Eberhardt Dodt, "Beiträge zur Elektrophysiologie des Auges. II. Mitteilung," *Von Graefes Archiv für Ophthalmologie*, CLIII [1952].)

Figure 10–7C (*Left*). Electrocochleogram

Forms of complex waves. (*Top to bottom*) stimulus wave made up of two simultaneous tones, 1,000 and 2,001⌐, forming a mistuned octave; the cochlear response produced by this sound (in the guinea pig); the cochlear response to a mistuned fifth (1,000 and 1,501⌐); the cochlear response to a mistuned fourth (900 and 1,201⌐).

(After E. G. Wever, *Theory of Hearing* [New York: John Wiley and Sons, 1949].)

Figure 10–8. Audiogram with Cochlear and Nerve Potential Recorded Simultaneously

Cathode ray pattern superimposes successive oscillations. Synchronization of responses of cochlea and a single neural element to various tones.

(After R. Galambos and H. Davis, "The Response of Single Auditory-Nerve Fibers to Acoustic Stimulation," *Journal of Neurophysiology*, VI [1943].)

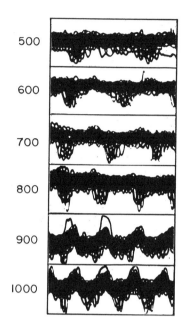

500

600

700

800

900

1000

The end organ of vision or hearing and the nerve that it activates are usually treated as two quite separate structures; but in reality the responses of the receptor and the nerve tract form one act, and just as in the muscle fiber, successive cycles are not absolute repetitions; the accompanying audiogram (Fig. 10–8), showing the response of cochlea and nerve, in unison, to tones of different pitches, shows the wide leeway of the minute acts concatenated into one responsive vibration.

Finally, the central nervous system—the spinal cord, brain stem, cerebrum and cerebellum—reflects and affects almost everything that goes on in the organism. In the higher animals, and especially in us, the cerebral cortex is the most active of all the vital organs, and its oscillations may be recorded even through the skull and scalp, whether the individual is awake or asleep. The little strip of electroencephalogram here reproduced (Fig. 10–9), showing a somewhat unusual pattern in a clinically normal person, is recorded simultaneously from six different leads. In all parts of the cortex, activity is continuous; acute stimuli only modify the spontaneous "firing" of cells in the brain. The encephalogram registers two blinks of the subject's eyelids. Note that the effect is stronger in the frontal and parietal lobes than in the occipital region that contains the visual area; evidently it is not the momentary elimination of light that records itself, but the act of blinking.

Not only the brain, but the whole organism "keeps going" even in repose. The first chart (Fig. 10–10) shows a simultaneous record of different functions in a sleeping subject. It exhibits a state of considerable inner activity; the sleeper is dreaming, as indicated by the rapid eye movements shown in the second tracing (*EM*). Respiration (*R*) de-

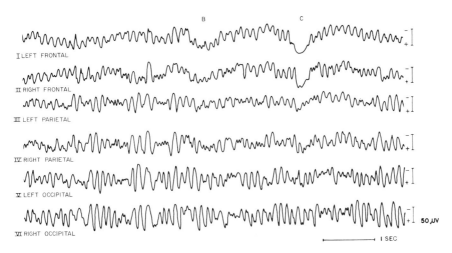

Figure 10–9. Electroencephalogram

Electroencephalogram of a twenty-nine-year-old male, clinically normal,
shows fairly continuous 10 per sec. activity in all leads mixed with
(unusual) high-voltage 6–7 per sec. activity. Blinks at *B* and *C*.

(From F. A. Gibbs and E. L. Gibbs, *Atlas of Electroencephalography*
[Cambridge, Mass.: Lew A. Cummings Co., 1941].)

Sleep; Rapid Eye Movements Associated With Dreaming.

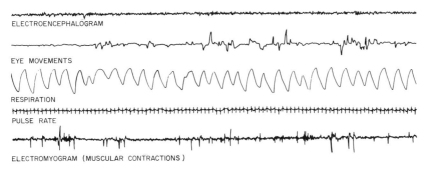

Figure 10–10. Five Constant Activities Recorded Simultaneously
in a Human Subject during Sleep

(Provided through the courtesy of Dr. Frederick Snyder)

velops some irregularities under these conditions, while the pulse rate (P) does not seem to be affected. The electromyogram (EMG) recorded from the chin shows markedly reduced muscle potentials associated with relaxation. This chart gives us an idea of the interactions of different functions during a short period in which a particular physiological state—dreaming—manifests itself.

The minute cycles that are concatenated and propagated by virtue of their own dialectical structures form continua which have their own larger oscillations, periods of high and low activation, motivated not so much by specific external influences as by the degree in which the body is receptive to such influences at all; that is, in the main, by its alternating states of sleep and wakefulness. The second chart (Fig. 10–11) presents something of a synoptic picture of simultaneous activities within the organism, on the larger scale of diurnal rhythms. It is based on computer analysis of minute-to-minute data simultaneously recorded with an eight-channel telemetry system on a normal human subject during several days and nights. The upper curve shows the frequency of an electroencephalogram which gives an indication of the brain activity. The basal skin resistance (second curve) is representative of the tonus of the autonomic nervous system. This is followed by respiration and pulse rate. The two lowest curves, surface temperature and core temperature (rectal), provide information on the heat exchange between the inner part (core) of the body and the surface (shell); this exchange is related to the metabolism.

Viewing the series of records in reverse, one gets a general impression of the progressive elaboration of vital functions: the metabolic cycles that are the most essential processes, then the circulatory mechanism that feeds the system and clears away its dross, the ventilation that implements its rapid and ubiquitous oxidation, then the autonomic system of nervous control of involuntary organic acts, and finally the activity of the central nervous system. This highest mechanism is the pacemaker in the major cycle of sleep and wakefulness, which all the large diurnal rhythms approximately follow—the ebb and tide that punctuate the stream of activity, and make each day physiologically one act.

If, now, we consider that a record from one muscle fiber represents thousands, each act in each fiber being slightly different from all others in that same fiber or in others; that a record from one nerve cell is one of many million, all spontaneously active, even if not specially activated, at every moment of an animal's life; and that no records are presented here from activities of chemical transformation, protoplasmic streaming, cell migration, division and aggregation, or the myriad effects of larger

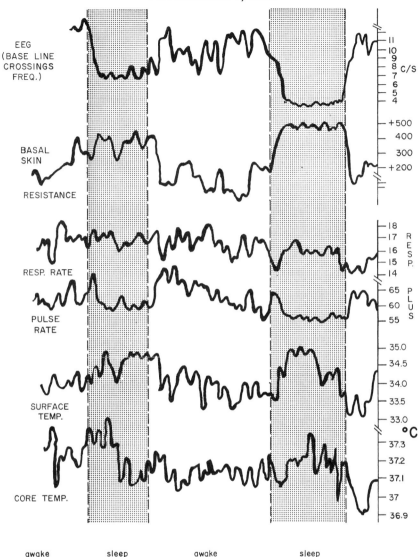

Figure 10–11. Large Diurnal Oscillations Imposed
on Continuous Organic Activities

(Provided through the courtesy of Dr. Karl Ernst Schaefer.)

changes such as locomotion, shifts of attention, etc., we may gather
something like a true impression of the complexity of life. Those larger
rhythms that rise and fall with our sleeping and waking, moving and

[393]

eating, are superimposed on the myriad microscopic activities and unconscious repetitive pulses that compose the organic matrix. Little we know
how every breath we draw reverberates through the whole vast dynamism
of interacting cycles, the internal situation of the next moment of life,
that motivates the next breath.

The development of special organs in all metazoa (and of quite
specialized organelles even in some protozoa[48]) has long harassed biologists seeking only natural, i.e., causal, explanations of apparently purposeful directedness in nature. The generally accepted view—the only
one, as far as I know, set up uncompromisingly against all kinds and
shades of teleology—is the neo-Darwinian, according to which chance
mutations constantly produce new forms, most of which are deleterious
if not lethal, and therefore are bred out again, while a few improve their
possessors' chances of survival and procreation, and consequently are
established and increased in the gene pool. That such a formative process
takes place is almost certainly true, and easy enough to accept as the
basic pattern of a universal life stream that consists of individuations—
numberless, ubiquitous, competing individuations, which naturally take
their forms in any way that the forces around them permit, exercising
all the functions they can, and surviving if, and as long as, the essential
functions are possible. Note, however, that this constant interplay of
forces, which makes shifting obstacles and openings for each individual
so that variously equipped organisms are differentially brought to grief,
is not a mechanism; the frequent references, in the literature, to the
"mechanism of selection" bear witness to the beguiling influence of the
term "natural selection," which seems to refer to an act, or at least a
function, of some specific power.[49] "Natural selection" is a historical
pattern, not a mechanism; it is the pattern of the natural history of life.

[48] T. M. Sonneborn, in the second part of his aforementioned article: "Does
Preformed Cell Structure Play an Essential Role in Cell Heredity?," gives a fairly
detailed account of the mechanisms in the cellular cortex of *Paramecium*; in two
other papers, "Beyond the Gene" and "Developmental Mechanisms in Paramecium"
(1957), he discussed the several types of nuclei and cytoplasmic inclusions contained
in these single cells.

[49] Gerhardt von Bonin, in "Types and Similitudes. An Enquiry into the Logic
of Comparative Anatomy" (1946), p. 196, speaks of "certain forces such as Natural
Selection" molding the shape of organisms. That is somewhat like speaking of "certain substrates such as Matter." Julian Huxley, in *Evolution, the Modern Synthesis*,
p. 474, writes: "R. A. Fisher has aptly said that natural selection is a mechanism
for generating a high degree of improbability."

In this pattern such forms as can arise and can coexist do so. This fact was expressed in the medieval doctrine of the "Great Chain of Being," which asserted that God created as many creatures as could logically coexist, and which Leibniz reinterpreted, in qualitative instead of quantitative terms, as "the best of possible worlds." In our own perspective, it appears that far more beings are engendered on earth than can coexist; also, some that could not continue even without competition, being internally deficient, non-viable.

Now, the study of evolution has been concerned, so far, almost entirely with the conditions of coexistence. Sir Julian summed it up fairly, I think, when he said, "The comparative study of the reagent—the varying, evolving organisms: the comparative study of the medium—the graded, fluctuating environment: and the comparative study of their interaction—the processes of selection and their consequences: it is along some such lines as these that the evolutionary text-book of the future must be written."[50] The characteristic concept here is the first one—the organisms, collectively, treated as "the reagent," instead of as so many agents. Then the question arises, of course, of how traits can be selected for survival that have to reach a fairly high development in phylogeny before they serve a purpose, and why their early stages, which (barring drastic steps of "macroevolution," as most Evolutionists do) must have endured for ages, were not eliminated as useless instead of being allowed to differentiate, step by step, until they were ready to function. Useless organs are supposed to be bred out; but while some structures not needed in a given environment do appear to atrophy or mutate to vestigial forms, like the eyes of cave fishes and the manatee's hind limbs, others seem to be quite needlessly developed and retained, as for instance the larval gills and adult lungs of the hellbender (*Cryptobranchus alleganiensis*), which breathes entirely through its skin.[51] True

---

[50] *Op. cit.*, p. 130.

[51] According to one simple but reliable account, "Hellbenders stay in the water all their lives. Although they have lungs they do not use them, but rely chiefly on their skins for respiration; their gill-slits which open into the sides of the throat do not seem to have any function" (Ann H. Morgan, *Field Book of Ponds and Streams* [1930], p. 352).

Recently, in a paper entitled "The Respiratory Function of Gills in the Larvae of Amblystoma punctatum" (1963), E. J. Boell, P. Greenfield and B. Hille stated that the gills of these larvae, too, have at most an emergency function, as their removal makes no difference to respiration under normal conditions and very little under low oxygen.

A truly bizarre example of a gratuitous organ (if such it be) is recorded by F. W. R. Brambell and D. H. S. Davis in their paper, "Reproduction of the Multimammate Mouse (*Mastomys erythroleucus* Temm.) of Sierra Leone" (1941),

neo-Darwinians are all too often driven to the question-begging assertion that such traits must bestow some advantage, or perhaps used to do so.

As soon as one regards evolution to be a pattern of acts rather than of the anatomical changes that form the record of acts, many such difficulties may be resolved by detailed study of the act form itself and the structured dynamisms it imposes wherever life is in progress. In the first place, random mutation is much easier to imagine (if not to understand) as a deviation from a long-perpetuated activity than as a "loss" of a chemical constituent. How was the constituent lost from the one cell that carried the parental genome? Nothing needs to have been lost; an activity has changed, perhaps during rejuvenation, in the crisis of contraction into a free cell, or perhaps long before that, before meiosis. That the new pattern of activity set up in a mutant locus should then continue itself throughout the new life and through subsequent generations is not surprising, in view of the general tendency of cyclic acts to prepare the optimum situation for their repetitions.

It may be, however, that long before the effects of a changed enzyme-generating process (such as the genes seem to actuate) comes to phenotypical expression, and affects the life chances of the whole organism and hence its own future, it has to run the gantlet of all the previously established and harmonized activities transferred in the rest of the germ plasm, which have to allow and often support its novel performances, especially where the mutation provides a new capacity, as in the cases of molds (*Neurospora*) observed by Horowitz, Raistrick and others, which produced mutants that could synthesize substances the ordinary strain had to find ready-made.[52] These mutants had probably arisen long before exhaustion of the substances in question made their activity evident by leaving them as the sole survivors on the depleted base. But no one knows what other internal processes, controlled by other genes, perhaps whole teams of them, were involved in the new process; or how many other mutants could potentially have made even greater advances, save that the active mechanisms of the otherwise normal *Neurospora* did not accommodate their vagaries. Perhaps this sort of internal discipline is what gives living process an appearance of heading in a constant direc-

---

p. 2: "The normal female, whether adult or immature, invariably possesses a conspicuous and well-developed prostate gland, which has the appearance of being actively functional."

[52] A highly interesting theory of the expansion of somatic acts to cover many steps between the impulse and the consummation was proposed by N. H. Horowitz in 1945, and has since then been tested and validated by several biologists; but the full significance of that insight will be more obvious in a later connection (the growth of the ambient), so their work will be discussed there.

tion—the evolutionary path, which is not straight, yet as a whole seems to hold a general course.[53] It is a form of pression that would rarely eliminate organisms, i.e., only in the unusual situations in which a mutation might be buffered out that would have saved the whole system; but it is probably a constant controlling influence on the expression of genes, and governs unknowable numbers that consequently never come to light at all. They may, however, be carried along in the gene pool of the stock for ages, and constitute the immeasurable capacity for new developments that seems to be stored in every individual matrix.[54]

The amount of modification of genetic expressions by internal adjustments, before ecological pressions have had their turn, and the direction thus given to all incipient processes by the organized acts already going on in a living system, is what L. L. Whyte in his recent little book, *Internal Factors in Evolution*, has termed "internal selection." His all-important point is that the control of gene expressibility, at this juncture in the history of a hereditary line, lies with what he calls the "coördinative conditions (C.C.)" prevailing in the cell, the interactions of all the mechanisms within it; and that this internal selection is not a struggle among rival genes for space and nutrition to exist, but an automatic assimilation of all available mechanisms that can work together to the integral system.[55]

What Mr. Whyte most emphatically insists on is that this differential admission of novel functions to the system is "a new[56] non-Darwinian mechanism of selection" (pp. 86, 87, 91 and elsewhere). Oddly enough,

---

[53] George Gaylord Simpson, in *The Meaning of Evolution* (1949), entitled his third chapter "Oriented Evolution."

[54] Cf. P. B. Medawar, *The Uniqueness of the Individual* (1957), p. 15: "The genetical mechanism is such that there are deep resources of hidden variation, of possible animals only awaiting the occasion to become real."

[55] Cf. Whyte, *op. cit.*, pp. 75–77: "In 1881 Wilhelm Roux extended the Darwinian conception of the struggle of individual organisms to a struggle of parts within the organism, using as the title of an essay: *Der Kampf der Teile im Organismus.* . . . In 1896 August Weismann, in his essay *On Germinal Selection*, adopted Roux's idea, called it Intra-selection, and gave it a new emphasis by applying it to the germ cells and the 'determinants' which stabilize hereditary characters. . . . [But] such intra-selection must not be considered as an anticipation of the internal selection with which we are here concerned. For Roux and Weismann were both thinking of a *struggle* between parts, i.e. a competition for space and nutrition in which the relative success of one part implied the relative failure of others. Nothing could be more different from an internal selection in which the criterion is the capacity of all the parts to fit together to form a single *coordinated functioning unit*. . . . Only when the internal functional coordination for some reason fails does the Roux-Weismann competitive struggle of the parts become a reality, and then we have cancer, not ordered differentiative growth or evolution."

[56] Meaning, of course, a newly discovered one.

[397]

its main difference from the Darwinian "mechanism of selection" is that it really involves a mechanism, and really is a process of selection. New, incipient activities are selectively entrained or incorporated, perhaps rearranged and incorporated, or overridden and suppressed by the established rhythms of the whole hereditary mechanism of the cell, when the time comes for the mutant gene to take effect. Sometimes its influence is too disharmonious to be fitted in and too strong to be suppressed; then the mechanism breaks down, i.e., the mutation is lethal. If, however, the change is accommodated and the organism survives to maturity, its whole activity is somewhat altered, and remains so if it is handed on in the tiny sample packages that are the mutant being's germ cells. In all its future cells the mechanism is modified by the mutation.

This sort of "internal selection" of what a new individual is going to present for ecological shaping at all, becomes even clearer if "the C.C.," meaning "coordinative conditions," be replaced by "the C.A.," or "co-ordinative activities." That the physiological activities in cells, tissues and organs constitute very exact mechanisms need not be argued with anybody who has even a popular idea of vital processes. But anyone who has more than such a bowing acquaintance with gene-controlled activities may also be led to grave skepticism about Mr. Whyte's sanguine hope that the C.C. will soon be amenable to mathematical description and calculation;[57] as C.A., they present all the disturbing features of acts —potentiality, motivation, impulse, rise and consummation and decline, rhythmic interactions, expansion and contraction, prepatterns and realization, integration with or without potential resegregation, metamorphosis, entrainment of rhythms and shifts of whole functions from one pattern into another, or from one organ to another.

Yet these earliest conditions, viewed as acts, give one a truer idea of the advance of life than the usual formulation of evolution theory, because in such an intrinsically active and self-harmonizing zygote there is a source of energy to push against extrinsic pressions, and to produce acts characteristic of the agent even when they are deeply impressed with the marks of ecological circumstance (Fig. 10–12). Contracted to a chemical tension pattern in the germ cell, the inherited rhythms con-

---

[57] Cf. Whyte, *op. cit.*, p. 35: "The C.C. are the expression of geometrical, 3ᴰ, or perhaps kinematic rules determining the necessary 3ᴰ or spatio-temporal network of the relations of the atoms, ions, molecules, organelles, etc. in a viable organism." Then after much interesting discussion, on p. 105: "The C.C. define the primary features of biological organization. . . . This covers a vast field of problems which are of great complexity when analyzed in the traditional manner. But there are grounds for expecting that they can be resolved by an elegant method, perhaps by one set of C.C."

Figure 10–12. Water Buttercup (*Ranunculus aquatilis* L.)
with Submerged, Aerial, and Transitional Leaves

*acts characteristic of the agent even when they are deeply impressed
with the marks of ecological circumstance*

(W. Keble Martin, *The Concise British Flora in Colour* [New York: Holt, Rinehart
and Winston, Inc., 1965]. Copyright © 1965 by George Rainbird. Reproduced by
permission of Holt, Rinehart and Winston, Inc.)

tinue, and as the new individual is synthesized by its own metabolic
procedures, its acts expand in every way possible to find consummation.
This process may attain fantastic complexity as integration of impulses,
reinforcement, elaboration of subacts, preparation of situations that motivate new acts and assimilation of concomitant acts in progress are
built up in the embryo of a higher organism.[58]

One principle, frequently stressed in the works of Evolutionists, is
that needless structures are bred out of a strain in the course of its

[58] One of the most striking examples of preparation by organisms for the continuation of their own activity is provided by the slime mold, *Dyctyostelium discoideum*,
which already commanded our attention (Chapter 9, pp. 347–48) by virtue of a
different peculiarity of its primitive but highly original way of life. A substance
known as acrasin that enables the cells to move in their amoeboid phase is generated
by themselves; as J. T. Bonner graphically describes it, "somehow they exude the
track and then walk on it, as the carpet of a Persian king is rolled before him"
(*Morphogenesis*, p. 174).

adaptation to an environment that makes them unnecessary. Such a statement is probably a dangerous generalization; a deleterious development is usually bred out by ecological screening ("selection against" it), but a great many harmless ones, of little or no apparent use, are carried along for whole biological ages without paying for themselves as they go. One could, indeed, set up a principle of tolerance, as reasonably as the principle of elimination of needless structures.[59] Tolerance is a large and rather neglected factor in evolution. Since useless forms play no noticeable part in the economy of organisms they are easily overlooked; but their existence is really significant for a theory of life, both for what it tells us about origins and for the unexpected role it is apt to play in phylogenetic crises.

The origin of useless or at least unnecessary processes is not harder to understand than that of useful ones: both arise from the matrix of acts which, in its turn, has arisen from chemical complexities in past conditions which we can only reconstruct by surmise. Each germ of life is a packet of potential impulses; each moment of its development is a configuration of occurring impulses seeking actualization among the basic acts already in process of realization. In this swarm of variously proceeding activities of a metazoan embryo each one persists as long as it can, each act that comes to consummation preparing further acts, so that complicated act structures record themselves in the formation of living cells. Some of these structures interact at the cellular level and form tissues, sometimes of great complexity—tubules, sheaths, laminae, cysts, threads and bundles—by chemical syntheses apparently catalyzed at a lower level, in cells, through the never interrupted activities of the chromosomes carried over from the parent strain. Unless these formative processes are stopped by the interference of others, by lack of implementing metabolites, or by destructive extrasystemic forces, they go on to form organs of still greater complexity than their component tissues.

It is here that a larger form of action arises. Over and above the microscopic cyclic acts of chemical transformation, which have already become staggered so that one set of changes supervenes on another (i.e., tissue differentiation on cell formation), acts that involve a whole area, and finally become localized in one of those specialized organs, begin to

---

[59] Cf. A. A. Boyden, "Comparative Evolution with Special Reference to Primitive Mechanisms" (1953), pp. 22–23: "Today after a thousand million years or more of evolutionary selection and fixation there is much wastage of protoplasmic materials concerned in evolutionary processes while the main patterns of organization are perpetuated."

occur in relative independence of the metabolic rhythms.[60] In the course of ontogenesis those larger acts take shape where and as they can, find their track in the living web of rhythms and expand as far as the developing anatomy permits. Again, the heart presents one of the best examples. It begins to form early in embryonic life, apparently serving no purpose until the incipient vascular system is ready to act with it. In the earliest phases, however, a characteristic function of periodic contraction, the so-called "pulse," appears in many evolving tissues, some of which will cease to exhibit it later, while others will join the cardiac development, so their rhythms will become entrained by larger ones and finally by the circulatory pulse.[61]

The "beat" that precedes heart formation illustrates a basic characteristic of organic function, namely, that its integrated activities are often detectable before their special mechanisms have even begun to appear. Nothing could demonstrate more aptly the primacy of acts in biological existence, and their gradual concentration in those regions of an organism where they can expand, dominate and integrate most fully. This order of development, from differentiating function to specialized location (tissue determination) and finally specialized form (cell determination), has been noted many times by embryologists. Charles Manning Child remarked, fifty years ago, "that differences in reaction or in capacity to react very commonly exist in different parts even before visible differentiation occurs, or in cases where it never occurs."[62] More recent and more circumscribed researches are corroborating that statement. Thus A. M. Dalcq, commenting on the unity of the nervous system, said: "Curiously enough, at the start that unity is not so much spatial as functional.... The nervous system does not really originate from a unique and continuous layer of cells."[63] And Clifford Grobstein, in an article based on his experimental studies of development in young embryos,[64] concluded that "when nervous tissue 'self-differentiates' ... the cells themselves have not yet acquired fixity of type as nerve cells. ...

[60] Cf. R. G. Harrison, "On the Origin and Development of the Nervous System Studied by the Methods of Experimental Embryology" (1935), p. 156: "The embryo carries on the ordinary functions of organisms, such as respiration and metabolism, but the peculiarly developmental processes are apart from or superimposed upon these and are for the most part continuously changing and irreversible."

[61] An excellent account of this functional development is given by J. D. Ebert, R. A. Tolman, A. M. Mun and J. F. Albright, in "The Molecular Basis of the First Heart Beats" (1955), passim.

[62] Senescence and Rejuvenescence (1915), p. 46.

[63] "Form and Modern Embryology" (1951), p. 104.

[64] "Tissue Disaggregation in relation to Determination and Stability of Cell Type" (1955), p. 1102.

some stabilization at the tissue level seems to precede stabilization at the cell level."[65]

It seems that organs differ in the degree of perfection they have to reach before assuming their specialized functions; these functions, meanwhile, are developing to whatever extent they can in other, often much less facile, ways. Sometimes the early ways are quite efficient until changes in other aspects of the growing system frustrate them or render them inadequate; as placental feeding and oxygenation are well suited to the embryonic phase of life, but with the ending of that phase they are automatically abolished. Where the ending is abrupt, the new vital acts have to be prepared to a degree that permits their consummation after a very few, if any, ineffective impulses. A few unsuccessful spasms of the breathing apparatus are all the neonate can afford without suffering permanent effects of anoxia, if not death. Feeding by behavioral means instead of purely organic acts has much more leeway, but also, in most mammals and birds, less perfectly developed means. The heart action is already established, and so is that of the kidneys.[66] The mouth reflexes are complete, the simple digestive apparatus is adequate. Only the forebrain seems to be still non-participant in the expanding behavior,[67]

[65] This observation is corroborated by a report in a more recent paper, Rita Levi-Montalcini's "Growth and Differentiation of the Nervous System" (1963), p. 263: "The band of thickened ectoderm along the middorsal line of the chick embryo has folded into a neural groove and the neural groove has changed in turn into a neural tube, before the cylindrical-shaped cells which form the tube show any sign of their essence." Also on the chick embryo, cf. J. A. A. Benoit, "Le rôle de l'hypophyse dans le développement des embryons de Vertèbres amniotes à sang chaud" (1962), p. 320: "Certaines méthodes expérimentales . . . ont montré en outre que les signes d'activité fonctionelle sont précoces et précèdent les signes morphologiques. Chez l'embryon de poulet en particulier l'élaboration de l'hormone thyréotrope et la sensibilité de la thyroïde à cette hormone sont établies 3 à 4 jours avant l'apparition des granulations glycoprotidiques et de la colloïde."

[66] See Davenport Hooker, *The Prenatal Origin of Behavior*, p. 111: "It has been generally assumed that, because the placenta carries out for the embryo and fetus the functions of lungs, urinary system, and the absorptive aspects of the digestive system, there is no need for these organs to begin function in the developing organism until shortly before or at the beginning of the viable period. Actually, these organs exhibit function of a sort as soon as they are capable of it. The kidneys of the fetus begin to function at an early age. . . . Alimentary canal function also occurs in mammalian fetuses. There is considerable doubt as to when various digestive juices are formed, but none regarding the facts of swallowing, peristalsis, and defecation."

[67] See, e.g., G. H. Glaser, "The Neurological Status of the Newborn: Neuromuscular and Electroencephalographic Activity" (1959), p. 174: "At birth the human infant has a well-developed brain up to at least the mid-brain level but not the higher cortical regions."

From the behavioral side we have a few reports, such as that of J. M. Nielsen and

though its basic, typically nervous rhythm—the infant sleep or "theta" rhythm—is in progress.

All the structures that go into action only at birth have developed, meanwhile, without performing their functions; the trophic activities that engendered them went on because they could, because their impulses were able to express themselves. In the enormous weaving advance of inductions, inhibitions, mutual pressions and compromises, not only the essential processes, immediately maintaining or simply enlarging the organism, manage to hold their own, but also a vast number of others that proliferate and differentiate in gene-determined (i.e., inherited) ways, formed by millions of years of gradual elaboration.

The point of all these embryological observations is that a great amount of differentiation and differential growth is neither needed nor promptly suppressed by essential activities in the actualization of a genetic impulse pattern, but tolerated. Organs develop when and where they can; and the competence of even mature bodies for such development is sometimes surprising.[68] As structures take shape, they offer opportunities to vital impulses to express themselves as they never did before, and total organic acts, such as the digestion of a meal, supervene upon the perpetual microscopic activities of life.

As George Gaylord Simpson has repeatedly pointed out, life, growth and development are essentially opportunistic in their progress. That is

---

R. P. Sedgwick, "Instincts and Emotions in an Anencephalic Monster" (1949), of what acts can take place in total absence of the forebrain. The infant those two authors were able to study during the ten weeks of its life showed most of the neonate responses, i.e., to falling, roughness, hunger, and also comfort, and was capable of feeding and of controlled posture. It seemed to lack sight and hearing— understandably, as its cerebral agenesis included even the thalami.

[68] See, for instance, Charles Huggins, "The Composition of Bone and the Function of the Bone Cell" (1937), p. 134: ". . . bone has been reported in nearly every mammalian tissue, most frequently in association with Ca deposits in dead tissue." And further, p. 135: ". . . the bladder of the dog and some other mammals is lined with an osteogenic membrane and the reason why bone does not form in this organ under natural states may be ascribed to a nearly perfect defense mechanism of the sub-epithelial fibroblasts. [A good example of repression where a development is not tolerable.] It is interesting that the whole of the bladder wall is not provided with this defense since the outer layers will ossify when the mucosa is transferred from the inside of the bladder to the outside. . . . The osteogenic epithelia modify certain fibroblasts to make bone cells and from then on their hereditary features are changed to subserve a new function."

Wolfgang Bargmann, in *Vom Bau und Werden des Organismus* (1957), p. 95, remarks that bone-forming processes do not cease as the organism matures and ages: "Umkonstruktionen mit *Abbau von Osteonen* und *Neubau von Lamellen* . . . finden [an gewissen Stellen] während des ganzen Lebens statt."

because every act requires some implementation, and any impulse shifts about—or perhaps, is shifted about with the flow of its situation—until it finds a medium for its expression. This medium may be simply a material to act upon, as a push requires some resistant matter in order to be a "push," and drinking entails the presence of a liquid that is imbibed; or it may involve a constellation of other acts, which may be in progress or ready to be induced, to support the actualization of the impulse in question. In the first case it is convenient to speak of the "substrate," in the second of the "means," or (in advanced forms) the "mechanism," of the act.

Now, it is a familiar fact (every housewife knows it) that wherever a substrate is given, some living thing is likely to attack it. Bacteria, molds, worms, flies will batten on any suitable organic matter available to them, and where grain is stored, rodents will find it. A coral island rising far out in the ocean becomes colonized by terrestrial plants, as soon as littoral seaweeds and flotsam have formed enough soil on its shores to support a few seeds that have survived passage through the digestive tracts of migrant birds. Bermuda is said to have had at least seventeen species of herbs and trees when it was discovered. All these instances are simple examples of exploiting a substrate. But the opportunism of agents, especially animals, goes much further than that, seizing on every implementation prepared by their own acts or by events in their ambient, and in that way concerting their hereditary activities in relation to original or gradually resultant conditions. That is the active process of adaptation, which goes on constantly in the frame of imposed limitations and the filter of ecological competition.

Something similar happens in the course of a creature's internal development. Wherever a tissue complex is formed, some elaborate hereditary impulse enlists its mechanical or chemical possibilities, i.e., imposes a vital function on it. Just how such a larger activity then induces sundry trophic impulses to develop the organ further in a functional direction would have to be learned in each case; but that it generally does so is evident. The activity and its mechanism evolve together. In some cases, the final development or the maintenance of the organ seems to depend, at a particular juncture, on the exercise of its function, without which agenesis or atrophy will set in;[69] but in other cases, especially in lower

[69] See, for instance, K. L. Chow, A. H. Riesen and F. W. Newell, "Degeneration of Retinal Ganglion Cells in Infant Chimpanzees Reared in Darkness" (1957), and Lars Gyllensten and Torbjörn Malmfors, "Myelination of the Optic Nerve and Its Dependence on Visual Function—A Quantitative Investigation in Mice" (1963). In both of these studies it was found that it was primarily, if not exclusively, the

animals, no stoppage occurs at the time when the young mechanism normally would go into operation. The morphological process is completed, and the organ can wait a long time—if not indefinitely—for implementation of its special action.[70]

The "preadaptation" of organs formed long before they are destined to serve any purpose has puzzled generations of economic-minded naturalists. For orthodox Selectionists it is an article of faith that every persistent structure must have survival value, and in a rather wide sense this is probably true (though, as far as I know, no survival value has ever been claimed for the vermiform appendix to the caecum, which shows no sign of disappearing from our anatomy). E. S. Goodrich asserted the extreme form of this tenet when he wrote: "In the evolution of an organ by natural selection every stage must be useful, and it is often difficult to picture the intermediate conditions."[71] If organs evolved wholly "by natural selection" this would, indeed, be difficult; but if they are found to evolve ontogenetically by hereditary impetus in a tolerant organism and take on whatever functions they are fit to perform, they may long serve only vital activities, perhaps even only temporary ones, dispersing or retreating with maturation of the organism, and so become very little exposed—if at all—to ecological competition, which in animal life at least is met largely by behavorial development. The tendency of acts to be entrained or assimilated by stronger or more extensively organized ones makes early internal activities of differentiation and growth prone to advance in harmony with the vital round, and be drawn into its repetitive acts so that these make use of the newly emerging tissues and complexes and the substrates they incidentally supply; that is to say, gene-borne impulses to structural elaboration tend to be subjected to the

---

retinal ganglion cells that either atrophied (in the apes) or failed to finish their differentiation (in the mice) without light. But whether this is due to lack of visual function or of photic influence on neural development seems to me still an open question here.

Bernhard Rensch, in "Die phylogenetische Abwandlung der Ontogenese" (1954), p. 105, writes: "Die definitive Struktur der Lunge . . . kommt bei niederen Amphibien erst durch den Einfluss des Ein- und Ausatmens zustande. Die Differenzierung unterbleibt, wenn die Tiere künstlich an der Lungenatmung gehindert werden."

70 See Leonard Carmichael, "The Development of Behavior in Vertebrates Experimentally Removed from the Influence of External Stimulation" (1926). The vertebrates were amphibian larvae, *Rana sylvatica* and *Ambystoma punctatum*, kept under chloretone anesthesia until five days after the controls had begun to swim. Their locomotor organs developed perfectly, and within an hour or less after the animals were placed in ordinary water, they swam as well as the controls.

71 *Living Organisms* (1924), p. 141.

opportunism of the basic life-supporting acts, and modeled into smoothly working mechanisms.[72]

This view of organogenesis receives strong support from the changes of function which many hereditary anlagen undergo in the course of ontogenesis, and the great variety of ultimate developments they exhibit in the phyletic spectra of their adaptive radiations.[73] In embryonic and fetal life some organs serve temporarily for other acts than they will negotiate later; and people who think of them only in relation to their final forms and uses wonder how they can "know" how to assume those forms long before it is time to launch on their predestined activities. Embryologists, however, have long been aware that many tissues and even highly articulated structures are put to one use after another, with successive changes of detail as their ever-changing total situation represses one activity and motivates another. Perhaps the best-known example of such shifts in the functional character of organs during their fetal development is the formation of the mesonephros first as an excretory organ, which acts in that role while the metanephros, the true kidney, is taking shape, and then, in male birds and mammals, gradually transforms its tubules into ductules of the epididymis, participating in the differentiation of the sex organs. But there are many more such functional changes, less evident anatomically. Sometimes even one and the same action has different effects in different stages.

It is an old belief that fetal organisms "practice" acts for future use: limb movements, pulmonary movements, pecking or sucking movements.[74] But the course of "preadaptation" of impulses and their expres-

---

[72] Cf. G. E. Coghill's paper, "Growth of a Localized Functional Center in a Relatively Equipotential Nervous Organ," p. 1091: "Local functional centers are not added by accretion to an already functioning mechanism, but they grow up within that mechanism and are normally held under its control." See also above, p. 305n, for another relevant passage from the same article.

[73] An astounding example is the modification of the arachnid hip (itself not quite easy to picture) into mouth parts. See Alfred Kästner's article, "Die Hüfte und ihre Umformung zu Mundwerkzeugen bei den Arachniden. Versuch einer Organgeschichte" (1931).

[74] This interpretation is still common in clinical and experimental literature. See A. Gerber, "Die frühesten psychischen Regungen des Embryos" (1949), p. 64: "Aus den gesamten Beobachtungen scheint hervorzugehen, dass weitaus die meisten, kurz nach Geburt ausgeübten Bewegungen im Mutterleib oder im Ei intensiv 'geübt' worden sind: Zweieinhalbmonatige Föten des Menschen zeigen schon einen deutlichen Greifreflex der Hände, und auch bei den Beinen zeigt sich etwas später ein deutlicher Klammerreflex."
With reference to an x-ray observation of an operatively aborted human fetus within its extraembryonic membranes, he reports (p. 65): "Dabei zeigten Magen und Darm, vor allem aber die Lungen, eine starke Verschattung. Der Zweck dieser

sions seems to be less rational and much more adventurous than that; for in their earliest stages they are tolerated, and expand and differentiate because nothing stops them, but as soon as they exert any appreciable effects of their own they tend to assume organic functions, often ancillary to other processes of prenatal development.[75] If they did not, they probably could not continue. Leslie B. Arey, in the introductory discussion at the beginning of his *Developmental Anatomy*,[76] comments on the functional lability of incipient organs, saying: "Certain ancestral organs abandon their original embryonic function, yet are retained and utilized for new purposes (e.g., mesonephric tubules and ducts become the permanent sex canals of the male). Other parts make their appearance, only to change at once into quite different structures (e.g., gill pouches into thymus and parathyroids); since these are necessary organs it is understandable why in this instance the embryonic pouches appear even though they are never respiratory in function" (p. 8). Why the necessity of endocrine glands should call for the original appearance of gill pouches is problematical; but why the inherited impulses to form gill pouches should start on their traditional road to actualization, only to be quickly diverted to other superacts, in a system where their continuity to full expression in gill formation and gill action is jeopardized by the prior oxygen supply through the placenta, is comprehensible.

It is also fairly obvious that with the adaptive changes effected in the stock by ecological competition the period of direct expression of such

---

'feuchten Ur-Inhalation' ist noch unklar. . . . Möglicherweise liesse sich aber der Vorgang verstehen, wenn man ihn als eine Art Vorübung für Trinken und Atmen im Mutterleib auffasste. *Saugen*: Auch hier sind fötale Vorübungen nachweisbar. . . ."

[75] A considerable body of factual material on this subject may be found in an article by the Russian B. S. Matveyev, relating these ontogenetic phenomena to the phylogenetic doctrine of the ethologist and ornithologist A. N. Sewertzov. One of the interesting observations in this paper is that fetal limb movements, whatever their motivation, have a prenatal function that is quite unrelated to their postnatal purposes. "The autonomous movements of the extremities of the fetus," says Matveyev, "are significant not as 'preparation' for movement after birth, but as an important motor reaction, raising the gaseous interchange in the body of the fetus."

A more dramatic kind of readaptation along the road from the zygote to maturity is the transformation of larval appendages into all sorts of special organs, of which he says: "The most striking examples are to be found in invertebrates, with their abrupt occurrence of metamorphosis. In insects the walking appendages of caterpillars are transformed into organs of the adult forms with completely different functions: in some, they become organs of the sexual apparatus, in others, organs for seizing food, in still others, organs of taste" ("The Transformation of Function in the Ontogeny of Animals" [1957]; unpublished translation by Judith Clifford).

[76] 6th ed., 1954.

ancient impulses would be more and more curtailed, the shift into the contemporary processes of development made ever earlier as the old activities become less and less congruent with the new over-all tendencies, until the original nature of the hereditary impulse is completely masked in its actualization. So the same principle, the continuity of impulse patterns through geologic ages of unbroken life, through alternate contractions into infinitesimal germs and expansions into new individuations, operates in phylogenetic as in ontogenetic progression.[77] The causes of evolution lie in the dynamic properties of acts and act-engendered entities. From the old, wonderfully versatile gill structures, for instance, all sorts of tissues have arisen in the mammalian and avian orders, in place of the gills that could not develop in their changed situations: not only the glands, mentioned by Arey, from the anlagen of the pouches, but the meatus of the ear instead of the ancient gill slits, jaws and ossicles from the arches, and supporting tissues of the throat from what used to make up various parts of the aquatic breathing apparatus.[78] As the germ plasm has an unbroken history of millions of years, so has every aspect of its expansions into individuations, though many of these are overlaid beyond recognition by the effects of past contingencies—mutations, inhibitions, distortions and admixtures of new impulses to the primordial ones. There is probably no new impulse that can establish itself without enlisting the support of a primitive one. Yet new impulses must arise with the growing cell assembly constituting

[77] This concept of the larger life process receives strong support from the ripe reflections of the geneticist C. E. McClung, who wrote, in a brief paper entitled "Evolution of the Germplasm" (1941): "The germplasm may be defined as 'the temporal record of racial experience' [p. 59].

"Each type of cell becomes highly specialized through modification of its cytosome, while its nucleus may lose its reproductive power or even disappear completely. . . . Only in the germ cells are nuclear potentialities retained. Retention of this power by the germ cells is made certain by various isolation devices which remove them from participation in differentiation during development.

"A germ cell of one individual generation . . . becomes detached and forms a complete organism of the next generation. By some insulating device the germinal elements within the gonad do not participate in somatic processes, but merely perpetuate themselves. On being freed from this inhibition they are free from the limited role of mere germ cells and may perform, through their descendants, all somatic activities. The germplasm, the record of activities to be repeated, is contained in all cells—in somatically included germ cells, there inhibited from full expression; in differentiated somatic cells, limited largely to a single expression; and is completely lost only in cells about to disappear" (p. 61).

[78] Paraphrased from Wolfgang Bargmann, *Vom Bau und Werden des Organismus*, p. 75. Bargmann closes his account with the words: "Diese Aufzählung erfasst nur einen Teil der Bildungen, die sich aus dem Anlagekomplex Kiemenapparat herleiten."

higher animal forms, in which vaster numbers of interfaces and phasic boundaries produce more energy for contractile functions and nervous discharges.[79]

If one thinks of life as a constant building up and expression of impulses, the oft-remarked fact that in development large acts of functional character frequently are established and quite precisely fixed prior to the subacts which afterward seem to compose them also presents no mystery. Acts take place as best they can; where the implementing elements in their situation are barely sufficient, the actualizations of pressing impulses will be crude and relatively simple, but if they reach their consummation the tension from beginning to end resolves in a way that invites repetition, as completed acts appear to do generally. The details are local and lesser expressions, which may have much play in the large act and many possible ways of merging their impetus and consummations in its passage.[80] Consequently the large, main functions in an organism are carried out at all levels of simplicity or refinement,[81] while their subordinate articulations proceed sometimes to higher perfection and sometimes to excess, or again in such a slack fashion that the complex remains primitive.

This capacity to fulfill the vital requirements accurately in many casually varied ways has an interesting parallel in the realm of behavioral acts; there, a similar (perhaps selfsame) directive tendency of a well-formed, deeply motivated act over chance-guided subacts which arise from the momentary situations created as its superior tension builds up to actuality is the physiological prerequisite for the formation and holding of purposes. The reappearance of basic principles that lie in the act form itself, on different levels of life and in many different contexts, goes through the whole realm of biological sciences, and extends to

---

[79] The idea that interfaces and phase boundaries generate the energy for vital activity and animal behavior was expressed by D'Arcy Thompson, *On Growth and Form*, Vol. I, p. 449, and has been repeatedly asserted since.

[80] A perfect instance is presented by Coghill, in a passage already cited in Chapter 8. See p. 269n.

[81] The earliest forms of the basic animal functions—alimentation, excretion, reaching out and retreat, as well as the simplest form of procreation, occur in amoebae, where no special organs are permanently articulated, yet the little drop of protoplasm, with the distinctness of cell wall, nucleus and cytoplasm as its only sign of any machinery, performs all those acts. At the other extreme of muscular movement by voluntary nervous control, it is still true that predictable exact form belongs to the large total act rather than its adaptive subacts; as O. Klemm has put it, "Die durchgliederte Bewegungsgestalt erhebt sich über das Einzeltum; denn das Ganze läuft genauer ab als seine Teile" ("Zwölf Leitsätze zur Psychologie der Leibesübungen" [1938], p. 391).

psychology and the historical disciplines. Such pervasive principles connect the phenomena of chromosome structure with the dynamics of ontogenesis, and further with behavior, feeling, nervous development, animal mentality and finally human mentality, or mind.

There are more of these implicit characters in the nature of acts, which explain some puzzling conditions met in the study of life, and especially of animal life. One has already been mentioned in connection with dormant germs: the contraction of acts to a minuscule scope in which change is suspended, and only tensions obtain. The most familiar instance of act suspension is the behavioral state known as "waiting." But waiting is not necessarily behavioral; it occurs in organic activities, too, and in cytological processes, and may well be the basis of the much-discussed mystery of "timing."[82] The actualization of an impulse does not necessarily collapse the moment it is not implemented. It may be suspended until the means of its continuance arrive, whereupon the act expands and accelerates again. The speeding and slowing of rhythms, such as breathing under different conditions of oxygen supply and of carbon dioxide concentration in the lungs, illustrate the automatic self-adjustment of vital acts, within fateful limits, to others which complete their promoting situation.[83] Entire suspension in mid-career is an extreme

---

[82] The principle of waiting is clearly exemplified in the conjoint actions of multienzyme systems, in which not the fastest but the slowest catalyst involved in a transformation is the "pacemaker," since chemical reactions are not driven by successive impulses, but require their own exact times, so that complex cycles are possible only if the faster reactions can be suspended until the slowest is completed. Ephraim Racker has furnished a good instance in his article, "Multienzyme Systems" (1959), p. 239: "The continuous flow of glycolysis requires 12 enzymes, three cofactors, inorganic phosphate and an appropriate ionic environment containing Mg and K. . . . The substrate is acted upon by a series of enzymes in a sequential order. The compound passes from enzyme to enzyme undergoing specific alterations of its structure at every step. When a steady state is established the rate of the overall process is governed by the rate of the slowest reaction. Then each step proceeds at the same rate." Cf. also Hudson Hoagland, *Pacemakers in relation to Aspects of Behavior* (1935), chap. iv, especially p. 35.

[83] See J. S. Haldane, *Organism and Environment as Illustrated by the Physiology of Breathing* (1936), pp. 7–9: "In ordinary breathing the ventilation of the lungs is such as to keep the percentage of carbon dioxide approximately constant in the air which is in close contact with the blood in the . . . alveoli of the lungs." So "a rise of about 0.2 percent in the alveolar CO_2 percentage is sufficient to double the breathing, while a fall of 0.2 percent produces cessation of breathing. . . . The quantity of CO_2 brought to the lungs by the blood is constantly varying in accordance with varying states of bodily activity. . . . it is evident that regulation of the oxygen percentage is involved in the regulation of the CO_2 percentage. The net result is that both the percentage of oxygen and that of CO_2 in the alveolar air are very constant, in spite of great changes in the amount of oxygen consumed and CO_2 given off by the body."

form; ordinarily, waiting for concomitant acts to develop and furnish substrates or means from moment to moment is an easy and continual practice, a characteristic of acts. In higher forms, where mental processes become prominent, it is one of the constant modifiers of their psychical phases, and sometimes a keenly felt act in its own right.

Like all basic principles, the rate control of acts extends over the entire gamut of vital events, from the very beginning. The action of genes seems to be very largely concerned with determining the tempo of ontogenetic process,[84] and perhaps of lifelong activities, such as metabolism and cerebral rhythms, as well. To say that acts are "encoded" in genes as biochemical patterns means, of course, that the first, i.e., direct, influence of genes must be chemical; and the translation of dynamic phenomena, such as rhythms, into chemical transformation rates is one of the most efficient methods of tracing them back to their genetic determiners.[85] The chemical phases of these basic acts which eventuate in growth, cell movements, embryonic organization, nervous and somatic and behavioral patterns, are of an almost unsurveyable intricacy; one has to isolate the smallest possible complex to track the genetic interactions that finally, by a contrapuntal balance of inductions and inhibitions,

[84] See Huxley, *Evolution, the Modern Synthesis,* pp. 73–74: ". . . wild-type characters are usually much less modifiable by changes in environment than are those determined by mutant genes. . . . The explanation is based on the fact that genes are in most cases concerned with the rates of processes [further reference is made here to Richard Goldschmidt's *Physiological Genetics* (1938)]. The curves expressing the rate of the processes are in general obliquely S-shaped, tending to a final equilibrium position. In wild-type genes, the flattening is normally completed before the imaginal state (or corresponding definitive stage) is reached, whereas in the majority of mutant genes, the curve is still in the ascending phase at this stage."

G. R. de Beer, in "Embryology and Evolution," also remarks on the rate-controlling function of genes: ". . . the qualities of certain results of development (such as male structures as contrasted with female structures in *Lymantria*, or the colour of the eyes in *Gammarus*) is controlled by the relative speeds at which processes of differentiation occur. These speeds have been shown to be the expression of quantitative properties of reaction-rate on the part of the genes concerned" (p. 59).

[85] In the course of a conference on nerve impulse held at Princeton in 1953, Dr. Hudson Hoagland remarked: "Suppose one does find a rhythmic process in the sense organ. It always seems to me one has to go back to the continuous chemical events that set up these things, in terms of the concept of relaxation oscillators. There is no escape from the view that there are continuous chemical processes taking place, which develop frequencies at a rate proportional to some function of their speed; so the fact that one does find an oscillating process in the receptor merely throws it back one more step toward the ultimate chemical kinetics of what is happening" (see David Nachmanson [ed.], *Nerve Impulse* [1954], pp. 202–3).

guide the synthesis of a single metabolite or effect a single function such as cell respiration.[86]

One further characteristic of acts should be mentioned here, though its full importance will emerge only in a later part of this essay. It is, again, a trait of major importance on at least two far-separated biotic levels, gene activity and behavior, respectively, and perhaps on others. That is the power of subacts, sometimes highly integrated complexes of subacts, to segregate from superacts in which they may have developed as original articulations, and to become involved in different superacts. On the genetic level this separability appears as the famous Mendelian law of segregation; and if so-called "genetic information" is indeed the hereditary continuance of activities, the segregation of subacts is well attested by Mendel's discovery. A detail of action developed in one particular strain of a plant or animal stock may in this way become an element in the future of another strain belonging to the same large stream of life.

In behavior, meanwhile, the same principle underlies the important phenomenon of repertoire. It has been recognized for some time that animal behavior is based on a fundamental repertoire of motions, postures and orientations out of which an individual's instinctive and adaptive acts seem to be compounded. Such repertoires have not been investigated with anything approaching the patience and effort that have been spent on the elements of inherited competence. The best observations of behavioral units, as far as I know, have been made in connection with instinctive actions[87] and with prenatal and neonatal movements, both animal and human (contrary to the usual order, in one of these categories—early infant behavior—the human material seems to be the best observed and recorded).

---

[86] A description of such a smallest complex is given in R. P. Levine's *Genetics* (1962), pp. 167–68, in what he calls "a brief outline," which I abbreviate further, only to show that interactions are involved in the most elementary processes: "In *E. coli*, for example, the utilization of lactose is dependent upon the action of four linked genes ($y+$, $z+$, $o+$, and $i+$). The $y+$ and $z+$ genes determine the structure of two enzymes essential for the conversion of lactose into glucose and galactose. . . . Both the $y+$ and $z+$ genes are necessary for lactose utilization. The action of both $y+$ and $z+$, however, is regulated by the $i+$ gene, and in the presence of this gene the synthesis of the two enzymes is repressed. They will not be synthesized unless the medium contains a specific class of compounds that can *induce* enzyme formation and thus overcome the repressive effect of $i+$." The fourth gene, $o+$, affects the expressibility of $y+$ and $z+$ under still further specialized conditions.

For the source of the statement see F. Jacob and J. Monod, "The Elements of Regulatory Circuits in Bacteria" (1963).

[87] See especially Nikko Tinbergen's excellent book, *The Study of Instinct* (1951).

These are undoubtedly not all the principles of act structure that set the phenomena of life apart from those of inorganic nature; more intimate study is fairly certain to reveal other aspects that are similarly grounded and just as ubiquitously expressed. But the ones so far discussed should serve to support the claim implicit in the heading of the present chapter, and explicitly made in its pages—namely, that "Evolution" is essentially an evolution of acts. It is acts that grow and continue and rejuvenate, and that may come to an end, from which there is no rebirth—extinction.

Besides the basic principles of dynamic structure, there are also functional properties of a less pervasive sort that characterize the ever-growing activity of life on earth: occasional major changes in the implementing mechanisms of some acts, changes which arise from no single sequence of events, but from summations, convergences and (chiefly, perhaps) from overdevelopment of successful functions that leads to excessive complication of structures and steps. Such eventualities beget crises in the existence of the stock that encounters them; but if it survives it usually makes an evolutionary advance, which may even be spectacular.

Finally, over and above the widely spaced yet systematically probable shifts of functions from old means to new and readier ones, there are some extremely rare and intrinsically unpredictable changes, not in patterns, but in the quality of events. They are unpredictable because until they have once occurred they are completely unimaginable: changes of phase in causally unbroken continuities. This is a large and problematical subject, by no means beyond its speculative, hypothetical stage; but its promise of generative ideas for the understanding of mental phenomena is great. And not only that; some light on the philosophical problems of genuine novelty and its emergence seems to come from this direction, too. So a further, searching pursuit of the natural history of acts may be expected to open the approach to a conception of mind and of its origin from the beginnings of life.

11

The Growth of Acts

THE MANIFOLD biological principles that lie in the formal structure and relations of acts operate throughout the realm of animate nature. We have considered them, so far, chiefly on the level of intraorganic activities, i.e., of cells, tissues and organs, reaching back sometimes to subcellular processes, which appear still to reflect the characteristic act form, and occasionally casting forward to the larger sphere of behavioral acts, or acts of the agent as a whole, that exhibit the same form no less clearly, with all its implications. In this chapter we shall be more closely concerned with behavior, and the relation of organisms, particularly animal organisms, to their *milieux externes*, the counterparts of their overt acts; though the phenomena that will chiefly engage us at the end of it and thereafter—the psychical phases into which some acts may go—are the most strictly intraorganic, i.e., covert, events in the world.

Despite the conservative, equilibrating principle of homeostasis, which is apparent especially in the inhibition of enzyme syntheses by an accumulation of their own products, tending to hold organisms in a steady state,[1] acts do grow, and their growth has several distinct dimensions

---

[1] This subject has been largely neglected in the foregoing chapters, because it is well known, but its importance is so great that some references, at least, are in order, for instance: Paul Weiss, "Self-Regulation of Organ Growth by Its Own Products"; H. J. Vogel, "Control by Repression" (1961); S. Spiegelman's paper, "Physiological Competition as a Regulatory Mechanism in Morphogenesis," especially pp. 143 ff, on the inhibitory influence of adult tissues placed among presumptive or adolescent ones. Spiegelman recognizes the autonomous nature of acts, held in check by the presence of a dominant (i.e., previously developed) complex; so he holds that the formation of a new hydranth on a coelenteron, which occurs when an older one in the immediate vicinity is removed, "is mainly due to the removal of an inhibitory regulative activity. It is not necessary to postulate any specific stimulations due to cutting, though such cannot of course be ruled out" (p. 144).

Compare also E. N. Willmer, *Cytology and Evolution* (1960), p. 156, where the same principle is declared to govern the activities of free-living cells: "Among the protozoa there are many examples where it has been shown that the by-products poured

(though, like all vital phenomena, these may have intimate interrelations). Acts grow in scope, in complexity, and in intensity, according to (1) their chances of implementation; (2) their organizing propensities, which depend largely on the opportunities they create for subacts to develop, and for lesser acts in progress to become entrained; and (3) the energy of their original motivation, which may be greatly enhanced by confluent impulses in the course of actualization. Each of these modes of increase may reach its own kind of limit, where it can develop no further in the same pattern, so a crisis occurs; the creature's activity undergoes a radical change, as the same essential impulse finds a different road to consummation. For instance, if the substrate for vital activities, let us say for feeding, becomes inadequate to the needs of the increasingly active organism, the feeding mechanism becomes overloaded and strained, and some other potential activity which had been inhibited by the vigor of the older one may expand in competition, and presently perform much or even all of the food intake. There is, then, a shift from one process to another in the history of the stock; perhaps from the holophytic to a holozoic form of nutrition.

The growth of acts obviously leads to tensions and inadequacies which, in turn, produce changes of relative opportunity for different ways of exploiting the environment, of keeping the organism intact, and of procreating, perpetuating the stock. Through all these changes, the vital activities which manage to persist continue to grow, and as they do so, continue to alter the ambient of the undying stock so that each future generation is born to an infinitesimally changed situation: a more pronounced development of one or another inherited action or means to action, a quicker synthesis of some enzymes, a more photosensitive eye-spot, and so forth. A living stock is not only an "open" system drawing on its environment, but also a hysteretic one, constantly preparing its own next condition and action; its repetitions are never exact, because the very impulse of an act is affected by what lies behind it, and no two acts—not even two cycles in one rhythmic series—have identical histories; just as no two numbers in the natural number series have the same class of antecedents. At each procreative juncture the conditions change noticeably, not necessarily by the rare act of mutation, but simply by the disturbance of the genetic pattern; in the case of mating organisms

---

into the culture medium of an organism will inhibit the activity of like organisms but favor the activity of other types." The passage is incidental to a discussion of colonial organisms, and continues: "This sort of relationship between anterior and posterior cells is obviously of great potential importance."

The list of relevant titles could be carried to great length.

(even without sexual distinction), by the confluence of their heritages. The purely cumulative changes in the course of an individual's rhythmic activity are far below the level of observability—indeed, even of imaginability—though in simple reason they are easy enough to demonstrate; but statistically they may be apparent, at least in larger organisms and in the long run from immaturity to old age. The implementation of greater, more elaborated and more integrated acts proceeds on the autocatalytic principle, as steady activities inscribe their form on metabolized substrate materials, building cells and other mechanisms, tissues and higher structures. The increasingly embodied and differentiated matrix motivates more and more concerted and potent impulses that tend to rhythmic, even if greatly variable, repetitions. So the forms of life develop wherever they can, on the basis of ancestral impulses, acts and physical records.

Since the hierarchy of structures in animal organisms—cells built out of elaborately patterned nucleoproteins that present as functional units, tissues composed of differentiated cells, organs of integrated tissues which may be of various origin—thus parallels the increasing intricacy and impetus of vital activities, it seems reasonable, at first sight, to postulate one geminate systematic order of complexity for structures and processes. But the order is not as simple as it looks; very advanced organic complexes may come to serve quite elementary purposes, and vice versa; and this phenomenon, which might, offhand, be viewed as an awkward exception to the general parallelism, is actually an important feature in the evolutionary advance of animal life. So, from the beginning, one cannot expect to match up complexity of structure directly with level of activity.[2]

A major instance in point is the occurrence of behavior, or overt holistic action, in the zoological realm. Since special internal processes, such as circulation of fluids carrying metabolites in solution, arise with the articulation of tubules or other differentiated tissues, and larger,

---

[2] J. W. Buchanan has pointed out the frequent discrepancy between levels of organic complexity and the fixity of function which, in general, marks the more advanced forms, so fragments of them cannot de-differentiate and be reorganized into new total individuals. In "Intermediate Levels of Organismic Integration," pp. 49–50, discussing the relations of procreative patterns—fission, the breaking off of bodily parts, budding, or gametic reproduction—to powers of regeneration from various bodily parts, he says: "But when one compares these properties among organisms it is found impossible to arrange a sequence of increasing structural stability which in any close degree parallels the commonly accepted sequence of increasing complexity of structure. . . . The generalization that structurally simple forms are also structurally labile is essentially correct but it is also true that much more highly complicated animals exhibit structural lability."

cumulative kinetic patterns with the appearance of special organs, one might expect the gathering internal activities to break over into action of the organism as a whole, i.e., into behavior, at some advanced phyletic stage. But that is not the order we find. Behavior is elementary in the animal kingdom. The simplest animals we know—unicellular, free-living amoebae—already perform unmistakably behavioral acts. According to two competent investigators, "Many of the larger amoebae respond to light and to mechanical and chemical stimuli; they pursue and capture prey, often highly differentiated multicellular animals about their own size."[3] Yet their internal structure is almost entirely transient, constantly created and dissolved by the changing phases of the protoplasmic stream, so that phase boundaries sometimes take the place of permanent membranes setting functional units apart. A. R. Moore, in *The Individual in Simpler Forms* (1945), said of this transitory type of organization: "The plasmodium and the amoeba are remarkable for their versatility, their ability to do many things with the same substance. One cell or part of it may at one moment act as a receptor and at another as an effector; one cell or part may serve either locomotion or digestion" (p. 22). There are no fixed internal organs, though the vacuoles, ectoplasmic tube and endoplasmic stream are visible enough; those are dynamic structures. Nuclei, mitochondria and Golgi bodies are the amoeba's nearest approach to internal organs,[4] and the provenance of these structures is so different from that of true organs that the comparison is more fanciful than scientific. For the rest, the cytoplasm seems to be totipotential within the range of amoeboid action.

Here, however, we meet again with a principle already familiar in the context of ontogenetic acts, namely, the appearance of functions before any visible differentiation of permanent special mechanisms to perform them. As with the heartbeat, oxygen exchange or other vital function in embryonic higher organisms, so with the over-all responses of these lowly little drops of protoplasm: before there is any fixed musculature, the organism as a whole goes into action at the touch of nutritive substances, the impingement of light rays, or when stimulated by gradients of tem-

[3] Robert D. Allen and Ronald Cowden, "Syneresis in Ameboid Movement: Its Localization by Interference Microscopy and Its Significance" (1962).

[4] Even such organelles may be essentially functional; thus A. V. Grimstone, in "Fine Structure and Morphogenesis in Protozoa" (1961), p. 117, says of the parabasal sacs of *Trichonympha*: "Irrespective of the mechanism of sac-formation, . . . it is clear that the parabasal body is normally in a state of dynamic equilibrium, with the rates of loss and gain of sacs approximately equal. It is not a static organelle, therefore, but a steady-state system, through which there is a constant flow of membranes. This concept is potentially applicable to any Golgi body."

perature or of chemical conditions in the surrounding waters. Its actions are simple, but sufficient for life to be sustained to the crisis of rejuvenation, i.e., to continue the stock. The slowness of its progress and smallness of its ambient, due to the limited methods at its disposal, are compensated by an advantage deriving from the very same condition that sets those behavioral limits, that is, from its lack of internal structure: a high degree of plasticity, which lets it change its form to withstand the ineluctable changes of its *milieu externe*. This plasticity gives not only the amoebae, but most (if not all) protozoans, a mode of evasion and toleration to make up for their smallness and simplicity.[5]

The one permanently structured organ in these briefly individuated little sol-gel systems is the cell membrane that divides them from their environment. Even in the most protean, multipotential and internally unfixed beings, the periphery has to have some constant identity to meet that most elementary need; and, moreover, it has to have some complexity of structure, because it is not only the dividing wall, but also the frontier of exchange between the organism and its external world. It has to be semipermeable, and that in a specially adapted fashion, to let some molecules go in and others out, and bar still others altogether.[6] The importance of the periphery holds for all creatures, no matter what their degree of complexity, size or behavioral freedom; and especially in free-living animals, which may enter new environments, the structural articulation of the surface membranes plays a leading part in the evolution of their organization. So it is not surprising that even in the protozoans any advance in complexity of the creature as a whole involves,

---

[5] G. H. Beale, in *Genetics of Paramecium Aurelia*, has commented on the extraordinary power of these protozoans to change their antigen complexes in meeting new conditions, which must mean the possession of genes ready for expression under all sorts of circumstances, many of which the species would never encounter in nature. See his statement given in Chap. 9, n. 23.

Similarly, L. L. Woodruff attests the variable nature of infusorians, in "The Protozoa and the Problem of Adaptation" (1924), p. 54, when he says: "Indeed, hardly a single species is strictly monomorphic, and many possess life histories which afford a varied pageant of dissimilar forms . . . determined largely, if not entirely, by the conditions of the environment. . . ."

Some detailed descriptions of extreme changes are given by Libbie Hyman in "The Transition from the Unicellular to the Multicellular Individual."

[6] Cf. H. Kacser, "The Kinetic Structure of Organisms" (1963), p. 26: "Kinetically speaking an organism is characterized by the set of reactions which take place within its boundaries but which do not take place outside. These boundaries are . . . selective toward the exit and entry of molecules. . . . Whatever the detailed mechanism, it is clear that the boundary of an organism acts essentially catalytically, for it is a selective action allowing some out of a large number of translational processes to take place."

above all, a great elaboration of the enclosing membrane, which may be granulated, ciliated, divided into layers, perforated by pores and tubules and fibers that turn one kind of terminus inward into the cytoplasm, and a differently qualified ending outward toward the environment; the outermost layer, and even the deeper ones, may be spun out into fine hairs which still have internal structures, differentiated perhaps to the molecular level. To appreciate the extent of the increase in surface complexity even with but moderate steps in evolutionary advance, one need only compare the relatively simple plasmalemma of an amoeba with the intricate cell membrane of a paramecium, set with rows of locomotory cilia and with defensive trichocysts.[7]

In higher animals the outer surface becomes specialized as an extremely complicated *integumentum commune* with many mutually adjusted, if not quite inseparably integrated, functions. From the standpoint of act development it is of inexhaustible interest, for its essential function is a dialectical one—to hold the organism apart from its environment, and at the same time to conjoin the two, and mediate their interchanges. This dual operation is reflected in all aspects of its structure, its variegation, its ontogenetic development and tendency to specialization. The protective function is expressed primarily in the epidermis of flattened cells that become keratinized, successively die, and are sloughed off, to be replaced from below, from the dermis, the "true skin." But its other, opposite properties, namely, its conjoining, mediating actions, are involved with its many detailed subordinate forms, which amount to millions of distinct and self-contained organs—sweat glands, sebaceous glands, hair follicles—as well as minute piloerector muscles, and above all the countless nerve endings, largely unmyelinated, with a variety of sensory functions. In its exposed position, the skin has grown to be the most adaptable of all organs, able to meet the multifarious and oftentimes sudden changes of conditions in the surrounding world. Even its protective action is implemented by elaborate structures that work in several different ways, mechanical, electrical and chemical.[8] The glands,

---

[7] For a detailed description of the microanatomy of these unicellular animals see Grimstone's paper, quoted above, "Fine Structure and Morphogenesis in Protozoa"; also T. M. Sonneborn, "Does Preformed Cell Structure Play an Essential Role in Cell Heredity?," especially pp. 167–73, for a more general account. By way of comparison, C. F. Ehret and E. L. Powers, "The Cell Surface of *Paramecium*" (1959), and A. Bairati and F. E. Lehmann, "Structural and Chemical Properties of the Plasmalemma of *Amoeba proteus*" (1953).

[8] A fair idea of these various means may be gathered from the following passage in William Montagna's *The Structure and Function of Skin* (1962), p. 100: "The proteinous-lipid product of the epidermis, the stratum corneum, is tailored in every

which secrete some of its shielding substances, also keep open the passages through which moisture and gases may enter, and through which some metabolic waste products may be eliminated. Such processes are most important in the skin of batrachians, which serves in large measure as a respiratory organ, supplementing the action of the lungs. But in every animal the external surface, be it a cell membrane, a squamous integument, a complex color-changing skin or a hairy one, is at once the creature's permanent separation from its surroundings and its organ of contact with them; a means of division, and of continuity.

Here we find in nature the principle of composition by division that is fundamental in art, where it is universally recognized as one of the essential characteristics of "living form." Plastic space is made by divisions that are conjunctions; and plastic space is the image of vital space, the space of an agent's being and ambient.[9] Nowhere in the inanimate world are cleavages organized to be connectives; but as every volume in pictorial, sculptural or architectural space serves to create the spatial unity in which it is defined, so every actual living form—every individual in its shell, hide, or containing membrane—creates its ambient, the world with which it has contact, through that same separating surface.

The protective character of peripheral activities is expressed in an intricate fabric of vital tissues and residual, deposited ones, closely related in conjoint guardianship of the vulnerable living system beneath; but

---

detail to protect the body against its environment. . . .

"An electronegatively charged horizontal field between the stratum granulosum and the stratum corneum is believed to repel anions, and to prevent cations from penetrating deeper. This field, a layer one or two cells deep, is thought to be the physical and psychological barricade against the penetration of substances and to be the 'barrier layer' of the epidermis. . . .

"However, it may be misleading to refer to a single structure as the barrier layer, since the entire epidermis is a bulwark against penetration. The cutaneous surface is coated with a complex layer of lipids and organic salts, liberated by the keratinizing cells and secreted by the sebaceous and sweat glands. . . . This layer, which has a pH of 4.5 to 6.0, is called the 'acid mantle,' and is said to have antifungal and antibacterial properties.

". . . The acid mantle, then, is the first barrier against invading agents. If agents should get past it, they must go through the stratum corneum, which is an effective filtering system in its own right; since the interstices between its cells become progressively smaller in the deeper parts, they probably serve as physical and chemical traps. . . . If microorganisms should pass through the barrier layer, the living epidermis is equipped with agents which often arrest their progress."

[9] Cf. above, Chapter 7, p. 205; see also *Feeling and Form*, chap. iv, *passim*. It was the dialectic nature of lines and planes encountered in that study that led me to notice its counterpart in actual living forms.

their negotiative nature leads to far greater developments, some of which are concentrated in small intensively differentiated areas: the external organs of sense. Through the sense organs, events in the world outside the animal's skin can influence the events inside without penetrating its barriers. Light, and other rays, which do traverse such divisions as cell walls and delicate multicellular skins, become differentially distributed as they do so, and their many effects are concentrated in some spots and deflected in others, filtered and analyzed and perhaps intrasystemically recombined. Photosensitive tissues, which respond to changes of light, are prone to organize themselves into "eyespots" on the bodies of almost all primitive metazoa, some protozoa and even some flagellates generally classed as protophyta.[10] But the photosensitivity of many animals—and by no means only unicellular ones—is not bound to such local spots, even though these may be well-developed eyes. In some creatures, pigments seem to negotiate photic reception, yet entirely unpigmented forms also react to light. There are organisms which, we are told, "have beautifully developed eyespots (Astasia ocellata, some species of Polytoma), but never react to light, while flagellates of the genus Chilomona, for instance, have no pigmented spot, but are distinctly photosensitive."[11]

A similar irregularity marks the phylogenetic relationships of the function and organ of response to sound. Hearing is generally, perhaps always, derived from the perception of more massive vibrations, which has been sometimes classed as touch, and sometimes—more aptly, I think—as a special "vibratory sense."[12] In lower animals—fish and amphibians—we have no way of knowing whether earth-conducted or water-conducted vibrations, which lie in man's auditory range, are experienced as sound or as vibrant contacts. Often the skin as a whole seems to be a highly multipotential sense organ; as Marcel Monnier has observed, "The sense organs of primitive fish are often very little differentiated and only vaguely separated from each other. At this stage, the ambient is

[10] See Rudolf Braun, "Der Lichtsinn augenloser Tiere" (1958), pp. 306–7: "Unter den Geisseltierchen (Flagellata) . . . haben . . . viele pigmentierte Augenflecken. Manche von ihnen (und das sind merkwürdigerweise pflanzliche Flagellaten) tragen über diesen Pigmentflecken sogar lichtstrahlensammelnde Linsen."

[11] Ibid., p. 307.

[12] David Katz has conducted the most extensive experiments as well as phylogenetic speculations on this subject. In The Vibratory Sense (1930), p. 90, he said: "My numerous experiments have led me to support the point of view that in the sense of vibration we have to do with a completely independent sense which has a correspondingly specific nervous apparatus." (This apparatus, it appears from further discussion, is conceived as cerebral in man.) And further, p. 95: "According to my hypothesis the ear is to be understood as a vibratory organ of the highest perfection." This statement is surely beyond dispute.

still experienced, so to speak, by way of the whole body surface. The external sense organs have all potentialities; they serve not only for perception of tactual stimuli, but occasionally also chemical, acoustical, and even optical ones. The taste receptors are not always specialized and located exclusively in the mouth. One finds them elsewhere, too, for instance in the vibrissae.... Finally, like the senses of taste and of hearing, so the faculty of vision in some creatures is not yet altogether differentiated from touch. Thus the whole surface may be photosensitive, so no special organ of sight is required. It has even been proposed that the pigmentary apparatus of the frog's skin may be a residuum of such optical potentiality."[13]

The peripheral surface of an organism develops from the very beginning in a systemically uncontrolled situation; as soon as a tissue is exposed to the inorganic world (which may have other organisms in it, that are, however, no part of the individual in question), that tissue meets with emergencies. Contacts with extraneous substances are radical changes of situation for active tissue, in which its activity is usually increased, sometimes in directions that were not apparent at all theretofore, and sometimes in its normal course. That is what it means to say that foreign substances tend to be inducing.[14] Changes of situation motivate acts. The outermost parts of any organism consequently are kept in constant and very variegated action, and such action records itself in a great complex of highly developed structures, mutually fitting and fusing as best they can all the time, and differentiating by virtue of their native impetus to continue in their respective actions and the diversity of pressions and opportunities each one encounters. The outer surface is geared to meet emergencies, and motivated all the time to specialize, diversify and intensify its growth and other functions.[15] It is always in a state of development, not unlike a young, individuating organism; consequently, even in the highest animal forms and at full maturity, the skin is a neoteinic structure.[16]

[13] "Aufbau und Bedeutung der Sinnesfunktionen" (1951), p. 16.

[14] This condition may be seen *in vitro* as well as *in vivo*.

[15] See, in corroboration, Dorothy Price (ed.), *Dynamics of Proliferating Tissues* (1958), p. 33: "Dr. Matolsky reported that, during renewal of the epidermis, the epidermal cells pass through various cytomorphic and physiologic alteration. The basal cells give rise to spinous cells, and the granular cells are converted into greatly flattened, inactive horny cells. The uppermost cornified cells gradually scale off and are lost at the skin surface."

[16] Cf. William Montagna, *The Structure and Function of Skin*, p. 426: "Most of the skin appendages show constant change in growth and differentiation similar to that of embryonal systems. . . . With a few notable exceptions, then, adult skin, in spite of apparent specialization, remains largely uncommitted; it behaves in the

The great evolutionary importance of this constant and inveterate activity of the peripheral surface in animals is that it probably engenders the first acts of such intensity that they enter a psychical phase, a moment of intraorganic appearance as sensation. There are several reasons to believe that sensory acts are the earliest ones to be felt, perhaps in a wholly indefinite psychical instant, at the highest concentration of impulse coming to expression under extreme contrary pressions, either sudden, or built up gradually to a maximum. Such fleeting moments of something felt would be unlikely to have any determinate qualitative character. Neither could they carry "information." They would presumably occur as the organism goes into action motivated by an acute situation; but it may have made comparable responses to "stimuli," received by proto-sensory organs, at a slightly lower pitch of activity countless times before the process culminated in some faint little gleam of feeling.[17] That momentary event, in itself, may have no perceptible effect on the organism's behavioral acts (wherefore we cannot know when and where it may occur); but it would be prone to prepare its own repetitions, so that psychical phases might become more frequent and finally common in some kinds of situation.

Whenever it was that the first of these completely intraorganic act phases occurred, there was a turning point in natural history, as I already remarked in the first chapter (compare p. 32). At such points, true novelties emerge from conditions that did not presage them, though retrospectively they may be seen to have set the stage for them. A truly novel phenomenon is one that could not have been imagined or conceptually constructed before the first instance of its kind had occurred. It has, therefore, the semblance of a *saltus naturae*; but it is possible, with some philosophical patience and thought, to treat such genuine novelties as emergent presentations instead of resorting to new metaphysical nou-

---

adult as it did in the embryo. The patterns of growth and differentiation leave one with the basic impression that all the cells of the cutaneous system are essentially *equipotential* and that true differentiation does not take place in the skin, but only *modulation*."

[17] The fact that sensory organs may operate without being felt is attested where we can check it, that is, in human experience. So Monnier (*op. cit.*, p. 18) says: "Wir finden in unserem Körperinnern primitive Sinnzellen, deren Funktion an die Sinnesfunktion der Amöbe erinnert. Sie vermögen nur ganz elementare Energieformen zu werten: chemische Reize aus dem Blut oder mechanische Reize, wie Aenderungen des Blutdruckes. Diese Reize werden von speziellen Sinnzellen in der Gefässwand, vor allem im Glomus caroticum und Sinus caroticus, aufgenommen, ohne dass wir uns dessen bewusst sind."

mena, and thereby hold to the unity of nature that underlies the possibility of any natural science.

If sensibility is indeed the earliest kind of feeling, it does not, of
course, have distinctive character as such in its most primitive setting;
nothing could feel "external" to a sensing creature before something also
felt "internal." For long ages, scattered tiny psychical episodes might
mark the most intense acts of some animals in those sudden developments of situation that are called "stimulations," without changing the
normal round of actions in which the agents periodically meet such
situations. But with the increase of acts which at some point in their
passage enter a psychical phase, a creature's behavioral actions fall under
the influence of its felt encounters and become organized to anticipate
repetitions of such episodes; more and more, then, behavior—the acts
of an organism as a whole in relation to extraorganic conditions—comes
to be guided and developed by feeling, which at this level had best be
termed "awareness." It is presumably momentary, and at the end of a
stimulated response may be simply extinguished. Between stimulations
there may be no feeling at all; even the response may not be a felt act,
but a disappearance of feeling, as the concerted excitation subsides and
falls below the threshold of psychical presentation. Such considerations
give some likelihood to the supposition that the first felt acts were
sensory.

With the concertment of responsive behavior, however, the counter-
activity gradually attains its psychical levels, too. At this juncture a basic
division of felt action takes place, the division into what is felt as impact
and what is felt as autonomous act: objective feeling, or sensibility, and
subjective feeling, or emotivity (cf. Chapter 1, p. 31). This dichotomy is
probably unimportant in the lowest forms of life in which it exists at all,
and it certainly attains its greatest significance only on the human level;
but it has some very interesting aspects in the middle ground of animal
development, too, especially in the problematical realm of instinct,
which will engage us in later chapters. At present let it only be said in
passing that our persistent failure to understand the actual, progressive
motivation of elaborate instinctive acts, such as the building of an
oriole's nest, is largely due to a preconceived notion of instinct as something automatic, not involving any mental acts; and this preconception
rests, in turn, on the identification of mentality with intelligence.[18] As

---

18 A typical expression of these tacit assumptions occurs on the very first page of
R. W. G. Hingston's *Problems of Instinct and Intelligence* (1928), in the words:
"Do animals low in the scale of nature reason? How do they solve their problems
of existence? Are they mere thoughtless automatons, or are they intelligent beings

soon as one classes as "mental" all acts which have a psychical phase, i.e., a felt passage, however faint or brief, the whole basis of animal mentality broadens; and instead of seeking out tiny bits of resemblance to human knowledge or reason in it, one is led to search the human mind for little-known elements of feeling, directly available, though rarely noticed, to furnish cues and hints for a hypothetical construction of subhuman motivations. But this large topic must bide its time.

Our concern just now is with the growth of acts beyond the development of the matrix itself and its internal functions, that is, with the growth of behavior. The chief characteristic of behavior is the massive release of energy, not at regular intervals, as when chemical or electrical forces, constantly engendered, periodically reach a threshold and motivate cyclic functions, but as much larger impulses to movements of the whole animal, or of its appendages, in its surroundings.[19] Such acts also

---

like ourselves?" After relating some of the more spectacular facts of insect activity, he closes a chapter on "The Perfection of Instinct" by saying: "These acts, though apparently so wonderfully prudent, are in reality thoughtless and blind. The knowledge behind them is quite unconscious. . . . Yet see to what remarkable perfection this blind unconscious impulse can attain." The fallacy that blocks scientific imagination in probing the phenomena of instinct is primarily the belief that knowledge must be involved in them somehow, which raises the unanswerable questions about "unconscious knowledge" and "Nature's foresight," and prompted Hingston to say, "We find ourselves against a brick wall" (p. 248), and "Certain acts of behaviour are beyond explanation. We call in the aid of an unknown sense" (p. 249). Only once, as a passing thought, he says, "Perhaps it is some kind of inherited feeling" (p. 261); but this possibility is immediately discarded again as offering no explanation.

[19] The sources of energy for behavioral acts are still in the realm of conjecture, but some very interesting conjectures are occasionally offered with respect to particular mechanisms. Dr. Hallowell Davis and collaborators, for instance, have speculated on the origination of the "cochlear microphonic" potential which implements the act of hearing. At the Cold Spring Harbor Symposium of 1952, Dr. Davis reported:

"The structures of the cochlear partition show surprisingly large resting DC potentials. . . . Briefly, if we take the perilymph of the scala vestibuli as reference point the endolymph in scala media shows a positive potential of 50 to 100 mV. The solid structures of the organ of Corti and of the stria vascularis show a negative potential of 30 to 50 mV. The spiral ligament, on the other hand, and the perilymph of the scala tympani are at very nearly the same potential as the perilymph of scala vestibuli. . . . 100 mV or more is a large potential, however, and we accept Békésy's view that these DC potentials constitute a pool of biological electric energy that is maintained by the metabolism of the tissue" (Hallowell Davis, I. Tasaki, and R. Goldstein, "The Peripheral Origin of Activity, with Reference to the Ear" [1952], p. 144).

See also Hallowell Davis, "Some Principles of Sensory Receptor Action" (1961), for the function of dendrites in deriving electrical from chemical processes.

arise, just as the organic activities do, out of the matrix; no external event can cause them except through its influence on the situation of the agent, in which external and internal elements intersect and interact. The change of situation effected by a single, acute external event may be so sudden (even though it be transient and trivial) that the motivated behavior seems to be directly caused by the "stimulus" (which psychologists today like to call the "input"); yet the most careful experiments indicate a much more complicated causal relationship. Their results really fit better into the conceptual frame of acts and motivating situations, than into the simpler frame of stimulus-response concepts which underlie the hypotheses of the reflex arc, and of the instinctive "releaser mechanism" triggered by a "sign stimulus," according to Lorenz and Tinbergen.

The experiments here referred to were actually made on neural rather than overt (muscular) responses; but since the former, especially when elicited by external stimuli, are normal commencements of behavioral responses, the observations recorded and discussed in connection with them by the experimenters are surely not irrelevant here. In the course of the Princeton Conference on Nerve Impulse already referred to above, Dr. Hartline declared: ". . . the response to light of the visual receptor is sluggish. You put a flash of light in and nothing happens for quite an appreciable length of time; then comes the response. . . . the significant point is how *long* this delay can be. You put in light, and whatever happens to a molecule of rhodopsin, it is hard to believe that the quantum is knocking around inside of it for a second, or several tenths of a second, before it does anything. . . . In response to a short flash of light, only a millisecond or so in duration, the electrical response begins abruptly only after the lapse of a period of time that may be as long as several tenths of a second. Not only that, but this reaction time is composed of two parts. . . . First, there is an interval during which effects of light at a certain fixed intensity are cumulative. After that, the response is fully determined, but there is as yet no sign of it. So that there is a delay, and during part of this delay the initial part of the response process is out of reach of the stimulus."[20]

The response that is "fully determined" even while "there is as yet no sign of it" is certainly the functional counterpart to the structural phenomenon of "prepattern" (cf. Chapter 9, pp. 334–35). But what is happening while the initial part of the act—the part that should receive and transmit the causal influence—is "out of reach of the stimulus"? It

[20] David Nachmanson (ed.), *Nerve Impulse*, pp. 44–45. Cf. Chapter 10 above, p. 411.

is taking shape at the impulse level, under the influence of the motivating situation altered by the stimulus, and it takes tenths, rather than thousandths, of a second to work itself out before it comes to expression as the effect of the external cause. Dr. George Wald seems to have seen Dr. Hartline's findings in this light when he interposed the comment: "Is it fair to state that all you have just said, with which I am in complete agreement, is a development of the thesis that it is not the first attack of the quantum of light on rhodopsin that produces the observed electrical effects? That is just a model. The action currents are caused by something deeper; it is something that happens later" (p. 45). And similarly, further on, when he said in the same discussion: "We must go more deeply into the excitatory mechanism to find the process responsible for the electrical variations that are observed" (p. 48).

That behavioral acts have the characteristic form of "pre-set," gathering impetus, consummation, and cadence or "finish," does not need to be argued at this point. The act form is most obvious in overt performances, and only found to be exemplified by intraorganic events upon attentive observation, comparison and generalization. But when it is found throughout all organisms, something momentous happens to the whole panorama of biological facts, from the chemistry of protoplasm to the psychology of man: they are seen to be of one piece, no matter how far apart in its vast structure. "Organic form" then appears in nature as it appears in art, and no matter how much scientific analysis may fragment it, every part still reflects and represents the whole; the image of life is restored, and all the models that mimic vital mechanisms in simpler terms do not destroy it, because the structures which their lifeless parts represent are clearly conceived as something else than ultimate changeless parts; they are conceived as tension patterns expressed in substance, which hold their form by a staggering complex of rhythmicized acts. Such dynamic patterns are not parts, but elements. Down to the structure of protein molecules, they determine the nature and potentialities of living matter.

As elements in a virtual form (i.e., in a work of art) grow out of other elements, so in an actual living form acts are made by and from other acts. This has already been discussed, in the previous chapter, on the somatic level, where that growth is gradual, and records itself in the deep, progressive differentiation of functioning tissues, each change adapted to the situation that implements it. Internal adaptations are constant, reciprocal interactions in the manifold courses of growth and pacing of functions, which lead to such perfect fitting of articulated forms against or around one another that it is often impossible to tell, without any knowledge of their history, whether an elaborate organ which they finally constitute has evolved by differentiation of previously homogeneous tissue or by integration of convergent embryonic processes. Every act within an individual has to get out of the way of other acts which, nevertheless, are making its situation and perhaps implementing its advance to consummation.

The parallels between vital processes on different levels of organization lead us at this point to an interesting question about a fundamental principle of life: nothing less, indeed, than the question of adaptation to a normal environment. Paul Weiss pointed out that adaptedness is a condition and adaptation a process.[21] By the most stable parts of an organism, such as the skeleton, internal adaptedness may be reached during youthful growth and prevail with so little active maintenance that forms like the interdigitating sutures of the skull, and perhaps the shapes of all the bones (though probably not of the joints), are really adapted to their places once and for all. But for the most part what goes on within the organism is a perpetual adaptation of every active tissue to its own neighborhood, dealing constantly, from one functional round to another, with the opportunities and pressions encountered there.

The theory of evolution, as it is most widely understood and accepted today, is built on the well-established premise that every being is adapted to its hereditary environment, and that all its internal rhythms—sleeping and waking, mating and procreating, as well as its visceral functions— have been originally paced by the great, unchanging or slowly changing cycles of day and night, tides and seasons. So it has come as a mystifying surprise to physiologists that under experimental conditions of isolation with no time indicators whatever, a person soon abandons the standard common multiple of all his periodic activities, the twenty-four-hour day, and slips into a schedule closer to a twenty-five-hour total round.[22]

There have been few speculations, so far, about the possible cause of this departure from the obviously advantageous rhythm adapted to the world-old twenty-four-hour solar cycle. The fact, though quite certain, is but recently established, and may still be too new to have received any serious theoretical interpretation. The usual way to explain non-adaptive traits in organisms is by a hypothetical assumption that they were quite properly established by "natural selection" under past conditions, to which, consequently, their possessors are still adapted. But there is no likelihood that the days of earth were twenty-five hours long for the greater part of hominid evolution; also, many animals—finches, for

[21] See Chapter 10 above, p. 365.
[22] See Jürgen Aschoff and Rütger Wever, "Spontanperiodik des Menschen bei Ausschluss aller Zeitgeber" (1962).
The whole of Vol. XXV of the *Cold Spring Harbor Symposia on Quantitative Biology* (1960), to which Drs. Aschoff and Wever both contributed, is devoted to the topic of "circadian" (circadiurnal) rhythms.

instance—seem to have a spontaneous cycle of less than twenty-four hours.[23] And, furthermore, if our ancestors or theirs had been adapted by evolutionary forces to a solar day of half a billion years ago, why have further generations been exempt from those forces ever since? This question applies equally, of course, to all explanations of non-adaptive forms or acts as anachronisms. Only quick changes of circumstance should engender misfitted strains—such as the pale strains of *Biston* in industrial "blacklands"—which are still in process of mutation or extinction.

The real significance of the twenty-five-hour organic cycle in man is, I think, something quite different, something that is relevant to our general concept of evolutionary pressions. We are wont to look for adaptedness in animal structures and functions, and certainly we find it in the countless vital mechanisms which have taken shape and survived in ecological competition to the present day. But adaptedness is a result and tangible record of adaptation; and a great deal of adaptation goes on without leaving any such material evidence. Adaptation is a process, a characteristic of acts, which constitutes the leeway of life; it goes on all the time; and where it is not difficult for a being to adapt its activities to given circumstances, any closer adaptation of its structure to its normal environment would give it no particular advantage over its rivals. It might, indeed, be slightly detrimental, because every refinement of one function in an organism is bought at the price of some other development, so that perfection *praeter necessitatem* is quite generally no asset. That is probably why a great many "pre-fittings" in nature are not very close. They are adequate, and if they are adequate there is no evolutionary pressure toward their further specialization. The adaptedness of organisms to their ambients is important in inverse proportion (very roughly speaking) to their powers of current and constant adaptation to circumstances.[24]

[23] See Aschoff, "Circadian Rhythms in Man" (1965), p. 1427.

[24] The term "adaptability" has two meanings which have rarely, if ever, been formally distinguished: one may be called "genetic adaptability," consisting in mutability of a stock with considerable strength to produce viable mutations, i.e., a wide potential adaptive radiation; the other, which I would call "functional adaptability," is the leeway of individual action in a situation. Some species of *Biston* are genetically more adaptable than others which produce no melanic mutants to take advantage of the new dark backgrounds, and which consequently tend to become eliminated. See D. F. Owen, "Industrial Melanism in North American Moths" (1961), p. 232.

G. F. Gause has noted a similar roughly inverse proportion between the adaptedness of species to their environments and their genetic adaptability (see "The Relation of Adaptability to Adaptation" [1942]). Unfortunately his use of words departs from the precise meanings pointed out by Weiss, so in his vocabulary "adaptation" must be understood to mean "adaptedness."

Such powers of adaptation which never inscribe themselves in the fossil record are, of course, part of a creature's adaptedness, and presumably develop by differential survival just as anatomical advantages do. This explanation of the "circadian" rhythms of men and animals, approximately but not exactly coincident with the astronomical twenty-four-hour cycle, has been proposed by Dr. Aschoff, one of its chief scientific investigators, in the reflective observation: "The adaptive significance of circadian rhythmicity is that it enables the organism to master the changing conditions of a temporally programmed world—that is, to do the right thing at the right time. This could be achieved, to some extent, by an exogenous rhythm. But by developing a self-sustained oscillation of approximately the same frequency as that of the environment, the organism, in its own organization, anticipates the respective states which will enable it to react properly to the environmental conditions which will ensue—it is prepared in advance."[25]

The interesting bearing of functional adaptability on evolution theory is that it sets an automatic limit to some progressions of genetic change toward bodily specialization, by abrogating the need, and therefore quickly reducing the survival value, of the latter's further increase where active adjustment can take over. This consideration throws some light on the twenty-five-hour intrinsic cycle of human activities as a discrepancy that has never been bred out, because it does not matter to man's viability. It is also possible that it has persisted because there are no variations from it, no mutants with a naturally shorter cycle, to win out over the "wild type" that has to adjust its rhythms to the solar day. Mutant strains can only develop where they occur in the first place. If they are non-existent or excessively rare, and functional adaptability makes the immutable stock viable anyway, the discrepancy may continue through geologic ages without being eliminated.

But if the length of the human organic round was not established from its very beginning by the most fixed and all-dominating periodicity of the earthly environment, the day, what set its paces originally so they coincide in approximately twenty-five-hour total cycles? Perhaps no environmental influence at all, but the rhythms that were paramount in the beginning, and changed only with internal developments of catalytic enzymes: the periods of chemical turnover, integrated in the metabolic acts. All the periodicities in a stock must have grown from these smallest and oldest time units in the course of phylogenesis. In the twenty-five-hour cycle of man we have another instance of that persistent constitutional standard, the whole act, unchangeably itself no matter how

25 "Circadian Rhythms in Man," p. 1431.

masked, that we have encountered before in the intrinsic rhythms of the heart, which yield to a pacemaker but reappear when that modifier is removed.

On the fundamental metabolic and trophic activities, the functional patterns of the special life-sustaining organs are superimposed in the course of embryonic development, and form a system of larger internal acts of the organism. Some of these processes have only intraorganic conditions to meet, but some have to negotiate more uncontrolled situations, because the organs in which they occur have some contact with the extraorganic environment through openings in the body wall. Heart and kidneys are safely buried, but the digestive organs and especially the breathing apparatus receive some foreign substances directly, so their procedures must be variable and quickly adaptable within fairly wide limits. They have, indeed, some behavioral powers through close association with the musculature. Breathing can be suspended for brief intervals, swallowing for much longer ones; coughing, sneezing and vomiting are defensive acts originating at more central loci.

The great organs of behavioral action in the metazoa, however, are the skeletal muscles with their peripheral adjunct organ, the skin. Their one inherent special activity is contraction, which goes on persistently without particular occasion and may fall to a very low level—so low that its constant presence was a discovery once upon a time, the discovery of "muscle tonus."[26] It has its rise and fall, which is slow, like organic changes. But superimposed on this basic activity is another system of contractile acts, characterized by massive impulses, very rapid rise and strong tendencies to entrain other acts or to become entrained, to wait for implementing acts that are forming and to suppress and supersede incipient obstructive acts.[27] These acute motions of the organism in its

---

[26] See Chapter 9 above, p. 327. This tonus-maintaining minute motion pattern is slightly enhanced in the oculomotor muscles during fixed vision, so the stimulus pattern of seen objects is imposed on it, according to Marcel Monnier: ". . . les globes oculaires ne sont jamais dans un état de fixité absolue; ils exécutent de minimes mouvements à une fréquence de 10 à 100 par seconde. Dans ces conditions, la rétine sur laquelle se projette l'image de l'objet est animée de mouvements fins, continuels. Le contour de l'image sur la rétine est animé, lui aussi, de mouvements continuels et les récepteurs rétiniens sur lesquels il se projette sont excités davantage que les récepteurs adjacents, sollicités par des parties plus neutres de l'image" ("L'organisation des fonctions psychiques à la lueur des données neurophysiologiques" [1951], p. 11).

[27] Cf. J. R. Hunt's Presidential Address to the American Neurological Association, "The Static and Kinetic Systems of Motility" (1920), pp. 353–54: "I believe that motility consists of 2 components, each being represented throughout the entire efferent nervous system by separate neural mechanisms, which are mutually coopera-

milieu externe constitute its behavior. They are not only motivated by the sudden changes in its ambient that are caused by fortuitous events in the outside world, but they also bring more such changes about, often without any regard to its own life rhythms. When an animal moves, all its spatial relations to things around it are altered;[28] the changes may bring it in contact with food, or into temperatures, waters or lairs that permit its vital activities to proceed more freely than before; then its energies will be spent in feeding, growing and perhaps procreating there. Or, movements—and especially locomotion—may take it to destruction. Short of that, inadequate means of life will continue to motivate its movements even to exhaustion unless its needs are assuaged somewhere.[29]

The potentialities that lie in the kinetic properties of protoplasm, first expressed in the flagellate movements that are common to a great many

---

tive and yet physiologically and anatomically distinct. . . . One of these components of motility is the movement proper, which is subserved by the *kinetic system.*

"The other represents the more passive form of contractility, which we recognize in tonus, posture, attitude, and equilibrium. This static function of the efferent system is subserved by separate neuromuscular pathways, which may be termed the *static system.*"

After briefly sketching the evolution of motile structures up to the appearance of muscular tissue, Dr. Hunt continued: "The first development is the unstriped muscle which, according to some authorities, is present as low in the scale as sponges (Parker), and is constant in all higher forms. The quality of its movement is slow and plastic. Striated muscle . . . makes its appearance in worms and mollusks and constitutes the skeletal musculature of vertebrates. The quality of its movement is quick and elastic" (p. 355).

[28] Cf. R. S. Woodworth, *Dynamics of Behavior*, p. 33, where he remarks that the classic stimulus-response formula, S → O → R (stimulus → organism → response), creates the impression "that the unit of behavior terminates in R, the motor response. Actually R itself produces an effect . . . not usually lost in the environment without any back action on the organism. The effect stimulates the organism; the motor response generates new stimuli; and these, rather than the muscular act, are the terminus, the destination, the goal of the S–R unit." (Here the "unit of behavior"—the act—is clearly a link in a rhythmically continuous series.)

Similarly, R. W. White, in "Motivation Reconsidered: The Concept of Competence," p. 322, has pointed out that all movements of an agent in space cause changes in his ambient: "Change in the stimulus field . . . happens when one is passively moved about, and it may happen as a consequence of random movements without becoming focalized and instigating exploration."

[29] L. L. Woodruff (*op. cit.*) recognized this principle of behavior at its lowest level: "Very generally stated, conditions injurious to the animal induce changes in behavior, and these changes subject it to new conditions. Just so long as the undesirable conditions prevail, these changes continue. In other words, they cease when conditions are attained . . . which no longer afford the unfavorable stimulus in question" (p. 62).

unicellular organisms (plant-like as well as animal-like ones) below the
level of separation into two "kingdoms" of life, are subsequently realized
in the animal kingdom to such a degree that their development has
become a major principle of metazoan evolution. The progressive increase and organization of contractile tissues has, in fact, given behavior
the leading role among survival mechanisms in all the higher animals,
and expanded the range of their actions so enormously that entirely new
ecological pressures have become paramount in shaping their adaptive
modifications. The vast numbers of interfaces and phase boundaries in
metazoan organisms, producing myriads of electrical potentials, probably provide the massive discharges that allow large units such as
visceral or skeletal muscles to contract. The general absence of behavioral
acts in the botanical realm and the anomaly of their limited, exceptional
occurrence might then be accounted for by the simpler chemistry of
plants, and their fewer internal tensions. It is reasonable to suppose
that these are normally just able to provide energy for the vital functions,
but might nevertheless permit, in specialized forms, concentrated local
developments that could build up sufficient potentials to engender
facultative impulses. In animals, however, the evolutionary trend to progressive chemical complexity, catalytic actions and energy production
has made for really incalculable possibilities of overt behavior.

A radical departure from the limited round of animal activities that
are still typified by the oyster, fast in its anchored shell, would be impossible, of course, without further, equally radical, innovations in the
anatomy and functional pattern of the advancing stock. Any great increase in the scope and strength of movement without concomitant
developments of protective or restraining mechanisms would soon bring
the phylogenetic adventure to an end. An example of that dangerous
discrepancy is furnished by our own infants between one and two years
of age, when they are able to walk and run, seize things and topple them,
but have not yet developed a proportionate awareness of imminent
situations. So, since many very mobile animals have evolved and survived to their present state without a baby-sitter, their increasing mobility must have been always counterbalanced by sufficient other changes
to meet the emergencies it created.

The first such change was probably the high elaboration and activation of the skin which, as previously said, in many animals is the source
of all their external sensory organs. But a protective and reactive—even
perceptive—outer surface is not enough to safeguard the unity and coherence of a being of large impulses that eventuate in overt acts. Many
of its muscles are too remote from the periphery to be under its control.

[434]

In very small organisms there may be some direct transference of impulses from one somatic cell to another,[30] but with increasing distances this could not be effective. Something more made integrated behavior possible—something derived from the same embryonic source as the skin, that is, from the ectoderm: the nervous structures. These penetrate the interior of the living system and extend the acts of specialized epidermal loci along sinuous courses to stations of similar specialization, forming either nerve nets, or long fibers and synaptic plexuses of a central nervous system, established in the gastrula stage of the higher animals and culminating in a brain.[31]

[30] See, for instance, G. F. Coghill's account in *Anatomy and the Problem of Behaviour*, Lecture II, titled "Dynamic Antecedents of Neural Mechanisms," p. 39: "The embryo is perfectly integrated before it has a nervous system. . . . Between this earlier non-nervous and the later nervous condition there appears superficially to be a hiatus in development, a point where one condition ends abruptly and something entirely different takes place." There is, in fact, a shift of functions. In the preneural stage the basic factor of organization is polarity.

Also E. N. Willmer (*op. cit.*), p. 231: "Each cell modifies its immediate environment and creates a special set of conditions for its next-door neighbour, which in turn can alter the conditions further for the first cell, so that groups of cells necessarily tend to behave differently from single cells. . . ." And O. Loewi, "The Ferrier Lecture on Problems Connected with the Principle of Humoral Transmission of Nervous Impulses" (1935), p. 306: "Within the cell-complex of Metazoa which have no nervous elements, the relations between the cells obviously can be of no other than a chemical nature."

[31] The peripheral sense organs, including the cutaneous ones, as well as limb buds and other effector anlagen, seem to be directly involved in the development of their respective innervations. Rita Levi-Montalcini, in her essay, "Growth and Differentiation of the Nervous System," p. 279, broaches a problem she refers to as "one of the most discussed problems of neuroembryology—the mechanism of control exerted by peripheral effectors and receptors on their associated nerve centers." In the conclusion of her own very interesting discussion of that unsolved problem, she writes (p. 292): "Under normal conditions, it seems conceivable that the peripheral end organs release small quantities of growth factors and that these factors utilize the nerve fibers as channels of diffusion to the associated nerve centers."

Cf. also R. G. Harrison, "On the Origin and Development of the Nervous System Studied by the Methods of Experimental Embryology" (1935), p. 172: "The ganglia of the lateral line and the ear develop in close association with the sensory epithelia. . . ." And later, p. 184: "There is even a strict selectivity among sensory nerves . . . of the tongue, where the trigeminus forms general sensory endings and the facial and glossopharyngeal run to the taste buds. It seems necessary to assume some specific reaction between each kind of end organ and its nerve. . . ."

One of the most interesting and baffling relationships, however, is mentioned by Montagna (*op. cit.*), pp. 10–11: "Mentally deficient, epileptic, and insane persons often show marked deviations from normal trends in the patterns of dermatoglyphics. . . . Perhaps dermatoglyphics are affected by the same agencies which impair the nervous system."

The receptors of the skin have not only to respond quickly, but to transform and propagate their responses to abruptly engendered conditions. From the very beginnings of their specialization, therefore, they develop blocking and straining mechanisms directed outward, and paths of impulse conduction directed inward and connected sometimes with the focal points of activation of the contractile tissues. Consequently the skin is closely related to the nervous system even at the highest level of neural development. In ontogenesis, the neural plate as well as the epidermis is formed from the ectoplasm.[32] Phylogenetically, the organs of hearing, wherever they occur, are formed from the skin, especially from its complex surface layer, and in low forms of life eyes are of epidermal (or analogously outermost) origin. Only with the higher development of the brain is the photosensitive part of the eye derived from the diencephalon (sclera and cornea are still developed from the skin). In a larger scheme of life, however, the organ of vision may still be viewed as a highly specialized center in the single structural unit, the ectodermal component of the animal body, which comprises skin, nerves and receptors.

The matrix of impulses in so complex an individual, and even at some lower levels of organization, is so dense that every impulse meets some competition; consequently wherever an actualization occurs there has been an option which the actualization has decided. This fundamental structure of animate process is what makes life irreversible. Most options are decided almost instantaneously and by millions, but the fact that potential conflicts lie everywhere indicates that options belong to the very nature of acts (and, perhaps, mark the faint line between proto-acts —act-like but independent physicochemical events—and true, or motivated, acts). The optional character of life is so pervasive that it presents not as a structural feature but as a quality; one of those elements of which we have as yet made no tangible or even mathematical model, but which is projected in works of art, and is perceived qualitatively as "vitality."

Out of the flood of unfelt options arise the larger ones that resolve themselves in behavioral acts. The emergence of motility in the course of ontogenesis is closely linked with the differentiation and maturation of the nervous system, beginning in the larva or foetus (as the case may

[32] The skin of vertebrates is a composite organ of double origin, for its deeper portion, the dermis, is formed from the embryonic mesoderm. But this is essentially a supporting and uniting structure; the differentiating and principally active portion is the epidermis. See Montagna, *op. cit.*, p. 122. The epidermis is even more closely related ontogenetically to the nervous system in that its pigment cells arise not from the general ectoderm but from the neural crest.

be), and proceeding in all the higher animals to such intimate conjunction of the musculature with the nervous system that the latter becomes completely dominant over the contractile mechanisms, although originally the muscle tissue itself may have considerable power of autonomous movement.[33]

One of the notable features of central nervous systems is their division into two sets of fibers and associated "centers," sensory and motor. This differentiation is the anatomical expression of an early division of animal acts into two principal classes: acts precipitated by a "stimulus," i.e., by an acute change of situation, which usually arise from peripheral impulses, and, if they have a psychical phase, are generally felt as impacts; and those which arise from interior impulses, motivated by more gradual situational changes, and, if felt, are normally felt as autonomous acts. The interplay of these two impulse patterns in the single dynamic system, the agent, is the foundation of behavior.[34]

Behavioral responses that are appropriate to the situation in which they occur may be induced by external receptor activities that do not rise above the psychical threshold, i.e., by sensory impressions that are not felt. Sir Henry Head described the adaptive movements of decerebrate experimental animals which, presumably, are not conscious of either their perceptions or their responses, and declared: "These [movements] reach a marvellous complexity and perfection in the decerebrate animal. Here the nature of the movement of ear and tongue depends on the

[33] So says Davenport Hooker, in *The Prenatal Origin of Behavior*, pp. 5–6: "The early activity of the heart in vertebrates has been demonstrated to be myogenic in nature in many forms, that is to say, the rhythmic contractility of its musculature is a property inherent in the muscle cells themselves. Only later in embryonic development do nerves, not yet developed when the cardiac beat begins, control the activity of the heart." Also, of the forming skeletal muscles in killifish (p. 11): "Coghill divided the early activity of *Fundulus* into five phases, beginning (phase A) with the first localized myogenic contractions of myotomes which became progressively active spontaneously in a general cephalocaudad direction. The wave of myogenic activity, as it spread caudalward in phase B, was shortly replaced (phase C) at the cephalic end of the body by another wave, this one of responses to external stimulation, the initial sensitivity becoming manifest in the area anterior to the eye. In phase D, only the tip of the tail and the pectoral fin still show localized spontaneous activity, which disappears in the last phase (E)." And Hooker quotes Coghill as saying, "The neurogenic system chases the myogenic system off the end of the tail."

[34] Cf. Hooker, *ibid.*, p. 100: ". . . there is no evidence that behavior patterns are built up from proprioceptive reflexes. On the contrary all available evidence indicates that overt behavior is primarily the response of the organism as a whole to its external environment." This does not mean, of course, that proprioceptive acts play no role in behavior, but only that "reflexes" triggered by such acts—for instance, response to internal pain or muscular fatigue—are not its basic elements.

form and quality of the stimulus rather than on its strength. . . . If we did not know that the whole of the brain had been removed, we should say that the actions of the decerebrate animal were directed by consciousness. It initiates no movement spontaneously, but purposive adaptation is evident in every response. . . . Thus, every one of these actions may be purely automatic. Their formal recognition, if present, would be accessory to a mode of behaviour, which can be carried into execution without it."[35]

I am not at all sure that these acts have no psychical character. In a human agent they presumably would not; but then, they also presumably could not take place. The anencephalic infant mentioned in the previous chapter had only indiscriminate reflexes, and had it survived the chances are small that it would have adjusted them in a telic fashion to differential stimuli. The responses of the mutilated animals had been learned, in each case, while the agent was intact. How much can a decerebrate dog or cat learn?

One thing, however, was made evident in Sir Henry's experiments and many similar ones since his day: there is no such entity as "consciousness" directing the course of behavior. "Consciousness" is not an entity at all, let alone a special cybernetic mechanism. It is a condition built up out of mental acts, essentially a qualitative aspect shared by all the mental acts of a particular life episode; this quality may change, either slowly as it does with oncoming drowsiness, or abruptly as in awaking suddenly; it may be altered by physical influences, drugs, atmospheric changes, excess or lack of sensory impingements and also by cerebral acts. If all acts of the organism decline in intensity below the threshold of feeling, "consciousness" disappears, as for instance under a total anesthetic.[36]

It is fairly certain that behavior arises in ontogenesis from many sources, just as organic functions do. The patterns of their respective developments are, indeed, very similar. Just as organs gradually assume their characteristic activities and adjust them to the conditions imposed by the rest of the system at each moment of their continuous history, so the organism as a whole performs all the acts its situation permits, fitting them into the conditions of its ambient. It develops motions of distinct and integrated form. A stronger and more concerted act may entrain

[35] "The Conception of Nervous and Mental Energy" (1923–24), pp. 132–34.

[36] This proper use of the term "consciousness," as in the phrase "alterations of consciousness," was pointed out to me by Dr. Karl E. Schaefer. See his article, "Die Beeinflussung der Psyche und der Erregungsablaüfe im peripheren Nervensystem unter langdauernder Einwirkung von 3% CO_2" (1949) and other, related, experimental studies (mainly in *Pflügers Archiv*), as well as the symposium edited by him, *Environmental Effects on Consciousness* (1960).

previously articulated ones of small scope, or override them, or force their shift to an altogether different consummation, just as maturing organic processes do among themselves. Some of the latter, e.g., the activity of the lungs, verge on behavioral action from the beginning. The skeletal muscles gradually, in postnatal life, attain a limited amount of control over respiratory cycles. This is a major step in establishing the power of vocalization in many animals. Other prenatally prepared patterns of motility are clearly recognizable in the subsequent instinctive acts that arise, in most cases, with as true a systemic timing as the staggered fetal organic functions.[37]

The basis of instinctive behavior is not primarily the desire to cause changes in the environment, together with some unlearned knowledge of how to do it, but the consummation of impulsive acts; and the whole gamut of articulated, essentially repeatable, overt actions of which an animal is capable is its repertoire. Highly elaborate acts are synthetic products of the instinctive repertoire, formed under the pressions or implementations, encouragements or discouragements of the outside world. Since behavior always does change the agent's ambient in some way, it automatically achieves good or bad situations for further acts; and such results quickly motivate acts incorporating favorable results in their normal consummations. The result need not therefore be preconceived by the animal; it is anticipated in that the tension set up in the impulse is not resolved until the practical change occurs, or some other change that takes its place. Beasts and even men may not know that the aim of an act in progress has shifted. A creature restlessly seeking a particular food where it has encountered such a prize before may find something else that consummates the act of hunting, and come to rest.

All practical animal performances, as well as the passes and posturings exhibited in play, are composed of elements originally developed from spontaneous expressions and adapted to the objectively perceived world. In this relationship, of course, the receptors of the body surface are crucial centers of action and, like all highly active organs, undergo adaptive specialization and increasing refinement. The spiral organ of Corti with its myriad hair cells in the inner ear, continued by a similarly spiraled nerve tract leading into its associated brain centers, and the retinal

---

[37] G. E. Coghill, referring to the beginning of motility in *Amblystoma* embryos, says: "During this early period of behaviour all muscular contractions begin in the head region and progress caudad: the individual performance recapitulates the history of its development" (*Anatomy and the Problem of Behaviour*, p. 6).

Ludwig von Bertalanffy, in *Das Gefüge des Lebens*, p. 145, remarked "dass die Wesenszüge des Instinkts geradezu verblüffend mit den Gesetzmässigkeiten der Embryonalentwicklung übereinstimmen."

organs refined to such a point that one quantum of light is enough to activate one rod, are familiar examples, yet no matter how familiar they may be, they never cease to impress one by their wonder.

The process of sense organ specialization is much like that of internal organ development: a complex that has reached a high articulation by virtue of its microambient becomes involved in one organic activity after another, so that its earlier functions are often—I would venture to say, usually—not primitive stages of its ultimate specialty at all. A new situation causes it to realize some quite unforeshadowed potentiality. It is often surprising to learn what ends a receptor has served in its phyletic past. Sometimes it continues to do so even after a newer function has become paramount, so its previous ones may go unrecognized in future. The non-visual activities of the eye are a striking instance. Its primitive function seems to have been light reception and propagation in connection with metabolic and trophic processes. Parietal eyes,[38] some ocelli[39] and very probably the eyespots on the backs of certain slugs are held by many physiologists to serve in such capacities rather than for vision. The mammalian eye still carries on this sort of function, through fibers that leave the optic nerve tract to enter the hypothalamus and in some cases terminate in the hypophysis, with somatic effects on skin and hair pigmentation.[40]

[38] Cf. S. G. Mygind, "The Function of the Parietal Eye" (1949). According to this author, the parietal eye found in some abyssal fishes and ancient reptiles "was never destined for sight," but served in some animals as a static organ, orienting the creature dorsoventrally, and in others as a light receptor without informative uses. George Wald and J. M. Krainin make similar observations in "The Median Eye of *Limulus*: An Ultraviolet Receptor" (1963), p. 1011: "It has been suggested . . . that median eyes may stimulate hormone production or secretion in the central nervous system and that in insects they may modulate the level of brain activity." And later (p. 1116): "The *Limulus* ocellus is an organ designed primarily for the reception of ultraviolet light, its sensitivity in the visible seeming negligible by comparison." Yet these experienced biologists do not overlook the fact that the organ does have at least two kinds of receptors, and add (p. 1117): "On the other hand, its relatively insensitive receptor units for the visible spectrum may serve quite different and important functions."

[39] L. J. Cloudsley-Thompson, in *Rhythmic Activity in Animal Physiology and Behaviour*, p. 169, recounts how Janet E. Harker showed, in 1954 to 1956, that the diurnal rhythms of the cockroach, *Periplaneta americana*, are controlled by a hormone acting on, or working through, the subesophageal ganglia of the central nervous system, and that "she suggested that the neurosecretory hormone whose periodic secretion is responsible for the activity rhythm, might be secreted by the suboesophageal ganglion under the influence of stimulation by light operating through the ocelli."

[40] Eliott B. Hague, in a medical study entitled "Uveitis; Dysacousia; Alopecia; Poliosis and Vitiligo" (1944), p. 532, p. 13, makes the general observation: "Func-

Many developments of this general sort will concern us in later chapters, dealing with the evolution of voluntary muscles, the shift of their control, and therewith of the options they express, to predominantly mental act complexes, the increasing effects of behavior on structure and correlatively increasing opportunities for variegated behavior, the rise and intensification of psychical phases and their role in the motivation and shaping of voluntary acts. Evolving processes are forever differentiating, so that distinct functional elements and associated structural articulations constantly arise. These follow their specialized courses with gathering impetus until they reach the limits set by the rest of the system, or until their own activities become overtaxed or overelaborated; then there is often a radical change of activities as another mechanism, which has formed just a little more slowly but with similar potentialities, takes over the function.

Such changes constitute crises in the course of life, phyletic or individual. Biological crises are not necessarily precipitated by external events or abnormal inward conditions, though of course they may arise that way; but ordinarily they occur as normal crests in the wave-like acts of vital development. For almost every physiological mechanism that achieves a high degree of differentiation, there are other commencements of potential structures, large impulses that could use the same opportunities for expression, but remain in abeyance or are diffusely enacted through other channels as long as the previously started rhythm is progressing freely; but if that rhythm flags for any serious reason, an easier and freer act complex, fitting into the same total organic pattern, is apt to be ready to displace it. This sort of functional shift has already come to our attention as a systematic phenomenon in ontogenesis. It also occurs in the wider sphere of behavior with the maturation of each individual. Few people realize what radical changes in the enlistment of muscles and other organs occur in the course of learning such acts as walking and jumping. Richard Voigt, in a study of the motor and sensory elements involved in a leap from a starting line to a goal, found that they were quite different for trained and untrained leapers: in an experiment comparing leaps performed in light and in darkness, respectively (the goal, in the latter case, a luminous line), he found that the trained athlete executes the leap with more assurance and ease in light, because he takes the total act for granted and makes corrections of small

---

tionally there is a relation of the retina, via the optic tracts, the hypothalamus and the activity of the pituitary, as shown by experimental work on amphibians, hens and other vertebrates." His own researches, guided by such findings, were in the human sphere.

aberrations in passage, whereas the untrained person feels the unity and ease of the act in darkness, where he concentrates solely on the goal.[41] The shift from one mechanism to another is still more pronounced in the development of higher abilities, such as reading;[42] while basic acts that are generally not even recognized as performances, acts of perception, seem to undergo similar changes in the course of special training, which lets one conjecture that such shifts occur quite normally in the course of maturation, too.[43] Here, as elsewhere, the characteristics of growth and progressive articulation, functional integration, and adaptation that belong to physiological processes reappear, sometimes in larger and more striking form, as typical properties of behavioral acts.

Throughout the evolution of an animal stock its activities grow in extent, diversity and intensity. With such increase, of course, the ambient of its individuated lives widens and becomes richer in opportunities for action. All life is opportunistic; every extension of the ambient— which is not only *Lebensraum,* but also the spectrum of sensory impressions, the range of awareness of external conditions as substrates or means of action—is also a further extension of behavior. Growth is the perpetual trend of life; the material self-enlargement of organisms is only one manifestation of it. Acts and ambients grow and diversify, reintegrate and shift to higher levels, together. That is the course of evolution. The power to negotiate a larger and more "difficult" ambient is often taken as the measure of evolutionary advance.[44] This is, of course, only

[41] "Über den Aufbau von Bewegungsgestalten" (1933), pp. 22–23. The whole article is of major importance for the analytic understanding of the act forms.

[42] H. Goepfert, "Beiträge zur Frage der Restitution nach Hirnverletzung" (1922): "Während der Geübte in Gesammtinnervationen liest, erfolgt das Lesenlernen des Kindes in geteilten Innervationen. Der geübte Leser . . . nimmt gleichzeitig, *simultan,* einzelne Worte und ganze Wortgruppen als einheitliche, zusammenhängende *Gestalten* wahr."

[43] Samuel Renshaw, in "The Visual Perception and Reproduction of Forms by Tachistoscopic Methods" (1945), says that O. D. Knight, in a similar study, showed "that the perceiver who approaches virtuosity . . . differs from the novice in the essential respect that he has come to utilize an entirely different set of functions. Imagery and the truly visual phases of the process tend to disappear. The emphasis shifts to the motor aspects of the whole perceptual act. . . . Exposure time, brightness, contrast level, size, blur, or clearness diminish in importance. As in other forms of manipulatory skills . . . where the spastic, disjoined tension movements of the untrained become replaced by the smooth, ballistic movements executed in the absence of antagonistic effector contractions, so it is with form perception. Seeing, to be effective, must be unitary, coherent, and fluent" (p. 218).

[44] See in this connection J. Z. Young, "The Evolution of the Nervous System and of the Relationship of Organism and Environment," pp. 181–82: "It has long been recognized that the organization of living systems is not in all cases equally

one aspect of "higher" activity, and essentially a physiological one; for, as J. Z. Young remarked in discussing the advantage bestowed on some animals by the development of swift locomotion, "We must beware of regarding increase of speed, or any other single process, as a necessary and constant feature of biological advance."[45] In human life the specialization of the neopallium has certainly motivated other than geographical extensions of the ambient, though just at present we are dizzily demonstrating its success in fantastic feats of locomotion.

Scientists have rarely treated growth as an equally basic principle in all vital processes, because it is not easily definable in terms that apply with proper exactitude to phenomena other than increasing physical dimensions. But in the created forms of art, all motion is growth; all spaces are "expanses," all elements grow until they check one another or intersect, or fuse, as the case may be. This pervasive semblance is fundamental in the artistic projection of life. Uexküll quotes K. E. von Bär's saying, " 'Animals and plants develop in the manner of a melody.' "[46] Melodies always move, and their advance is growth.

But growth—in nature as in art—is not always expansive. Acts may grow in intensity, where more and more diversification in a limited compass implements their progress, while they gather impetus by entrainment of other incipient acts, pooling of impulses. Such acts finally break over into the purely intraorganic phase of being felt.

---

complex, but that in some sense we can recognize 'higher' and 'lower' organisms. . . . Some organisms have acquired in the course of their evolution the capacity to maintain themselves in environments which are inaccessible to others, they can maintain a greater difference between themselves and their surroundings." This requires special abilities, and a higher degree of activity than existence in "easy" environments. "Thus freshwater crustacea maintain a concentration of salts inside their bodies higher than that outside, by means of a process which involves so much work that they require approximately twice as much oxygen per gramme per hour as do their marine relatives" (p. 183).

Monnier, in "Aufbau und Bedeutung der Sinnesfunktionen," p. 17, notes the intimate correlation between acts and ambients: "Die mächtige Erweiterung eines aktiv beherrschten Lebensraumes ist nur durch eine hochgradige Spezialisierung der Sinnesorgane möglich. . . ."

Cf. also M. F. Glaessner, "Evolutionary Trends in Crustacea (Malacostraca)" (1957), especially pp. 181–83, on the widening of the ambient with one "spectacularly successful" trend, and A. Tumarkin's broadly conceived, careful study of amphibian phylogeny, "On the Evolution of the Auditory Conducting Apparatus: A New Theory Based on Functional Considerations," pp. 231 ff.

[45] *Op. cit.*, p. 190.

[46] *Umwelt und Innenwelt der Tiere*, p. 28. Raymond Ruyer, in his *Éléments de psycho-biologie*, p. 114, speaks of a smoothly progressing actualization of structured impulses as "*actualisation mélodique.*"

This is not a shift of functions, but the emergence of an entirely new phenomenon, "feeling" in the broadest sense, or consciousness. It is a crisis in natural history as great as the emergence of life from physico-chemical processes; the emergence of a novel quality in the evolutionary course of life. This momentous crisis may not have been a "crisis" in the ordinary sense of a single, more or less cataclysmic, event, but a vastly distributed, protracted process taking eons to develop. As it did so, how-ever, "life" in another than physical sense originated with it—"life" as the realm of value. For value exists only where there is consciousness. Where nothing ever is felt, nothing matters. Biologists speak of "survival value" with relation to organisms to which they do not impute any felt acts, but the term is really inexact, and reflects the natural and world-old belief that all individuals not only automatically strive, but desire, to live.

But this is all anticipation of later chapters. At present our concern is with the growth of behavioral acts and its harmony with all other evolu-tionary patterns, as far as we can survey them, from the simplest forms to the most elaborately organized. The significance of mental acts in this connection is that they prepare very rapid advances in behavior and concomitant growth of animal ambients, and themselves have such power of growth and such flexibility in mutual adaptation that they shift from one course or one mechanism to another, and build up a whole functional response system, which we refer to as the agent's mentality. The study of these processes belongs to psychology; but the basis of that study is the conception of living form and function, which takes one from biology into psychology, and further into the strictly human part of that discipline, the investigation of mind and all its reaches and expressions.

For in one primate stock—the hominid stock—all the developments of special talents seem to have tended in one general direction, which was toward cerebral activity; and at some fateful juncture in the history of that genus, there occurred a shift—doubtless long-prepared—of a sys-temically and practically unimportant cerebral function that took all related ones along, somewhat as the turn of a milometer that has reached 99,999.9, starting with the decimal figure, replaces one 9 after another by 0, until at the left end of the line a new space is filled by a "1": 100,000.0. Such was the great shift, the shift from animal to human estate, that initiated the development of mind. To reconstruct in theory what probably or even conceivably happened in actuality is still a large undertaking, to which the next part of this essay will be devoted.

Bibliography

ABERCROMBIE, M., HICKMAN, C. J., AND JOHNSON, M. L. A *Dictionary of Biology*. 2d rev. ed.; 1st ed., 1951. Harmondsworth: Penguin Books, 1954.

ADRIANI, BRUNO. *Problems of the Sculptor*. New York: Nierendorf Gallery, 1943.

ALLEN, ROBERT D., AND COWDEN, RONALD R. "Syneresis in Ameboid Movement: Its Localization by Interference Microscopy and Its Significance," *Journal of Cell Biology*, XII (1962), 185–89.

ANAND, B. K., AND SUA, S. "Circulatory and Respiratory Changes Induced by Electrical Stimulation of Limbic System (Visceral Brain)," *Journal of Neurophysiology*, XIX (1956), 393–400.

ANDERSEN, WAYNE V. "A Neglected Theory of Art History," *The Journal of Aesthetics and Art Criticism*, XX (1962), 389–404.

ANDERSON, NORMAN. "Cell Division. Part One: A Theoretical Approach to the Primeval Mechanism, the Initiation of Cell Division, and Chromosomal Condensation," *The Quarterly Review of Biology*, XXXI (1956), 169–99.

AND'YAN, A., LISSAK, K., AND MARTIN, I. "New Data on the Mechanism of Action of Sympathomimetic Substances, in Particular Alendrine" [in Russian], *Fiziologicheskii Zhurnal S.S.S.R.*, XXXIV (1948), 727–38. English abstract, here used, in *Journal of Mental Science*, XCVI (1950), 357.

ANFINSEN, CHRISTIAN B. *The Molecular Basis of Evolution*. New York: John Wiley, 1959.

ANSCHÜTZ, G. "Zur Frage der musikalischen Symbolik," *Die Musik*, XX (1928), 113–18.

AREY, LESLIE B. *Developmental Anatomy*. 6th ed. Philadelphia and London: W. B. Saunders, 1954.

ARGELANDER, ANNELIES. *Das Farbenhören und der synästhetische Faktor der Wahrnehmung*. Jena: Fischer, 1927.

ARISTOTLE. *De Poetica*. Translated by Ingram Bywater. Oxford: Clarendon Press, 1924.

ARNHEIM, RUDOLF. *Art and Visual Perception: A Psychology of the Creative Eye*. Berkeley: University of California, 1954.

———. "Perceptual Abstraction and Art," *Psychological Review*, LIV (1947), 66–82.

ARRHENIUS, SVANTE. *Worlds in the Making: The Evolution of the Universe*. English translation by H. Borns. New York and London: Harper, 1908.

[445]

ASCH, SOLOMON E. "A Perspective on Social Psychology," in *Psychology: A Study of a Science*, Vol. III. Edited by Sigmund Koch. New York: McGraw-Hill, 1959, pp. 363–83.

ASCHOFF, JÜRGEN. "Circadian Rhythms in Man," *Science*, CXLVIII (1965), 1427–32.

————, AND WEVER, RÜTGER. "Spontanperiodik des Menschen bei Ausschluss aller Zeitgeber," *Die Naturwissenschaften*, XLIX (1962), 337–42.

ASHBY, W. R. "Principles of Self-Organizing Dynamic Systems," *Journal of General Psychology*, XXXVII (1947), 125–28.

AUGIER, E. *La mémoire et la vie. Essai de défense du mécanisme psychologique*. Paris: Alcan, 1939.

BABCOCK, HARRIET. *Time and the Mind*. Cambridge, Mass.: Sci-Art Publishers, 1941.

BACH, C. P. E. *Versuch über die wahre Art das Klavier zu spielen*. Reprint from the 2d ed., Berlin, 1759 and 1762; 1st ed., 1753. Leipzig: C. F. Kahnt, 1925.

BAILEY, O. T. "Levels of Research in the Biological Sciences," *Philosophy of Science*, XII (1945), 1–7.

BAIRATI, A., AND LEHMANN, F. E. "Structural and Chemical Properties of the Plasmalemma of *Amoeba proteus*," *Experimental Cell Research*, V (1953), 220–33.

BALDWIN, ERNEST. "Rigidification in Phylogeny," in *Perspectives in Biochemistry*. Edited by Joseph Needham and David E. Green. Cambridge: University Press, 1937, pp. 99–107.

BARFIELD, OWEN. *Poetic Diction, a Study in Meaning*. London: Faber and Gwyer, 1928.

BARGMANN, WOLFGANG. *Vom Bau und Werden des Organismus*. Hamburg: Rowohlt, 1957.

BARR, A. H. *Cubism and Abstract Art*. New York: The Museum of Modern Art, 1936.

BARRETT, WILLIAM. *Irrational Man: A Study in Existentialist Philosophy*. Garden City, N.Y.: Doubleday, 1958.

BARTLETT, F. C. *Remembering*. Cambridge: University Press, 1932.

BASTIAN, H. *The Beginnings of Life*. New York: Appleton, 1872.

BAYER, RAYMOND. "Essence du Rythme," *Revue d'esthétique*, VI (1953), 277–90.

BEADLE, G. W. "Genes and Biological Enigmas," *Science in Progress*, VI (1949), 184–249.

BEALE, G. H. *Genetics of Paramecium aurelia*. Cambridge: University Press, 1954.

BECKING, GUSTAV. *Der musikalische Rhythmus als Erkenntnisquelle*. Augsburg: Dr. B. Filser, 1928.

BENEVISTE, S., CARLSON, H. B., COTTON, J. W., AND GLASER, N. "The Acute Confusional State in College Students: Statistical Analysis of Twenty Cases," *Journal of Psychology*, XLVIII (1959), 271–77.

BENNINGHOVEN, ERICH. "Der Geist im Werke Hindemiths," *Die Musik*, XXI (1929), 718–23.

BENOIT, J. A. A. "Le rôle de l'hypophyse dans le développement des embryons de Vertèbres amniotes a sang chaud," *L'année Biologique*, LXVI (1962), 297–325.

BERNARD, CLAUDE. *La science expérimentale*. Paris: Baillière, 1878.

BERNARD, ÉDOUARD. *Essai de la philosophie musicale*. Paris: Gavot, 1916.

BERRILL, N. J. "Salpa," *Scientific American*, CCIV (1961), 150–60.

BERTALANFFY, LUDWIG VON. *Das Gefüge des Lebens*. Leipzig and Berlin: Teuber, 1937.

BETHE, ALBRECHT, AND SCHAEFER, HANS. "Erregungsgesetze einer Blinkschaltung in Vergleich zu denen biologischer Objekte," *Pflügers Archiv für die gesamte Physiologie*, CCXLIX (1948), 313–38.

BICKFORD, R. G., MULDER, D. W., DODGE, H. W., JR., SVIEN, H. J., AND ROME, H. P. "Changes in Memory Function Produced by Electrical Stimulation of the Temporal Lobe in Man," in *The Brain and Human Behavior*. Edited by Harry Solomon, Stanley Cobb and Wilder Penfield. Baltimore: Williams & Wilkins, 1958.

BIEDERMAN, J. C. *Art as the Evolution of Visual Knowledge*. Red Wing, Minn.: by the author, 1948.

BLANCHARD, F. B. *Retreat from Likeness in the Theory of Painting*. New York: Columbia University Press, 1945.

BLOSSFELDT, KARL. *Urformen der Kunst*. 5th ed.; 1st ed., 1928. Berlin: Ernst Wasmuth, 1953.

BLUM, H. F. "On the Origin of Self-Replicating Systems," in *Rhythmic and Synthetic Processes in Growth*. Edited by D. Rudnick. Princeton: University Press, 1957, pp. 155–72.

———. *Time's Arrow and Evolution*. Princeton: University Press, 1951.

BOELL, E. J., GREENFIELD, PHYLLIS, AND HILLE, BERTIL. "The Respiratory Function of Gills in the Larvae of Amblystoma punctatum," *Developmental Biology*, VII (1963), 420–31.

BÖLSCHE, WILHELM. *Der Termitenstaat: Schilderung eines geheimnisvollen Volkes*. Stuttgart: Kosmos, Gesellschaft der Naturfreunde, 1931.

BÖRNSTEIN, WALTER. "Der Aufbau der Funktionen in der Hörsphäre," *Abhandlungen aus der Neurologie, Psychiatrie, Psychologie und ihren Grenzgebieten*, LIII (1930), 1–126.

BONIN, GERHARDT VON. "Types and Similitudes. An Enquiry into the Logic of Comparative Anatomy," *Philosophy of Science*, XIII, No. 3 (1946), 196–202.

BONNER, J. T. *Morphogenesis: An Essay on Development*. Princeton: University Press, 1952.

———. *Size and Cycle: An Essay on the Structure of Biology*. Princeton: University Press, 1965.

———. *The Evolution of Development*. Cambridge: University Press, 1958.

———. "The Role of Toxic Substances in the Interactions of Higher Plants," *Botanical Review*, XVI (1950), 51–65.

BOOLE, GEORGE. *An Investigation of the Laws of Thought, on Which Are Founded the Mathematical Theories of Logic and Probabilities*. London: Walton and Maberly, 1854.

BORING, EDWIN G. *A History of Experimental Psychology*. 2d ed. New York: Appleton-Century-Crofts, 1950.

BORISSAVLIÉVITCH, MILOUTINE. *La Science de l'harmonie architecturale.* Paris: Fischbacher, 1925.

BORODIN, GEORGE. *This Thing Called Ballet.* London: Macdonald, n.d.

BOSANQUET, BERNARD. *History of Aesthetic.* New York: Macmillan, 1892.

———. *The Principle of Individuation and Value.* London: Macmillan, 1912.

———. *Three Lectures on Aesthetic.* London: Macmillan, 1915.

BOULE, MARCELLIN. *Fossil Men.* Edinburgh: Oliver and Boyd, 1923.

BOURGUÈS, LUCIEN, AND DÉNÉREAZ, ALEXANDRE. *La musique et la vie intérieure.* Paris: Alcan, 1921.

BOWIE, HENRY P. *On the Laws of Japanese Painting.* New York: Dover, 1951.

BOYDEN, A. A. "Comparative Evolution with Special Reference to Primitive Mechanisms," *Evolution,* VII (1953), 21–30.

BRACHET, JEAN. "Les Acides Nucléiques et l'Origine des Protéines," *Proceedings of the International Symposium on the Origin of Life on the Earth, Moscow, 1957.* Edited for the Academy of Sciences of the U.S.S.R. by A. I. Oparin *et al.* New York: Pergamon, 1959, pp. 361–67.

BRAIN, WILLIAM RUSSELL. "The Cerebral Basis of Consciousness," *Brain,* LXXIII (1950), 465–79.

BRAMBELL, F. W. R., AND DAVIS, D. H. S. "Reproduction of the Multimammate Mouse (*Mastomys erythroleucus* Temm.) of Sierra Leone," *Proceedings of the Zoological Society of London, Series B,* CXI (1941), 1–11.

BRAUN, RUDOLF. "Der Lichtsinn augenloser Tiere," *Umschau,* LVIII (1958), 306–9.

BRITSCH, GUSTAF. *Theorie der bildenden Kunst.* 2d ed.; 1st ed., 1926. Munich: Bruckmann, 1930.

BRODA, E. "Die Entstehung des dynamischen Zustandes," *Proceedings of the International Symposium on the Origin of Life on the Earth, Moscow, 1957.* Edited for the Academy of Sciences of the U.S.S.R. by A. I. Oparin *et al.* New York: Pergamon, 1959, pp. 334–43.

BROWN, G. BALDWIN. *The Art of the Cave Dweller.* London: John Murray, 1928.

BROWN, MORTIMER. "On a Definition of Culture," *Psychological Review,* LX (1953), 215.

BRUNSWIK, EGON. *The Conceptual Framework of Psychology.* Vol. I, No. 10, of *The International Encyclopedia of Unified Science.* Chicago: University of Chicago Press, 1952.

BUCHANAN, J. W. "Intermediate Levels of Organismic Integration," *Biological Symposia,* VIII (1942), 43–65.

BUCHLER, JUSTUS. *Toward a General Theory of Human Judgment.* New York: Columbia University Press, 1951.

BUCK, PERCY C. *The Scope of Music.* 2d ed. London: Oxford University Press, 1927.

BÜCHNER, LUDWIG. *Kraft und Stoff.* Frankfurt am Main: Meidinger, 1855. Translated into English as *Force and Matter* in 1864.

BULLOCK, T. H. "The Origins of Patterned Nervous Discharge," *Behaviour,* XVII (1961), 48–59.

————, AND TERZUOLO, C. A. "Diverse Forms of Activity in the Somata of Spontaneous and Integrative Ganglion Cells," *The Journal of Physiology,* CXXXVIII (1957), 341–64.

BURNET, F. M. *Virus as Organism.* Cambridge, Mass.: Harvard University Press, 1945.

BURNETT, ALLISON L. "The Maintenance of Form in Hydra," *Society for the Study of Development and Growth. Symposium,* XX (1962), 27–52.

BURR, H. S. "Some Experiments on the Transplantation of the Olfactory Placode in Amblystoma, I: An Experimentally Produced Aberrant Cranial Nerve," *Journal of Comparative Neurology,* XXXVII (1924), 455–79.

BUTLER, J. A. V. *Man Is a Microcosm.* New York: Macmillan, 1951.

BUYTENDIJK, F. J. J. *Algemene Theorie der Menslijke Houding en Beweging als Verbinding en Tegenstelling van de physiologische en de psychologische Beschouwing.* Antwerp: Standaard Boekhandel, 1948.

CAMPBELL, C. A. *On Selfhood and Godhead.* The Gifford Lectures. London and New York: Allen & Unwin and Macmillan, 1957.

CAMPBELL-FISHER, IVY G. "Aesthetics and the Logic of Sense," *Journal of General Psychology,* XLIII (1950), 245–73.

————. "Intrinsic Expressiveness," *Journal of General Psychology,* XLV (1951), 3–24.

————. "Static and Dynamic Principles in Art," *Journal of General Psychology,* XLV (1951), 25–55.

CARMICHAEL, LEONARD. "The Development of Behavior in Vertebrates Experimentally Removed from the Influence of External Stimulation," *Psychological Review,* XXXIII (1926), 51–58.

CARNAP, RUDOLF. *The Logical Syntax of Language.* 1st ed. (German), Vienna, 1934. London: Kegan Paul, Trench, Trubner, 1937.

CARPENTER, WILLIAM BENJAMIN. *Principles of Mental Physiology.* 1st ed. London: H. S. King, 1874.

Cassell's New French Dictionary. New York: Funk and Wagnalls, 1934.

CASSIRER, ERNST. *Individuum und Kosmos in der Philosophie der Renaissance. Studien der Universität Warburg,* Vol. X. Leipzig: Teubner, 1927.

————. *Philosophy of Symbolic Forms.* Translated by R. Manheim. New Haven: Yale University Press, 1953–57.

CHILD, CHARLES MANNING. *Individuality in Organisms.* Chicago: University of Chicago Press, 1915.

————. *Senescence and Rejuvenescence.* Chicago: University of Chicago Press, 1915.

CHOW, K. L., RIESEN, A. H., AND NEWELL, F. W. "Degeneration of Retinal Ganglion Cells in Infant Chimpanzees Reared in Darkness," *Journal of Comparative Neurology,* CVII (1957), 27–41.

CLEMENTS, GRACE. "An Abstraction Is a Reality," *California Arts and Architecture* (May, 1944), 42–43.

CLOUDSLEY-THOMPSON, L. J. *Rhythmic Activity in Animal Physiology and Behaviour.* New York: Academic Press, 1961.

CLUTTON-BROCK, A. "The Function of Emotion in Painting," *Burlington Magazine,* XII (1907–8), 23–26.

[449]

COGHILL, G. E. *Anatomy and the Problem of Behaviour.* London: Cambridge University Press, 1929.

———. "Growth of a Localized Functional Center in a Relatively Equipotential Nervous Organ," *Archives of Neurology and Psychiatry,* XXX (1933), 1086–91.

CONRAD, KLAUS. "New Problems of Aphasia," *Brain,* LXXVII (1954), 491–509.

COOPER, J. B. "An Exploratory Study of African Lions," *Comparative Psychology Monographs,* XVII, Series 90, No. 7 (1942), 1–48.

CORREDOR, J. M. *Conversations with Casals.* Translated by A. Mangeot. New York: Dutton, 1957.

DALCQ, A. M. "Form and Modern Embryology," in *Aspects of Form.* Edited by Lancelot Law White. New York: Pellegrini & Cudahy, 1951, pp. 91–120.

DALTON, H. C. Introduction to "Molecular Events in Differentiation Related to Specificity of Cell Type," *Annals of the New York Academy of Sciences,* LX (1955), 967.

DANCKERT, WERNER. *Ursymbole melodischer Gestaltung.* Kassel: Bärenreiter, 1932.

DARLINGTON, C. D. *The Evolution of Genetic Systems.* Cambridge: University Press, 1939.

DARWIN, CHARLES. *Insectivorous Plants.* New York: Appleton, 1897.

———. *The Origin of Species by Means of Natural Selection.* London: John Murray, 1859.

DAVIS, HALLOWELL. "Some Principles of Sensory Receptor Action," *Physiological Reviews,* XLI, No. 2 (April, 1961), 391–416.

———, TASAKI, I., AND GOLDSTEIN, R. "The Peripheral Origin of Activity, with Reference to the Ear," *Cold Spring Harbor Symposia on Quantitative Biology,* XVII (1952), 143–54.

DAWSON, A. B. "Cell Division in Relation to Differentiation," *Second Symposium on Development and Growth,* Ithaca, N.Y. (1940), pp. 91–106.

DE BEER, G. R. "Embryology and Evolution," in *Evolution.* Edited by G. R. de Beer. Oxford: Clarendon Press, 1938, pp. 57–78.

DELGADO, J. M. R., ROBERTS, W. W., AND MILLER, N. S. "Learning Motivated by Electrical Stimulation of the Brain," *American Journal of Physiology,* CLXXIX (1954), 587–93.

DÉRIBÉRÉ, MAURICE. *Images étranges de la nature.* Paris: Editions de Varennes, 1951.

DEUTSCH, K. W. "Mechanism, Organism and Society; Some Models in Natural and Social Science," *Philosophy of Science,* XVIII (1951), 230–52.

DEWEY, JOHN. *Art as Experience.* New York: Putnam's, 1934.

———. *Reconstruction in Philosophy.* Boston: Beacon Press, 1919.

DE ZAYAS, MARIUS, AND HAVILAND, P. B. *A Study of the Modern Evolution of Plastic Expression.* New York: "291," 1913.

DICKINSON, GEORGE. "Aesthetic Pace in Music," *Journal of Aesthetics and Art Criticism,* XV (1957), 311–21.

DIETRICH, FRITZ. *Musik und Zeit; eine musikmorphologische Skizze.* Kassel: Bärenreiter, 1933.

DOMBROWSKI, H. J. "Lebende Bakterien aus dem Paläozoicum," *Biolo-*

gisches Zentralblatt, LXXXII (1963), 477–84.

DORCHESTER, D., JR. "The Nature of Poetic Expression," *Poet Lore*, V (1893), 81–90.

DOTY, R. W., AND BOSMA, J. F. "An Electromyographic Analysis of Reflex Deglutition," *Journal of Neurophysiology*, XIX (1956), 44–60.

DREIER, KATHERINE. " 'Intrinsic Significance' in Modern Art," in *Three Lectures on Modern Art*. New York: Philosophical Library, 1949.

DRIESCH, HANS. *Die Biologie als selbstständige Grundwissenschaft*. Leipzig: Engelmann, 1893.

———. *Der Vitalismus als Geschichte und als Lehre*. Leipzig: Barth, 1905.

———. *Die "Seele" als elementarer Naturfaktor. Studien über die Bewegungen der Organismen*. Leipzig: Engelmann, 1903.

———. *Philosophie des Organischen. Gifford Vorlesungen, gehalten an der Universität Aberdeen in den Jahren 1907–1908*. 2 vols. Leipzig: Engelmann, 1909.

———. *The Crisis in Psychology*. Princeton: University Press, 1925.

D'UDINE, JEAN [pseud. of Albert Cozanet]. *L'Art et le geste*. Paris: Alcan, 1910.

DUVE, HELMUTH. "Das Bewegungsprinzip in der Skulptur," *Zeitschrift für Aesthetik und allgemeine Kunstwissenschaft*, XXII (1928), 438–44.

EBERT, J. D. *Interacting Systems in Development*. New York: Holt, Rinehart and Winston, 1965.

———, TOLMAN, R. A., MUN, A. M., AND ALBRIGHT, J. F. "The Molecular Basis of the First Heart Beats," *Annals of the New York Academy of Sciences*, LX (1955), 968–85.

ECCLES, J. C. *The Neurophysiological Basis of Mind*. Oxford: University Press, 1953.

EDMONDS, E. M., AND SMITH, M. E. "The Phenomenological Description of Musical Intervals," *American Journal of Psychology*, XXXIV (1923), 287–91.

EDWARDS, JOHN S. "Insect Assassins," *Scientific American*, CCII (1960), 72–78.

EHRET, C. F., AND POWERS, E. L. "The Cell Surface of *Paramecium*," *International Review of Cytology*, VIII (1959), 97–133.

EISLER, MAX. "Das Musikalische in der bildenden Kunst," *Zeitschrift für Aesthetik und allgemeine Kunstwissenschaft*, XX (1926), 317–22.

ELIOT, T. S. *On Poets and Poetry*. New York: Farrar, Straus and Cudahy, 1957.

———. *Selected Essays, 1917–1932*. New York: Harcourt, Brace, 1932.

———. *The Sacred Wood: Essays on Criticism and Poetry*. London: Methuen, 1920.

ELLSON, D. G. "Linear Frequency Theory as Behavior Theory," in *Psychology: A Study of a Science*, Vol. II. Edited by Sigmund Koch. New York: McGraw-Hill, 1959, pp. 637–62.

———. "The Application of Operational Analysis to Human Motor Behavior," *Psychological Review*, LVI (1949), 9–17.

EMIROGLU, FÜRUZAN, AND WINTERSTEIN, HANS. "Über den Ruhetonus des Nervensystems," *Pflügers Archiv für die gesamte Physiologie*, CCLXII (1956), 405–12.

EMPSON, WILLIAM. *Seven Types of Ambiguity*. London: Chatto and

Windus, 1930.

Enenkel's English-Italian Dictionary. Philadelphia: McKay, 1908.

Estève, J. "La pensée musicale," *La Revue Musicale,* VII (1926), 236–41.

Fauré-Fremiet, Philippe. *La re-création du réel et l'équivoque.* Paris: Alcan, 1940.

———. *Pensée et re-création.* Paris: Alcan, 1934.

Fawcett, E. D. *The Individual and Reality.* New York: Longmans, Green, 1909.

Feininger, Andreas. *The Anatomy of Nature.* New York: Crown, 1956.

Fergusson, Francis. *The Idea of a Theater.* Princeton: University Press, 1949.

Fiedler, Konrad. "Moderner Naturalismus und künstlerische Wahrheit" (1881), in *Konrad Fiedlers Schriften über Kunst.* Edited by Hermann Könnerth. Munich: Hirzel, 1913–14.

———. *Über den Ursprung der künstlerischen Thätigkeit.* Leipzig: Hirzel, 1887.

Fisher, Sara C. "The Process of Generalizing Abstraction and Its Product, the General Concept," *Psychological Monographs,* XXI, No. 2 (1916).

Fisher, William A. "What Is Music?," *Proceedings, Music Teachers' National Association,* XXII (1927), 9–21.

Focillon, Henri. *The Life of Forms in Art.* Translated from the 2d French ed. by C. B. Hogan and G. Hubler; 1st French ed., 1934. New York: Wittenborn, Schultz, 1948.

Foppa, Klaus. "Motivation und Bekräftigungswirkungen im Lernen," *Zeitschrift für angewandte Psychologie,* X (1963), 646–59.

Forti, Edgard. "La nature de l'émotion," *Revue de Metaphysique et Morale,* XLIII (1936), 359–84.

Fowler, F. G., and Fowler, H. W. *American Oxford Dictionary.* New York: Albert and Charles Boni, 1931.

Francastel, Pierre. "Espace génétique et espace plastique," *Revue d'esthétique,* I (1948), 349–80.

———. "Naissance d'un espace: mythes et géometrie au Quattrocento," *Revue d'esthétique,* IV (1951), 1–45.

———. *Peinture et société; naissance et destruction d'un espace plastique de la Renaissance au cubisme.* Lyon: Audin, 1952.

Francé, Raoul H. *Die Pflanze als Erfinder.* 7th ed. Stuttgart: Franckh'sche Verlagsbuchhandlung, 1920.

Freud, Sigmund. "Über den Gegensinn der Urworte," *Gesammelte Werke,* VIII (1909–13). London: Imago Publishing Co., 1943, pp. 214–21.

Friedmann, Hermann. *Die Welt der Formen. System eines morphologischen Idealismus.* 2d ed.; 1st ed., 1925. Munich: Beck, 1930.

Fry, Roger. *Transformations.* London and New York: Chatto and Windus and Brentano, 1926.

———. *Vision and Design.* London: Chatto and Windus, 1925.

Fuortes, M. G. F. "Integrative Mechanisms in the Nervous System," *The American Naturalist,* XCIII (1959), 213–24.

Gardner, Helen. *Art through the Ages.* New York: Harcourt, Brace, 1926.

Garnett, Edward (ed.). *Conrad's Prefaces to His Works.* London: J. M. Dent, 1937.

Bibliography

GAUSE, G. F. "The Relation of Adaptability to Adaptation," *Quarterly Review of Biology*, XVII (1942), 99–114.

GEHRING, ALBERT. *The Basis of Musical Pleasure, together with a Consideration of the Opera Problem and the Expression of Emotions in Music*. New York: Putnam, 1910.

GERBER, A. "Die frühesten psychischen Regungen des Embryos," *Revue suisse de Psychologie et de psychologie appliquée*, VIII (1949), 61–67.

GERNET, JACQUES. "L'expression de la couleur en chinois," in *Problèmes de la couleur*. Edited by Ignace Meyerson. Paris: S.E.V.P.E.N., 1957.

GIEDION, S. *The Eternal Present: The Beginnings of Art*. New York: Pantheon, 1962.

GLAESSNER, M. F. "Evolutionary Trends in Crustacea (Malacostraca)," *Evolution*, XI (1957), 178–84.

GLASER, G. H. "The Neurological Status of the Newborn: Neuromuscular and Electroencephalographic Activity," *The Yale Journal of Biology and Medicine*, XXXII (1959), 173–91.

GODDARD, JOSEPH. *The Deeper Sources of Beauty and Expression in Music*. London: Wm. Reeves, n.d.

GOEPFERT, H. "Beiträge zur Frage der Restitution nach Hirnverletzung," *Zeitschrift für die gesamte Neurologie und Psychiatrie*, LXXV (1922), 411–59.

GOETHE, JOHANN WOLFGANG VON. *Sämtliche Werke*. Stuttgart: Cotta, 1866–68.

GOLDSCHMIDT, RICHARD B. "Chromosomes and Genes," *Cold Spring Harbor Symposia on Quantitative Biology*, XVI (1951), 1–11.

GOLDWATER, ROBERT, AND TREVES, MARCO. *Artists on Art*. New York: Pantheon, 1945.

GOODDY, WILLIAM. "On the Nature of Pain," *Brain*, LXXX (1957), 118–31.

———. "Sensation and Volition," *Brain*, LXXII (1949), 312–39.

GOODRICH, E. S. *Living Organisms*. Oxford: Clarendon Press, 1924.

GOODWIN, BRIAN C. *Temporal Organization in Cells. A Dynamic Theory of Cellular Control Processes*. London and New York: Academic Press, 1963.

GOTTSCHALK, L. A., GLESER, G. C., AND HAMBRIDGE, G. "Verbal Behavior Analysis," *A.M.A. Archives of Neurology and Psychiatry*, LXXVII (1957), 300–11.

GRAF, MAX. *Die innere Werkstatt des Musikers*. Stuttgart: Enke, 1910.

GRANICK, S. "Inventions in Iron Metabolism," *The American Naturalist*, LXXXVII (1953), 65–75.

GREGORY, W. K. *Evolution Emerging: A Survey of Changing Patterns from Primeval Life to Man*. 2 vols. New York: Macmillan, 1951.

———. "Polyisomerism and Anisomerism in Cranial and Dental Evolution among Vertebrates," *National Academy of Sciences of the U.S.A., Proceedings*, XX (1934), 1–9.

GRIMSTONE, A. V. "Fine Structure and Morphogenesis in Protozoa," *Biological Reviews of the Cambridge Philosophical Society*, XXXVI (1961), 97–150.

GROBSTEIN, CLIFFORD. "Tissue Disaggregation in Relation to Determination and Stability of Cell Type," *Annals of the New York Academy of Sciences*, LX (1955), 1095–1107.

GROOS, KARL. *Die Spiele der Thiere.* Jena: Fischer, 1896. Translated into English as *The Play of Animals* (New York: D. Appleton, 1898).

GUGGENHEIM, LOUIS. *Phylogenesis of the Ear.* Culver City, Calif.: Murray & Gee, 1948.

GUTHRIE, E. R. "Association by Contiguity," in *Psychology: A Study of a Science,* Vol. III. Edited by Sigmund Koch. New York: McGraw-Hill, 1959, pp. 158–95.

GUYER, MICHAEL F. *Animal Biology.* Rev. ed.; 1st ed., 1931. New York: Harper, 1937.

GYLLENSTEN, LARS, AND MALMFORS, TORBJÖRN. "Myelination of the Optic Nerve and Its Dependence on Visual Function—A Quantitative Investigation in Mice," *Journal of Embryology and Experimental Morphology,* II (1963), 255–66.

HAECKEL, ERNST. *Die Welträtsel.* Bonn: E. Strauss, 1899.

———. *Kunstformen der Natur.* 2d ed. abridged; 1st ed., 1899. Jena: Bibliographisches Institut, 1924.

———. *Natürliche Schöpfungsgeschichte.* 1st ed. Berlin: Reimer, 1868.

HAGUE, ELIOTT B. "Uveitis; Dysacousia; Alopecia; Poliosis and Vitiligo," *Archives of Ophthalmology,* XXXI (1944), 520–38. (Footnote reference is to an expanded reprint, separately paged.)

HALDANE, J. S. *Organism and Environment as Illustrated by the Physiology of Breathing.* New Haven: Yale University Press, 1936.

HALLIBURTON, W. D. *Handbook of Physiology.* 13th ed. Philadelphia: Blakiston, 1917.

HAMBURGER, VICTOR, AND BALABAN, MARTIN. "Observations and Experiments on Spontaneous Rhythmical Behavior in the Chick Embryo," *Developmental Biology,* VII (1963), 533–45.

HARBURGER, W. *Form und Ausdrucksmittel in der Musik.* Stuttgart: J. Engelhorns Nachfolger, 1926.

HARKER, JANET. "Diurnal Rhythms in the Animal Kingdom," *Biological Reviews of the Cambridge Philosophical Society.* XXXIII (1958), 1–52.

HARLOW, H. F. "Love in Infant Monkeys," *Scientific American,* CC (1959), 68–74.

———, AND STAGNER, R. "Psychology of Feelings and Emotions. II. Theory of Emotions," *Psychological Review,* XL (1933), 184–95.

HARRIS, J. DONALD. *Some Relations between Vision and Audition.* Springfield, Ill.: Thomas, 1950.

HARRIS, ROY. "The Basis of Artistic Creation in Music," in *The Bases of Artistic Creation.* Edited by M. Anderson, R. Carpenter and R. Harris. New Brunswick, N.J.: Rutgers University Press, 1942.

HARRISON, R. G. "On the Origin and Development of the Nervous System Studied by the Methods of Experimental Embryology," *Proceedings of the Royal Society of London, Series B,* CXVIII (1935), 155–96.

HARTMANN, EDUARD VON. *Philosophie des Unbewussten.* Berlin: C. Duncker, 1869.

HARTSHORNE, CHARLES. "The Intelligibility of Sensations," *The Monist,* XLIV (1934), 161–85.

Bibliography

————. "The Monotony Threshold of Singing Birds," *The Auk*, LXXIII (1956), 176–92.
————. *The Philosophy and Psychology of Sensation*. Chicago: University of Chicago Press, 1934.
HASKINS, CARYL P. *Of Societies and Men*. New York: Norton, 1951.
HAUSER, ARNOLD. *The Social History of Art*. London: Routledge and Kegan Paul, 1951.
HAYES, J. S., RUSSELL, W. M. S., HAYES, CLAIRE, AND KOHSEN, ANITA. "The Mechanism of an Instinctive Control System: A Hypothesis," *Behavior*, VI (1954), 85–119.
HAYES, WILLIAM C. *The Scepter of Egypt; a Background for the Study of the Egyptian Antiquities in the Metropolitan Museum of Art*. New York: Harper and the Metropolitan Museum of Art, 1953.
HEAD, HENRY. "The Conception of Nervous and Mental Energy," *British Journal of Psychology (Section, General)*, XIV (1923–24), 126–47.
————. "The Physiological Basis of the Spatial and Temporal Aspects of Sensation," *Proceedings of the Aristotelian Society*, Suppl. Vol. II (1919): *Problems of Science and Philosophy*, pp. 77–86.
Heath's New German and English Dictionary. New York: Funk and Wagnalls, 1939.
HEBB, D. O. *The Organization of Behavior: A Neuropsychological Theory*. New York and London: John Wiley and Chapman & Hall, 1949.
HECHT, SELIG. "Energy and Vision," *Science in Progress*, Series 4 (1945), pp. 75–97.
HEINE, HEINRICH. *Werke*. Leipzig: Walzel, 1911.
HEINRICH, F. "Die Tonkunst in ihrem Verhältnis zum Ausdruck und zum Symbol," *Zeitschrift für Musikwissenschaft*, VIII (1925–26), 66–92.
HEISENBERG, WERNER. *Philosophic Problems of Nuclear Science*. New York: Pantheon, 1952.
HERON, PATRICK. *The Changing Forms of Art*. New York: Noonday Press, 1955.
HERRICK, C. J. *Introduction to Neurology*. Philadelphia and London: W. B. Saunders, 1931.
————. *The Evolution of Human Nature*. Austin: University of Texas Press, 1956.
HEYER, G. R. "The Psycho-Somatic Unity—A Few Practical Inferences," *Journal of Mental Science*, LXXXIV (1938), 1069–71.
HILDEBRAND, ADOLF. *The Problem of Form in Painting and Sculpture*. New York: G. E. Stechert, 1932.
HILL, W. F. "Activity as an Autonomous Drive," *Journal of Comparative and Physiological Psychology*, XLIX (1956), 15–19.
HINGSTON, R. W. G. *Problems of Instinct and Intelligence*. New York: Macmillan, 1928.
HOAGLAND, HUDSON. *Pacemakers in Relation to Aspects of Behavior*. New York: Macmillan, 1935.
HOCKING, W. E. *The Self, Its Body and Freedom*. The Terry Lectures (1928). London: H. Milford, Oxford University Press, 1928.
HOESLIN, I. K. VON. "Die Melodie als gestaltender Ausdruck seelischen

Lebens," *Archiv für die gesamte Psychologie*, XXXIX (1920), 232–68.
HOFMANN, HANS. *Search for the Real.* Andover, Mass.: The Addison Gallery of American Art, 1948.
HOLMES, S. J. *Organic Form and Related Biological Problems.* Berkeley: University of California Press, 1948.
HOLST, E. VON, AND SAINT PAUL, URSULA. "Vom Wirkungsgefüge der Triebe," *Naturwissenschaften*, XVIII (1960), 409–22.
HOLTFRETER, JOHANNES. "Some Aspects of Embryonic Induction," *Tenth Symposium on Development and Growth*, Ithaca, N.Y. (1951), pp. 117–52.
HOLZKAMP, KLAUS. "Ausdrucksverstehen als Phänomen, Funktion und Leistung," *Jahrbuch für Psychologie und Psychotherapie*, IV (1956), 297–323.
HOOKER, DAVENPORT. *Evidence of Prenatal Function of the Central Nervous System in Man.* The James Arthur Lecture on the Evolution of the Human Brain, 1957. New York: American Museum of Natural History, 1958.
———. *The Prenatal Origin of Behavior.* Porter Lectures, Series 18. Lawrence: University of Kansas Press, 1952.
HOROWITZ, N. H. "Genetic and Nongenetic Factors in the Production of Enzymes by Neurospora," *Tenth Symposium on Development and Growth*, Ithaca, N.Y. (1951), pp. 47–62.
———. "On Defining Life," *Proceedings of the International Symposium on the Origin of Life on the Earth*, Moscow, 1957. Edited for the Academy of Sciences of the U.S.S.R. by A. I. Oparin *et al.* New York: Pergamon, 1959, pp. 106–7.
———. "On the Evolution of Biochemical Syntheses," *Proceedings of the National Academy of Sciences* (1945), pp. 153–57. Reprinted in *Great Experiments in Biology.* Edited by Mordecai L. Gabriel and Seymour Fogel. Englewood Cliffs, N.J.: Prentice-Hall, 1955, pp. 297–300. (References are to the reprint.)
———, BONNER, D., MITCHELL, H. K., TATUM, E. L., AND BEADLE, G. W. "Genic Control of Biochemical Reactions in Neurospora," *The American Naturalist*, LXXIX (1945), 304–39.
HÜLLER, J. A. "Abhandlung von der Nachahmung der Natur in der Musik," in *Historisch-kritische Beyträge zur Aufnahme der Musik*, Vol. I. Edited by F. W. Marpurg. Berlin, 1754.
HUGGINS, CHARLES. "The Composition of Bone and the Function of the Bone Cell," *Physiological Review*, XVII (1937), 119–43.
HUNGERLAND, HELMUT. "The Concept of Expressiveness in Art History," *The Journal of Aesthetics and Art Criticism*, III (1944), 22–28.
HUNT, C. C. "On the Nature of Vibration Receptors in the Hind Limb of the Cat," *Journal of Physiology*, CLV (1955), 175–86.
HUNT, J. R. "The Static and Kinetic Systems of Motility," *Archives of Neurology and Psychiatry*, IV (1920), 353–69. Presidential Address to the American Neurological Association.
HUSMANN, HEINRICH. "Der Aufbau der Gehörswahrnehmungen," *Archiv für Musikwissenschaft*, X (1953), 95–115.

HUXLEY, ALDOUS. *The Doors of Perception.* New York: Harper, 1954.

HUXLEY, JULIAN. *Evolution, the Modern Synthesis.* New York and London: Harper, 1943.

HYMAN, LIBBIE. "The Transition from the Unicellular to the Multicellular Individual," *Biological Symposia,* VIII (1942), 27–42.

JACOB, F., AND MONOD, J. "The Elements of Regulatory Circuits in Bacteria," in *Biological Organization at the Cellular and Supercellular Level.* Edited by R. J. C. Harris. UNESCO Symposium at Varenna, September, 1962. London and New York: Academic Press, 1963, pp. 1–24.

JACQUES, E. F. "The Composer's Intention," *Proceedings of the Musicological Association,* XVIII (1891–92), 35–45.

JAMES, D. G. *Skepticism and Poetry: An Essay on the Poetic Imagination.* London: Allen and Unwin, 1937.

JAMES, WILLIAM. *Essays in Radical Empiricism.* New York: Longmans, Green, 1912.

———. *The Principles of Psychology,* Vol. I. New York: Henry Holt, 1890.

JANCKE, H. "Das Spezifisch-Musikalische und die Frage nach dem Sinngehalt der Musik," *Archiv für die gesamte Psychologie,* LXXVIII (1931), 103–84.

JASPERS, KARL. *Reason and Existenz.* New York: Noonday Press, 1955.

JONAS, HANS. "Motility and Emotion. An Essay in Philosophical Biology," *Proceedings of the Ninth International Congress of Philosophy,* VII (1953), 117–22.

JONES, M. R. (ed.). *Nebraska Symposium on Motivation.* Lincoln: Nebraska University Press, 1961.

JONES, ROBERT EDMOND. *The Dramatic Imagination.* New York: Duell, Sloan & Pearce, 1941.

JUNG, CARL GUSTAV. *Naturerklärung und Psyche. Synchronizität als ein Prinzip akausaler Zusammenhänge.* Zürich: Rascher, 1952. Published in English as *The Interpretation of Nature and Psyche* (New York: Pantheon, 1955).

———. *Psychologische Abhandlungen.* Vol. II: *Über die Energetik der Seele und andere Psychologische Abhandlungen.* Zürich: Rascher, 1928.

KACSER, H. "The Kinetic Structure of Organisms," in *Biological Organization at the Cellular and Supercellular Level.* Edited by R. J. C. Harris. UNESCO Symposium at Varenna, September, 1962. London and New York: Academic Press, 1963, pp. 25–41.

KANDINSKY, WASSILY. *Point and Line to Plane.* 1st ed. (German), 1926. New York: Museum of Non-Objective Painting, 1947.

KANT, IMMANUEL. *Kritik der Urteilskraft,* 1790.

KÄSTNER, ALFRED. "Die Hüfte und ihre Umformung zu Mundwerkzeugen bei den Arachniden. Versuch einer Organgeschichte," *Zeitschrift für Morphologie und Ökologie der Tiere,* XXII (1931), 721–58.

KATZ, DAVID. *The Vibratory Sense.* University of Maine Studies, Series II, No. 14. Orono: Maine University Press, 1930.

KAVANAU, J. L. "A Theory of Causal Factors in the Origin of Life," *Philosophy of Science,* XII (1945), 190–93.

KEPES, GYORGY. *Language of Vision*. Chicago: Theobald, 1944.

KIENZL, WILHELM. "Die Werkstätte des Komponisten," *Miscellen. Gesammelte Feuilletons und Aufsätze über Musik, Musiker und musikalische Erlebnisse*. Leipzig: Matthes, 1886.

KLAGES, LUDWIG. *Der Geist als Widersacher der Seele*. Leipzig: J. A. Barth, 1929.

KLEE, PAUL. *The Thinking Eye*. Translated from the German, *Das bildnerische Denken* (1956), by R. Manheim. New York and London: George Wittenborn and Lund Humphries, 1961.

KLEMM, O. "Zwölf Leitsätze zur Psychologie der Leibesübungen," *Neue psychologische Studien*, IX, No. 4 (1938). Monograph.

KOCH, SIGMUND. "Epilogue to Study I," in *Psychology: A Study of a Science*, Vol. III. Edited by S. Koch. New York: McGraw-Hill, 1959, pp. 729–88.

KOEHLER, WOLFGANG. *Gestalt Psychology*. New York: Liveright, 1929.

KOZLOFF, L. M. "Virus Reproduction and Replication of Protoplasmic Units," in *Dynamics of Growth Processes*. Edited by E. J. Boell. Princeton: University Press, 1954.

KRAUSS, STEPHAN. "Untersuchungen über Aufbau und Störung der menschlichen Handlung," *Archiv für Psychologie*, LXXVII (1930), 649–92.

KRUEGER, FELIX. "Das Wesen der Gefühle. Entwurf einer systematischen Theorie," *Archiv für die gesamte Psychologie*, LXV (1928), 91–128.

KURTH, ERNST. *Musikpsychologie*. Berlin: Hesse, 1930.

LABAN, RUDOLF VON. *Die Welt des Tänzers; fünf Gedankenreigen*. Stuttgart: W. Seifert, 1920.

LA BARRE, W. *The Human Animal*. Chicago: University of Chicago Press, 1954.

LAIRD, JOHN. *Problems of the Self*. The Shaw Lectures of 1914. London: Macmillan, 1917.

LANGER, SUSANNE K. *Feeling and Form*. New York: Scribner's, 1953.

———. *Philosophical Sketches*. Baltimore: Johns Hopkins, 1962.

———. *Philosophy in a New Key*. Cambridge, Mass.: Harvard University Press, 1942.

———. *Problems of Art*. New York: Scribner's, 1957.

LASHLEY, K. S. "The Problem of Cerebral Organization in Vision," in *Visual Mechanisms*. Edited by H. Klüver. Lancaster, Pa.: Jacques Cattel, 1942, pp. 301–22.

LE CORBUSIER [C. E. JEANNERET-GRIS]. *Toward a New Architecture*. Translated into English from the 13th Fr. ed. by F. Etchells. New York: Payson and Clark, n.d.

LEDERBERG, JOSHUA. "Cell Genetics and Hereditary Symbiosis," *Physiological Review*, XXXII (1952), 403–30.

LEGENDRE, R. Note, *La Nature*, LXXV (1947), 270.

LELE, P. P., AND WEDDELL, G. "The Relationship between Neurohistology and Corneal Sensibility," *Brain*, LXXIX (1956), 119–54.

LENZEN, VICTOR F. *Procedures of Empirical Science*, Vol. I, No. 5, of *International Encyclopedia of Unified Science*. 1st ed. Chicago: University of Chicago Press, 1938.

LEPPIK, E. E. "Phyllotaxis, Anthotaxis and Semataxis," *Acta Biotheoretica*, XIV (1961), 1–28.

Bibliography

LEVI-MONTALCINI, RITA. "Growth and Differentiation of the Nervous System," in *The Nature of Biological Diversity*. Edited by J. M. Allen. New York: McGraw-Hill, 1963, pp. 261–95.

LEVINE, R. P. *Genetics*. New York: Holt, Rinehart and Winston, 1962.

LEWISOHN, LUDWIG. *The Permanent Horizon; a New Search for Old Truths*. New York and London: Harper, 1934.

LIDDELL, H. S. "Adaptation on the Threshold of Intelligence," *Adaptation*. Edited by J. Romano. Ithaca: Cornell University Press, 1949, pp. 53–76.

LILLIE, R. S. *General Biology and Philosophy of Organism*. Chicago: University of Chicago Press, 1945.

LINDSLEY, D. B. "Psychophysiology and Motivation," in *Nebraska Symposium on Motivation*. Edited by M. R. Jones. Lincoln: Nebraska University Press, 1957, pp. 44–108.

LIVINGSTON, ROBERT B. "Perception and Commitment," *Bulletin of the Atomic Scientists*, XIX (1963), 14–18.

LLOYD, FRANCIS E. *The Carnivorous Plants*. Waltham, Mass.: Chronica Botanica Co., 1942.

LOCKE, JOHN. *An Essay Concerning Human Understanding*. London: Th. Basset, 1690.

LOEB, LEO. *The Biological Basis of Individuality*. Springfield, Ill.: Thomas, 1945.

LOESER, J. A. *Die psychologische Autonomie des organischen Handelns. Abhandlungen zur theoretischen Biologie*, Vol. XXX. Berlin: Bornträger, 1931.

LOEWI, O. "The Ferrier Lecture on Problems Connected with the Principle of Humoral Transmission of Nervous Impulses," *Proceedings of the Royal Society of London, Series B*, CXVIII (1935), 299–316.

LONDON, I. D. "Psychologists' Misuse of Auxiliary Concepts of Physics and Mathematics," *Psychological Review*, LI (1944), 266–303.

———. "Some Consequences for History and Psychology of Langmuir's Concept of Convergence and Divergence of Phenomena," *Psychological Review*, LII (1945), 170–88.

LORENTE DE NÓ, RAFAEL. "Anatomy of the Eighth Nerve," *Laryngoscope*, XLII (1933), 1–38.

———. "Cerebral Cortex: Architecture, Intracortical Connections, Motor Projections," in J. F. Fulton, *Physiology of the Nervous System*. 2d ed. rev., 1938. New York: Oxford University Press, 1943.

LORENZ, KONRAD. "The Comparative Method in Studying Innate Behaviour Patterns," in Society for Experimental Biology, *Physiological Mechanisms in Animal Behavior*. New York: Academic Press, 1950, pp. 221–68.

LOTKA, A. J. *Théorie analytique des associations biologiques*. Paris: Hermann, 1934.

LUMHOLTZ, CARL. "Symbolism of the Huichol Indians," *Memoirs of the American Museum of Natural History*, III (1907), 1–228.

MACLEOD, J. J. R. *Physiology in Modern Medicine*. St. Louis: C. V. Mosby, 1935.

MALRAUX, ANDRÉ. *The Psychology of Art*. Vol. I: *Museum without Walls*. Bollingen Series, No. XXIV. Translated by S. Gilbert. New York: Pantheon, 1949.

MARSHALL, W. H., AND TALBOT, S. A. "Recent Evidence for Neural Mechanisms in Vision Leading to a General Theory of Sensory Acuity," in *Visual Mechanisms*. Edited by H. Klüver. Lancaster, Pa.: Lancaster Press, 1942, pp. 117–64.

MASLOW, A. H. "A Theory of Human Motivations," *Psychological Review*, L (1943), 370–96.

MATVEYEV, B. S. "The Transformation of Function in the Ontogeny of Animals," *Zoologicheskij Zhurnal*, XXXVI (1957), 4–25.

MAYER, ERNST. *Dialektik des Nichtwissens*. Basel: Verlag für Recht und Gesellschaft, 1950.

McCLINTOCK, BARBARA. "Chromosome Organization and Genic Expression," in "Genes and Mutations," *Cold Spring Harbor Symposia on Quantitative Biology*, Vol. XVII. Lancaster, Pa.: Science Press, 1952, pp. 13–47.

McCLUNG, C. E. "Evolution of the Germplasm," in *Cytology, Genetics, and Evolution*. Edited by M. Demerec et al. Philadelphia: University of Pennsylvania Press, 1941, pp. 59–66.

McDOUGALL, WILLIAM. *Body and Mind: A History and Defense of Animism*. New York: Macmillan, 1911.

McMULLIN, MICHAEL. "The Symbolic Analysis of Music," *The Music Review*, VIII (1947), 25–35.

MEAD, G. H. *Mind, Self and Society from the Standpoint of a Social Behaviorist*. Chicago: University of Chicago Press, 1934.

MEDAWAR, P. B. *The Uniqueness of the Individual*. New York: Basic Books, 1957.

MERSMANN, H. "Versuch einer musikalischen Wertästhetik," *Zeitschrift für Musikwissenschaft*, XVII (1935), 33–47.

MEYER-ABICH, ADOLF. "Organismen als Holismen," *Acta Biotheoretica*, XI, No. 2 (1955), 85–106.

MEYERSON, IGNACE. *Les Fonctions psychologiques et les oeuvres*. Paris: Librairie philosophique J. Vrin, 1948.

MILLER, NEAL E. "Liberalization of Basic S-R Concepts: Extensions to Conflict Behavior, Motivation, and Social Learning," in *Psychology: A Study of a Science*, Vol. II. Edited by Sigmund Koch. New York: McGraw-Hill, 1959, pp. 196–292.

MIRABENT, F. "Émotion et Compréhension dans l'expérience du beau," *Revue d'esthétique*, I (1948), 250–73.

MITTASCH, ALWIN. *Wilhelm Ostwald's Auslösungslehre*. Heidelberg: Springer, 1951.

MITTENZWEI, KUNO. "Über abstrahierende Apperzeption," *Psychologische Studien*, II (1907), 358–492.

MOHOLY-NAGY, LASZLO. *Vision in Motion*. Chicago: Theobald, 1947.

MONNIER, MARCEL. "Aufbau und Bedeutung der Sinnesfunktionen," in *L'organisation des fonctions psychiques*. Edited by Marcel Monnier. Neuchatel-Suisse: Griffon, 1951.

———. "L'organisation des fonctions psychiques à la lueur des données neurophysiologiques," in *L'organisation des fonctions psychiques*. Edited by Marcel Monnier. Neuchatel-Suisse: Griffon, 1951, pp. 5–15.

MONTAGNA, WILLIAM. *The Structure and Function of Skin*. 2d ed. New York and London: Academic Press, 1962.

MONTAGUE, C. E. A Writer's Notes on His Trade. Garden City, N.Y.: Doubleday Doran, 1930.

MOORE, A. R. The Individual in Simpler Forms. Eugene, Ore.: University Press, 1945.

MOORE, JOHN A. Principles of Zoology. New York: Oxford University Press, 1957.

MORGAN, ANN H. Field Book of Ponds and Streams. New York and London: Putnam's, 1930.

MORLEY, THOMAS. Plaine and Easie Introduction to Practical Musicke. London: P. Short, 1597.

MORRIS, DESMOND. The Biology of Art. New York: Knopf, 1962.

MOULYN, ADRIAN C. "Mechanisms and Mental Phenomena," Philosophy of Science, XIV (1947), 242–53.

MÜNZ, BERNHARD. Hebbel als Denker. Munich: Georg Müller, 1913.

MUENZINGER, K. F. "The Need for a Frame of Reference in the Study of Behavior," Journal of General Psychology, L (1954), 227–36.

MUNN, N. L. "Pattern Brightness Discrimination in Raccoons," Journal of Genetic Psychology, XXXVII (1930), 227–36.

MURRAY, J. A. H. A New English Dictionary. Oxford: Clarendon Press, 1888.

MURRY, J. MIDDLETON. The Problem of Style. 1st ed., 1922. London: Oxford Paperbacks, 1960.

MYGIND, S. G. "The Function of the Parietal Eye," Acta Neurologica et Psychiatrica, XXIV (1949), 607–28.

NACHMANSON, DAVID (ed.). Nerve Impulse. Transactions of the Fourth Conference, March 4, 5 and 6, 1953, Princeton, N.J. New York: Josiah Macy, Jr., Foundation, 1954.

NAFE, JOHN. "The Psychology of Felt Experience," American Journal of Psychology, XXXIX (1927), 367–89.

NEGUS, V. E. The Comparative Anatomy and Physiology of the Larynx. London: Heinemann Medical Books, 1949.

NEURATH, OTTO. Foundations of the Social Sciences. Vol. II, No. 1, of International Encyclopedia of Unified Science. Chicago: University of Chicago Press, 1944.

New Biology #16. Baltimore: Penguin, 1954.

NEWELL, N. D. "Biology of the Corals," Part II of "The Coral Reefs," Natural History, LXVIII (1959), 227–35.

NICHOLS, GEORGE E. "The Terrestrial Environment in Its Relation to Plant Life," in Organic Adaptation to Environment. Edited by M. R. Thorpe. New Haven: Yale University Press, 1924.

NIELSEN, J. M. "Epitome of Agnosia, Apraxia, and Aphasia with Proposed Physiologic-Anatomic Nomenclature," Journal of Speech Disorders, VII, No. 2 (1942), 105–41.

NIELSEN, J. M., and SEDGWICK, R. P. "Instincts and Emotions in an Anencephalic Monster," Journal of Nervous and Mental Disease, CIX (1949), 387–94.

NOHL, HERMANN. Die aesthetische Wirklichkeit: eine Einführung. Frankfurt am Main: Schulte-Bulmke, 1935.

———. Stil und Weltanschauung. Jena: E. Diederichs, 1920.

———. "Die mehrseitige Funktion der Kunst," Deutsche Vierteljahr-

schriften für Literaturwissenschaft und Geistesgeschichte, II (1924), 79–92.

NUTTALL, GEORGE H. F. "Symbiosis in Animals and Plants," *The American Naturalist,* VII (1923), 449–75.

OLSON, EVERETT C. *Origin of Mammals Based upon Cranial Morphology of the Therapsid Suborders.* New York: The Geological Society of America, 1944.

OPARIN, A. P. *Origin of Life.* Translated by S. Morgulis. New York: Macmillan, 1938. 1st ed. (Russian), 1936.

OSTOW, MORTIMER. "Entropy Changes in Mental Activity," *Journal of Nervous and Mental Diseases,* CX (1949), 502–6.

OSTWALD, WILHELM. *Julius Robert Mayer über Auslösung.* Edited by A. Mittasch. Weinheim: Chemie, 1953.

OWEN, D. F. "Industrial Melanism in North American Moths," *American Naturalist,* XCV (1961), 227–33.

PALÁGYI, MELCHIOR. *Ausgewählte Werke.* Vol. II: *Wahrnehmungslehre.* Leipzig: Barth, 1925.

PANTIN, C. F. A. "Behavior Patterns in Lower Invertebrates," *Symposia of the Society for Experimental Biology,* IV (1950), 175–95.

PARKER, DE WITT. *The Analysis of Art.* New Haven: Yale University Press, 1926.

————. *The Self and Nature.* Cambridge, Mass.: Harvard University Press, 1917.

PARKER, H. *The Nature of the Fine Arts.* London: Macmillan, 1885.

PARSONS, J. H. "Adequate Stimuli (with Special Reference to Cutaneous Nerves)," *Brain,* LXXII (1949), 302–11.

PARSONS, TALCOTT. "Psychological Theory in Terms of Theory of Action," in *Psychology, A Study of A Science,* Vol. III. Edited by S. Koch. New York: McGraw-Hill, 1959, pp. 612–711.

PAUL, HERMANN. *Deutsches Wörterbuch.* 4th ed. Edited by K. Euling. Halle a/S: Max Niemeyer, 1935.

PAULI, W. F. *The World of Life: A General Biology.* Boston: Houghton Mifflin, 1949.

PAULIAN, R., AND CORIC, F. "Les staphylins, commensaux des mañans," *La Nature,* LXXV (1947), 53–44.

PEIRCE, C. S. *Collected Papers,* Vol. I. Cambridge, Mass.: Harvard University Press, 1931.

PENFIELD, WILDER. "Mechanisms of Voluntary Movement," *Brain,* LXXVII (1954), 1–17.

————. "Some Observations on the Functional Organization of the Human Brain," *Proceedings of the American Philosophical Society,* XCVIII (1954), 293–97.

PERSSON, P. *Beiträge zur indogermanischen Wortforschung.* Uppsala: Almqvist and Wiksell, 1912.

PERRY, R. B. *The Moral Economy.* New York: Scribner's, 1909.

PETERS, R. S. "Motives and Causes," in Aristotelian Society Supplementary Volume, *Men and Machines* (1952), pp. 139–62. Symposium paper.

————. *The Concept of Motivation.* New York: Humanities Press, 1958; London: Routledge and Kegan Paul, 1960.

PETRINOVICH, L., and BOLLES, R. "Delayed Alternation: Evidence for

Symbolic Processes in the Rat," *Journal of Comparative and Physiological Psychology*, L (1957), 363–65.

PETRUNKEVITCH, ALEXANDER. "Environment as a Stabilizing Factor," in *Organic Adaptation to Environment*. Edited by M. R. Thorpe. New Haven: Yale University Press, 1924, pp. 67–110.

PFUETZE, P. E. *The Social Self*. New York: Bookman Associates, 1954.

PHILIPPE, JEAN. *L'image mentale (évolution et dissolution)*. Paris: Alcan, 1903.

PHILLIPS, JOHN. "Succession, Development, the Climax, and the Complex Organism: An Analysis of Concepts," *Journal of Ecology*. Part I, XXII (1934), 554–71; Part II, XXIII (1935), 488–508.

PIAGET, JEAN. *The Psychology of Intelligence*. Translated by M. Piercy. Original French edition, *La psychologie de l'intelligence* (1947). New York: Harcourt, Brace, 1950.

PICK, JOSEPH. "The Evolution of Homeostasis: The Phylogenetic Development of the Regulation of Bodily and Mental Activities by the Autonomic Nervous System," *Proceedings of the American Philosophical Association*, XCVIII (1954), 298–303.

PIERCE, J. S. "Visual and Auditory Space in Baroque Rome," *Journal of Aesthetics and Art Criticism*, XVIII, No. 3 (1959), 55–67.

PIRIE, N. W. "Chemical Diversity and the Origins of Life," *Proceedings of the International Symposium on the Origin of Life on the Earth, Moscow, 1957*. Edited for the Academy of Sciences of the U.S.S.R. by A. I. Oparin *et al*. New York: Pergamon, 1959, pp. 76–83.

PIRRO, ANDRÉ. *L'esthétique de Jean-Sebastien Bach*. Paris: Fischbacher, 1907.

PITTENDRIGH, C. S., AND BRUCE, V. C. "An Oscillator Model for Biological Clocks," in *Rhythmic and Synthetic Processes in Growth*. Princeton: University Press, 1957, pp. 75–109.

PLATT, JOHN R. "Properties of Large Molecules That Go Beyond the Properties of Their Chemical Sub-Groups," *Journal of Theoretical Biology*, I (1961), 342–58.

POSTMA, C. *Plant Marvels in Miniature*. New York: John Day, 1961.

PRALL, DAVID. *Aesthetic Analysis*. New York: Crowell, 1936.

PRATT, MARJORY B. *Formal Designs from Ten Shakespeare Sonnets*. Privately printed, 1940.

PRICE, DOROTHY (ed.). *Dynamics of Proliferating Tissues*. Report of the Conference on Dynamics of Proliferating Tissues held at Upton, N.Y., 1956. Chicago: University of Chicago Press, 1958.

PRINGLE, J. W. S. "On the Parallel between Learning and Evolution," *Behaviour*, III (1951), 174–215.

PROCHNOW, OSKAR. *Formenkunst der Natur*. Berlin: Ernst Wasmuth, 1934.

PROTHRO, E. T., AND KEEHN, J. D. "Stereotypes and Semantic Space," *Journal of Social Psychology*, XLV (1957), 197–209.

PRUCHE, BENOIT. *Existentialisme et acte d'être*. Paris: Arthaud, 1947.

PUBOLS, B. H., JR. "Jan Swammerdam and the History of Reflex Action," *American Journal of Psychology*, LXXII (1959), 131–35.

RACKER, EPHRAIM. "Multienzyme Systems," *The American Naturalist*, XCIII (1959), 237–44.

RAPHAEL, MAX. *Prehistoric Cave Paintings.* Bollingen Series, No. IV. New York: Pantheon, 1945.

RASHEVSKY, NIKOLAS. *Mathematical Biology of Social Behavior.* Chicago: University of Chicago Press, 1951.

———. *Mathematical Biophysics: Physicomathematical Foundations of Biology.* Chicago: University of Chicago Press, 1938.

———. *Mathematical Theory of Human Relations: An Approach to a Mathematical Biology of Social Phenomena.* Bloomington, Ind.: Principia Press, 1947.

READ, HERBERT. *Icon and Idea. The Function of Art in the Development of Human Consciousness.* The Charles Eliot Norton Lectures, 1953. Cambridge, Mass.: Harvard University Press, 1955.

———. *The Meaning of Art.* 1st ed., 1931. London: Faber & Faber, 1936.

———. *The True Voice of Feeling. Studies in English Romantic Poetry.* New York: Pantheon, 1953.

RENSCH, BERNHARD. "Die phylogenetische Abwandlung der Ontogenese," in *Die Evolution der Organismen.* 2d ed. Edited by G. Heberer. Stuttgart: Gustav Fischer, 1954, pp. 103–30.

———. *Psychische Komponenten der Sinnesorgane; eine psychophysische Hypothese.* Stuttgart: Thieme, 1952.

RENSHAW, SAMUEL. "The Visual Perception and Reproduction of Forms by Tachistoscopic Methods," *Journal of Psychology,* XX (1945), 217–32.

RETI, RUDOLPH. *The Thematic Process in Music.* New York: Macmillan, 1951.

RIBOT, T. *Essai sur l'imagination créatrice.* 1st ed.; 6th ed., 1921. Paris: Alcan, 1900.

RICHARDS, I. A. *Principles of Literary Criticism.* 2d ed. New York: Harcourt, Brace, 1926.

RIGNANO, EUGENIO. *Biological Memory.* Translated by E. W. MacBride. Original Italian ed., 1925. New York: Harcourt, Brace, 1926.

RODMAN, SELDEN. *Conversations with Artists.* New York: Capricorn Books, 1961.

ROSE, S. M. "Cellular Interaction during Differentiation," *Biological Reviews of the Cambridge Philosophical Society,* XXXII (1957), 351–82.

ROSENBLUETH, A., WIENER, N., AND BIGELOW, J. "Behavior, Purpose and Teleology," *Philosophy of Science,* X (1943), 18–24.

ROYCE, JOSIAH. *The World and the Individual.* New York and London: Macmillan, 1900.

RUSKIN, JOHN. *The Elements of Drawing in Three Letters to Beginners.* London: G. Allen, 1889.

RUSSELL, BERTRAND. "Mysticism and Logic," in *Mysticism and Logic, and Other Essays.* New York: Longmans, Green, 1918. (Originally published in *The Hibbert Journal,* XII (1914), 780–803.)

———. "Philosophy of Logical Atomism," *The Monist,* XXVII (1918), pp. 485–527.

———. *The Analysis of Mind.* London and New York: George Allen & Unwin and Macmillan, 1921.

———. "The Study of Mathematics," *New Quarterly* (November, 1907).

Reprinted in *Philosophical Essays* (London, 1910) and in *Mysticism and Logic* (New York, 1918).

———. "The Ultimate Constituents of Matter," *The Monist*, XXV (1915), pp. 399–417.

RUSSELL, CLAIRE, AND RUSSELL, W. M. S. "An Approach to Human Ethology," *Behavioral Science*, II (1957), 169–200.

RUSSELL, E. S. *The Directiveness of Organic Activities*. Cambridge, Mass.: Harvard University Press, 1945.

RUSSELL, W. M. S., MEAD, A. P., AND HAYES, J. S. "A Basis for the Quantitative Structure of Behaviour," *Behaviour*, VI (1954), 153–206.

RUTZ, OTTMAR. *Musik Wort und Körper als Gemütsausdruck*. Leipzig: Breitkopf and Härtel, 1911.

RUYER, RAYMOND. *Éléments de psycho-biologie*. Paris: Presses Universitaires de France, 1946.

SACHS, CURT. *The Commonwealth of Art: Style in the Fine Arts, Music and the Dance*. New York: W. W. Norton, 1946.

———. *World History of the Dance*. Translated by B. Schönberg from *Eine Weltgeschichte des Tanzes*. New York: W. W. Norton, 1937.

SAKHAROV, ALEKSANDR. *Réflexions sur la danse et la musique*. Buenos Aires: Viau, 1943.

SANDER, FRIEDRICH. "Über Gestaltqualitäten," *Bericht des 8. internationalen Kongresses*, 1927.

SANTAYANA, GEORGE. "Apologia pro Mente Sua," in *The Philosophy of George Santayana*. Edited by Paul Schilpp. 2d ed. New York: Tudor, 1951.

———. *The Life of Reason: Or, the Phases of Human Progress*, Vol. IV: *Reason in Art*. New York: Scribner's, 1905.

———. *The Sense of Beauty*. New York: Scribner's, 1896.

SCHAEFER, KARL ERNST. "Die Beeinflussung der Psyche und der Erregungsabläufe im Peripheren Nervensystem unter langdauernder Einwirkung von 3% CO_2," *Pflügers Archiv für die Gesamte Physiologie*, CCLI (1949), 716–25.

SCHAEFER, KARL ERNST (ed.). *Environmental Effects on Consciousness*. New York: Macmillan, 1960.

SCHAFFNER, J. H. "Orthogenetic Series Involving a Diversity of Morphological Systems (Studies in Determinate Evolution, II)," *Ohio Journal of Science*, XXIX (1929), 45–61.

SCHENKER, HEINRICH (ed.). *Das Meisterwerk in der Musik, ein Jahrbuch*. Munich: Drei Masken Verlag, 1925–30.

———. *Neue musikalische Theorien und Phantasien*. Vol. III: *Der freie Satz*. Vienna and Leipzig: Universal-Edition, 1935.

———. *Tonwille*. (A journal irregularly published by the author, and devoted to his own work, during the 1920's.)

SCHILLER, J. C. FRIEDRICH VON. "Über naive und sentimentalische Dichtung" (1795–96), in *Werke*, Vol. VII. Meyers Klassiker-Ausgabe. Leipzig: Bibliographisches Institut, n.d., pp. 441–549.

SCHLEGEL, RICHARD. "Atemporal Processes in Physics," *Philosophy of Science*, XV (1948), 25–35.

SCHMEER, W. *Farbenmusik und die Übernahme der Gesetze der Musik in alle Farbenkünste; mit Schlüssel und Farbakkordgreifer.* Nuremberg: Zeiser, 1925.

SCHMITT, FRANCIS O. "The Macromolecular Assembly—A Hierarchical Entity in Cellular Organization," *Developmental Biology*, VII (1963), 549–59.

SCHOENBERG, ARNOLD. *Style and Idea.* New York: Philosophical Library, 1950.

———. *The Structural Functions of Harmony.* New York: W. W. Norton, 1954.

SCHOENHEIMER, R. *The Dynamic State of Bodily Constituents.* Cambridge, Mass.: Harvard University Press, 1942.

SCHOPENHAUER, ARTHUR. *Die Welt als Wille und Vorstellung* (1818). (Translated into English as *The World as Will and Idea*, 1883.) *Schopenhauers sämmtliche Werke*, Vols. II and III. Leipzig: F. A. Brockhaus, 1919.

SCHOTTLÄNDER, RUDOLF. "Recht und Unrecht der Abstraktion," *Zeitschrift für philosophische Forschung*, VII (1953), 220–34.

SCHUBERT, ANNE. "Drawings of Orotchen Children and Young People," *Journal of Genetic Psychology*, XXXVII (1930), 232–44.

SCHUMANN, ROBERT. *Gesammelte Schriften*, Vol. I. Leipzig: Breitkopf and Härtel, 1914.

SCHWARZ, GEORG. "Über Handlungsganzheiten und ihre Bedeutung für die Rückfälligkeit," *Psychologische Forschung*, XVIII (1933), 143–90.

SCHWEITZER, ALBERT. *J.-S. Bach, le musicien-poète.* Leipzig: Breitkopf and Härtel, 1905. Translated into English by Ernest Newman as *J. S. Bach.* New York: Macmillan, 1955.

SEGOND, JOSEPH. "Esthétique de la mobilité," *Revue d'esthétique*, I (1948), 40–56.

SELBACH, H. "Das Kippschwingungsprinzip in der Analyse der vegetativen Selbststeuerung," *Fortschritte der Neurologie, Psychiatrie und ihrer Grenzgebiete*, XVII, No. 3 (1949), 129–50; XVII, No. 4 (1949), 151–69.

SELINCOURT, BASIL DE. "Music and Duration," *Music and Letters*, I (1920), 286–93.

SEMON, RICHARD WOLFGANG. *Die Mneme als erhaltendes Prinzip im Wechsel des organischen Geschehens.* Leipzig: W. Engelmann, 1911.

SESSIONS, ROGER. "The Composer and His Message," in *The Intent of the Artist.* Edited by Augusto Centeno. Princeton: University Press, 1941, pp. 101–34.

SHALER, N. S. *The Individual: A Study of Life and Death.* New York: Appleton, 1900.

SIEVERS, EDWIN. *Demonstration zur Lehre von den klanglichen Konstanten in Rede u. Musik.* n.p.: no publisher, 1913.

SIMMEL, GEORG. "Michelangelo," in his collected essays, *Philosophische Kultur.* Leipzig, 1911.

SIMONS, ELWYN L. "The Early Relatives of Man," *Scientific American*, CCXI, No. 1 (1964), 50–62.

SIMPSON, GEORGE GAYLORD. *The Meaning of Evolution.* New Haven: Yale University Press, 1949.

————, PITTENDRIGH, COLIN S., AND TIFFANY, LEWIS H. *Life: An Introduction to Biology.* New York: Harcourt, Brace and World, 1957.

SINNOTT, EDMUND W. *The Problem of Organic Form.* New Haven: Yale University Press, 1963.

SITTARD, J. "Die Musik im Lichte der Illusionsaesthetik," *Die Musik*, II (1903), 243–52.

SKEAT, W. W. *A Concise Etymological Dictionary of the English Language.* New and corrected impression; originally published, 1911. Oxford: Clarendon Press, 1924.

SMITH, MAPHEUS. "An Approach to the Study of the Social Act," *Psychological Review*, XLIX (1942), 422–40.

SMUTS, JAN C. *Holism and Evolution.* New York: Macmillan, 1926.

SONDHI, K. C. "The Biological Foundations of Animal Patterns," *Quarterly Review of Biology*, XXXVIII (1963), 289–327.

SONNEBORN, T. M. "Beyond the Gene," *American Scientist*, XXXVII (1949), 33–59.

————. "Developmental Mechanisms in Paramecium," *Growth*, XI (7th Symposium) (1957), 291–307.

————. "Does Preformed Cell Structure Play an Essential Role in Cell Heredity?," in *The Nature of Biological Diversity.* Edited by J. M. Allen. New York: McGraw-Hill, 1963, pp. 165–221.

SORANTIN, ERICH. *The Problem of Musical Expression: A Philosophical and Psychological Study.* Nashville, Tenn.: Marshall and Bruce, 1932.

SOURIAU, ÉTIENNE. "L'art chez les animaux," *Revue d'esthétique*, I (1948), 217–49.

SPEMANN, HANS. *Embryonic Development and Induction.* New Haven and London: Yale University Press and H. Milford, 1938.

SPIEGELMAN, S. "Physiological Competition as a Regulatory Mechanism in Morphogenesis," *Quarterly Review of Biology*, XX (1945), 121–46.

SPITZER, LEO. *Linguistics and Literary History.* Princeton: University Press, 1948.

STERN, PAUL. "Über das Problem der künstlerischen Form," *Logos*, V (1914–15), 165–72.

STRACHE, WOLF. *Forms and Patterns in Nature.* New York: Pantheon, 1956.

STUMPF, CARL. "Erscheinungen und psychische Funktionen," *Abhandlungen der königlich-preussischen Akademie der Wissenschaften*, Abhandlung IV (1906), pp. 1–40.

STURTEVANT, A. H. "The Evolution and Function of Genes," in *Science in Progress*, VI (1949), 250–65.

SULLIVAN, L. H. *Kindergarten Chats.* Originally published serially, 1901. New York: Wittenborn and Schultz, 1947.

SWINTON, WILLIAM E. *The Dinosaurs; a Short History of a Great Group of Extinct Reptiles.* London: T. Murby, 1934.

SYDOW, ECKART VON. *Form und Symbol.* Potsdam and Zürich: Müller & Kiepenheuer and Orell Füssli, 1928.

SZASZ, T. S. *Pain and Pleasure.* New York: Basic Books, 1957.

SZENT-GYÖRGYI, A. VON. "Oxydation and Fermentation," in *Perspectives in Biochemistry.* Edited by Joseph Needham and David E. Green. Cambridge: University Press, 1939, pp. 165–74.

TEILHARD DE CHARDIN, PIERRE. *La place de l'homme dans la nature: Le groupe zoologique humain.* Paris: Michel, 1956.

THIESS, FRANK. *Der Tanz als Kunstwerk.* 3d ed. Munich: Delphen-Verlag, 1923.

THOMPSON, D'ARCY. *On Growth and Form.* 2d ed.; 2 vols. 1st ed., 1917. Cambridge: University Press, 1942.

THURSTONE, L. L. "The Vectors of the Mind," *Psychological Review,* XLI (1934), 1–32.

TINBERGEN, NIKKO. *The Study of Instinct.* Oxford: Clarendon Press, 1951.

TOCH, ERNST. *The Shaping Forces in Music.* New York: Criterion Music Corp., 1948.

TOLMAN, E. C. "Principles of Purposive Behavior," *Psychology: A Study of a Science,* Vol. II. Edited by Sigmund Koch. New York: McGraw-Hill, 1959, pp. 92–157.

TORREY, T. W. "Organisms in Time," *Quarterly Review of Biology,* XIV (1939), 275–88.

TOVEY, DONALD FRANCIS. "Music," in *The Encyclopaedia Britannica.* 11th ed.

TRAXEL, WERNER. "Über Dimensionen und Dynamik der Motivierung," *Zeitschrift für experimentelle und angewandte Psychologie,* VIII (1961), 418–28.

TRENDELENBURG, W. E. T. *Einheitlichkeit und Anpassung in den Lebens-erscheinungen.* An academic speech. Berlin, 1929.

TUMARKIN, A. "On the Evolution of the Auditory Conducting Apparatus: A New Theory Based on Functional Considerations," *Evolution,* IX (1955), 221–43.

TWITTY, V. C., AND NIU, M. C. "The Motivation of Cell Migration, Studied by Isolation of Embryonic Pigment Cells Singly and in Small Groups *in Vitro*," *Journal of Experimental Zoology,* CXXV (1954), 541–73.

UEXKÜLL, JAKOB VON. *Umwelt und Innenwelt der Tiere.* Berlin, 1909.

UREY, H. C. "On the Early Chemical History of the Earth and the Origin of Life," *Proceedings of the National Academy of Sciences, U.S.A.,* XXXVIII (1952), 351–63.

URMSON, J. O. "Motives and Causes," *Proceedings of the Aristotelian Society,* Supplementary Volume XXVI (1952), 179–84.

VIAL, JEANNE. *De l'être musical.* Neuchatel: de la Baconnière, 1952.

VINCI, LEONARDO DA. *A Treatise on Painting.* Translated anonymously, 1721. Translated by John Francis Rigaud, London, 1802, 1877. Two-volume edition, 1956. Vol. I, translated and annotated by A. P. Macmahon. Introduction by Ludwig H. Heydenreich. Vol. II, facsimile of *Codex Vaticanus* MS. Princeton: University Press, 1956.

———. *Das Buch von der Malerei,* 1882. Vols. 15–16 of *Quellenschriften für Kunstgeschichte und Kunsttechnik des Mittelalters u. der Renaissance.* Translated by H. Ludwig. Contains Italian text of *Libro di Pittura,* Codex Vaticanus No. 1270.

VISCHER, FRIEDRICH T. *Kritische Gänge. Neue folge.* 1st ed., 1861–63. Stuttgart: Cotta, 1873.

Bibliography

VISCHER, ROBERT. *Über das optische Formgefühl. Ein Beitrag zur Aesthetik*. Leipzig: n.p., 1873.

VOGEL, H. J. "Control by Repression," in *Control Mechanisms in Cellular Processes*. Edited by David Bonner. New York: Ronald Press, 1961, pp. 23–65.

VOGEL, P. "Beiträge zur Physiologie des vestibulären Systems beim Menschen," *Pflügers Archiv für die gesamte Physiologie*, CCXXX (1932), 16–32.

VOIGT, RICHARD. "Über den Aufbau von Bewegungsgestalten," *Neue psychologische Studien*, IX (1933), 5–32.

VOLKELT, JOHANNES I. *Das aesthetische Bewusstsein; Prinzipienfragen der Aesthetik*. Munich: Beck, 1920.

WADDINGTON, C. H. *The Nature of Life*. London: Allen and Unwin, 1961.

WALD, GEORGE, AND KRAININ, JAMES M. "The Median Eye of *Limulus*: An Ultraviolet Receptor," *Proceedings of the National Academy of Sciences, U.S.A.*, L (1963), 1011–17.

WALDE, ALOIS. *Vergleichendes Wörterbuch der indo-germanischen Sprachen*. Edited by S. Pokorny. 2 vols. and index vol. Berlin and Leipzig, 1930.

WALLASCHEK, RICHARD. "On the Origin of Music," *Mind*, XVI (1891), 375–86.

WALLON, HENRI. *De l'acte à la pensée: Essai de psychologie comparée*. Paris: Flammarion, 1942.

WARDEN, C. J., JENKIND, T. N., AND WARNER, L. H. *Comparative Psychology*. Vol. II: *Plants and Invertebrates*. New York: Ronald Press, 1935–40.

WARDLAW, C. W. "The Floral Meristem as a Reaction System," *Proceedings of the Royal Society of Edinburgh*, LXVI (1955–57), 394–408.

WATSON, JOHN B. *Behavior, An Introduction to Comparative Psychology*. New York: Holt, 1914.

———. *Psychology from the Standpoint of a Behaviorist*. Philadelphia and London: Lippincott, 1924.

WATTERSON, R. L. (ed.) *Endocrines in Development*. Report on the Shelter Island Symposium of 1956. Chicago: University of Chicago Press, 1959.

WEBER, C. O. "Homeostasis and Servo-Mechanisms for What?," *Psychological Review*, LVI (1949), 234–37.

Webster's New World Dictionary of the American Language. Cleveland and New York: World, 1960.

WEIDLÉ, WLADIMIR. "Biologie de l'art," *Diogène*, XVIII (1957), 6–23.

WEISS, PAUL. "Self-Differentiation of the Basic Patterns of Coordination," *Comparative Psychology Monographs*, XVII (1941), 1–96.

———. "Self-Regulation of Organ Growth by Its Own Products" (abstract only), *Science*, CXV (1952), 487–88.

———. "The Biological Basis of Adaptation," in *Adaptation*. Edited by J. Romano. Ithaca: Cornell University Press, 1949, pp. 1–22.

———. "The Compounding of Complex Macromolecular and Cellular Units into Tissue Fabrics," *Proceedings of the National Academy of Sciences, U.S.A.*, XLII (1956), 789–830.

————. "The Problem of Cell Individuality in Development," in *Biological Symposia*. Edited by Jacques Cattell. Lancaster, Pa.: Jacques Cattell, 1940.

WELLEK, ALBERT. "Die Aufspaltung der 'Tonhöhe' in der Hornbostelschen Gehörpsychology und die Konsonanztheorie von Hornbostel und Krueger," *Zeitschrift für Musikwissenschaft*, XVI (1934), 481–96, 537–53.

WERNER, HEINZ. "Microgenesis and Aphasia," *Journal of Abnormal and Social Psychology*, LII (1956), 347–53.

WETZEL, N. C. "On the Motion of Growth. VIII. The Connection between Growth and Heat Production in the Amphibian, *Bufo vulgaris*, from Fertilization to Metamorphosis," *Proceedings of the National Academy of Sciences, U.S.A.*, XX (1934), 183–89.

WEVER, E. G. *Theory of Hearing*. New York: Wiley, 1949.

WHEELER, R. H. *The Laws of Human Nature*. London: Nisbet, 1931.

WHEELER, WILLIAM M. "On Instinct," *Essays in Philosophical Biology*. Cambridge, Mass.: Harvard University Press, 1939.

WHITAKER, D. M. "Physical Factors of Growth," *Second Symposium on Development and Growth*, Ithaca, N.Y. (1940), pp. 75–88.

WHITE, R. W. "Motivation Reconsidered: The Concept of Competence," *Psychological Review*, LXVI (1959), 297–333.

WHITEHEAD, A. N. *Process and Reality. An Essay in Cosmology*. New York: Macmillan, 1929.

WHYTE, LANCELOT LAW. *Internal Factors in Evolution*. New York: Braziller, 1965.

WIENER, NORBERT. *Cybernetics: Or Control and Communication in the Animal and the Machine*. New York and Paris: John Wiley and Hermann, 1948.

WILLIAMS, R. C. "Relations between Structure and Biological Activity of Certain Viruses," *Proceedings of the National Academy of Sciences, U.S.A.*, XLII (1956), 789–830.

WILLMER, E. N. *Cytology and Evolution*. New York and London: Academic Press, 1960.

WILSON, G. A. *The Self and Its World*. New York: Macmillan, 1926.

WITT, P. N. "Spider Webs and Drugs," *Scientific American*, CXCI, No. 6 (December, 1954), 80–86.

WITTGENSTEIN, LUDWIG. *Tractatus Logico-Philosophicus*. London: Kegan Paul, Trench, Trubner, 1922.

WITTMANN, J. "Die physiognomische Urbedeutung des Wortes und das Problem der Bedeutungsentwicklung," in *Bericht über den XIII Kongress der deutschen Gesellschaft für Psychologie*. Jena: Fischer, 1934.

WÖLFFLIN, HEINRICH. "Über den Begriff des Malerischen," *Logos*, IV (1913), 1–7.

WOLF, HELMUT. "Über die Genauigkeit der Herztätigkeit in der Tierreihe," *Pflügers Archiv für die gesamte Physiologie*, CCXLIV (1941), 181–204.

WOODRUFF, L. L. "The Protozoa and the Problem of Adaptation," in *Organic Adaptation and the Environment*. Edited by M. R. Thorpe. New Haven: Yale University Press, 1924, pp. 45–66.

WOODWORTH, R. S. *Dynamics of Behavior.* New York: Henry Holt, 1958.

WORRINGER, WILHELM. *Abstraktion und Einfühlung. Ein Beitrag zur Stilpsychologie.* Munich: R. Piper, 1908. Translated into English as *Abstraction and Empathy.* London: Rutledge and Kegan Paul, 1953.

WRIGHT, SEWALL. "Genes as Physiological Agents; General Considerations," *American Naturalist,* LXXIX (1945), 289–301.

YOUNG, J. Z. *Doubt and Certainty in Science.* The B.B.C. Reith Lectures, 1950. Oxford: Clarendon Press, 1951.

————. "The Evolution of the Nervous System and of the Relationship of Organism and Environment," in *Evolution.* Edited by G. R. de Beer. Oxford: Clarendon Press, 1938, pp. 179–204.

YOUNG, P. T. *Emotion in Man and Animal.* New York: Wiley, 1943.

————. *Motivation and Emotion: A Survey of the Determinants of Human and Animal Activity.* New York: Wiley, 1961.

————. *Motivation of Behavior: The Fundamental Determinants of Human and Animal Activity.* New York: Wiley, 1936.

————. "The Role of Hedonic Processes in Motivation," in *Nebraska Symposium on Motivation.* Edited by M. R. Jones. Lincoln: Nebraska University Press, 1955, pp. 193–238.

ZALOKAR, MARKO. "Ribonucleic Acid and the Control of Cellular Processes," in *Control Mechanisms in Cellular Processes, A Symposium.* Edited by David M. Bonner. New York: Ronald Press, 1961, pp. 87–140.

ZAMBIASI, GIULIO. "Intorno alla misura degli intervalli melodici," *Rivista Musicale Italiana,* VIII (1901), 157–77.

ZIETZ, KARL. "Gegenseitige Beeinflussung von Farb- und Tonerlebnissen," *Zeitschrift für Psychologie und Physiologie der Sinnesorgane,* CXXI (1931), Abt. I, 257–356.

ZUCKERKANDL, VICTOR. *Sound and Symbol.* New York: Pantheon, 1956.

Index

Index

Lorenz, Konrad, 290, 376n, 427: quoted, 376n
Lotka, A. J., 15n, 40, 263 and n: quoted, 40n
Ludwig, H., 166n
Lumholtz, Carl, 169, 170: quoted, 169n

M

Macleod, J. J. R., 386
McClung, C. E., quoted, 408n
McDougall, William, 275n, 299n
Macmahon, A. P., 190n
McMullin, Michael, quoted, 155
Macroevolution, 395
Macromolecular Assembly, 325
McTeer, William, 24n
Magdalenian art, 133n
Magic: in art, 131, 239, 243
Maillol, Aristide, 96
Malmfors, Torbjörn, 404n
Malraux, André, quoted, 227n
Manet, Edouard, 121n
Mantelhandlung, 372–73
Maringer, Johannes, 132
Marshall, W. H., quoted, 291n
Martin, I., quoted, 259n
Martin, W. Keble, 399
Maslow, A. H., quoted, 276n, 279–80
Mathematical expression, 80
Matolsky, 423n
Matrix: of life processes, 328–29
Matveyev, B. S., quoted, 407n
Mayer, Ernst, 10n
Mayer, Julius Robert, 275n, 284 and n
Mead, A. P., 298 and n: quoted, 293, 294, 295–97, 299n
Mead, G. H., 308n
Mechanical brains, 272n, 305, 320n, 359
Mechanisms, 270–75, 357–60, 371, 394, 397–98, 404–5, 427
Medawar, P. B., quoted, 397n
Memories, organic, 199
Mendel, Gregor Johann, 358, 412
Mendelssohn, Felix, 208n
Mentality, animal and human, 229. *See also* Mind
Mercadier, 206n
Mersmann, H., quoted, 209n
Metamorphosis, 374
Metaphor, sensuous, 186–93, 202–3, 213, 230
Metaphysical Subject, 13–14, 15, 307, 308

Methodology, prescriptive, 36–37
Meyer-Abich, Adolf, 263n, 275
Meyerson, Ignace, quoted, 305n
Michelangelo, 238
Mill, John Stuart, 287
Miller, Neal E., 277, 321n: quoted, 44, 45, 46
Mimesis, 172
Mind, 7, 8, 19, 29–30, 51–53, 55, 58, 77–78, 130n, 147, 200, 202, 205, 285–86, 307, 310, 312, 314
Mirabent, F., 140n: quoted, 121n
Mittasch, Alwin, 284n
Mittenzwei, Kuno, 156
"Mobility," 201
Models, 59–60, 63, 67–69, 95, 97, 355, 359
Mondrian, Piet, 86n
Monnier, Marcel, quoted, 422–23, 424n, 432n, 443n
Monod, J., 412n
Montagna, William, 436n: quoted, 420n, 421n, 423n, 424n, 435n
Montague, C. E., quoted, 119n
Moore, A. R., 360n: quoted, 384n, 418
Moore, John A., 349
Moore, O. K., 35n
Morgan, Ann H., quoted, 395n
Morgan, C. T., 268n
Morley, Thomas, 190
Morphogenesis, 347
Morris, Desmond, 137n
Moss, 289n
Motif, 201, 202, 220
Motion: melodic, 234, 236; virtual, 160
Motivation, 46, 220, 275, 277–81, 283, 287–90, 297, 315, 369, 371, 379, 416
Motives, 220, 275–77, 281, 290
Moulyn, Adrian C., quoted, 270n
Movement: in art, 160, 201, 230–31
Movements, 265–67, 275
Mozart, Wolfgang Amadeus, 88, 209n
Muenzinger, K. F., quoted, 25n
Mun, A. M., 401n
Munn, N. L., quoted, 24n
Münz, Bernhard, 117n, 242n: quoted, 113n
Murray, J. A. H., 191, 192n, 193, 279n
Murry, J. Middleton, 118n, 119n, 225: quoted, 118
Music, 66, 82–84, 88, 155, 161–64, 180, 190, 203, 215–17, 231–37
Musical space, 231–37

Index

"Phase beauty," 211–13, 214
Philippe, Jean, 63, 90, 91, 93–94, 95, 98, 99: quoted, 93, 99n
Phillips, John, 337n
Photography, 231
Phylogenesis, 383, 431
Physics: modified concepts in, 271–72
Physiology, animal, 4n
Physical science, 33–34, 38
Physicalism, 11, 16
Piaget, Jean, 149n
Picasso, Pablo, 86, 97n, 188, 208n
Pick, Joseph, quoted, 384n
Pictet, F. J., 382
Pierce, J. S., quoted, 160n
Pirie, N. W., quoted, 259n, 321–22, 353
Pirro, André, 82n, 83n
Pittendrigh, Colin S., 326n, 348n: quoted, 383n
Plants: behavior of, 3–4, 352, 362–66, 369; chemistry of, 434
Plato, 17, 308
Platt, John R., quoted, 271–72
Poetry, 102n, 125–27, 182–83, 210, 238, 241
Pokorny, Julius, 191n
Pollock, Jackson, quoted, 145n
Portmann, Adolf, 69
Postma, C., 244, 330
Potentiality, 206
Powers, E. L., 420n
Prall, David, 180
Pratt, Marjory B., 183, 184, 185: quoted, 185n
Prepatterns, 334–35, 427
Pression, 370, 373, 376, 397
Price, Dorothy, quoted, 379n, 423n
Pringle, J. W. S., 295n: quoted, 383n
Prochnow, Oskar, 244n
Projection: in art, 73–106, 107–11, 137, 200, 208, 213–14
"Projektion," 95n
Prothro, E. T., 40n
Proto-acts, 273, 274, 337, 436
Protozoa, 338, 365–66
Proust, Marcel, 120
Pruche, Benoit, 264n
"Psychic factor," 17–18
Psychical phase, 22, 29
Psychoanalysis, 7, 23
Psychology, 3–4, 8, 16–17, 23, 34, 48, 55, 59, 65, 280, 287: motivation, 280

Psychophysical causality, 203
Psychophysical entities, 5, 9, 14, 15
Psychophysical organism, 200
Pubols, B. H., Jr., 327n

Q

Quality, artistic, 106, 107, 121, 124, 126, 127, 158

R

Racker, Ephraim, quoted, 410n
Rainer, 183n
Raistrick, Harold, 396
Raphael, Max, 134–35, 137n: quoted, 134–35
Rashevsky, Nikolas, 41–42: quoted, 42
Rationality, 148, 150
Rationalization, 148
Read, Sir Herbert, 121–24, 125–27, 128, 134, 140, 185n: quoted, 121n, 122n, 123, 126 and n
Reagent, 395
Réalisation, 63
Reality: concepts of, 8; conceptual, 290; "neutral," 14; phenomenal, 290
Realization: in art, 95, 179
Reasoning, 146: and animals, 425n
"Re-creation," 100
Reflexes, 13, 269, 270, 272n, 286n
Relationships: in art, 200–6
"Releaser mechanism," 427
Rembrandt, 86, 97n, 121n, 208n
Renaissance, 308
Renoir, Pierre Auguste, 121n
Rensch, Bernhard, 29n: quoted, 405n
Renshaw, Samuel, quoted, 442n
Repertoire: of animals, 412, 439
Repetition: as design, 227–28
Representation: in art, 86, 87, 94–97, 166
Respighi, Ottorino, 138
Responses, 24, 44, 287, 290, 427–28, 437–38
Reti, Rudolph, 197n
Rhythms, 204–5, 213, 214, 228, 274, 323, 324, 341, 383, 384–86, 392–94, 401, 429–33: "circadian," 429–30, 431; dialectical, 324–25; diffusion, 333; simultaneous, 333–34; spatial, 230–31
Ribot, T., 91–93, 99

Mind: An Essay on Human Feeling
Volume I
by Susanne K. Langer

Designer:	Edward King
Typesetter:	Monotype Composition Company, Inc.
Typefaces:	Electra, Text; Palatino, Display
Printer:	Universal Lithographers, Inc.
Paper:	Warren 1854
Binder:	Moore & Co., Inc.
Cover Material:	Holliston Record Buckram
Manuscript Editor:	Jean Owen